Apoptotic Chromatin Changes

Gáspár Bánfalvi

Apoptotic Chromatin Changes

Springer

Prof. Gáspár Bánfalvi
University of Debrecen
Institute of Biology and Ecology
Debrecen
Hungary

ISBN 978-1-4020-9560-3   e-ISBN 978-1-4020-9561-0

DOI 10.1007/978-1-4020-9561-0

Library of Congress Control Number: 2008940586

© Springer Science+Business Media B.V. 2009
No part of this work may be reproduced, stored in a retrieval system, or transmitted
in any form or by any means, electronic, mechanical, photocopying, microfilming, recording
or otherwise, without written permission from the Publisher, with the exception
of any material supplied specifically for the purpose of being entered
and executed on a computer system, for exclusive use by the purchaser of the work.

Cover design: Boekhorst Design b.v.

Printed on acid-free paper

9 8 7 6 5 4 3 2 1

springer.com

Preface

The writing of this book is based on: 1. earlier experience writing textbooks for biology students with a university-level background in biology and biochemistry with many (>200) figures in these books (in Hungarian), 2. and on the necessity to present university lectures as power point presentations to catch the interest of students. The author realizes that young readers who were grown up in an information society are relactant to read too much unless they have to take exams. Even then they prefer books which contain illustrations for a better understanding. In collaboration with colleagues, referees and members of the publishing staff an extensive set of photographs and illustrations were collected to provide a graphic follow-up to the text. Further aids for the student, instructor and the curious reader are provided by summaries, extensive sets of readings and references for each chapter, a glossary of the terms, list of abbreviations and a DVD with a red-blue eyeglass to visualize three dimensional chromatin structures.

The reader could ask: Why another book on apoptosis? The answer to the question is related to the definition of the process. The term apoptosis has been introduced to describe typical morphological changes leading to controlled self-destruction of cells. The first demonstrated biochemical feature of this type of cell death was internucleosomal fragmentation, which was occasionally preceded by the generation of large DNA fragments. It turned out that cytoplasmic changes may take place in cells lacking nuclei, indicating that apoptosis may occur without endonuclease mediated DNA degradation while reproducible proteolytic cleavage was observed, pointing to the importance of proteases in the induction and completion of apoptotic cell death. However, the most typical large-scale morphological changes of apoptosis at the chromatin level could not be visualized so far. The book contains reproducible fluorescent chromatin structures some of them subjected to Computer Image Analysis (CIA method) to visualize not only large-scale chromatin changes but also those structures which were invisible for the human eye, but traceable with the aid of a computer.

As apoptosis has been extensively reviewed in several other books this book will discuss primarily the morphological aspects of apoptosis. Chapter 1 will focus on the evolutionary aspects of the genetic material, Chapter 2 deals with the structure of DNA, Chapter 3 is devoted to chromatin structure, Chapter 4 to apoptotic chromatin changes in healthy cells and Chapter 5 to genotoxic chromatin changes.

Attachment

In the attachment under the back cover you will find a DVD entitled: "Chromatin structures under normal and apoptotic conditions" by G. Nagy, G. Pinter, G. Trencsenyi, D. Rozsa and G. Bánfalvi. The DVD contains: 1. All figures with a better resolution including colour figures and the plectonemic mechanism of *Drosophila* chromatin condensation, 2. Chromatin image analysis: traveling inside the interphase chromosome, 3. 3D visualization of chromatin structures (colour anaglyph cardboard eyeglass attached), 4. Apoptotic dance (danse macabre), Bonus executables (secret hidden content).

Acknowledgements

It is with pleasure to acknowledge the contribution of colleagues, who generously offered their comments and suggestions and artwork to improve the quality of the book and reviewed parts of the text or the manuscript.

In particular I wish to thank the reviewers for their helpful comments:

Prof. Laszlo Nagy
University of Debrecen
Prof. Peter Nemeth
University of Pecs

University of Debrecen   Gáspár Bánfalvi
Debrecen, Hungary
October, 2008

Contents

1 DNA Empire 1
  Summary 1
  Historical Events Leading to the Recognition of Core Processes of
    Genetic Information 1
    Jacob, "The First Geneticist" 1
    Mendel, "The Father of Modern Genetics" 2
    Darwin, "The Most Dangerous Man" 2
    Schleiden and Schwann: Living Units Are Cells 3
    Why Are Cells So Small? 4
    Is Life on Earth Possible Without Cells? 7
    Why Is It Difficult to Imagine Cellular and Subcellular Proportions? 7
  RNA World... 8
    Pre-RNA World 9
    Additional Arguments Supporting RNA World Hypothesis 9
  Basic Principles of the Transfer of Genetic Information 12
  Core Information Processes Belonging to the DNA Empire 16
    When Did DNA Evolve? 16
    Molecular Oxygen Serving the Gradual Transition from RNA World to
      DNA Empire 19
    Saturation of Seawater with Oxygen and Development of Oxygen
      Atmosphere 19
    Advantage of Oxygen Production 20
    Oxygen the Janus Faced Molecule 20
    Protection of Anaerobes from Oxygen Toxicity 22
    Free Radical Formation During the Synthesis of Deoxyribonucleotides .. 22
  Cellular Responses to DNA Damages 24
    Conceptual Changes in Gene Stability 24
    Primary Information on Cytotoxicity Can Be Obtained at DNA Level ... 24
    Detection of Genotoxic Changes at the Structural Level of DNA 25
  Conclusions 26
  References 27

vii

viii — Contents

2 Structural Organization of DNA ... 31
Summary ... 31
Building Blocks of Nucleic Acids ... 31
   Structure of Nucleotides ... 31
   Nucleic Acid Bases ... 32
   Sugar Component of Nucleotides ... 39
   Phosphate as Nucleotide Component ... 49
   Bond Types in Nucleotides ... 49
   Nomenclature of Bases, Nucleosides and Nucleotides ... 50
   Metabolism of Nucleotides ... 50
   Regulation of Nucleotide Biosynthesis ... 57
   Degradation of Nucleotides ... 57
   Biosynthesis of Coenzymes ... 57
   Tissue Specific Purine Synthesis ... 59
   Functions of Nucleotides ... 59
DNA Structure ... 62
   Structural Levels of DNA ... 63
   Primary Structure ... 63
   Secondary DNA Structures ... 63
   Tertiary Structure of DNA (Topology) ... 77
   Thermodinamic Aspects of Supercoiling ... 79
   Topoisomerases ... 80
   Models of Gyrase Action ... 82
   Topology of Eukaryotic DNA ... 85
   Supranucleosomal (Chromosomal) Organization of DNA ... 87
   Early Chromatin Models ... 87
   Solenoid versus Zig-Zag Model ... 88
   Chromosomes of Animal Cells ... 93
   Temporal and Spatial Order of Gene Replication ... 112
   DNA Is Replicated and Repaired in Several Subphases in S-Phase ... 113
References ... 114

3 Chromatin Condensation ... 125
Summary ... 125
Importance of Chromatin Condensation ... 125
Active and Inactive Chromatin ... 126
   Euchromatin and Heterochromatin ... 127
   Histone Code Hypothesis ... 127
Chromosome Arrangement in the Nucleus ... 129
Models of Chromosome Condensation ... 129
Chromatin Folding in the Interphase Nucleus Is Poorly Understood ... 135
Chromatin Models ... 136
Review of Methodologies to Follow Chromosome Dynamics ... 136
Methods to Monitor Chromosome Condensation During the Cell Cycle ... 139

Centrifugal Elutriation 139
Permeable Cells 140
Technical Limitations to Visualize Large Scale Chromatin Structures ... 141
Reversible Permeabilization 141
Isolation of Interphase Chromosomes: Visualising "Babies"
Before Birth 142
Visualization of Intermediates of Chromatin Condensation in CHO Cells .. 142
Biotinylation Interpheres with Chromatin Folding 142
Decondensed Chromatin Structures After Biotinylation 144
Globular, Supercoiled, Fibrous, Ribboned Structures
upon DNA Biotinylation..................................... 146
Chromatin Condensation of Non-biotinylated DNA Studied
in Synchronized Cells.. 148
Decondensed Chromatin Structures in Cells Synchronized in Early S
Phase (2.0 – 2.5 C) 149
Transition from Veiled to Ribboned Structures in Cells Synchronized
in Early Mid S Phase (2.5–3.0 C) 150
Linear Connection of Chromosomes in Condensing Interphase
Chromatin in Cells Synchronized at Late Mid S Phase (3.0–3.5 C) ... 151
Distinct Forms of Early Chromosomes in Cells Synchronized
at the End of S Phase (3.5–4.0 C) 153
Chromatin Image Analysis 153
Linear Connection of Condensing Chromosomes in Nuclei During Cell
Cycle.. 155
Linear Connection of Isolated Chromosomes....................... 155
Linear Connection of Chromosomes Inside the Nucleus 157
Common Pathway of Chromatin Condensation in Mammalian Cell Lines .. 158
Decondensed Chromatin Structures at the Unset of S Phase (2.0–2.25
C Value) 161
Supercoiled Chromatin in Early S (2.25–2.5 C)..................... 161
From Veiled to Fibrous Structures Later in Early S Phase
(2.5–2.75 C) 161
Transition to Ribboned Chromatin Structures in Early Mid S Phase
(2.75–3.0 C)... 161
Chromatin Bodies After the Mid S Phase Pause (3.0–3.25 C) 162
Early Chromosomes Later in Mid S Phase (3.25–3.5 C)............... 162
Early Elongated Forms of Chromosomes (3.5–3.75 C)............... 162
Final Stage of Chromosome Condensation (3.75–4.0 C) 162
Chromatin Condensation in Resting Cells and Chemically Induced
Tumours of Rat Hepatocytes 164
Transmission of Hepatocellular Tumour with the HeDe Cell Line 164
Decondensed Chromatin Structures in Resting Hepatocytes 165
Transition from Round Chromatin Bodies to Linear Chromosomes in
Murine preB and CHO Cells 166
Cytometric Analysis of Synchronized preB Cell Populations 168

x   Contents

Chromatin Structures Excluded from Nuclei of preB Cells 169
Condensation of Round Shaped Interphase Chromosomes in preB Cells . 170
Karyotype of Early Interphase Chromosomes 170
Linear Connection of Chromosomes Excluded from Nuclei
   of CHO Cells 175
Condensation of Interphase Chromosomes in CHO Cells. 176
Structure of Interphase Chromosomes in *Drosophila* Cells 180
Synchronization of Drosophila Cells at Low Resolution of Elutriation . . . 180
Cytometric Analysis of Drosophila Cells Synchronized at Low
   Resolution of Elutriation 181
Chromatin Condensation in Synchronized *Drosophila* Cells 182
Chromatin Rodlets in Nuclei of Drosophila Cells 184
Linear Arrangement of Drosophila Chromosomes 185
Interphase Chromosomal Forms in Nuclei of Drosophila Cells 185
Intermediates of Chromatin Condensation in Drosophila Cells at High
   Resolution of Synchronization 187
High Resolution of Synchrony of Elutriated Fractions and DNA
   Synthesis in Synchronized Cells 188
Major Steps and Structures of Chromatin Condensation in Drosophila
   Cells 189
Building Units of Drosophila Chromosomes 191
Folding of Nucleosomal Chromatin String in Drosophila Cells 192
Plectonemic Model of Chromatin Condensation in Drosophila Cells 193
References 196

4   Apoptosis 203
Summary 203
History 203
Increased Scientific Interest to Understand Apoptosis. 206
Relationship of Apoptosis to Genetic Communication 208
Destruction of Biological Information 208
Have You Ever Seen Apoptosis? 209
What Triggers Apoptosis? 211
Apoptosis (Type I Programmed Cell Death) and Necrosis 212
Characterization of Apoptosis 212
Necrosis 218
Induction of Apoptosis 220
Apoptosis Inducing Signals 229
Chemical Inducers of Apoptosis 229
Apoptotic Pathways. 237
Irradiation and Stress-Induced Apoptosis 238
Induction of Nucleases 240
Developmentally Induced Cell Death 241
Granzyme Mediated Apoptosis 241

Induction of Apoptosis Through Death Receptors 243
Caspase-Dependent Initiation of Apoptosis and Necrosis 248
p53 in Normal and Deregulated Cancer Cells 251
Apoptosis Triggered by Chemical DNA Damage 258
Retinoids as Apoptotic Agents 259
Trombospondin Induced Apoptosis in Angiogenesis 259
Antiapoptotic Pathways 260
Reverse Effects of Apoptosis-Inducing Factor (AIF) in Apoptosis and
in Cell Survival 260
Akt Pathway Promotes Cell Survival 261
Inactivation of p53 in Malignant Transformation of Cells 261
Viral Oncoproteins Inactivate p53 262
Wild-Type Level and Overexpression of p53 263
Antitapoptotic Pathways by Growth Factor Activation 264
Stress Induction of Heat Shock Protein Regulation 264
Alternative Mechanisms to Suppress Apoptosis by Cytokines 265
Nicotinic Acetylcholine Receptors Against Apoptosis 267
Trefoil Factors for Mucosal Healing 267
Transgenic Mice for Apoptotic Research 268
Apoptosis Protocols 269
Staining Dead Cells 269
TUNEL Assay 269
ISEL (*In Situ* End Labeling) (LEP) 270
DNA Laddering 270
Immunological Detection of Low Molecular Weight DNA 270
Annexin V Analysis 270
Fluorescence-Activated Cell Sorting (FACS) 271
Light Scattering Flow Cytometry 271
Comet Assay 271
Apoptosis Proteins 272
Detection of Mono- and Oligonucleosomes 272
Anti-Fas mAb 272
p53 Protein Analysis 272
Caspase Staining Kits 273
References 273

5 Apoptotic Chromatin Changes 293
Summary 293
Genotoxic Agents 293
Changes in DNA Structure Caused by Genotoxic Agents 294
Chromatin Changes upon Genotoxic Treatment 297
Chemically Induced Chromatin Changes 297
Heavy Metal Induced Cytotoxicity 301

Relationship Between Replicative and Repair DNA Synthesis
in Undamaged Cells During the Cell Cycle 301
Cadmium as a Genotoxic and Carcinogenic Agent 302
Effect of Ionizing Radiation on Chromatin Structure 326
Gamma Irradiation-Induced Apoptosis in Murine preB Cells 326
Gamma Irradiation-Induced Apoptosis in Radiation Resistant Human
Erythroleukemia K562 Cells 334
Preapoptotic Chromatin Changes Induced by Ultraviolet B Irradiation
in Human Erythroleukemia K562 Cells 346
Summarizing the Apoptotic Changes of Genotoxic Treatments 356
References 358

Abbreviations 365

Glossary 379

Index 411

Chapter 1
DNA Empire

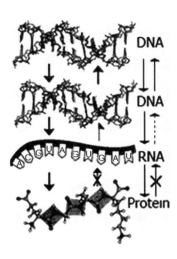

Summary

Different functions of ribonucleotides provide additional arguments that RNA preceded DNA. It is proposed that the DNA Empire came into being after photosynthesizing algae provided an oxygen rich environment. Oxygen the "by-product" of photolyis contributed to the RNA => DNA transition. The reductive power generated by the photolyis produced more carbohydrates than the photosynthesizing algae needed and provided a carbohydrate rich environment for microbial symbiotic communities in the "local primordial soup". The discovery of the structure of the double stranded DNA led to the recognition of several basic principles of modern biology. Based on these rules the informational processes belonging to the DNA Empire have been compiled into a comprehensive hierarchically arranged system consisting of at least twelve processes known as the transfer of genetic information or genetic communication.

Historical Events Leading to the Recognition of Core Processes of Genetic Information

Jacob, "The First Geneticist"

Guesses related to the transfer of genetic information were based on observations and similarities among relatives, consequently these ancient perceptions are much older than the only one century old telecommunication. The results of these ancient observations are known as plant improvements and domestication of animals.

The first documented evidence of controlled animal propagation comes from the Bible (The book of Genesis 31). Jacob (meaning *deceiver*, who God later renamed as Israel, meaning *he struggles with God*, the second-born of the twin sons of Isaac and Rebekah and the grandson of Abraham and Sarah, Genesis 25:24–26), after serving as a shepherd for fourteen years asked his uncle Laban who was also his father-in-law, to pay his wages. It became about when Rachel had borne Joseph. Laban said to Jacob: "What shall I give you?" Jacob said, "You shall not give me anything. If you will do this one thing for me, I will again pasture and keep your flock: Let me pass through your entire flock today, removing from there every speckled and spotted sheep, and every black one among the lambs, and the spotted and speckled among the goats; and such shall be my wages (Genesis 30:31–32, New American Standard Bilde, 1995). After dividing the flock Jacob became exceedingly prosperous within a few years (Genesis 30 verses 41–42). His method was a stroke of genius. It was a kind of selective breeding. What determined the colour of offspring of the flocks was the characteristics of the black male that mated with the white female goats. Jacob's attention was drawn to the fact that all the white female goats which were mating with black males, gave birth to striped, speckled, and mottled offsprings. Based on this recognition Jacob allowed only the striped, speckled, and mottled males to mate with the white female goats of Laban, none of the rest. Soon Jacob became a rich man and Laban lost his flock.

Mendel, "The Father of Modern Genetics"

Compared to the ancient observations, the science of heredity (genetics) is a relatively young discipline. Gregor Mendel's (1822–1884) work became the foundation for modern genetics. The theories of heredity are based on his work with pea plants, which was so brilliant and unprecedented that it took thirty-four more years for the scientific community to realize its significance. Mendel recognized the basic laws of heredity: 1. hereditary factors do not combine, but are passed intact, 2. each member of the parental generation transmits only half of its hereditary factors to each offspring (with certain factors "dominant" over others); 3. different offsprings of the same parents receive different sets of hereditary factors.

Darwin, "The Most Dangerous Man"

In the book of "The Origin of Species" (1859), which is felt to be the major book of the 19th centure, Charles Darwin (1809–1882) provided evidence that all species have evolved over time from one or a few common ancestors through the process of natural selection. The theory of evolution took an even longer time to be accepted by the 1930s, but now it forms the basis of modern evolutionary theory. In its modified form, the theory of Darwin provides a unifying explanation for the diversity of life. By its unifying concept genetics became the most coherent system in natural science.

Misunderstood:

,,Decended from the apes! My dear, let us hope that it is not true, but if it is, let us pray that it will not become generally known" (The wife of the Bishop of Worcester).

,,You care for nothing but shooting, dogs, and rat-catching, and you will be a disgrace to yourself and all our family" (Drawin's father).

,,Poor man, he just stands and stares at a yellow flower for minutes at a time. He would be better off with something to do" (Darwin's gardener).

Darwin's theory is among the most important ones dealing with the historical perspectives of living organisms such as:

- the origin of Earth
- the origin of life
- the theory of Oparin
- the theory of inheritance by Charles Drawin.

The basic theories of modern biology are:

- Cellular theory (cell doctrine)
- Evolution
- Transfer of information

Schleiden and Schwann: Living Units Are Cells

The cellular theory concept was recognized in 1839 by Schleiden and Schwann and asserts that living beings are composed of similar units called cells. Before the discovery of the cell, it was not recognized that the building blocks of organisms are cells. The tenet that cells come into being by free-cell formation turned out to be wrong and the correct interpretation was enunciated by Ruldolph Virchow's powerful statement: "Omnis cellula e cellula" (All cells came from pre-existing cells). The idea of cell theory preceded other paradigms of biology including Darwin's theory of evolution (1859), Mendel's laws of inheritance (1865), and the establishment of comparative biochemistry (1940). We regard now the inheritable "elements" of Mendel as the genes, which are selected and allotted in each generation. The processes of transfer of genetic information became known after the discovery of

Fig. 1.1 The development of biology from cell theory to genomics and proteomics

Biology

- **Cell theory**
- **Evolution**
- **Transfer of information**

Cell theory
↓
Inheritance
↓
Genetic material
↓
Transfer of information
↓
Molecular biology
↙ ↘
Gene techniques   Protein techniques
↙ ↘
Genomics   Proteomics

DNA structure (Watson and Crick, 1953) in the 1960s. The era of molecular biology started after the discovery of restriction endonucleases in the 1970s (Fig. 1.1).

Schleiden and Schwann:

In 1838 after a dinner, Theodor Schwann and Matthias Schleiden were discussing their studies on cells. According to the legend Schwann was struck when he heard from Schleiden that plant cells with nuclei are similar to cells he had observed in animal tissues. They immediately went to Schwann's laboratory to compare animal and plant cells. Next year Schwann published his book on animal and plant cells (Schwann 1839), without acknowledging the contribution of Schleiden (1838).

Why Are Cells So Small?

Respiration in water takes place by diffusion. Taking a hypotethetical spherical organism with a diameter of 2 cm, an oxygen consumption rate of 1 μl/g and the diffusion constant of an animal tissue (1.1×10^{-5} atm^{-1} min^{-1}), to supply the entire organism with oxygen to the center by diffusion, 15 atm would be necessary. If we consider the required oxygen concentration of a smaller organism with a diameter of 2 mm, 0.15 atm partial oxygen pressure would be enough for its oxygen supply. Aerated water is in equilibrium with the atmosphere, which contains

Fig. 1.2 Cell size of prokaryotic and eukaryotic, including human cells

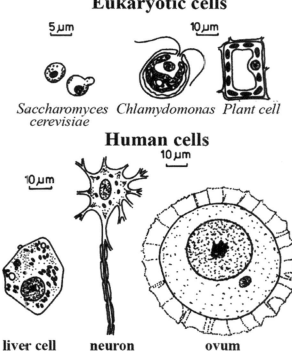

0.21 atm oxygen (21%, corresponding to 0.21 atm partial oxygen pressure), consequently the basic oxygen supply of smaller organism is feasible (Fig. 1.2). One can conclude from this hypothetical experiment, that an active organism to be supplied with oxygen by diffusion must be smaller than 1 mm (protozoans, flatworms) or have a much lower metabolic rate (e.g. jellyfish). Following the same logic cells inside tissues are normally smaller than unicellular organisms or individual cells (Schmidt-Nielsen, 1997). Cell sizes (Fig. 1.2) and the schematic view of tissue cells are shown in Fig. 1.3.

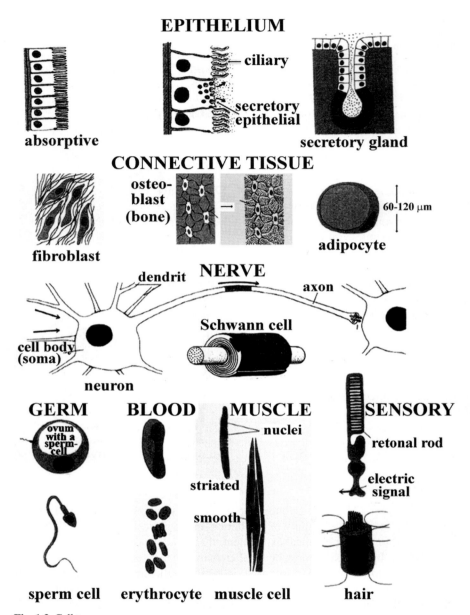

Fig. 1.3 Cell types

> **States of matter including plasma state**
> Life on Earth is in aqueous liquid state. All of the visible universe is in the so-called fourth state of matter beyond the familiar Solid - Liquid - Gas phases. In a solid, atoms and molecules are mostly arranged in regular crystal structures, in a liquid they are much more loosely bound, and in a gas they can move more freely. The fourth state of matter is named the plasma state. The plasma consists of hot, ionized gas. This state does not exist at the temperatures that are normal in our environment. As the atmosphere becomes thinner with altitude, hydrogen, helium, nitrogen and oxygen are ionized by ultraviolet rays and atoms become positively charged and they eject some of their electrons. This mixture of ionized atoms and electrons is called a plasma. The region in space around Earth where the atmosphere becomes rich in plasma is called the plasmasphere. Plasmas are the most common phase of matter. Essentially all of the visible light from space comes from stars, which are plasmas with a temperature such that they radiate strongly at visible wavelengths. Most of the ordinary matter in the universe is found in the intergalactic medium, which is a much hotter plasma, radiating primarily x-rays. The current scientific consensus is that about 96% of the total energy density in the universe is not plasma or any other form of ordinary matter, but a combination of cold dark matter and dark energy.

Is Life on Earth Possible Without Cells?

Viruses are constituted by two types of genetic material, DNA or RNA enwrapped in a protein capsule. They do not have membrane and cell organelles, neither an own metabolism. They can reproduce their genetic material only in a host cell. The life of viruses without cells would be impossible. Viruses are not living organisms, only infectious genetic agents.

Another non-cellular possibility would be plasma life, which however belongs to the science fiction category since the extreme conditions in plasma state contradict biological life. Since life in plasma state is impossible and we know only about cellular life the definitive answer to the above question is no.

Why Is It Difficult to Imagine Cellular and Subcellular Proportions?

The substructures of cells named organelles fulfil specific functions. Their size ranges between $2\,nm$ and $10\,\mu m$. Their function in eukaryotic cells is to create separate units (compartments) with different microenvironment, organization and macromolecular composition. Such organelles are the nucleus ($5–10\,\mu m$ in

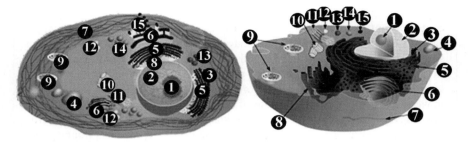

Fig. 1.4 Organelles of eukaryotic cells. 1. nucleolus, 2. nucleus, 3. ribosomes, 4. transmembrane protein, 5. rough endoplasmic reticulum, 6. Golgi apparatus, 7. cytoskeleton, 8. smooth endoplasmic reticulum, 9. mitochondrion, 10. centriole, 11. mitotic spindle, 12. vacuole, 13. peroxisome, 14. lysosome, 15. vesicle

diameter), mitochondrion ($\approx$1 μm long, 0.75 μm in diameter, but changes quickly its shape), metaphase chromosomes (2–10 μm long), ribosomes (20–25 nm) etc. (Fig. 1.4).

Larger intracellular particles can be separated by differential centrifugation using increasing centrifugal force. Since these sizes are deeply below the human dimensions it is difficult to imagine them. To come closer to human dimensions and give an idea of their proportions Table 1.1. shows not only the real sizes of an average eukaryotic cell and its organelles but also their proportions in metric dimensions by multiplying sizes by one million.

Table 1.1 Sizes of components of a typical eukaryotic cell

Cell component	Original size	Magnified size ($\times 10^6$)
Cell	30 μm	30 m (size of an amphitheatre)
Nucleus	5 μm	5 m (size of a theatre stage)
Chloroplast	5 μm	5 m
Mitochondrion	1 μm (length)	1 m
Chromosome	1.4 μm (diameter)	1.4 m
Ribosome	25 nm	25 nm (size of a ping-pong ball)
Hemoglobin	6.4 nm	6.4 mm (size of a pea)
DNA	2 nm (diameter)	2 mm
	μm–m (length)	m–1000 km
Aminoacid	0.6–1.2 nm	0.6–1.2 mm
Water molecule	0.4 nm	0.4 mm

$1 \text{ m} = 10^{-3} \text{ km} = 10^3 \text{ mm} = 10^6 \text{ μm} = 10^9 \text{ nm}$

RNA World

Contrary to the enormous variety of living organisms all living beings are regarded as descendants of a unique ancestor cell. Genetics was able to bring unity to the vast diversity of organisms by defining the principles of reproduction. The first major recognition was self duplication of DNA at the time of the discovery of double stranded DNA by Watson and Crick (1953). Francis Crick devoted most of his efforts between 1954 and the mid-1960s to studying the genetic coding problem, the action by which genes controlled the synthesis of proteins, the building blocks of life. His seminal discoveries took place in Cambridge at the heart of the by then defunct British Empire. Nevertheless, as a metaphor for DNA, the term DNA Empire is recommended, since the phrase RNA World is already occupied. The expression "RNA World" was first used by Walter Gilbert (1986), stating that self-replicating RNA-based systems would have arisen first, and DNA and proteins would have been added later. Central to this hypothesis is an RNA enzyme that replicates other RNA molecules (Cech, 1986). Francis Crick was the first who argued that RNA must have been the first genetic material able to act as a template and reproduce itself as an enzyme.

Ribozymes (RNA enzymes or catalytic RNAs) are RNA molecules that catalyse the hydrolysis of their own RNA phosphodiester bonds, or other RNAs. The first ribozymes were discovered by Thomas R. Cech, during his studies on RNA splicing in the ciliated protozoan *Tetrahymena thermophila* (Cech, 1986).

Pre-RNA World

The troublesome subunit in ribonucleotides is ribose which can be synthesized only in small quantities under prebiotic conditions, and has a short lifetime in water. Scientists dealing with prebiotic synthesis have suggested that the first RNA molecules probably did not contain ribose (Larralde et al., 1995) leading to the hypothesis that a pre-RNA World existed before the RNA World. Even if ribose would have been produced, it would be difficult to explain how a substantial amount of the short lived ribose could have been produced in the primordial soup. Ribose as a reducing sugar would have been quickly eliminated by reactions with amino acids in the soup, if there was such soup at all. So far no explanation has been given for the ribose enrichment by researchers. The best guess regarding an inreased sugar concentration is the production of carbohydrates by photosynthestic bacteria (Smith et al., 1969). Some of the glucose could have been oxidized directly in the pentose cycle leading to the conversion of C_3–C_7 sugars including ribose. The purine ring of purine nucleotides is assembled on ribose phosophate, while the ribose phosphate is aquired in a later step during the formation of pyrimidine nucleotides.

Additional Arguments Supporting RNA World Hypothesis

There are recently three arguments which speak for an early RNA World: (a) RNA has an enzymatic activity (Cech, 1986), (b) RNA enzymes (ribozymes) are similar to protein enzymes creating catalytic surfaces with complex secondary and tertiary structures. (c) RNA nucleotides are more readily synthesized than deoxyribonucleotides. Although, the majority of biologists accepts the RNA World theory, the minority sees no grounds for considering it established. To strengthen the feasibility that RNA existed before DNA further arguments are listed below:

1. The 2'OH group of ribose in RNA plays a direct role in self-splicing (Pley et al., 1994; Murray and Arnold, 1996).
2. Experiments carried out under different "prebiotic" conditions produced amino acids, sugars and ribonucleotides. Deoxyribose and its derivatives were not among them. Comparison of configurations and sugar pucker conformations of pentoses indicate that the β-anomer of D-ribose was not randomly selected, but the only choice, since β-D-ribose fits best into the structure of physiological forms of nucleic acids (Bánfalvi, 2006).
3. Most ribozymes work along with proteins and perform not only cutting but also pasting reactions.
4. Substrates of DNA replication are deoxyribonucleoside triphosphates (dNTPs). dNTPs are derived from precursor deoxyribonucleoside diphosphates (dNDPs), that are synthesized from ribonucleoside diphosphates (rNDPs), catalysed by ribonucleotide reductase (Reichard, 1993). Class I ribonucleotide reductase consists of two subunits (B1 and B2). Active B2 is formed from the apoprotein and Fe^{2+} in the presence of O_2 and a thiol (Thelander and Reichard, 1979). Further details of ribonucleotide reductase and superoxide dismutase reactions will be discussed in this chapter under the subtitle: Free radical formation during the synthesis of deoxyribonucleotides. There is a general agreement among scientists that molecular oxygen (O_2) was not present in the ancient reducing atmosphere and erose later during the evolution (Fig. 1.5).

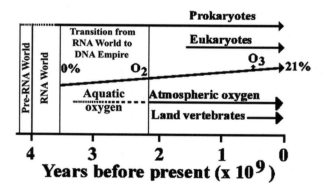

Fig. 1.5 The evolution of life on Earth. Earliest possible appearance of life took place at an estimated 4.2 billion years ago. Oxygen production started some 3.6 billion, atmospheric oxygen erose 2.2 billion, eukaryotes 1.5 billion, land vertebrates 500 million years ago. Pre-biotic RNA life could have taken place between 4.2 and 4.0 billion years ago. RNA World followed and lasted till an estimated 400 million years. Transition from RNA World to DNA Empire required the production of oxygen when it reached its saturation concentration in the oceans (∼3%), and appeared in the atmosphere. All living organisms were subjugated to the DNA Empire by around 2.2 billion years ago

RNA World 11

5. Adenosine nucleotides (but not deoxiadenosine nucleotides) are components of the three major coenzymes: Coenzyme A, NAD$^+$, FAD.
6. dATP is not involved in glycolysis which is regarded as one of the oldest metabolic pathways and is working under reducing conditions (without molecular oxygen). In this patway NAD and ATP are also present.
7. Cyclic AMP (cAMP) produced from ATP is an ancient "hunger signal" in all kingdoms of life. Cyclic dAMP is not known.
8. The major chemical energy currency of cells is adenosine triphosphate (ATP). dATP has no such function.
9. ATP is the immediate source of energy for locomotion, biosynthetic and transport processes. GTP powers the intracellular movements of macromolecules. In contrast to ATP, the only function of dATP and the other three dNTPs is that they are substrates for DNA synthesis.
10. Exergonic biochemical reactions result in energy being given off, while others require energy (endergonic). For both processes to be carried out efficiently, they must be "coupled". Coupled reactions involve ATP or other ribonucleoside triphosphates (aderylation, uridylation, guanylation, cytidylation of intermediates). Nucleotide derivatives are activated intermediates in many biosyntheses. Examples are: UDP-glucose, CDP-diacylglycerol which are precursors of glycogen and phosphoglycerids, respectively. S-adenosylmethionine carries an activated methyl group. There are no such activated deoxyribonucleotide derivatives.
11. Ribonucleotides serve as metabolic regulators. cAMP and cGMP are intracellular mediators (second messengers) in the action of many hormones (primary messengers). The substrate specity of adenylate cyclase is limited to ATP not to dATP.
12. Covalent modifications introduced by ATP alter the activities of many enzymes, as exemplified by the phosphorylation of glycogen phophorylase, glycogen synthase or adenylation of glutamine synthase.

These examples indicate that ATP and other ribonucleotides existed long before deoxyribonucleotides and took many additional functions during the early development of metabolism.

Major presumed transitory steps leading from RNA World to DNA Empire are summarized in Fig. 1.6. The evolving oxygen atmosphere produced by photosynthesizing bacteria made it possible to carry out radical reactions such as ribonucleotide reductase catalyzing the synthesis of deoxyribonucleotides. The first informational molecules synthesized by viral retrotranscription could have been RNA-DNA molecules from the mixed nucleotide pool (dNTP, rNTP). The first DNA probably contained dAMP, dGMP, dCMP and dUMP (U-DNA). Cytosine in the presence of molecular oxygen spontaneously deaminates at a perceptible rate to uracil. Uracil pairs with adenin in the next round of replication, leading to A-U mutation rather than the original G-C pair (Fig. 1.7). The presence of thymine instead of uracil in DNA permitted the repair of deaminated cytosine in the process known as base excision repair (BER).

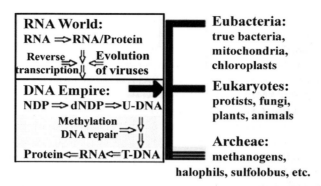

Fig. 1.6 Rise of DNA Empire and fall of RNA World. The left side summarizes major engagements in the battlefield. To the right: T-DNA, the winner took it all and reigns the Empire ever since. Abbreviations: U-DNA, dUMP containing DNA; T-DNA, dTMP conatining DNA

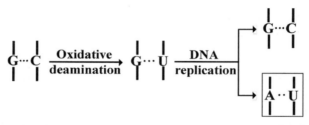

Fig. 1.7 The presence of molecular oxygen led to the oxidative deamination of cytosine to uracil. Uracil had to be replaced by thymine (5-methyl uracyl) to repair mutation. Mutated uracil could be distinguished from methylated uracyl and repaired by base excision repair (BER)

Basic Principles of the Transfer of Genetic Information

The principles of the transfer of genetic information led to the recognition of molecular processes belonging to the DNA Empire. First the principles will be reviewed briefly then the processes belonging to the transfer of genetic information will be summarized.

1. **The general idea** is expressed by one of the most important principles of modern biology stating that DNA makes RNA makes protein:

    DNA ===> RNA ===> Protein

    The extention of the general idea was suggested by Hunt et al. (1983):

    DNA => RNA => Protein => Cell => Organism

2. **Central dogma**. Francis Crick made two hypotheses by the end of the 1960s, the central dogma and the sequence hypothesis (Crick, 1970; Crick, 1968). Both ideas are accepted today as basic rules of molecular biology. The central dogma, states that once the information entered protein, it cannot get out again (Crick, 1970). In retroviruses RNA can be transcribed reversely to DNA. Reverse transcription was discovered by Temin and Baltimore (1972).

$$DNA \rightleftarrows RNA \rightarrow Protein$$

> **Lamarkism** (named after Lamarck, Jean-Baptiste de Monet). Lamarck was the first to use the word biology, in 1802. His idea that acquired traits are inheritable is known as Lamarckism and is controverted by Darwinian theory. Lamarckism was discredited after the 1930s, except in the Soviet Union, where it survived as Lysenkoism, and dominated Soviet genetics until the 1960s. An example of Lamarckism: Trees in the seashore blown constantly by the wind are bent. When their seeds are planted in an other soil far from the sea and come into leaf the young trees will become bent (not true). Michurin was one of the founding fathers of scientific agricultural selection. He worked on hybridization of plants of similar and different origins and made major contributions in the development of genetics. Michurin's theory of influence of the environment on the heredity was regarded as a variant of Lysenkoism and without his intention he was promoted as a Soviet leader in theory of evolution.
>
> The main argument against Lamarck's "soft inheritance" is that experiments do not support his "law of acquired traits" which are not inherited. Lamarckism contradicts the central dogma. The central dogma states that protein represent a "dead end street" and information cannot get out of it and cannot make RNA or DNA.

3. **Imperial idea** (named after the DNA Empire). Neither the general idea, nor the central dogma refers to the DNA ⇔ DNA transfer although, there are several such processes (DNA replication, recombination, mutation, repair). The imperial idea includes these processes, combines the general idea and the central dogma by stating that DNA makes DNA makes RNA makes protein, and occasionally RNA makes DNA in retroviruses (Fig. 1.8).

4. **Sequence hypothesis**. Crick came to the conclusion that the sequence in which the four bases of DNA were arranged encoded the instructions for building long chains made of a combination of the twenty protein building amino acids. Messenger RNA acts as an intermediate in conveying information from the sequence of nucleotides in DNA to the sequence of amino acids in the polypeptide chain.

5. **Enzyme catalysis**. Metabolic reactions are catalysed by specific enzymes. In the overwhelming majority of biological catalysis enzymes are proteins. Ribozymes are relics of the past, these are RNA molecules that can perform biological catalysis (Zaug and Cech, 1980).

6. **Amino acid sequence**. The action of an enzyme is determined by its amino acid sequence.

7. **The one gene – one enzyme theory** states that each gene in an organism controls the production of a specific enzyme (Beadle and Tatum, 1941). This hypothesis was later modified to the concept that one cistron controls the production of one polypeptide.

8. **The gene** is (a) the smallest informational unit of living organisms to be inherited, (b) a segment of DNA to be transcribed, (c) a DNA segment coding functional products (RNA, protein).

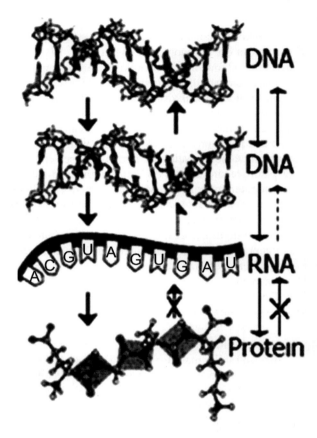

Fig. 1.8 The imperial idea: DNA makes DNA makes RNA makes protein. But protein never makes RNA, and RNA makes DNA only in retroviruses

9. **Codon**. Each amino acid is specified by a codon corresponding to the sequence of three bases in mRNA. mRNA is by definition positive (+) stranded RNA, transcribed from the negative (−) DNA strand.
10. **Gene-protein colinearity**. Francis Crick's seminal concept was that RNA and protein sequences are colinear with the genetic DNA information. The missing link in defining the relationship between genes and proteins was that the nucleotide sequence of genes corresponded to the amino acid sequence of respective proteins. Studies with bacteriophage T4 provided an approach that could be used to attack the genetic aspect of the colinearity problem without requiring the isolation of a gene or determining its nucleotide sequence (Benzer, 1957). Further evidences followed and supported the colinearity hypothesis (Sarabhai et al., 1964; Yanofsky et al., 1964).
11. **Genetic code**. The principles from 4 to 10 contributed to the recognition of the Genetic Code which establishes the relationship between genes and proteins. In 1961 Crick and Brenner described that the genetic code was to be read three bases at a time (Crick et al., 1961). At the same time the team of the American biochemist Marshall Nirenberg reached the same conclusion (Matthaei

et al., 1962). In the next few years, scientists in England and in the United States identified the triplets for all the twenty amino acids.

12. **RNA as the primary information**. RNA represents secondary information in cells and DNA viruses. RNA is the primary information in RNA viruses that do not have a DNA phase.

13. **Mutation** is a heritable change in the genetic material DNA and the major source of genetic variation which fuels evolutionary changes.

14. **DNA repair**. While some genetic variation must occur to adapt to environmental changes and to develop distinguishing characteristics, radical or inappropriate modification of the genetic material if unattended can lead to the development of diseases. Cellular DNA is the only molecule which can be repaired. Repair is contributed by the high fidelity of DNA polymerases to minimize mutation. Mitochondrial (maternal) DNA is not repaired, consequently the "molecular clock" of mitochondrial DNA is ticking faster than the clock of the cellular (chromosomal) DNA. Mitochondrial Eve, the matrilineal most recent common ancestor is believed by some to have lived about 140,000 years ago in Africa. The time she lived is calculated based on the molecular clock technique. The notion of a "molecular clock" is attributed to Zuckerkandl and Pauling (1965).

15. **Protein targeting**. Information is news, but news becomes information only when reaching its target. Proteins are synthesized either in free ribosomes or at the surface of the endoplasmic reticulum. The location of ribosomes determines the fate of the proteins synthesized by them. Proteins synthesized in free ribosomes remain either in the cytosol, are transported to the nucleus or to the mitochondria. Proteins produced by ribosomes on the endoplasmic reticulum are chaneled inside the endoplasmic reticulum and undergo core glycosylation. The majority of proteins moves further in vesicular form to the cys Golgi and from the trans Golgi apparatus to: (a) lysosomes for destruction, (b) to the cellular membrane, (c) to neighboring cells, or (d) is externalized by exocytosis.

16. **Apoptosis**. Since genetic information is neither material nor energy, the laws of material and energy constancy do not apply to information. Consequently, new biological information can be generated (e.g. recombination), destroyed (e.g. apoptosis) and fabricated (e.g. viruses). It is logical to assume that for the benefit of the organism damaged cells undergo self destruction in a process called apoptosis.

Although, it may sound non-scientific, for apoptosis the expression of Captain Spock is borrowed:
„The needs of the many outweigh the needs of the few or the one" quoted from Star Treck. But of course, Admiral Kirk proved him wrong, since sometimes the needs of the few or the one outweigh the needs of the many (e.g. the fertilized egg cell).

17. **Epigenetic processes** refer to features such as chromatin and DNA modifications that are stable over rounds of cell division but do not involve changes in the DNA sequence of the organism (Bird, 2007). These processes play a role in cellular differentiation, allowing cells to maintain different characteristics despite containing the same genomic material.

Core Information Processes Belonging to the DNA Empire

Based on the basic principles, the transfer processes of genetical information (Fig. 1.9) the transfer processes of genetic information have been summarized in a comprehensive hierarchical system (Fig. 1.10). The highest level of transfer of genetic information in this scheme is represented by recombination, the exchange of genetic material, referred to as DNA → DNA' hybridization of the parental strands which can be manifested occasionally by the birth of a genius or an individual with capabilities much below the average. The second level represents other DNA⇔DNA transfer processes. One step below the DNA level are processes related to RNA information. At the bottom of the hierarchy is the protein level.

There are further related processes such as cell division, cellular damages, gene regulation, genetic rearrangement, genetic selection, mutagenesis, DNA methylation, genetic determination, DNA packaging, postranscriptional, postranslational modifications, protein splicing, etc. which are not involved directly in the flow of genetic information.

When Did DNA Evolve?

To come to some conclusion regarding the age of DNA we have to return to the era of RNA World. Unfortunately there is no remnant or trace evidence of a precellular

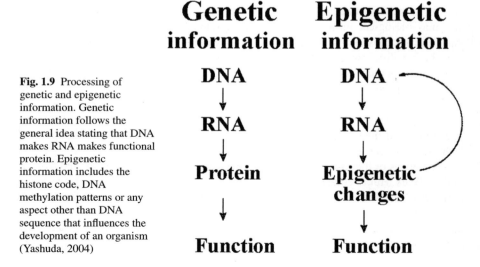

Fig. 1.9 Processing of genetic and epigenetic information. Genetic information follows the general idea stating that DNA makes RNA makes functional protein. Epigenetic information includes the histone code, DNA methylation patterns or any aspect other than DNA sequence that influences the development of an organism (Yashuda, 2004)

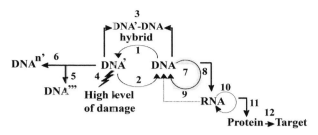

Fig. 1.10 Processes belonging to the intracellular transfer of genetic information. 1 – 7. DNA ⇔ DNA transfer processes. 1. Mutation: DNA => DNA'. 2. Repair: DNA' => DNA. 3. Recombination: crossover, gene conversion (DNA'-DNA hybridization). Recombination can take place intracellularly, but is mainly an intercellular process. 4. Apoptosis (programmed cell death, high levels of DNA damage). 5. Aging, several mutations, persistent DNA damage: DNA' => DNA'' => DNA'''. 6. Malignant transformation with multiple mutations: DNA => DNA$^{n'}$ (persistent DNA damage, many mutations, mutant p53). 7. DNA replication: DNA⇔DNA reduplication (high fidelity, HiFi process, 1:10^{10} miscopoorated deoxyribonucleotide). 8 – 9. DNA⇔RNA transfer. 8. Transcription: DNA => RNA (medium fidelity, MeFi process, 1:10^5 misincorporated ribonucleotide). 9. Reverse transcription: RNA => DNA (in retroviruses). 10. RNA replication: RNA⇔RNA (in RNA viruses). 11. Translation: RNA => protein (low fidelity, LoFi process, 1:10^4 misincorporated amino acid) 12. Protein targeting: information reaches intra- or extracellular destination

life today. It is even doubted that it ever existed. There seems to be a consensus at least among those scientists who accept the RNA World theory, that the time between its appearance and evolution into the simplest bacteria was relatively short. One supporter of the RNA World theory allows 400 million years between "The Rise and Fall" of the RNA World and by 3.6–3.8 billion years ago the DNA/protein Empire has already risen (Joyce, 1991). Others see evidence for a prokaryotic life within 100 million years (Gould. 1989). Based on fossil records it was suggested that protein based life existed as early as 3.6 billion years ago, leaving a short time to get life started (Crick 1993). The shape, the size of these microfossils, as well as other associated fossils, suggest that these are the remains of cyanobacteria. Chemical and radio-isotopic trace evidence of biologically processed carbon in Earth's oldest surviving rocks – as old as 3.85 billion years – suggest that self-replicating, carbon-based microbial life became well developed, confirming the idea that cellular metabolism existed before 3.8 billion years ago (Mojzis et al., 1996). By comparing the age of Earth (∼4.5 billion years) with the fossil evidence it was concluded that cellular organisms resembling bacteria existed by 3.6 billion years before the present, leaving a relatively short "window opportunity" for the RNA World as the Earth's surface was uninhabitable at the beginning due to the heat generated (Moore et al., 1993). The discovery of evidence for early life in rocks in Australia also suggests that simpler prokaryotic cells may have lived even earlier than 3.5 billion years ago (Brocks et al., 1999). By accepting the isotopic evidence that some kind of metabolic activity, primitive cellular life form, common ancestor existed 3.8 billion years ago, it is realistic to think that its complexity, both morphologically, genetically and metabolically could have evolved to the level of cyanobacteria 3.6 billion years ago. These data suggest that the RNA World existed earlier. It is now generally believed that the archaea and bacteria developed separately from a common ancestor less than 4 billion years ago. Non-oxygen-producing bacterial species, such as purple and green bacteria are the oldest photosynthetic bacteria. Blue-green algae (also called blue-green bacteria or cyanobacteria) are known from rocks that are 3.5 billion years old. Cyanobacteria were among the last prokaryotic bacterial groups to form.

The explanation that one cannot push closer the development of the RNA World to the age of Earth is that during the evolution of our solar system the Earth and the other inner planets were subjected to a great deal of bombardment by asteroids and comets between 4.2 and 4.3 billion years ago. This means that the RNA World could not have existed much after 4.0 and before 4.2 billion years ago as a primitive prebacterial life (Fig. 1.5). Since photosynthesis took place in an aqueous environment, the ocean must have been saturated first with oxygen before its accumulation took place in the atmosphere.

There is a gap between the appearance of prokaryotes and development of oxygen between 4.0 and 3.6 billion years ago. This gap is regarded as the RNA World. Based on Fig. 1.5. there might be some correlation between oxygen production and the decline of the RNA World.

Stromatolites are layered mounds, columns, and sheets found in the Archean rocks of Western Australia, dated 3.5 billion years old. They were originally formed by the growth of layer upon layer of *cyanobacteria*, a single-celled photosynthesizing microbe growing on the sea floor. Stromatolites became more common 2.5 billion years ago and gradually changed the Earth's reducing atmosphere to the present-day oxygen-rich atmosphere. These observations are in correlation with the modell of Russel and Hall regarding the emergence of life at warm submarine wents (Russel and Hall, 1997) and compatible with the theory of the last universal common ancestor (Luca) proposing a free-living, inorganically housed assemblage of replicable genetic elements, harbouring an RNA genome (Forterre et al., 2005). The evolution of enzymatic systems for biosynthetic processes enabled the first bacteria (eu-, and archebacteria) to escape from the hatchery to other places (Koonin and Martin, 2005) of the sea floor closer to the surface making photosynthesis possible.

The comparison of the origin of DNA replication in the three domains of life (bacteria, eukarya and archea) has shown that the core information processing machineries of archea are distinct from bacteria and are fundamentally related to those of eukaryotes (Bell and Jackson, 2001; Kelman and Kelman, 2003). To explain these major evolutionary transitions and the formations of the three domains of life theoretical considerations suggest that viruses played a critical role (Forterre et al., 2005; Forterre, 2006). It is assumed that the invention of DNA required enzymatic activities for the synthesis of DNA precursors, reverse transcription and the replication of DNA (Forterre et al., 2003).

To summarize the data related to appearance of macromolecular biosynthesis: earliest preRNA life could have started between 4.2 and 4.3 billion years ago. A strong candidate for a pre-RNA World informational molecule is a peptide nucleic acid, or PNA. It could have bases bound to a peptide-like backbone, in contrast to the sugar-phosphate backbone of RNA. At the time of appearance of cyanobacteria the atmosphere of Earth was reducing. RNA World started ≈ 4.0 billion years ago and could have been present till DNA emerged when oxygen appeared in the atmosphere as a by-product of photosynthesizing cyanobacteria. Geological and geochemical evidence indicate that the level of atmospheric oxygen was extremely low before 2.45 billion years ago, and that it had reached considerable levels by 2.22 billion years ago (Bekker et al., 2004). A corollary of the appearance of DNA hypothesis is that DNA could have developed only in oxidizing atmosphere when oxygen reached a critical level influencing most of the existing organisms at around 2.2 billion years ago. This estimate narrows the gap between the appearance of oxygen atmosphere and DNA.

Molecular Oxygen Serving the Gradual Transition from RNA World to DNA Empire

There is no doubt that bacteria played a major role in the events during the transition from RNA World to DNA Empire. No one disputes the fact that photosynthesis originated in bacteria, with some bacteria possessing photosystems that produced oxygen similar to how green plants do today. How did oxygen production influence the transition from RNA World to DNA Empire?

Saturation of Seawater with Oxygen and Development of Oxygen Atmosphere

To reach conclusion regarding the influence of oxygen on ancient life we discuss first the effect of oxygen on organisms. O_2 dissolved in water diffuses into cells at relatively low concentration. It is generally accepted that at the beginning of life, isoionic conditions existed, intracellular and extracellular concentrations were identical. The transition from aquatic to terrestrial mode of life is thought to have occured during the Devonian Era some 400 million years ago. The inner environment in which terrestrial vertebrates live is the blood and other body fluids ("milieu interior"). In spite of evolutionary divergence and different ways of osmotic regulation, land vertebrates have nearly identical ion concentrations in their blood. These animals maintained the osmolality of their inner environment (≈ 300 mOsm), while the osmolality of the present day ocean is more than three times higher (1.09 Osm). One osmole corresponds to the osmolality of 1 kg aqueous solution containing 6.0221415×10^{23} (Avogaro's number) dissolved particles. The present osmolality of sea reflects a long-term concentration process of salt in oceans (Bánfalvi, 1991). At the beginning of the photosynthesis the salinity of the ancient ocean was probably much lower than 300 mOsm. Since the solubility of gases is inversely proportional to the salinity of the solution, the primordial ocean due to its lower salt content could have dissolved more oxygen than the present day ocean. Atmospheric oxygen originates from the sea. The reducing atmosphere refers to the absence of oxygen, while some oxygen could have already been present in the seawater. The lack of oxygen also means that there was much more seawater than today. The increasing amount of oxygen in the atmosphere resulted in the gradual loss of water and an icreased land surface. To saturate the immense body of ancient sea with the oxygen must have been a long process before significant amount of oxygen appeared in the atmosphere. The slow saturation process of the sea followed by the gradual increase of oxygen concentration in the atmosphere allowed a relatively long transition period from RNA World to DNA Empire between 3.5 and 2.2 billion years ago.

Advantage of Oxygen Production

All air-breathing organisms get most of their energy by directing electrons from certain food molecules to O_2 in a process called aerobic respiration where O_2 picks up hydrogen atoms and becomes water. The process in which the electrons are "falling" down the energy hill and land on O_2 is called the electron-transport, allowing cells to make 18-times more ATP, than under anaerobic conditions.

Oxygen the Janus Faced Molecule

Oxygen as a by-product of photosynthesis has shown probably its worse side at the beginning of life. Oxygen oxidizes-SH groups to form S-S interactions. Aerobic cells have developed enzymes which reduce back disulfides (S-S) to thiols containing –SH groups. Other enzymes can also reduce components oxidized by O_2. Cells can tolerate O_2 primarily because the first of the two electrons required for reduction to H_2O is not easily picked up by oxygen in the aqueous environment. As a result some of the oxygen taken up by the cells is always converted to reactive oxygen species (ROS). ROS are either free radicals, reactive anions containing oxygen atoms, or molecules containing oxygen atoms that can either produce free radicals or are chemically activated by them. The most agressive radical oxygen species are hydroxyl radical (OH·), superoxide (O_2·⁻) anion, hydrogen peroxide (H_2O_2) and other molecules. Although, the hydroxyl radical often acts as a first step to remove air pollutants and is referred to as the "detergent" of the atmosphere, it has equally damaging effect in living organisms and is regarded as the most dangerous ROS species. The high reactivity of hydroxyl radical and that it cannot be detoxified by supreoxide dismutase makes this compound very dangerous. The only means to protect cellular structures from hydroxyl radical is their antioxidant capacity provided by gluthathione, NADPH and by the consumption of antioxidants mainly of plant origin. The hydroxyl radical can damage virtually all types of macromolecules including nucleic acids causing mutations, lipids (lipid peroxidation), amino acids and carbohydrates. Cell damage is also induced by other reactive oxygen species. Among these molecular species, the life span of OH· and

$O_2 \cdot^-$, which have an unpaired electron, is the shortest. OH· has the highest reactivity and it reacts with various molecules by a diffusion controlled rate. The main sources of ROS *in vivo* are aerobic respiration, peroxisomal β-oxidation of fatty acids, microsomal cytochrome P450 metabolism of xenobiotic compounds, stimulation of phagocytosis by pathogens or lipopolysaccharides, arginine metabolism, and tissue specific enzymes (Fiers et al., 1999; Nicholls and Budd, 2000). Malarial infection induces the generation of hydroxyl radical (OH·) in liver, responsible for the induction of oxidative stress and apoptosis (Guha et al., 2006).

During the development of oxygen atmosphere many mechanisms have been developed which protect cellular constituents from the potentially harmful effects of O_2. Aerobic organisms have enzymes which protect other enzymes and substances inside the cells from the potentially harmful effects of molecular oxygen. These enzymes are: superoxide dismutase, catalase, peroxidase.

Superoxide dismutase (SOD) provides the first line of defence against oxygen toxicity. Superoxide dismutase catalyses the dismutation by the disproportionation of two superoxide anions ($O_2 \cdot^-$) into hydrogen peroxide and molecular oxygen. The turnover of SOD is one of the highest among enzymes (10^6/sec) catalysing the reaction:

$$2O_2 \cdot^- + 2H^+ \rightarrow H_2O_2 + O_2$$

The highly toxic superoxide radical is converted to the less toxic hydrogen peroxide. Catalase provides the second defence line against the toxic intermediates of oxygen. It catalyses the inactivation of peroxide:

$$H_2O_2 + H_2O_2 \rightarrow O_2 + 2H_2O$$

Peroxidases represent the third defence line against the toxic intermediates of oxygen. Peroxidases are oxidoreductases which use H_2O_2 as electron acceptor for catalyzing different oxidative reactions. The overall reaction is as follows:

$$\text{Oxygen donor} + H_2O_2 \rightarrow \text{oxidized donor} + 2H_2O$$

Returning to the most important defence enzymes represented by superoxid dismutases, in all SODs the metal ion (M) catalyses dismutation of the superoxide radical through a cyclic oxidation-reduction mechanism:

$$M^{3+} + O_2 \cdot^- \rightarrow M^{2+} + O_2$$
$$M^{2+} + O_2 \cdot^- + 2H^+ \rightarrow M^{3+} + H_2O_2$$

Four classes of SODs are known, distinguished by the metal prosthetic group: Cu/Zn, Fe, Mn and Ni. Fe- and Mn-SODs constitute a structural family (Parker et al., 1987, Parker and Blake, 1988).

The appearence of molecular oxygen in the cells and the development of oxygen-defence mechanisms immediately suggests that there must be a close evolutionary relationship between them. This idea is supported by the fact that superoxide dismutase is widespread in nature, present in all aerobic organisms indicating that, in both prokaryotes and in eukaryotes, superoxide dismutase is an important component of the defences against oxygen toxicity (Gregory et al., 1974). Superoxide dismutase is also found in anaerobic bacteria to protect them from oxygen toxicity (Hewitt and Morris, 1975). Cu/Zn-SOD is found primarily in the cytosol of eukaryotic cells and in the stroma of chloroplasts. Mn-SOD is typical to prokaryotes and present in the matrix of mitochondria. Fe-SOD is characteristic to prokaryotes and to the chloroplasts of some eukaryotic families. Moreover, organ specific differentiation of peroxisomes may reveal further SOD variants, questioning the common endosymbiontic origin of mitochondria and peroxisomes. While mitochondrion is still regarded an endosymbiontic organ, peroxisomes might have also arisen from internal membrane systems.

Protection of Anaerobes from Oxygen Toxicity

In order to survive, the members of local primordial microbial communities had to cope with the toxic effects of molecular oxygen (O_2) which diffused inside the primitive prebiotic living forms. Biochemical and ultrastructural analyses suggest that the cytoplasmic membrane and the genetic material are the main sites damaged by singlet oxygen (metastable state of molecular oxygen). Primitive bacteria protected themselves with thinner (Gram positive) or thicker cell walls (Gram negative) against oxygen. The major difference between the two cell types is in the thickness of the peptidoglycan rather than the chemical makeup. The cell wall lies immediately external to the plasma membrane. Studies on cell samples exposed to singlet oxygen for different periods of time showed a drastic decrease in survival for *Streptococcus faecium* (Gram positive), while *Escherichia coli* (Gram negative) becomes sensitive only when the integrity of the outer membrane is altered (Valduga et al., 1993). To evade the toxic effects of oxygen flagella for locomotion anchored in the membrane allowed the bacterial cell to swim away from a highly toxic environment. Bacteroides are the most important obligately anaerobic and heterogenous Gram-negative bacilli implicated in anaerobic infections. Their taxonomy is under revision.

Free Radical Formation During the Synthesis of Deoxyribonucleotides

Deoxyribonucleotides are synthesized by a radical mechanism. Ribonucleotide reductase catalysis the synthesis of deoxyribonucleotides by a radical mechanism. The synthesis of deoxynucleotides takes place at the ribonucleoside diphosphate level.

The precursors (dNDPs) of substrates for DNA synthesis (dNTPs) are formed by the reduction of ribonucleoside diphosphates (rNDP) to deoxyribonucleoside diphosphates (dNDP).

$$rNDP + NADPH + H^+ \rightarrow dNDP + NADP^+ + H_2$$

The reaction mechanism is more complex than this equation would suggest since the electrons from NADPH are transferred to the substrate via a series of carriers (Reichard, 1993). Moreover, all known ribonucleotide reductases rely on a radical to achieve catalysis. The regeneration of the stable free tyrosyl radical requires molecular oxygen (O_2). The B2 subunits of ribonucleotide reductase in *Escherichia coli* contain stable tyrosyl radicals which are generated by an iron-sulfur center rather than an iron-oxygen center (Thelander and Reichard, 1979). The reduction of ribonucleotides to deoxyribonucleotides takes place by the abstraction of a hydrogen atom and the formation of free-radicals (Reichard and Ehrenberg, 1983). Three proteins are needed to catalyse the regeneration of the radical and restore enzymatic activity of ribonucleotide reductase. One of them is superoxide dismutase. Molecular oxygen is the electron acceptor in the one-electron oxydation of the tyrosine residue to radical cation. The supporting activity of superoxide dismutase in maintaining the catalytic activity of ribonucleotide reductase points to the possible role of molecular oxygen in the formation of deoxynucleotides and appearance of DNA.

One can argue that the assumption of ribonucleotide reduction requiring molecular oxygen is true for class I ribonucleotide reductases, but not for the class II (adenosylcobalamine-dependent) or class III (anaerobic) ribonucleotide reductases, which appear to be the evolutionary antecedents of the oxygen-requiring class I enzymes (Stubbe et al., 2001). The class I enzymes use a diferric cluster and a stable tyrosyl radical. The class II enzymes use adenosylcobalamin, a precursor to a 5′-deoxyadenosyl radical and cob(II)alamin. The class III enzymes function only under anaerobic conditions and utilize $[Fe_4S_4]^{+1}$ and S-adenosylmethionine (SAM) to generate a stable glycyl radical (Mulliez et al., 1999).

The similarities in the active site structures of the three classes of ribonucleotide reductases, and the similarities in their chemistry together provide strong support for the hypothesis that these proteins are evolutionarily related. They further point to the importance of the unique chemistry in the evolution of the different classes of ribonucleotide reductase from one precursor. Assuming that a ribozyme will not be capable of catalyzing such complex radical chemistry, then proteins must have preceded DNA in the transition from an RNA World to a DNA Empire (Stubbe, 2000). The complex pathway of adenosylcobalamin generation and the fact that S-adenosyl methionine is a poor adenosylcobalamin source makes it likely that class III ribonucleotide reductase could have been the progenitor enzyme (Reichard, 1997). The biosynthesis of S-adenosylmethionine is less complex, but cannot initiate a radical-dependent reaction and in addition requires the presence of the quite unusual Fe_4S_4 cluster, the availability of which could be seriously doubted. Although, we do not know the composition of the prebiotic soup and the biosynthetic pathways in it, but whenever oxygen is taken up by the cells it is always converted to reactive oxygen species (ROS). Consequently, the incidental ribonucleotide reductase reactions driven by class II or III enzymes were rapidly overwhelmed by the radical mechanism of class I enzymes in the presence of molecular oxygen, favouring the idea that oxygen might have been the major molecular player in the development and expansion of the DNA Empire.

Cellular Responses to DNA Damages

Conceptual Changes in Gene Stability

The paradigm shift from genes viewed as stable entities to genes as structures subject to environmental and metabolic perturbation led to the acknowledgement of repair mechanisms in the 1960s. In the last two decades a further conceptual change took place. It came from the recognition that repair in itself would not be enough to cope with DNA damage, thus alternative mechanisms are required to respond to DNA damage, such as the induction of cell cycle checkpoints and/or ultimately the induction of apoptosis. To maintain the integrity of the genome eukaryotic cells employ mechanisms, that temporarily block the cell cycle. These blocks are called checkpoints. The slowing down of the cell cycle and blocking of DNA replication and mitosis provides opportunities for DNA repair and induction of protein synthesis involved in these processes. In multicellular organisms a further mechanism is induced, namely apoptosis. The programmed cell death of a limited number of damaged cells serves the survival of the organism.

Primary Information on Cytotoxicity Can Be Obtained at DNA Level

On the top of the hierachical arrangement of processes involved in the transfer of genetic information the DNA $\Leftrightarrow$ DNA transfer processes (DNA recombination, replication) carry the primary information from one generation of cells to the next one. DNA $\Leftrightarrow$ DNA transfer is also responsible for maintaining the integrity of the genetic material (DNA replication and repair) (Fig. 1.10). DNA damage suppresses DNA replication to avoid mutagenic changes to be perpetuated in the genome in the next generation (Murray, 1992). Such control points or checkpoints are found at the borders of cell cycle phases (Hartwell and Weinert, 1989). Checkpoints not only ensure the completion of downstream events before further progress through the cell cycle is permitted but also offer possibility for DNA repair before mutations are fixed and replication could continue. Alternatively, DNA damage of multicellular organisms may lead to programmed cell death (apoptosis) to protect the organism by sacrifying damaged cells.

Detection of Genotoxic Changes at the Structural Level of DNA

Enzymatical Detection of Cellular Lesions: Too Late

Cellular lesions are related to macromolecular synthetic processes including the hiearchical flow of genetic information: DNA => RNA => protein. Biochemical parameters such as enzymatic measurements may serve as useful tools for monitoring environmental pollutions, but they are not early signs of cytotoxicity and represent late effects in the hierachy of transfer of genetic information. In the metabolism of xenobiotics the release of reactive intermediates causing genotoxic effects is followed by the activation of biotransformation enzymes. Organic phosphates used as insecticides or nerve gases in chemical warfare are inhibitors of acetylcholine esterases. The inhibition of cholinesterase activity is a warning signal of chemical pollution. The degree of genotoxic effects can be estimated by the increased activity of cellular enzymes (e.g. lactate dehydrogenase, glutamate dehydrogenase, glutamate-oxalacetate transaminase etc. in blood).

Early Detection of Cytotoxicity: at DNA Level

Enzymatic reactions could be replaced with assays at the level of the primary source of genetic information. Early detection of cytotoxicity at structural and functional level of DNA combined with high sensitivity are the expected advantages. Toxic treatment of cells increases the rate of mutations leading to a shift to the left:

$$\mathbf{DNA'''} <= \mathbf{DNA''} <= \mathbf{DNA'} <=> \mathbf{DNA} \text{ information}$$

where DNA', DNA'', DNA''' are increasingly mutated and DNA is the normal or repaired genetic information.

Cytotoxicity could be followed by methods such as: 1. measuring simultaneously the two different types of DNA synthesis (replicative and repair), 2. cytometric measurement of nascent DNA without detecting the bulk of cellular DNA, 3. following the chromatin condensation which is reflecting early toxic changes at the structural level, and 4. computer analysis of chromosome images being in different stages of condensation.

1. Oxidative damage changes the ratio of replicative and repair DNA synthesis. Reduced DNA replication and elevated repair synthesis are expected in cells after chemical treatment and upon irradiation. p53 is directly involved in base excision repair (BER) (Offer et al., 1999) and BER synthesis varies during the cell cycle (Offer et al., 2001). Toxic agents are expected to influence different types of repair synthesis at different stages of cell cycle.
2. The direct cytometric measurement of pg amounts of nascent DNA is regarded as a potential tool to indicate cell growth and cancer treatment. The essence of the cytometric method is: (a) Pulse labeling of replicative and repaired DNA in permeable cells in the presence of biotinylated nucleotide (biotin-11-dUTP). (b) Reversal of permeabilization to restore membrane integrity. Isolation

of nuclei and synchronization. (c) Immunofluorescent amplification of nascent and repaired DNA. (d) Flow cytometric measurement of replicative and repair synthesis. Cytometric analysis of DNA replication and DNA repair are potential indicators of genotoxic effects since toxic treatment blocks replication and increases repair DNA synthesis. Checkpoints at which inhibition and elevated repair synthesis takes place can be characterized by their C-values. The ability to visualize and to measure newly synthesized (relicative and/or repair) DNA without the interference by the bulk of cellular nucleic acids offers considerable advantage for cytometric analysis of replicating cells and for repair synthesis (Schweighoffer et al., 1991). Based on replicative and repair synthesis measured by cytometry cytotoxic measurements can be automated.

3. Highly purified fractions of cells and nuclei representing narrow cell cycle segments can be obtained by centrifugal elutriation (Bánfalvi, 2008). Temporally distinct forms of chromosome condensation obtained from synchronized populations of cells offer at least two lines of research. (a) Analysis of intermediates of chromatin condensation with respect to their temporal development in synchronized populations. (b) Analysis of intermediates of chromatin condensation using electron microscopy, confocal microscopy, fluorescent microscopy.

4. Computer analysis of chromatin structures: (a) Data obtained by microscopic imaging can be interphased with a digital network, data stored and analyzed by computer. Computer programmes allow a 2–3-fold magnification of microscopic images. (b) Software programmes adapted to distinguish between density and light intensity of a fluorochrome are used in the study of chromosome condensation. Compactness of chromatin can be colour coded and coloured condensing chromosomes analyzed in three dimensions. The software programmes of chromosome analysis are able to resolve structural differences invisible for the human eye. (c) Additional optical devices allow the 3-D analysis and a further 2–3-fold magnification (total 4000–9000-fold) of chromosomes contributing significantly to the analysis of chromosome aberrations at the chromatin level.

Conclusions

The RNA World theory is strengthened by additional arguments based on the functions of ribonucleotides. By accepting the RNA World theory a corollary of the hypothesis on the DNA Empire is that:

1. Oxygen played a crucial role in the formation of DNA.
2. Although, RNA was relatively stable under the reducing atmosphere, the free 2' OH group of ribose in ribonucleotides and the cis-diol at the 3'-end of RNA became extremely vulnerable when oxygen diffused into the cells.
3. Nucleotide bases are sensitive to the presence of oxygen. Bases containing amino groups (A, G, C) undergo spontaneous oxydative deamination (adenine→ hypoxantin; guanine → xanthin; citosine → uracil). The citosine → uracil (C → U) transition turned out to be one of the most dangerous mutations in DNA.

Without repair the mutated U would pair with A in the next generation and in the long run if unattended the G/C base pairs would have gradually turned to A/T pairs and the four letter genetic code reduced to two bases.

4. The evolution of the oxygen atmosphere brought two new inventions in the nucleic acid structure: a) the metylation of uracil to thymine distinguishing it from deaminated citosine, b) base excision repair, by cutting out first the mutated uracil from DNA, then the rest of the nucleotide and replacing it with citidylate.

5. PreRNA-World probably existed when the primitive life form started on Earth as early as 4.2 billion years ago.

6. RNA-World came into being when prokaryotes emerged around 4.0 billion years ago.

7. The rise of DNA Empire seems to be connected with the development of atmospheric oxygen nearly 2.2 billion years ago.

8. The DNA Empire can be summarized as a hierarcically arranged group of processes.

9. Different processes belonging to the DNA Empire can be selected as potential genotoxic indicators Early indicators are related to DNA ⇔ DNA transfer processes and to changes in the early stage of chromatin condensation.

Before going into details of chromatin condensation the next chapter deals with DNA structure from nucleotides to chromosomes.

References

Bánfalvi, G. (1991). Evolution of osmolyte systems. Biochem Educ. **19**, 136–139.

Bánfalvi, G. (2006). Why ribose was selected as the sugar component of nucleic acids. DNA Cell Biol. **25**, 189–196 (2006).

Bánfalvi, G. (2008). Cell cycle synchronization of animal cells and nuclei by centrifugal elutriation. Nature Protocols. **3**, 663–673.

Beadle, G.W. and Tatum, E.L. (1941). Genetic control of biochemical reactions in Neurospora. Proc Natl Acad Sci U S A. **27**, 499–506.

Bekker, A., Holland, H.D., Wang, P.L., Rumble III, D., Stein, H.J., Hannah, J.L., Coetzee, L.L. and Beukes, N.J. (2004). Dating the rise of atmospheric oxygen. Nature. **427**, 117–120.

Bell, S.D. and Jackson, S.P. (2001). Mechanism and regulation of transcription in archaea. Curr Opin Microbiol. **4**, 208–213.

Benzer, S. (1957). The elementary units of heredity. In The Chemical Basis of Heredity. Johns Hopkins University Press, Baltimore MD. pp. 70–93.

Bird, A. (2007). Perceptions of epigenetics. Nature. **447**, 396–398.

Brocks, J.J., Logan, G.A., Buick, R. and Summons, R.E. (1999). Archean molecular fossils and the early rise of eukaryotes. Science. **285**, 1033–1036.

Cech, T.R. (1986). A model for the RNA-catalysed replication of RNA. Proc Nat Acad Sci USA. **83**, 4360–4363.

Crick, F.H.C. (1968). The origin of the genetic code. J Mol Biol. **38**, 367–397.

Crick, F.H.C. (1970). Central dogma of molecular biology. Nature. **227**, 561–563.

Crick, F.H.C. (1993). In The RNA World. Gesteland, R.F. and Atkins, J.F. eds. Cold Spring Harbor Laboratory Press, p. xiv.

Crick, F.H.C., Barnett, L., Brenner, S. and Watts-Tobin, R.J. (1961). General nature of the genetic code for proteins. Nature. **192**, 1227–1232.

Fiers, W., Beyaert, R., Declercq, W. and Wandenabeele, P. (1999). More than one way to die: Apoptosis, necrosis and reactive oxygen damage. Oncogene. **18**, 7719–7730.

Forterre, P. (2006). The origin of viruses and their possible roles in major evolutionary transitions. Virus Research. **117**, 5–16.

Forterre, P., Filée, J. and Myllykallio, H. (2003). The Genetic Code and the Origin of Life. Ribas de Pouplana, L. ed. Springer Verlag, New York. p. 24.

Forterre, P., Gribaldo, S. and Brochier, C. (2005). Luca: The last universal common ancestor. Méd Sci, Paris. **21**, 860–865.

Gilbert, W. (1986). The RNA World. Nature. **319**, 616–618.

Gould, S.J. (1989). In Wonderful Life: The Burgess Shale and the Nature of History. W.W. Norton and Company. p. 58.

Gregory, E., Goscin, S. and Fridovich, I. (1974). Superoxide dismutase and oxygen toxicity in a eukaryote. J Appl Bacteriol. **117**, 456–460.

Guha, M., Kumar, S., Choubey, V., Maity, P. and Bandyopadhyay, U. (2006). Apoptosis in liver during malaria: Role of oxidative stress and implication of mitochondrial pathway. FASEB J. **20**, 1224–1226.

Hartwell, L.H. and Weinert, T. A. (1989). Checkpoints: Controls that ensure the order of cell cycle events. Science **246**, 629–634.

Hewitt, J. and Morris, J. (1975). Superoxide dismutase in some obligately anaerobic bacteria. FEBS Letts. **50**, 315–318.

Hunt, T., Prentis, S. and Tooze, J. (1983). In DNA Makes RNA Makes Protein. Elsevier Press, Amsterdam. 226–232.

Joyce, G.F. (1991). The rise and fall of the RNA World. New Biologist. **3**, 339–407.

Kelman, L.M. and Kelman, Z. (2003). Archaea: An archetype for replication initiation studies? Mol Microbiol. **48**, 605–615.

Koonin, E.V. and Martin, W. (2005). On the origin of genomes and cells within inorganic compartments. Trends in Genet. **21**, 647–654.

Larralde, R., Robertson, M.P. and Miller, S.L. (1995). Rates of decomposition of ribose and other sugars: Implications for chemical evolution. Proc Natl Acad Sci U S A. **92**, 8158–8160.

Matthaei, J.H., Jones, O.W., Martin, R.G. and Nirenberg, M.W. (1962). Characteristics and composition of RNA coding units. Proc Natl Acad Sci U S A. **48**, 666–677.

Mojzis, S.J., Arrhenius, G., Mckeegan, K.D., Harrison, T.M., Nutman, A.P. and Friend, R.L. (1996). Evidence for life on Earth before 3,800 million years ago. Nature. **384**, 55–59.

Moore, P.B., Gesteland, R.F. and Atkins, J.F. (1993). In The RNA World. Cold Spring Harbor Laboratory Press. p. 131.

Mulliez, E., Meier, C., Cremonini, M., Luchinat, C., Trautwein, A.X. and Fontecave, M. (1999). Iron-sulfur interconversions in the anaerobic ribonucleotide reductase from *Escherichia coli*. J Biol Inorg Chem. **4**, 614–620.

Murray, A.W. (1992). Creative blocks: Cell-cycle checkpoints and feedback controls. Nature. **359**, 599–604.

Murray, J.B. and Arnold, R.P. (1996). Antibiotic interactions with the hammerhead ribozyme: Tetracyclines as a new class of hammerhead inhibitor. Biochem J, Great Britain. **317**, 855–860.

Nicholls, D.G. and Budd, S.L. (2000). Mitochondria and neuronal survival. Physiological Reviews. **80**, 315–360.

Offer, H., Wolkowicz, R., Matas, D., Blumenstein, S., Livneh, Z. and Rotter, V. (1999). Direct involvement of p53 in the base excision repair pathway of the DNA repair machinery. FEBS Lett. **450**, 197–204.

Offer, H., Zurer, I., Bánfalvi, G., Rehak, M., Falcovitz, A., Milyavsky, M., Goldfinger, N. and Rotter, V. (2001). p53 modulates base excision repair activity in a cell cycle specific manner following genotoxic stress. Cancer Res. **61**, 88–96.

Parker, M.W., Blake, C.C., Barra, D., Bossa, F., Schinina, M.E., Bannister, W.H. and Bannister, J.V. (1987). Mitochondria and neuronal survival. Protein Eng. **1**, 393–397.

Parker, P.W. and Blake, C.C. (1988). Iron- and manganese-containing superoxide dismutases can be distinguished by analysis of their primary structures. FEBS Lett. **229**, 377–382.

References

Pley, H.W., Flaherty, K.M. and Mckay, D.B. (1994). Three-dimensional structure of a hammerhead ribozyme. Nature. **372**, 68–74.

Reichard, P. (1993). The anaerobic ribonucleotide reductase from *Escherichia coli*. J Biol Chem. **268**, 8383–8386.

Reichard, P. (1997). The evolution of ribonucleotide reduction. Trends Biochem Sci. **22**, 81–85.

Reichard, P. and Ehrenberg, A. (1983) Ribonucleotide reductase-a radical enzyme. Science. **221**, 514–519.

Russel, M.J. and Hall, A.J. (1997). The emergence of life from iron monosulphide bubbles at a submarine hydrothermal redox and pH front. J Geol Soc London. **154**, 377–402.

Sarabhai, A.S., Stretton, A.O.W., Brenner, S. and Bolle, A. (1964). Co-linearity of the gene with the polypeptide chain. Nature. **201**, 13–17.

Schleiden, M.J. (1938). Beitrage zue Phytogenesis. Arch. Anat. Physiol. Wiss. Med. **13**, 137–176.

Schmidt-Nielsen, K.S. (1997). Respiration in water. In Animal Physiology. Cambridge University Press. pp. 16–17.

Schwann, T. (1939). Mikroskopische Untersuchungen über die Übereinstimmung in der Struktur und dem Wachstum der Tiere und Pflanzen, Sander'schen Buchhandlung, Berlin.

Schweighoffer, T., Schweighoffer, E., Apati, A., Antoni, F., Molnar, G., Lapis, K. and Bánfalvi, G. (1991). Cytometric analysis of DNA replication inhibited by emetine and cyclosporin A. Histochemistry. **96**, 93–97.

Smith, D., Muscatine, L. and Lewis, D. (1969). Carbohydrate movement from autotrophs to heterotrophs in parasitic and mutualistic symbiosis. Biol Rev Camb Phil Soc. **44**, 17–90.

Stubbe, J. (2000). Ribonucleotide reductases the link between an RNA and a DNA world? Curr Opin Struct Biol. **10**, 731–736.

Stubbe, J., Ge, J. and Yee, C.S. (2001). The evolution of ribonucleotide reduction revisited. Trends Biochem Sci. **26**, 93–99.

Temin, H.M. and Baltimore. D. (1972). RNA-directed DNA synthesis and RNA tumour viruses. Adv Virus Res. **17**, 129–36. Review.

Thelander, L. and Reichard, P. (1979). Reduction of ribonucleotides. Annu Rev Biochem. **48**, 133–158.

Valduga, G., Bertoloni, G., Reddi, E. and Jori, G. (1993). Effect of extracellularly generated singlet oxygen on gram-positive and gram-negative bacteria. Journal Photochem Photobiol. *B*. **21**, 81–86.

Watson, J.D. and Crick, F.H.C. (1953). Molecular structure of nucleic acids. a structure for deoxyribose nucleic acid. Nature. **171**, 737–738.

Yanofsky, C., Carlton, B.C., Guest, J.R., Helinski, D.R. and Henning, U. (1964). On the colinearity of gene structure and protein structure. Proc Natl Acad Sci U S A. **51**, 266–271.

Yashuda, K. (2004). Biotecnology approach to determination of genetic and epigenetic control in cells. J. Nanobiotechnol. **2**, 11.

Zaug, A.J. and Cech, T.R. (1980). *In vitro* splicing of the ribosomal RNA precursor in nuclei of *Tetrahymena*. Cell. **19**, 331–338.

Zuckerkandl, E. and Pauling, L. (1965). Molecules as documents of evolutionary history. J Theor Biol. **8**, 357–366.

Chapter 2
Structural Organization of DNA

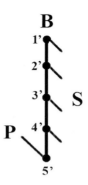

Summary

This chapter summarizes (a) the building blocks of nucleotides and in this context discusses separately the nucleic acid bases, the sugar and phosphate components, (b) the bond types in nucleotides, (c) the metabolism of nucleotides, (d) the lower levels of DNA structure (primary, secondary, tertiary, nucleosomal), (e) supranucleosomal organization of eukaryotic DNA, (f) chromosomes of animal cells, and (g) the temporal and spatial order of gene replication.

Building Blocks of Nucleic Acids

Structure of Nucleotides

The building blocks of nucleic acids, the nucleotides are complex compounds themselves consisting of 3 components: nucleic acid base, sugar (pentose) and 1–3 phosphates (Title Fig. 2.0, see above). Nucleic acids (RNA and DNA) are polymers made up of monomers called mononucleotide units. The monomers of deoxyribonucleic acid (DNA) are the deoxiribonucleotides while the building blocks of ribonucleic acid (RNA) are ribonucleotides. The skeletal model of a nucleotide shows that nucleotides consists of three parts: sugar (S), base (B) and 1–3 phosphates (P).

If a pentose (ribose or 2'-deoxyribose) is added to the nitrogen containing base, the resulting compound is named nucleoside, which however does not belong to the building blocks of nucleic acids. Depending on their phosphate content deoxyribonucleoside monophosphates (dNMPs), deoxyribonucleoside diphosphates (dNDPs) and deoxyribonucleoside triphosphates (dNTPs) are distinguished. The abbreviations of the corresponding ribonucleotides are NMP, NDP, NTP. Nucleoside triphosphates (NTPs) (Fig. 2.1a) are not to be confused with trinucleotides consiting of three nucleotides (Fig. 2.1b). The units in DNA are deoxyribonucleoside monophosphates (dNMPs: dAMP, dGMP, dCMP, dTMP). RNA contains ribonucleoside monophosphate units (NMPs: AMP, GMP, CMP, UMP).

G. Bánfalvi, *Apoptotic Chromatin Changes*, DOI 10.1007/978-1-4020-9561-0_2, 31
© Springer Science+Business Media B V. 2009

Fig. 2.1 The difference between nucleoside triphosphate and trinucleotide. (**a**) nucleoside triphosphate (NTP), (**b**) a trinucleotide consisting of three nucleotides (3 nucleoside monophosphates, NMPs)

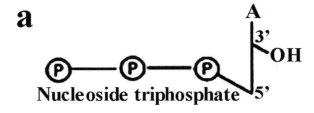

Nucleic Acid Bases

Types. There are two types of nucleic acid bases (Fig. 2.2): pyrimidines (cytosine, uracil, thymine) and purines (adenine, guanine). The repeating units of ...sugar-phosphate-sugar-phosphate... provide the backbone of nucleic acids, and the linear order of the four bases attached to the sugar units bring variety (information) into the structure. The basic compounds of the bases are the purine and pyrimidine rings. The purine ring itself is a pyrimidine derivative, the combination of a pyrimidine and an imidazole ring.

Aromatic Character, Planar Structure of Nucleic Acid Bases

X-ray diffraction experiments revealed that the aromatic pyrimidine ring is a planar structure. The puring ring has also an aromatic character with some folding between the two rings. The essence of the X-ray diffraction analysis is that repeating units of

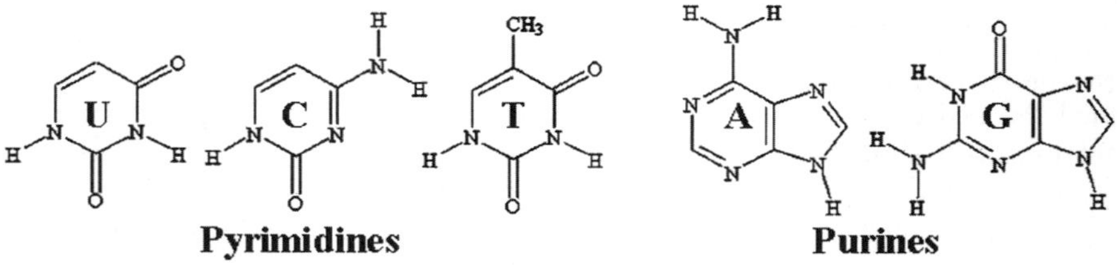

Fig. 2.2 Pyrimidine and purine bases. U, uracil; C, citosine; T, thymine; A, adenine, G, guanine

Building Blocks of Nucleic Acids

atomic or molecular units can be determined in a crystal, depending on the direction and degree of diffraction of X-rays passing through the crystal. Atoms of high electron density (e.g. heavy metals) show a high degree of diffraction, while the ability to diffract X-rays of low density atoms (e.g. hydrogen) is much less.

Free Nucleic Acid Bases

Free purine and pyrimidine bases are found only in traces in cells due to the enzymatic cleavage of nucleic acids to nucleotides. Free bases and nucleosides are converted back to nucleotides by means of the so called salvage reactions. Free purine and pyrimidine bases have a low solubility and exist in two tautomeric forms known as the oxo-enol and the amino-imino conversions (Fig. 2.3). As a general rule one can formulate that under physiological conditions (pH 7.4) the oxo and amino forms of tautomers are stable. Enol and imino tautomers are rare, their non-standard base pairing can lead to mutations.

During the tautomerism of nucleic acid bases the proton and electron shift does not take place between two carbon atoms (oxo-enol tautomerism), but between a carbon and a nitrogen atom (lactam – lactim tautomers) (Fig. 2.4).

Spontaneous oxydative deamination of cytosine into uracil occurs in DNA. This spontaneous mutation is corrected by the removal of uracil in the base excision repair. As opposed to uracil the spontaneous deamination of 5-methylcytosine to thymine in DNA cannot be corrected since the repair mechanism does not recognize thymine.

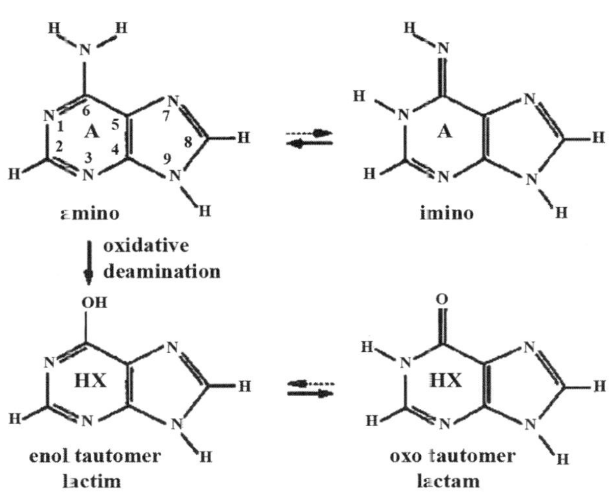

Fig. 2.3 Amino-imino tautomers of adenine and lactim–lactam tautomers of hypoxanthine. Spontaneous oxidative deamination of adenine (6-aminopurine) results in hypoxanthine (HX) and ammonia. The amino and oxo (lactam) tautomers dominate over the imino and enol (lactim) forms. Hypoxanthine is oxidized to xanthine then to uric acid (2,6,8 trioxypurine) by the catalytic action of xanthine oxidase. In humans and higher primates uric acid is the end product of purine catabolism

Fig. 2.4 Enol-oxo and lactim-lactam tautomers. Enol – oxo tautomerisms takes place between two carbon atoms (C-C). In the electron and proton shift of the lactim – lactam transition nitrogen and carbon atoms (N-C) are involved

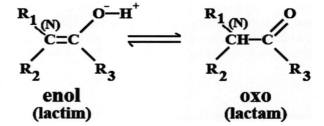

Light Absorption

The aromatic bond system of purine and pyrimidine bases of nucleic acids possesses a strong light absorption capacity in the 250–260 nm wavelenght band. This property is used for the estimation of the content of nucleic acid bases, nucleotides and nucleic acids. The light absorption maximum of aromatic amino acids (phenylalanine, thyrosine, tryptophane) is somewhat higher around 280 nm wavelength. The comparison of the light absorption curve of a protein solution with that of a DNA solution shows some overlapping (Fig. 2.5). A fast method to prove that the purified DNA solution does not contain protein impurity is to measure the optical densities (OD) at 260 and 280 nm. The DNA solution is protein free when the OD 260/280 ratio is higher than 1.8.

Apolar Interactions

In the double stranded DNA the bases are arranged above (below) each other in both strands forming week apolar interactions. The strength of these additive interactions should not be underestimated as they are almost as strong as the hydrogen bonds between the base pairs and contribute significantly to the stability of the double helical structure. An important difference between the two interactions is that while the hydrogen bonds connect the two strands, the hydrophobic forces act among the

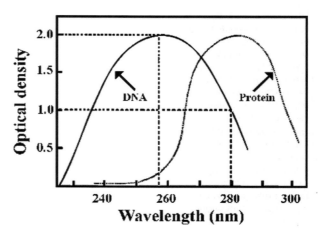

Fig. 2.5 Absorption spectra of DNA and protein solutions

Building Blocks of Nucleic Acids

bases of the same DNA strand. This so called stacking effect ceases when DNA is denatured and single DNA strands are randomly coiled. The stacking effect is involved in the maintainance of the secondary DNA structure.

Hypochrome, Hyperchrome Effect

The absorption of ultraviolet light is much smaller in a solution containing double stranded DNA than in single stranded DNA. The reason is that in the DNA duplex the base pairs above each other reduce the light absorption (hypochrome effect). After denaturing DNA (melting) much more bases are exposed to the ultraviolet light and the opposite hypechrome effect can be measured by a spectrophotometer. The effect of temperature on the light absorption of double stranded DNA is shown in Fig. 2.6.

Major, minor and rare bases. The distiction reflects their relative incidence in nucleic acids.

a. *Major bases*: uracil (only in RNA), thymine (only in DNA), adenine, cytosine and guanine. Functional goups of major bases are the amino and the oxo groups. The amino group behaves as a partial proton donor, the electron attraction of the oxo group is due to the electronegativty of oxygen. These substituents are participating in the H-bond formation between the base pairs (A–T, G–C) of the

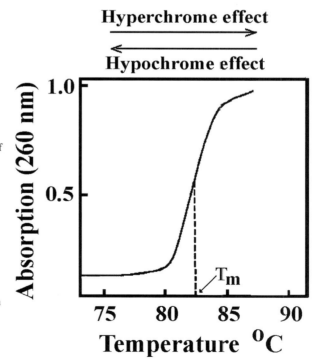

Fig. 2.6 The melting curve of double stranded DNA. The heating of dissolved double-stranded DNA (e.g. bacterial DNA) results in a sudden increase of light absorption at a relatively narrow zone of temperature (hyperchrome effect). The inflexion point of the transition is called the melting point (Tm). By following the light absorption of the denatured DNA in a cooling solution the opposite hypochrome effect can be measured

double stranded DNA. As already mentioned the function of the methyl group in thymine is to distinguish it from uracil. The spontaneous deamination of cytosine results in uracil. The DNA contains thymine to be able to recognize and repair deaminated cytosine (i.e. uracil) in DNA. RNA contains the "less expensive" uracil rather than thymine.

b. **Minor bases** are methyl derivatives such as 5-methylcytosine, 5-hydroxymethylcytosine, 6-methyadenine, 2-methylguanine, etc.

c. Among the **rare bases** are also many methyl derivatives, but other substitutions may also occur: 5,6-hydroxyuracil, 1-methyluracil, 5-hydroxymethyluracil, 2-thiouracil, N4-acetylcytosine, 2-methyladenine, 7-methyladenine, N6,N6-dimethyladenine, 1-methylguanine, 7-methylguanine, N2,N2-dimethylguanine, etc. Most of the rare bases are found in tRNA where their percentage may be more than 10%. So far several dozens of rare bases have beeen identified.

As the order of base pairs in various nucleic acids is different one cannot expect that the distribution of the four bases will be uniformly 25%. Actually, the ratio of base pairs (A–T/G–C) shows significant deviations in different species. To the contrary the A/T ratio similarly to the G/C shows equimolarity. The ratio of purines versus pyrimidines is always 1:1 irrespective of the change of the total amount of G+C versus A+T (Chargaff's rule).

Base Analogues

The best known purine analogues are the xanthine, hypoxantine derivatives and antimetabolite analogues.

1. **Xanthine derivatives**: caffeine, theophylline, theobromine (Fig. 2.7) are alkaloids. Alkaloids are organic, nitrogen containing bases mainly of plant origin exerting strong physiological effects, but amines produced by animals and fungi are also called alkaloids.

    Caffeine (1,3,7 trimethylxanthine) is most commonly consumed as infusions extracted from the beans of the coffee plant (*Coffea arabica*). Caffeine is a central nervous system stimulant, with the potencial to restore alertness. Beverages such as coffee, tea, soft drinks and energy drinks contain caffeine.

    Theophylline (1, 3 dimethylxanthine) originally extracted from *Thea synensis* (Chinese tea) is used as a drug in the therapy of respiratory diseases, primarily asthma. Theobromine (3,7 dimethylxanthine) is the primary alkaloid of the cacao plant (*Theobroma cacao*) found in cocoa and chocolate. It has a vasodilating effect in heart, brain, skin and kidney.

Fig. 2.7 Xanthine derivatives: caffeine, theophylline and theobromine. These purine analogue alkaloids are also of therapeutic importance

Caffeine **Theophylline** **Theobromine**

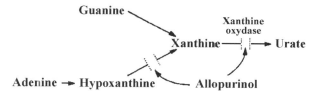

Fig. 2.8 Allopurinol is preventing the hypoxanthine → xanthine → uric acid transition. Allopurinol is a competitive inhibitor of the uric acid biosynthesis

2. *Hypoxanthine analogue.* In allopurinol (4-hydroxypyrazole pyrimidine) the positions of N7 and the C8 atoms have been converted. Allopurinol inhibits xanthine oxydase by preventing the formation of uric acid. In gout the serum urate concentration is high. Due to the low solubility of uric acid it precipitates in the intraarticular synovial liquid of joints causing arthritic pain. Xanthine oxidase is catalyzing the reaction of hypoxanthine → xanthine → uric acid, but in the presence of allopurinol the enzyme cannot dissociate from the product and a so called "suicide inhibition" takes place. The therapeutic effect of allopurinol is based on the inhibition of purine base degradation, preventing the formation of uric acid from purine bases and decreasing the gouty complaints (Fig. 2.8).

Gout used to be the joint disease of rich people. Wealthy people were often suffering from gout due to the fact that they were eating more meat and lived in big, cold castles and palaces. Among the royalties we find Septimus Severus, Roman Emperor; Queen Ann, Queen of England; George IV, King of England; Henry VIII, King of England; Frederic the Great, King of Prussia; Kublai Khan, Mongol Emperor; Louis XVIII, King of France; Matthias I Hunyadi, King of Hungary, also known as Matthias Corvinus. Because of his hunting habit King Matthias Corvinus (1458–90) used for wolf and bear hunting the Kuvasz dog (Ku Assa meaning "Dog of the Horse"). It is said that he kept several Ku Assa with him at all times for protection and their fur as a „hand-towel" after the copious meals. A gift of Kuvasz from the King was a special honor. Among the honored was Count Dracula, upon his release from prison.

Relationship between gout and apoptosis. Gout tophi are characterised by foreign body granulomas consisting of mono- and multinucleated macrophages surrounding deposits of monosodium urate microcrystals. Experimental data show that macrophages are continuously recruited into the gout tophi. *In situ* end-labelling of fragmented DNA demonstrated that CD68+ macrophages undergo apoptosis within gout tophi (Schweyer et al., 2004).

Piece of advice: Do not ride often your bicycle or motorcycle in cold weather, you may get gout.

Antimetabolites

Antimetabolites are similar to normal body molecules but slightly different in structure. Antimetabolites prevent the conversion of these normal substances (metabolites) and interfere with the normal functions of cells including cell growth and division. Although, there are also antibiotics (e.g. sulfanylamides) and folic acid analogues (e.g. metothrexate) among the antimetabolites, but most of the antimetabolites are purine and pyrimidine analogues used in cancer therapy. The best known feature of anti-metabolites is that they often inhibit DNA synthesis and growth of cancer cells which need to make and repair DNA to grow and multiply.

Azathioprine is an inactive base analogue and is converted in the organism to its active metabolites, the most important of which is 6-mercaptopurine. The drug is used in organ transplantation, autoimmune diseases and in Chron's bowel disease as an immunosuppressant.

Azidothymine (AZT) is an inhibitor of reverse transcriptase. The compound persists, after intracellular hydrolysis as phosphorylated or non-phosphorylated antiviral nucleoside. The drug is effective in preventing retroviral replication in HIV infections. Prolonged activity can be reached in combination with other antiviral agents.

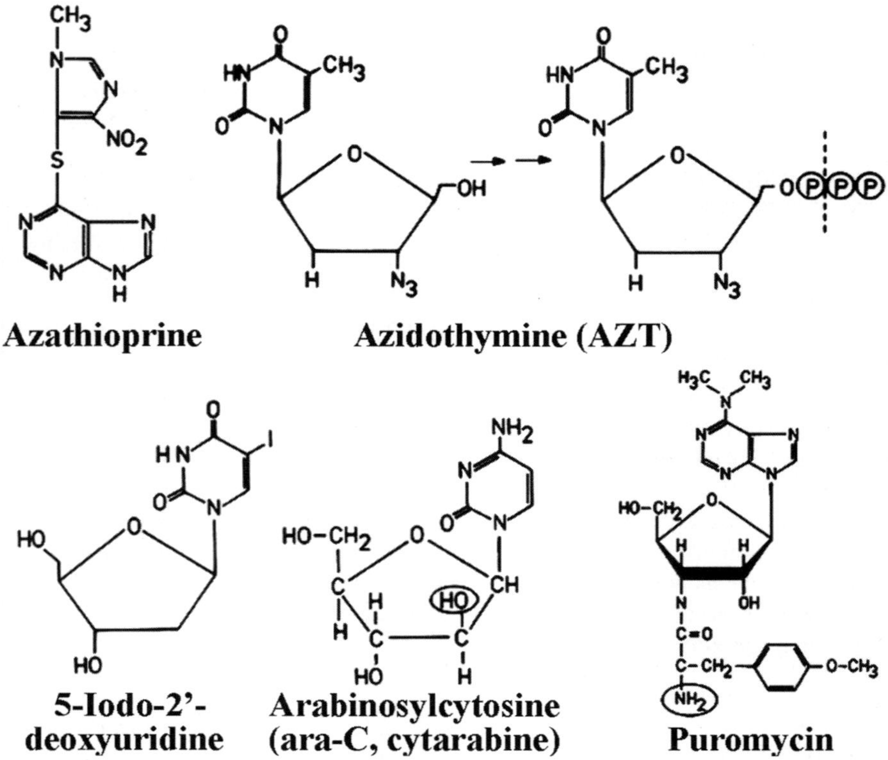

Fig. 2.9 Antimetabolites: base, nucleoside and nucleotide analogues. Base analogue: azathioprine. Nucleoside analogues: AZT, 5-iodo-deoxyuridine, ara-C, puromycine

Building Blocks of Nucleic Acids

5-iodo-2'-deoxyuridine is known to limit the multiplication of deoxyribonucleic acid (DNA) in the host cell, interfering with the formation of herpes viruses causing the "cold sore" on lips, skin and infections at venereal regions (Fig. 2.9).

Arabinosylcytosine (ara-C) is used to treat cancers (head, neck), acute leukemias, non-Hodgkin's lymphoma.

Puromycin is an aminonucleoside antibiotic, derived from the bacterium *Streptomyces alboniger*. Puromycin as an aminoacyl tRNA analogue nucleotide-antibiotic causes premature chain termination during protein biosynthesis. Since it is neither selective for prokaryotes, nor for eukaryotes it is a toxic substance.

Sugar Component of Nucleotides

Why Ribose Was Selected as the Sugar Component of Nucleic Acids

To understand the selection criteria why ribonucleotides have been preferred over other furanosyl derivatives one approach involves the synthesis of nucleotide and nucleoside derivatives to mimic the hypothetical process leading to the selection of RNA (Orgel, 1968; Usher, 1972; Usher and McHabe, 1976; Kierzek et al., 1982; Eschenmoser and Dobler, 1992. Eschenmoser, 1993; Sawai et al., 1996; Prakash et al., 1997). Alternatively, sugar-base connections of nucleotides have been chosen as molecular and graphical models (Calladine, 1982; Dickerson, 1982; Dickerson et al., 1982) to test which sugar and base would fit best in the helical structure of nucleic acids. For precise measurements computer models are the appropriate ones. For teaching purposes structural models of nucleic acids are convenient to use and to explain the primary, secondary and tertiary structures of nucleic acids. These demonstrations gave clues and led to the recognition why ribose was selected as the exclusive component of nucleic acids (Bánfalvi, 2006a).

The selection of ribo- and deoxyribonucleic acids as genetic systems belongs to the central elements of theories related to the origin of life. From the limited knowledge of emergence of RNA in a primordial system (Joyce and Orgel, 1993) it was conjectured that oligonucleotides derived from a library of structural alternatives involving furanosyl and pyranosyl derivatives of sugars (Eschenmoser, 1994). The comparison of pyranosyl derivatives with weaker (Hunziker et al., 1993; Ottig et al., 1993; Groebke et al., 1998) and stronger Watson-Crick pairing (A–T and G–C) (Pitsch et al., 1995; Bolli et al., 1997) indicates that originally base-pairing was not an important criterion for the selection of the sugar component. However, the increased number of atoms in pyranosyl derivatives causes steric hyndrance and unwanted interactions.

As far as the criteria for sugar selection are concerned consideration is given to: (a) chemical stability, (b) rigidity of nucleotides, (c) sugar puckering, (d) maintainance of steric freedom and (e) reducing nonbonding interactions to a minimum.

Molecular models were built in three dimensions using backbone models (the Blackwell Molecular Models) (Fletterick et al., 1985). These models were used for the visualization of pentoses and their basic conformational changes in the furanose

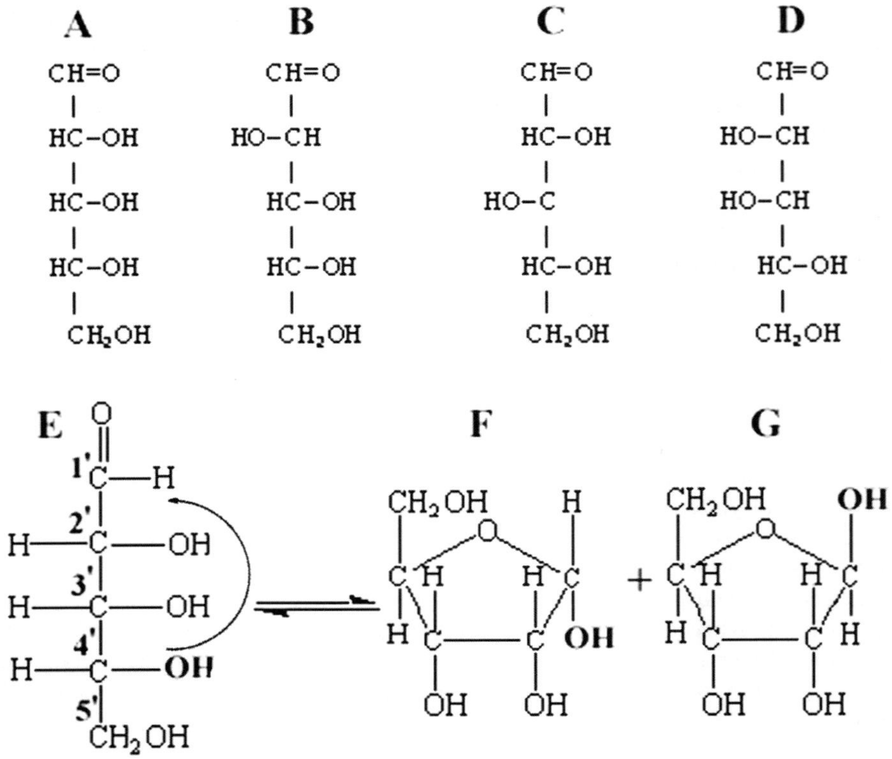

2 Structural Organization of DNA

Fig. 2.10 Open forms of pentoses and ring formation of ribose. Open forms of (**A**) D-ribose, (**B**) D-arabinose, (**C**) D-xylose, and (**D**) D-lyxose. (**E**) The $C_{1'}$ aldehyde and $C_{4'}$ hydroxyl group of D-ribose (*open form*) react with each other and form cyclic hemiacetal resulting in either (**F**) α (alpha) or (**G**) β (beta) anomers of D-ribose (Reproduced with permission of Bánfalvi, 2006a)

ring. Linear and circular forms of pentoses and the ring formation in aqueous environment are shown in Fig. 2.10.

Criteria for Sugar Selection

(*a*) *Chemical Stability of the Polynucleotide Chain*

The most effective way to reduce the number of potential chemical interactions is the reduction of substituents of sugar residues contributing to the stability of the polynucleotide chain. This is achieved by selecting pentoses (Fig. 2.10 A–D) rather than hexoses and selecting 2'-deoxyribose in DNA versus ribose in RNA. The 2'-hydroxyl group of ribose is responsible for the hydrolytic lability of RNA, providing a mechanism for numerous biological functions. Chemical cleavage studies show that the 2'- hydroxyl group stabilizes the sugar moiety in RNA towards oxidation relative to DNA (Thorp, 2000). The reactivity of ribose could have other evolutionary significance, such as the catalytic RNA (ribozyme) (Kruger et al., 1982; Guerrier-Takada et al., 1983). The catalytic

Building Blocks of Nucleic Acids   41

activity of RNA led to the concept known as RNA World (Gilbert, 1986; Sharp, 1985). However, the fast rates of decomposition of ribose and other sugars suggest that there might have been an earlier period before the RNA World. In this pre-RNA World living organisms contained a backbone different from ribose-phosphase and bases different from A, G, C, U. Alternatively, ribose could have been stored in its more stable prebiotic derivatives in the primitive ocean and converted back to ribose from ribulose by isomerization or from lactone with NADH (Larralde et al., 1995).

(b) *Rigidity of nucleotides*

When a pentose is dissolved in water, its linear form turns to a ring structure in a chemical reaction between the C_1' aldehyde in the open-chain form of pentose and the sterically favourable $C_{4'}$ hidroxyl grup to form an intramolecular hemiacetal (Fig. 2.10E). The hemiacetal formation of D-ribose is shown in Fig. 2.10 (E, F and G). Carbon-1', the carbonyl carbon atom in the open-chain form becomes an asymetric atom in the ring form. Two anomeric ring structures can be formed: α-D-ribose (Fig. 2.10F) and β-D-ribose (Fig. 2.10 G). The designation alpha (α) means that the hydroxyl group attached to $C_{1'}$ is below the plane of the ring; beta (β) means that the glycosidic hydroxyl group is above the plane of the ring. Rigidity is contributed by the ring formation in pentoses (Muller et al., 1990; Zubay, 1998). In addition, the $-C_{1'}-O_{4'}-C_{4'}-$ angle is less flexible than angles with carbon as a vertex (Levitt and Warshel, 1978). Conformations of Watson-Crick type polynucleotides are restricted to torsion angles with nucleotides in their most preferred "rigid" forms (Sasisekharan and Pattabiraman, 1978).

(c) *Sugar pucker conformation*

Five membered furanose rings, like six-membered pyranose rings are not planar. The planar furanose ring is energetically unfavourable since in this arrangement all torsion angles are $°0$ and substituents are fully ecclipsed. Small deviation of ring atoms from planarity is called minor, large deviation major puckering. Furanose rings can be puckered in two basic conformations: envelope and twisted or half-chair forms. The puckering of five membered rings containing heteroatoms is based on the concept of endocyclic pseudorotation originally described for cyclopentane (Klpatrick et al., 1947; Pitzer and Donath, 1959) and adapted to substituted furanosyl derivatives (Hall et al., 1970; Saenger, 1984).

Envelope form (E): four of the five atoms in the furanose ring of a nucleotide are nearly coplanar and the fifth is about 0.5 Å away from this plane. The structure resembles an open envelope with the back flap raised from the plane. The probability of being four atoms coplanar rather than three is lower, similarly to hit three rather than four winning numbers in the lottery. Consequently, the envelope form can be excluded as a conformer from the pentose structure.

Twisted form (T) three atoms ($-C_{1'}-O_{4'}-C_{4'}-$) are coplanar and the other two lie away on opposite sides of this plane. The atom which is above the plane is called *endo* and the one under the plane is in *exo* position. In projected formulas of pentoses in nucleotides the largest substituent at $C_{5'}$ carrying the primary hydroxylic group ($HOCH_{2-}$) and the β-glycosidic-bond at $C_{1'}$ are above the

Fig. 2.11 Sugar puckering affecting the sterical positions of the phosphoester and the glycosidic bonds in deoxyribonucleotides. (**A**) In $C_{2'}$ *endo* conformation (as in B-DNA) the $C_{2'}$ carbon atom is above the plane of the ring structure, while $C_{3'}$ is slightly under the plane. Functional groups at $C_{1'}$ and $C_{5'}$ are closer to each other than in (**B**) where the $C_{3'}$ carbon atom is in *endo* position (e.g. A-DNA). White circles indicate hydrogen, black spheres carbon and grey spheres oxygen atoms (Reproduced with permission of Bánfalvi, 2006a)

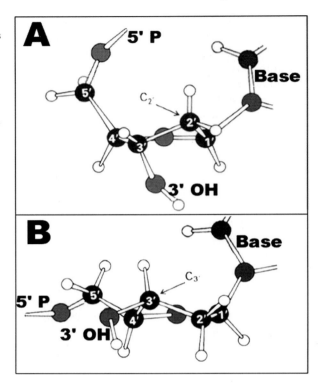

plane. The $C_{2'}$ or $C_{3'}$ atom in *endo* position is above, while the *exo* atom is slightly below the plane of the pentose ring (Fig. 2.11).

Endocyclic rotation: in cyclopentane the pucker rotates freely virtually without energy barriers, giving rise to infinite number in the conformational wheel (Altona and Sundaralingam, 1972). The presence of the heteroatom oxygen and the asymmetrical substitution in furanose create energy thresholds, which limit pseudorotation. The strong electronegativity of the heteroatom oxygen in furanose has a limiting effect on the free rotation of neighboring atoms. Consequently, these three atoms (-$C_{1'}$-$O_{4'}$-$C_{4'}$-) will be in one plane and adopt a 'rigid' $O_{4'}$ *endo* (*east*) sugar pucker.

The other two ($C_{2'}$ and $C_{3'}$) atoms are either in *endo* or in *exo* position. In nucleotides and nucleosides sugar puckering is limited to two twisted conformations: $C_{2'}$-*endo* major $C_{3'}$-*exo* minor (briefly $C_{2'}$-*endo*, or *south*) (Fig. 2.12a) and $C_{3'}$-*endo* major, $C_{2'}$-*exo* minor (briefly $C_{3'}$-*endo*, or north) forms (Fig. 2.12b).

Spectroscopic data indicate a rapid $C_{3'}$-*endo* ⇔ $C_{2'}$-*endo* interconversion. Pyrimidine nucleotides and ribonucleosides favour the $C_{3'}$-*endo* mode of puckering, while most (60–80%) of the purine derivatives occur in $C_{2'}$-*endo* puckering. In deoxyribonucleosides and deoxyribonucleotides $C_{2'}$-*endo* sugar puckering was observed (Altona, 1975; Davis, 1978; Bolli et al., 1997). $C_{2'}$-*endo* conformation is characteristic to the Watson-Crick B-DNA. The sugar

Building Blocks of Nucleic Acids   43

Fig. 2.12 Twisted conformations of sugar puckering in pentoses. (**a**) $C_{2'}$ *endo* conformation. (**b**) $C_{3'}$ *endo* conformation. For the sake of clarity hydroxy groups and hydrogen atoms have been omitted

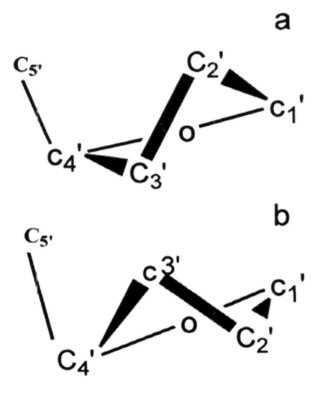

in A-DNA and in RNA is in $C_{3'}$-*endo* conformation. C-C torsion angles are maximally staggered in $C_{2'}$-*endo* or -*exo* and in $C_{3'}$-*endo* or -*exo* puckering. Sugar puckering markedly affects the orientation of the glycosidic bond and those of phosphodiester bridges. Besides the endocyclic rotation which is limited by the heteroatom oxygen, functional groups attached to furanose ring must have played an important role when ribose was selected as the sugar component of nucleic acids. This notion has been confirmed by synthetic nucleotides with modified sugars, which influence sugar puckering and when incorporated into RNA significantly contribute to its RNase resistance. These derivatives which do not exist in nature point to the importance of sugar puckering and to a strong selection against such derivatives.

Exocyclic rotation: In the furanose ring of pentoses: ribose, arabinose, xylose and lyxose the exocyclic rotation about the carbon $C_{4'}$-$C_{5'}$ bond allows $O_{5'}$ to assume three main conformations with staggered positions relative to the furanose ring. In the Klyne-Prelog notation anti (ap) denotes a torsion angle higher than high anti (ac) and high-*syn* (sc) is higher than *syn* (sp) (Prusiner and Sundaralingam, 1972). According to the IUPAC-IUB recommendation it is sufficient to define the torsion angle γ about the $C_{4'}$-$C_{5'}$ looking in the $C_{5'} => C_{4'}$ direction (IUB-IUFAC, 1983). The most likely conformation is *anti* (ap).

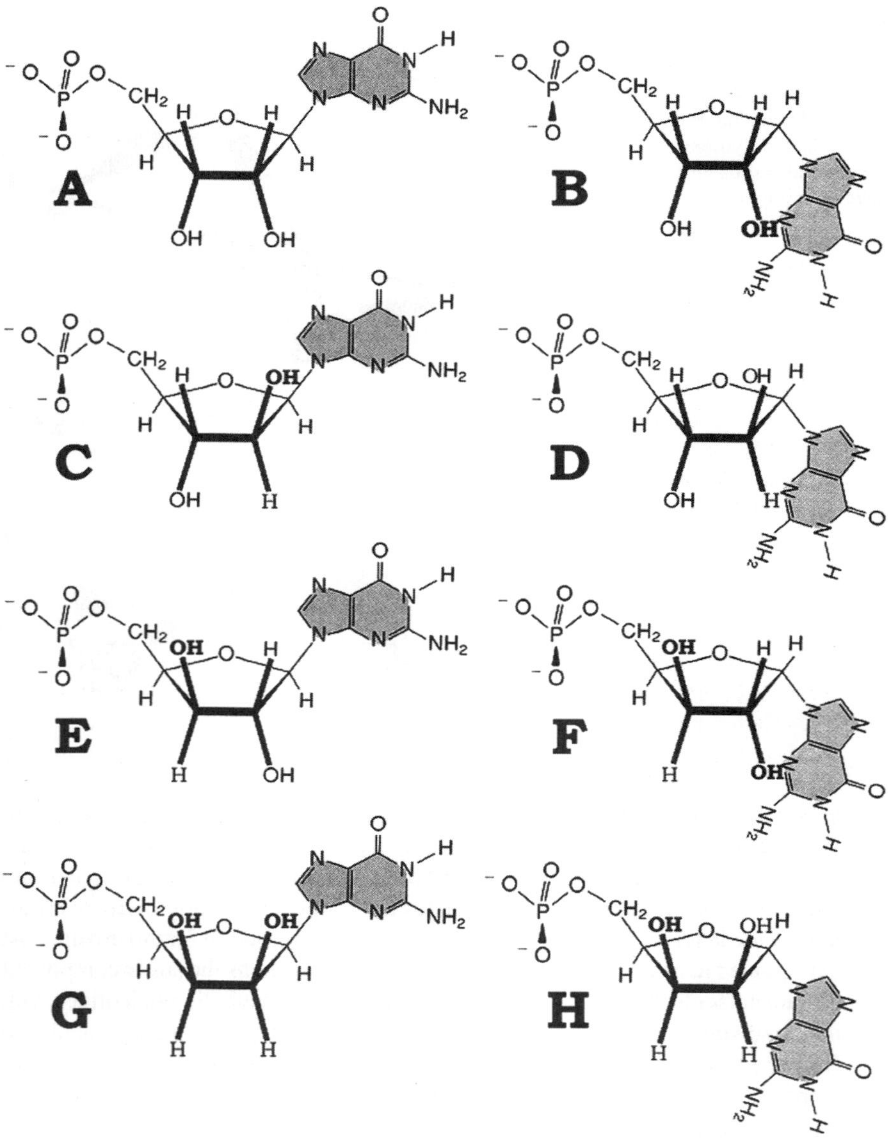

Fig. 2.13 Nucleotides containing alpha and beta anomers of different pentose sugars. Guanosine monophosphates of (**A**) β-ribonucleotide, (**B**) α-ribonucleotide, (**C**) β-arabinonucleotide, (**D**) α-arabinonucleotide, (**E**) β-xylonucleotide, (**F**) α-xylonucleotide, (**G**) β-lyxonucleotide, (**H**) α-lyxonucleotide (Reproduced with permission of Bánfalvi, 2006a)

Building Blocks of Nucleic Acids

(*d*) *Maintaining steric freedom about the glycosyl bond*

With respect to the base orientation about the glycosyl bond in β-nucleotides and β-nucleosides the base moiety relative to the sugar can adopt two extreme positions, $C_{1'}$-N_1 in pyrimidines and $C_{1'}$-N_9 in purines. The IUPAC-IUB nomenclature defines these two conformational ranges as *anti* and *syn*. In β-nucleotides the heterocyclic base is attached to the β-anomer $C_{1'}$ and is *anti* positioned (Fig. 2.13 A, C, E and G).

In α-nucleotides sugar puckering can be regarded as the mirror image of β-nucleotides. Generally the $C_{2'}$-*exo* and $C_{3'}$-*exo* forms are preferred in α-nucleotides (Suncaralingam, 1971). In α-nucleotides the heterocyclic base is attached to the α-anomer $C_{1'}$ and is *syn* positioned (Fig. 2.13 B, D, F, H). α-nucleotides are not found in natural polymeric nucleic acids, but do occur as constituents of smaller molecules in living organisms (Suzuki et al., 1967; Seto et al., 1972). Further examples include the α-nucleoside component of vitamine B_{12} (Hawkinson et al., 1970)

Naturally Occurring Nucleotides have β Configuration

In a B-DNA (Watson-Crick) type double helical structure (Fig. 2.14A) the replacement of one or two β-nucleotides by α-nucleotides (Fig. 2.14B, C) clearly demonstrates that bases in α-anomeric position are unable to base pair, eliminating the possibility of helix formation. Consequently, α-nucleotides could not be selected as sugar components of nucleic acids.

This notion is confirmed not only by the biosynthesis of purine nucleotides where the configuration of ribose at $C_{1'}$ is inverted to β in the second step of biosynthesis. The committed step in the *de novo* biosynthesis of purine ring is the formation of 5-phosphorybosylamine from 5-phophoribosyl-1-pyrophosphate. When the purine ring is assembled on ribose phosphate the amide group from the side chain of glutamine displaces the pyrophosphate. In this reaction the configuration at C-1 of 5-phophoribosyl-1-amine is inverted from α to β. The β-configuration resulting from the C–N glycosidic bond is characteristic to all naturally occuring nucleotides. The pyrimidine biosynthesis also involves 5'-phosphoribosyl-1'-pyrophosphate where ribose is initially an α-anomer. The α configuration of ribose is inverted to β during the formation of orotidylate from orotate (Stryer, 1995).

(*e*) *Reducing interactions of furanose ring substituents to minimum*

$C_{2'}$ and $C_{3'}$ inversions of pentoses are critical factors in sugar selection. Nuclear magnetic resonance (NMR) spectroscopy experiments confirmed that $C_{2'}$ and $C_{3'}$-substituents influence the $C_{3'}$-*endo* ⇔ $C_{2'}$-*endo* equilibrium. Substitutions of adenosine and uridine derivatives at $C_{2'}$ showed that the amount of the $C_{3'}$-*endo* conformer increased linearly with the electronegativity of the $C_{2'}$-substituent (Uesugi et al., 1979; Guschlbauer and Janlowski, 1980). These experiments indicate that the lack of the 2'-OH shifts the equilibrium to the $C_{2'}$-*endo* puckering in deoxyribonucleotides, deoxyribonucleosides and in B-DNA.

Ribonucleotides. That C_2 and $C_{3'}$ inversions are critical factors in sugar selection is demonstrated by showing the structural differences and sterical

Fig. 2.14 Double helix formation is prevented in DNA in the presence of α-D-deoxyribo-nucleotides. (**A**) Watson-Crick double helix formation in the presence of β-D-deoxyribonucleotides. (**B**) and (**C**) α-D-deoxyribonucleotides are not perpendicular to the axis of the helix and will prevent the hydrogen bonding between bases. (Reproduced with permission of Bánfalvi, 2006a)

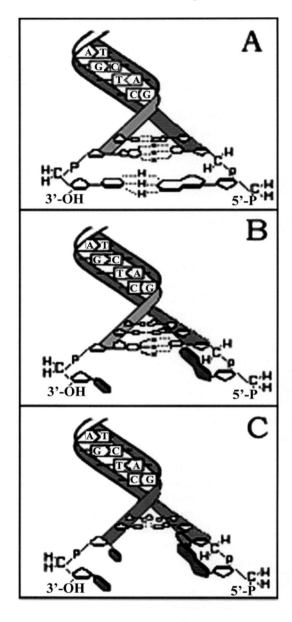

hyndrances in ribo-, arabino-, xylo-, and lyxonucleotides (Fig. 2.15 A–D). In ribonucleotides the bulky base is attached to the $C_{1'}$ carbon and the phosphate(s) at the other end of the sugar to the 5' end, while the smaller 2' and 3' hydroxyl groups are completely separated from them since they are located under the plane of the furanosyl ring. Free rotation of the subtituents is allowed above and

Building Blocks of Nucleic Acids   47

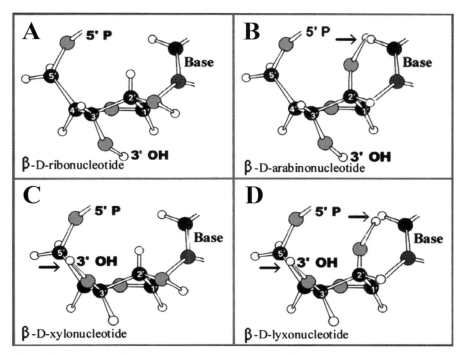

Fig. 2.15 Selection of ribose as the best fitting pentose in the nucleotide structure. Nucleotides contain β-D-pentoses. (**A**) Ribonucleotide: the bulky base at C$_{1'}$ and the phosphate group at C$_{5'}$ positions are far enough to be outside of the van der Waals distance. C$_{2'}$ and C$_{3'}$ hydroxyl groups are at the other side of the ring structure. Due to the C$_{2'}$-*endo* major and C$_{3'}$-*exo* minor conformations hydroxyl groups at these positions (C$_{2'}$ and C$_{3'}$) point to different directions allowing their free rotaion. (**B**) Arabinonucleotide: the C$_{2'}$ hydroxyl group is above the plane and within van der Waals reach with the base. Steric hyndrance due to the vicinity of these groups prevents free rotation as indicated by the arrow at the right upper corner. (**C**) Xylonucleotide: the proximity of C$_{3'}$ C$_{5'}$ substituents causes steric hyndrance indicated by the arrow. (**D**) Lyxonucleotide: all functional groups are above the plane in the sugar moiety. Steric incompatibility is due to the lack of space. Free rotation is hindered between the 5'-phosphate and 3'hydroxyl, and between 2'-OH and the base indicated by the two arrows. White rings indicate hydrogen, black spheres carbon and grey ones oxygen atoms (Reproduced with permission of Bánfalvi, 2006a)

under the plane of the ring. None of the substituents are within van der Waals distance (Fig. 2.15A).

Arabinonucleotides. In arabinose the point of configurational inversion is the C$_{2'}$ atom (Fig. 2.15B). The O$_{2'}$ hydroxyl on C$_{2'}$ is *cis*-oriented to the glycosyl C$_{1'}$-N link. The vicinity of O$_{2'}$-hydroxyl to the base causes steric hindrance, limiting the rotation of the O$_{2'}$-hydroxyl group and affecting base orientation about the glycosidic bond. Any change including nonbonding interactions and base propelling around the C$_{1'}$-N bond will influence proper base pairing with another helix. The removal of 2'-OH in 2'-deoxyarabinose would solve the problem posed by the steric hyncrance. Eventually 2'-deoxyarabinose is equivalent to 2'-deoxyribose and could be used as a building block in DNA but not in RNA since deoxyribonucleotides are formed from ribonucleotides at the ribonucleoside diphosphate level. The vicinity of the $C_{2'}$-OH group to the $C_{1'}$-β-OH in β-D-arabinose would be unfavourable for the glycosidic bond formation in arabinonucleotides. This gives an additional argument to the debate that the evolution of ribo- and potentially arabinonucleotides could have preceeded those of 2'-deoxynucleotides. In arabinonucleotides due to the presence of the bulky base at $C_{1'}$ access to the 2'-OH would be denied for the ribonucleotide reductase enzyme. In α-D-arabinonucleotides there is no such steric interference between the base coupled through the $C_{1'}$-N glycosidic bond and the $C_{2'}$-OH group since they are at different sides of the sugar ring. However, α-nucleotides would favour major $C_{2'}$-*exo* and $C_{3'}$-*exo* sugar pucker conformation which would be detrimental to helix formation in nucleic acids. That arabinose is suitable in certain structures is indicated by its presence in small natural molecules but not in nucleic acids. Due to its suitability the convenient synthesis of arabinonucleotide containing oligodeoxyribonucleotides is possible (Ozaki et al., 2000). Hybrids of RNA and arabinonucleic acid (ANA) were recently shown to be resistant to RNase H. Although RNase H binds to double-stranded RNA, no cleavage occurs with such duplexes. ANA/RNA hybrids may prove helpful in the design of future antisense oligonucleotide analogs (Denisov et al., 2001).

Xylonucleotides. They do not exist in nature. In xylose the configurational inversion takes place at the $C_{3'}$ carbon atom relative to ribose. The $C_{3'}$-OH would be the point of attachment to the next nucleotide (Fig. 2.15C). The ester bond at $C_{5'}$ would be the same but the bond angle at 3'-OH in the furanose ring would change by an angle of 108° due to the configurational inversion at this position. The polymerization of xylonucleotides would result in a zig-zag polymer and would eliminate the possibility of helix formation.

Lyxonucleotides. In lyxose the points of configuration are the $C_{2'}$ and $C_{3'}$ carbon atoms relative to ribose. All the bulky groups would be at one side of the planar furanose ring (Fig. 2.15D). These inversions combine the drawbacks mentioned with arabino- and xylonucleosides. No wonder that none of the lyxonucleotides is found in nature. The steric hyndrance of the nearby functional groups would not only prevent the formation of lyxonucleotides but also those of lyxonucleosides.

Structural considerations of 1. sugar pucker conformation, 2. steric analysis of anomeric inversion of glycosidic-OH, and phosphoester formation and 3. the configurational inversion of $C_{2'}$ and $C_{3'}$ carbon atoms in pentoses indicate that β-D-ribose is the exclusive pentose sugar that perfectly fits into nucleotides. As far as helical structures are concerned, β-D-ribose has been replaced by β-D-deoxyribose, since ribose does not fit into a B-DNA type helix because there is no sufficient room for the $C_{2'}$–oxygen atom (Dickerson, 1983). Bases in α-anomeric position are unable to base pair, eliminating the possibility of helix formation.

Phosphate as Nucleotide Component

Phosphate in the organism is represented either as a phosphate ion or as phosphoric acid. Of the physiologically important oxygenated acids of phosphorus:

- Orthophosphoric acid (H_3PO_4), occurs *in vitro* as phosphate (Pi)
- Metaphosphoric acid (HPO_3), does not exist as in the aqueous environment it would immediately form orthophosphoric acid: $HPO_3 + H_2O = H_3PO_4$
- Pyrophosphoric acid (PPi): $H_4P_2O_7 + H_2O \leftrightarrow 2H_3PO_4$ is the anhydride of orthophosphoric acids and hydrolyses to H_3PO_4.

Phosphate is the third component of nucleotides besides the base and sugar. Each nucleotide contains 1–3 phosphate units. The most important biological functions of phosphate are:

- phosphate buffer (NaH_2PO_4/Na_2HPO_4) does not contribute significantly to the stability of pH due to the low concentration of phosphate in the blood. One of the component of the phosphate buffer, the monobasic phosphate (NaH_2PO_4) is more acidic and is regarded as the weak acid of the buffer (HA). The other component, dibasic phosphate (Na_2HPO_4) is closer to the neutralized salt (Na_3PO_4), and is regarded as the salt component (A^-) of the buffer when the pH is calculated by the Henderson-Hasselbalch equation.

$$pH = pK_a + \log_{10} \frac{[A^-]}{[HA]}$$

- phosphorylated sugars are the key compounds of energy production (e.g. ATP),
- phosphorylated aminoacids are involved in signal transmission (serine, threonine, tyrosine),
- phospholipids are components of cellular membranes (phosphatides),
- phosphate is the building component of bones (calcium phosphate) and teeth (calcium fluorophosphate),
- phosphate is the component of nucleotides (NMP, NDP, NTP, dNMP, dNDP, dNTP).

Bond Types in Nucleotides

The bond types in nucleotides are shown in the skeletal structure of ATP (Fig. 2.16). The same bond types are present in GTP, CTP and UTP, with the exception that the bases are guanine, cytosine and uracil, respectively. Deoxyribonucleotides contain 2'-deoxyribose in place of ribose.

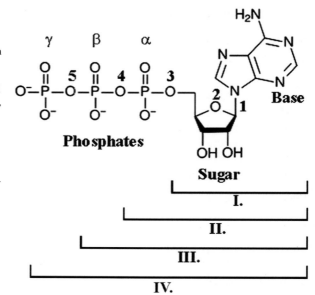

Fig. 2.16 Bond types in nucleotides. Numbers representing bonds: **1**. N-glycosidic bond between the base and sugar, **2**. cyclohemiacetal formed inside the sugar, **3**. ester bond between sugar and phosphate, **4** and **5**. anhydride bond between two phosphates. Phosphates: α, β, γ. Roman numerals: **I**. sugar + base nucleoside (adenosine). **II**, **III** and **IV** are nucleotides. **II**. nucleoside monophosphate (AMP), **III**. nucleoside diphosphate (ADP), **IV**. nucleoside triphosphate (ATP)

Nomenclature of Bases, Nucleosides and Nucleotides

The names of bases and their abbreviations, the names of ribonucleosides and deoxyribonucleosides as well as the names of ribonucleotides (ribonucleosides with one phosphate) and deoxyribonucleosides (deoxyribonucleoside monophosphates) are summarized in Table 2.1.

Metabolism of Nucleotides

De novo Biosynthesis

Based on didactic considerations the biosynthesis of nucleotides is divided to the following subchapters: the basic conception of biosynthesis (starting materials: aminoacids, carbamoil phosphate, pentoses), biosynthesis of purine and pyrimidine nucleotides, interconversion of nucleotides (transphosphorylation, synthesis of deoxyribonucleotides from ribonucleoside diphosphates, UTP → CTP, NDP → dNDP, dCMP → dUMP transitions).

Basic Conception of Biosynthesis

The general view of nucleotide biosynthesis is summarized in Fig. 2.17. The essence of these reactions is that the purine and pyrimidine ring formation is followed by the

Building Blocks of Nucleic Acids

Table 2.1 Nomenclature of bases, nucleosides and nucleotides

Base	Ribonucleoside (base + ribose)	Ribonucleotide (base + ribose + phosphate) (Ribonucleoside 5'-monophosphate) (NMP)
Adenine (A)	Adenosine	Adenosine 5'-monophosphate, AMP, adenylate*
Guanine (G)	Guanosine	Guanosine 5'-monophosphate, GMP, guanylate*
Cytosine (C)	Cytidine	Cytidine 5'-monophosphate, CMP, cytidylate*
Uracil (U)	Uridine	Uridine 5'-monophosphate, UMP, uridylate*
Base	Deoxyribonucleoside (base + deoxyribose)	Deoxyribonucleotide (base + deoxyribose + phosphate) (Deoxyribonucleoside 5'-monophosphate) (dNMP)
Adenine (A)	Deoxyadenosine	Deoxyadenosine 5'-monophosphate, dAMP, deoxyadenylate*
Guanine (G)	Deoxyguanosine	Deoxyguanosine 5'-monophosphate, dGMP, deoxyguanylate*
Cytosine (C)	Deoxycytidine	Deoxycytidine 5'-monophosphate, dCMP, deoxycytidylate*
Thymine (T)	Deoxythymidine	Deoxythymidine 5'-monophosphate, dTMP, deoxythymidilate*

Nucleoside diphosphates:
Abbreviated names:
   NDPs: ADP, GDP, CDP, UDP
   dNDPs: dADP, dGDP, dCDP, dTDP
Nucleoside triphosphates:
Abbreviated names:
   NTPs: ATP, GTP, CTP, UTP
   dNTPs: dATP, dGTP, dCTP, dTTP
*Anionic forms of phosphate esters under physiological conditions.

synthesis of ribonucleoside monophosphates (AMP, GMP, UMP). These are phosphorylated to nucleoside di- and triphosphates. Ribonucleoside diphosphates are converted to deoxyribonucleoside diphosphates by the catalytic activity of ribonucleotide reductase. Nucleoside triphosphates (NTPs, dNTPs) are obtained by the phosphorylation of nucleoside diphosphates (NDPs, dNDPs). Nucleoside triphosphates are the substrates for the biosynthesis of nucleic acids, activated metabolites and coenzymes.

Ribonucleotide Biosynthesis

Before deoxyribonucleotide biosynthesis the synthesis of the four ribonucleoside diphosphates are required (Fig. 2.18).

CTP is produced from UTP and glutamine in an ATP-dependent reaction catalysed by CTP synthase (Fig. 2.19). With the production of CTP all ribonucleotides are available for the synthesis of RNA, activated nucleotides and coenzymes.

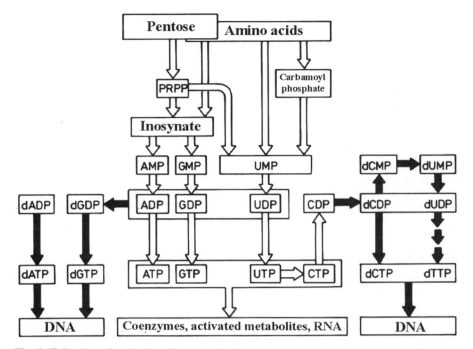

Fig. 2.17 Review of nucleotide biosynthesis. Purine nucleotide biosynthesis through phosphorybosylpyrophosphate (PRPP) results in inosynate (IMP). The synthesis of inosynate (IMP) nucleotide takes place on the ribose molecule. Different steps lead from IMP to AMP and GMP. The initial compounds of pyrimidine nucleotide biosynthesis are amino acids (aspartate, glutamine) and carbamoyl phosphate. The pyrimidine base is attached to the sugar after ribose is synthesized. After the phosphorylation of UMP all pyrimidine nucleotides are synthesized from UTP, the first step of which is UTP→CTP conversion. The dTTP formation is a rather expensive process involving CTP→CDP→dCDP→dCMP→dUMP→dUDP→→dTTP reactions. dTTP is the specific nucleotide which makes DNA synthesis quite expensive

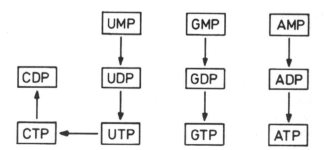

Fig. 2.18 Ribonucleoside di- and triphosphate synthesis from monophosphates. Nucleoside mono-, di-, and triphosphates can be reversibly interconverted by transphosphorylation

Building Blocks of Nucleic Acids

UMP + ATP ➜ UDP + ADP

(catalyzed by UMP kinase)

UDP + ATP ➜ UTP + ADP

(catalyzed by UDP kinase)

AMP + ATP ➜ 2ADP

(catalyzed by adenylate kinase)

Generally:

XMP +YTP ➜ XDP + YDP

XDP +YTP ➜ XTP + YDP

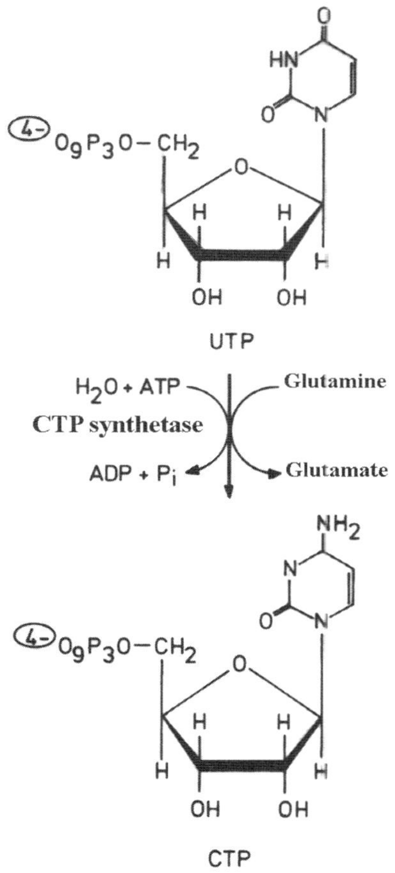

Fig. 2.19 CTP synthetase reaction. UTP →CTP conversion

Deoxyibonucleotide Biosynthesis

The NDP (CDP, UDP, GDP, ADP) conversion to dNDP is summarized schematically in Fig. 2.20. The reaction is catalysed by the ribonucleotide reductase enzyme.

The removal of oxygen from the 2'OH group of ribose is not a simple one. The complexity of the conversion is not reflected by the initial and terminal state of the reaction (Fig. 2.21). Thioredoxin or glutaredoxin are the hydrogen donors for ribonucleotide reductase, the crucial enzyme providing deoxyribonucleotides for DNA synthesis. Both small proteins can supply ribonucleotide reduction with electrons from NADPH by a specific mechanism (Thelander and Reichard, 1979; Holmgreen, 1989; Reichard, 1993). Even more surprising is that the B1 subunit of ribonucleotide reductase contains a free tirosyl radical and molecular oxygen

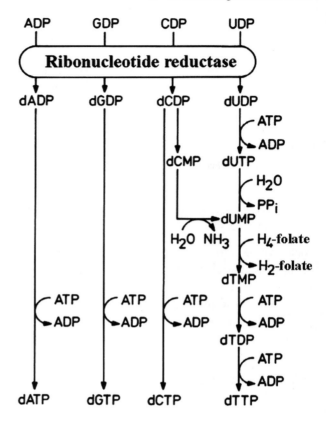

Fig. 2.20 Review of deoxyribonucleotide biosynthesis

is the electron acceptor during the tyrosil residue formation. For the regeneration of the free radical the enzyme superoxide dismutase (SOD) is needed. Superoxide dismutase is primarily found in aerob organisms. The formation of deoxyribonucleotides seems to be related to the presence of molecular oxygen. A corollary of this hypothesis is that the formation of DNA might have taken place when saturating

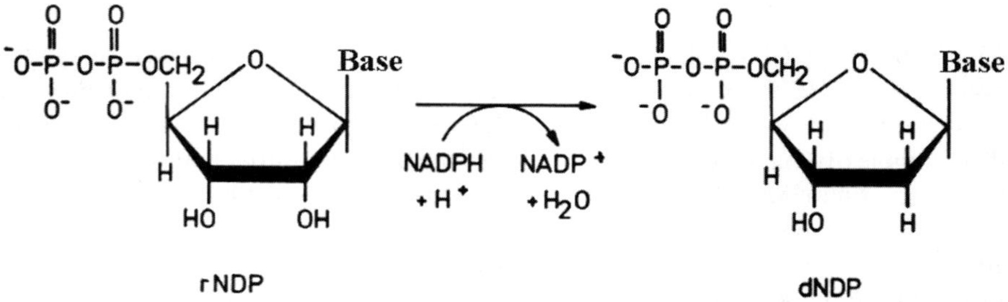

Fig. 2.21 The reaction catalysed by ribonucleotide reductase enzyme. In the reaction leading to the rNDP → dNDP conversion NADPH + H$^+$ is the coenzyme

Building Blocks of Nucleic Acids 55

Fig. 2.22 Review of dNTP formation

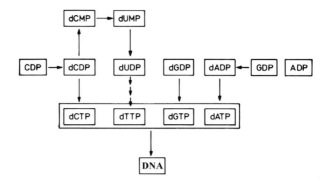

oxygen concentration was present in the sea and oxygen appeared in the atmosphere of the Earth more than 2 billion years ago.

dNDP – dTNP Conversion

Deoxyribonucleotide triphosphates are formed by the phosporylation of dNDPs (Fig. 2.21). The four dNTPs are the substrates for DNA synthesis (Fig. 2.22).

dUMP Formation

Thymidylate (dTMP) necessary for the formation of one of the four substrates of DNA synthesis is produced primarily from UMP in four steps:

$$UMP \rightarrow UDP \rightarrow dUDP \rightarrow dUMP \rightarrow dTMP$$

Altrenative pathways for dUMP production:

dUDP + ADP → dUMP + ATP (phosphorylation)
dUDP + ATP → dUTP → dUMP + PPi (phosphorylation and hydrolysis)
dCMP + H_2O → dUMP + NH_4^+ (deamination)

Thymidylate Formation (dUMP → dTMP Reaction)

RNA contains G, C, A and U bases, DNA contains T rather than U. Thymine (5-methyluracil) is formed in cells by converting uracil to 5-methyluracil. The methylation takes place at the level of dUMP by means of the thymidylate synthase enzyme (Fig. 2.23). The inhibition of thymidylate synthase is one of the key enzymes of tumour chemotherapy, as the disabling of tymidylate synthase acts specifically on DNA synthesis. The antimetabolite 5-fluorouracil after conversion to its nucleotide analogue mimics thymidylate as the compeptitive inhibitor of thymidylate synthase. Another possible point of attack is the dihydrofolate reductase enzyme. Methotrexate a specific inhibitor of dihydrofolate reductase by binding

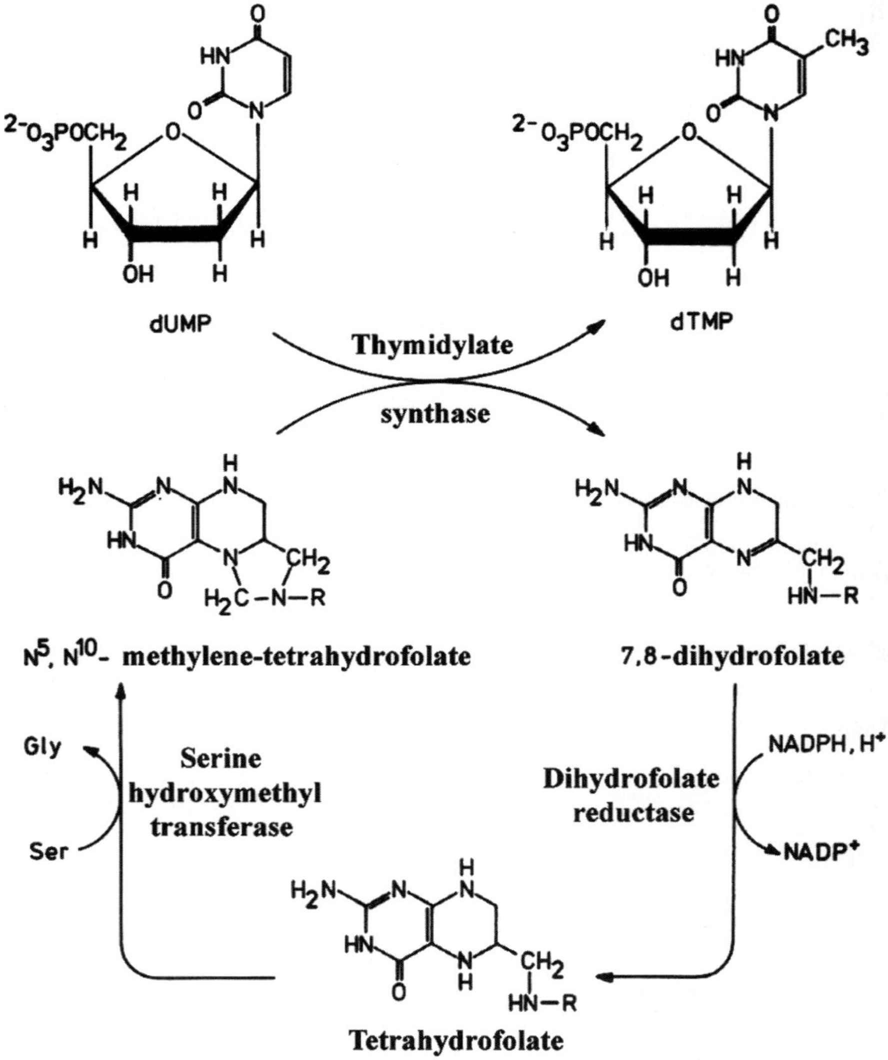

Fig. 2.23 Conversion of uridylate to thymidylate. The dUMP → dTMP transition is catalysed by thymidylate synthase. The methyldonor is N^5, N^{10}-methylene tetrahydrofolate. The methyl goup of tetrahydrofolate is obtained from the methyl "sponge" serine in the serine → glycine transformation

strongly to the enzyme reduces the amount of tetrahydrofolate and consequently the producation of dTMP.

In the dUMP → dTMP reaction the methyl group is mediated by tetrahydrofolate in a cyclic process, where an amino acid, serine is the ultimate methyl donor. Serine is representing the methyl pool in transmitting its methyl group via the serine – glycine transformation to tetrahydrofolate (Fig. 2.23).

Regulation of Nucleotide Biosynthesis

The best known regulation of *E. coli* pyrimidine nucleotide synthesis is feed back inhibition. UMP inhibits the initial step catalysed by carbamoyl phosphate synthase. CTP exerts it feed back inhibition on aspartate transcarbamoylase (catalyzing the transformation of carbamoyl phosphate to carbamoyl aspartate) and on CTP synthase catalyzing the UTP → CTP reaction.

The regulation of purine nucleotide synthesis is based on a similar strategy. End products of purine nucleotide biosynthesis inhibit the first steps of IMP synthesis. IMP itself has the same feed back effect on the committed step of its own synthesis. Cross regulation takes place by GTP inhibiting the production of AMP and ATP inhibiting the overproduction of GMP.

Degradation of Nucleotides

The catabolism of pyrimidine nucleotides results in acetylcoenzyme A (acetyl-CoA) and succinate. Three major steps of the degradation process are distinguished: 1. pyrimidine nucleotides loose their phosphate groups in a general nucleotidase reaction and are converted to nucleosides (cytidine, uridine, deoxythymidine), 2. the spontaneous oxydative deamination of cytidine is followed by the loss of phosphorylated pentose resulting in free bases (U, T), 3. finally after reduction of bases the ring is opened and uracil through the formation of β-alanine is converted to acetyl-CoA and thymine through aminoisobutyrate to succinyl-CoA.

Purine nucleotides are either deaminated and then dephosphorylated or vice versa. Nucleoside phosphorylase catalyses the removal of phosphorylated pentose. The bases are undergoing the hypoxanthine → xanthine → uric acid reactions (Fig. 2.8). Uric acid is the end product of purine degradation in man. Uric acid is also the main nitrogenous constituent of the urinary excreta of a group of lizards and snakes. In most mammals except higher apes the opening of the pyrimidine ring of urate (deprotonated uric acid) is catalysed by uricase (urate oxydase) and the more soluble allantoine is excreted. Allantoin is present in the urinary concretions of lizards and absent from that of snakes. Bony fish degrade allantoin to allantoic acid. Most of the fish, amphibians and molluscs excrete urea. Sea invertebrates, crustacens etc. metabolize urea to ammonia and carbon dioxide (Fig. 2.24).

Biosynthesis of Coenzymes

Coenzyme A (CoA) biosynthesis starts from panthothenate. Panthothenic acid is not produced by animals, it has to be present in the food as vitamin. There are five steps of the synthetic process which are not detailed.

Flavine adenine dinucleotide (FAD). The initial compound for the biosynthesis is riboflavine (vitamine B_2), which is first phosphorylated then adenylated with ATP.

Nicotinamide adenine dinucleotide (NAD^+). In the biosynthesis of NAD the aminoacid triptophane is converted in seven steps to hydroxyanthranylate, in

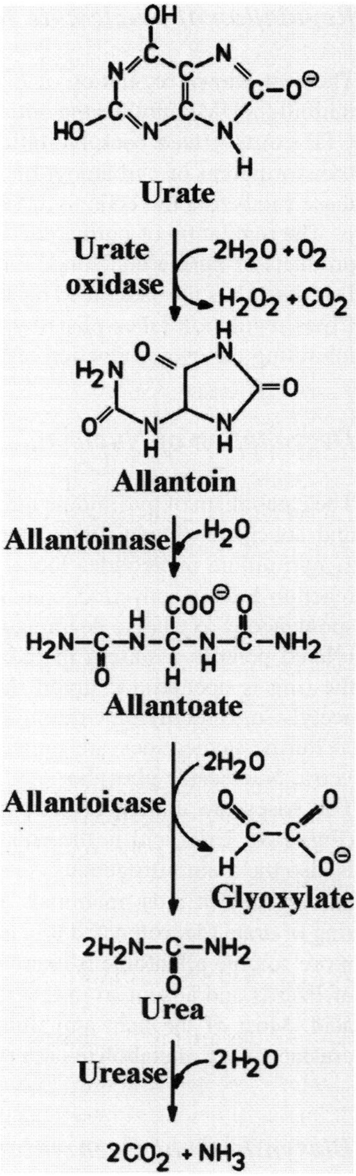

Fig. 2.24 Degradation of uric acid

further two steps to quinolinate, by phosphoribosylated nicotinate through deamido-nicotinamide mononucleotide and deamido-NAD to NAD^+. An alternative possibility is the phosphorybosylation of nicotinamide to form nicotinamide mononucleotide (NMN) and adenylation.

Building Blocks of Nucleic Acids 59

NAD$^+$ and FAD are carriers of reducing equivalents playing important role in oxydoreductive reactions and in energy production in mitochondria where hydrogen is oxidized to water. NAD$^+$ is not to be confused with nicotinamide adenine dinucleotide phosphate (NADP$^+$), which is used in those anabolic reactions where NADPH serves as a reducing agent in biosynthetic processes such as fatty acids and nucleic acids. NAD$^+$ and NADP$^+$ are not only coenzymes that function in oxidation-reduction reactions, but are also regarded as metabolites of ATP.

Tissue Specific Purine Synthesis

The major organ of purine nucleotide biosynthesis is the liver. Those tissues which lack purine synthesis utilize purine bases and nucleosides obtained from the liver in a process called salvage. In red blood cells and polymorphonuclear leukucytes the phosphoribosyl aminotransferase reaction does not work, thus their exogeneous purine supply is necessary. Similarly, the brain has also lowered phosphoribosyl transferase level. This enzyme is catalyzing phosphoryl transfer reactions on phosphoribosyl pyrophosphate, an activated form of ribose-5-phosphate.

The degradation of nucleic acids and the reutilization of nucleobases and nucleosides is summarized in Fig. 2.25. Endonucleolytic activity produces oligonucleotides, which are further hydrolysed by phosphodiesterases to nucleotide building blocks. Nucleotidases break down nucleotides to nucleosides, and phosphorylases to nucleobases. Nucleobases can also be generated from nucleotides in the phosphorybosyltransferase reaction. Bases follow the pathway of purine and pyrimidine degradation. Salvage reactions are working in the opposite direction, nucleobases and nucleosides can be converted to nucleotides. The thick lines in Fig. 2.25 indicate the reutilization of bases and nucleosides known as salvage reactions.

Functions of Nucleotides

Nucleotides are present in several biochemical reactions such as:

1. activated precursors of DNA and RNA biosynthesis (dNTP, NTP),
2. building blocks of nucleic acids (dNMP, NMP)
3. activated precursors of biosynthetic processes:

    – UDP-glucose (precursor of glycogen synthesis),
    – CDP-diacylglycerol (precursor of phophoglyceride biosynthesis)
    – S-adenosylmethionine (carrier of active methyl group)

Examples of metabolic activation:

    – Phosphatidic acid + CTP $\rightarrow$ CDP-diacylglycerol + CMP + PPi
    – Fatty acid + ATP $\rightarrow$ Acyl-AMP + PPi

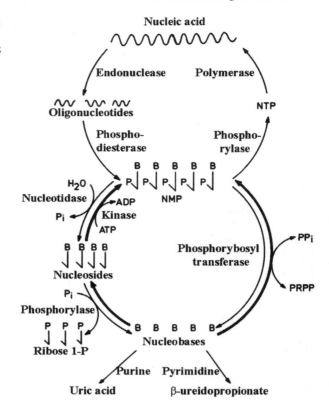

Fig. 2.25 Review of nucleic acid degradation. Abbreviations: P, phosphate; PPi, pyrophosphate; PRPP, phophoribosyl pyrophosphate; NMP, ribonucleoside monophosphate

- Aminoacid + ATP → Aminoacyl-AMP + PPi
- Glucose 1-P + UTP → UDP-glucose + PPi (glycogen synthesis)
- Glucose 1-P + ATP → ADP-glucose + PPi (starch biosynthesis)
- Glucose 1-P + TTP → TDP-glucose + PPi (rhamnose synthesis)
- Glucose 1-P + GTP → GDP-glucose + PPi (cellulose synthesis)

4. Universal energy currency in biosynthetic processes (ATP)
5. Components of coenzymes (NAD$^+$, FAD, CoA)
6. Metabolic regulation (cAMP as an ancient hunger signal)
7. Removal of extrahepatic ammonia (purine nucleotide cycle). Briefly: the urea cycle is not working in extrahepatic tissues and ammonia is removed through the aspartate – purine cycle. This cycle is connecting the urea cycle to the citrate cycle in the metabolic cycle system (metabolic clockwork) summarized in Fig. 2.26 (bold faced letters in the figure legend).

Building Blocks of Nucleic Acids 61

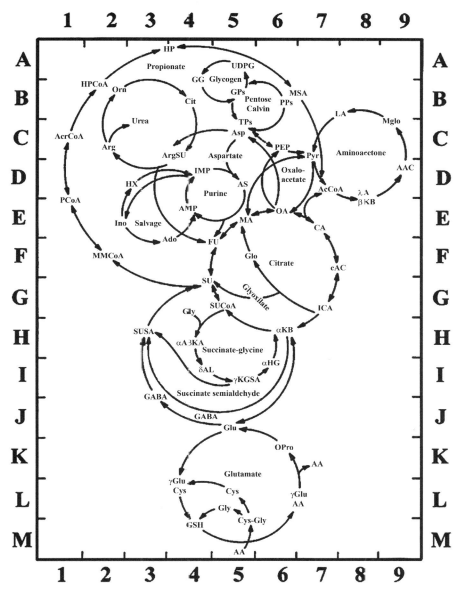

Fig. 2.26 Schematic view of the metabolic clockwork. Metabolic cycles contain the names of their metabolites in abbreviated form. Abbreviations and cycles in alphabetical order are listed below. Positions of cycles are indicated in parenthesis. Note: not all intermediates in a cycle may be shown. One-headed arrows indicate unidirectional, two headed arrows bidirectional reactions. 1. Adenylosuccinate cycle: AS, adenylosuccinate; FU, fumarate; MA, malate; OA, oxaloacetate; Asp, aspartate (C–F, 5–6). 2. Aminoacetone cycle: AcCoA, AcetylCoenzyme A; α-A-β-KB, α-amino-β-ketobutirate; AAc, aminoacetone; MGlo, methyl-glyoxalate; LA, lactic acid; Pyr, pyruvate (C–D, 7–9). 3. γ-Aminobutyrate cycle: α-KG, α-ketoglutarate; Glu, glutamate; GABA, γ-aminobutyrate;

62   2 Structural Organization of DNA

DNA Structure

There exist two types of macromolecules in nature: (a) non-informational biopolymers consisting of repetitive units and/or branched structures such as polysaccharides, peptidoglycans, glycoproteins, poly ADP-ribose, and (b) informational macromolecules which are heteropolymers carrying primary genetic information (nucleic acids) and secondary genetic information (proteins).

In animal and plant cells, DNA is concentrated into a specialized body termed nucleus. The nucleus of eukaryotic cells contains several or many nuclear chromosomes. The total amount of DNA per nucleus is a variable quantity in different species of eukaryotic cells. Many species have a much larger DNA content than the minimum in any class with a variation in genome size of about 10-fold. The total length of DNA in a single human somatic cell has been calculated to be about 1.74 m which is equivalent to about 3.2×10^9 base pairs corresponding to a molecular weight of 2.3×10^{12} dalton (Da). In human cells there are 46 chromosomes each with an average length of 4 cm DNA.

The bacterial genome consists of a closed circle of double-stranded DNA molecule, enormous compared to the size of the cell itself. The contour length of

---

Fig. 2.26 (continued) SUSA, succinate semialdehyde; SU, succinate; FU, fumarate; MA, malate; OA, oxaloacetate; CA, citrate; cAC, *cis*-aconitate; iCA, isocitrate (G–J, 3–6). 4. Aspartate cycle: OA, oxaloacetate; Asp, aspartate; ArgSU; argininosuccinate; FU, fumarate; MA, malate (C–F, 3–6). 5. Aspartate-purine cycle: Asp, aspartate; AS, adenylosuccinate; FU, fumarate; MA, malate; OA, oxaloacetate (C–F, 5–6). 6. Calvin cycle: PPs, pentose phosphates; TPs, triose phosphates (B–C, 5–6). 7. Citrate cycle: OA, oxaloacetate; CA, citrate; cAC, *cis*-aconitate; iCA, isocitrate; α-KG, α-ketoglutarate; SU-CoA, succinylCoA; SU, succinate; FU, fumarate; MA, malate (E–G, 5–7). 8. Glutamate (Meister) cycle: Glu, glutamate; γ-GluCys, γ-glutamyl cysteine; GSH, glutathione; Cys, cysteine; Gly, glycine; AA, aminoacid; OPro, 5-oxoproline (K–M, 4–6). 9. Glycogen cycle: GPs, glucose phosphates; GG, glycogen (A5–B5). 10. Glyoxylate cycle: OA, oxaloacetate; CA, citrate; cAC, *cis*-aconitate; iCA, isocitrate; Glo, glyoxylate; SU, succinate; FU, fumarate; MA, malate (E–G, 5–7). 11. γ-Ketoglutarate semialdehyde cycle: γ-KGSA, γ-ketoglutarate semialdehyde; SUSA, succinate semialdehyde; SU, succinate; SUCoA, succinylCoA; Gly, glycine; α-Aβ-KA, α-amino β-ketoadipate; δ-AL, δ-aminolevulinate (H–I, 3–5). 12. Oxaloacetate cycle: OA, oxaloacetate; PEP, phosphoenolpyruvate; Pyr, pyruvate (C–E, 6–7). 13. Pentose cycle: GPs, glucose phosphates; PPs, pentose phosphates; TPs, triose phosphates (B5–6). 14. Propionate cycle: SU, succinate; MMCoA, methylmalonylCoA; PCoA, propionylCoA; AcrCoA, acrylylCoA; HPCoA, hydroxypropionylCoA; HP, hydroxypropionate; MSA, malonate semialdehyde; AcCoA, acetylCoA; CA, citrate; cAC, *cis*-aconitate; iCA, isocitrate; α-KG, α-ketoglutarate; SUCoA, succinylCoA (A–G, 1–7). **15. Purine (nucleotide) cycle: IMP, inosine-5'-monophosphate; AS, adenylosuccinate; AMP, adenosine-5'-monophosphate (D–E, 4–5).** 16. Pyruvate (pyruvate-malate) cycle: Pyr, pyruvate; OA, oxaloacetate; MA, malate (C–E, 5–7). 17. Salvage cycle I: AMP, adenosine-5'-monophosphate; IMP, inosine-5'-monophosphate; Ino, Inosine; Ado, adenosine (D–E, 3–4). Salvage cycle II: IMP, inosine-5'-monophosphate; Ino, inosine; HX, hypoxanthine (D–E, 3–4). 19. Succinate-glycine cycle: SUCoA, succinylCoA; Gly, glycine; α-Aβ-KA, α-amino β-ketoadipate; δ AL, δ-aminolevulinate; γKGSA, γ-ketoglutarate semialdehyde; α-HG, α-hydroxyglutarate; α-KG, α-ketoglutarate (H–I, 5–6). 20. Succinate semialdehyde cycle: α-KG, α-ketoglutarate; SUSA, succnate semialdehyde; FU, fumarate; MA, malate; OA, oxaloacetate; CA, citrate; cAC, *cis*-aconitate; iCA, isocitrate (G–J, 3–6). 21. Urea cycle: Orn, ornitine; Cit, citrulline; ArgSU, argininosuccinate; Arg, arginine (B–C, 3–4). (Reproduced with permission of Bánfalvi, G. 1994)

DNA Structure

Escherichia coli DNA for example, is about 1.36 mm corresponding to 4×10^6 nucleotide base pairs, with a thickness of 20 Å (1 Angstrom: 1.0×10^{-10} m) and a molecular weight of 2.8×10^9 Da. The bacterial chromosome is tightly and compactly looped in the nuclear zone called the mesosome. In addition to the relatively large chromosome, the bacterial cell may contain small circular duplex DNA molecules called plasmids, resembling viral DNA in size, ranging from 5 to 100 MDa. Some of the plasmids, called episomes, become incorporated into the host-cell chromosome.

Structural Levels of DNA

In discussing the structural organization of DNA it is convenient to refer to different levels such as primary, secondary, tertiary and higher (chromosomal) organization of DNA.

Primary Structure

A single DNA chain is a long, threadlike molecule made up of a large number of building units (dNMPs). The primary structure includes all covalent bonds of the molecule similarly to the linear array of amino acids in the primary structure of proteins. The backbone of the primary structure of DNA consists of deoxyriboses linked by phosphodiester bridges. The phosphodiester bonds are formed between the 3'-hydroxyl and 5'-phosphate groups of the successive sugar molecules (Fig. 2.27).

The ten rotational angles of the primary structure allow a flexible backbone (Fig. 2.28). The backbone of DNA is constant throughout the molecule, the variable part of DNA giving individuality to the chain is its sequence of bases. Single stranded DNA is folded into a random coil. For random coil DNA sequences in which the DNA is in single-stranded state and no intra- or inter-strand hydrogen bonds or base stacking are present, the proton chemical shifts of a specific nucleotide depend not only on the type of its nearest neighbours, but also its next nearest neighbours (Lam et al., 2002).

Secondary DNA Structures

It is worth mentioning in advance that the dividing line between the secondary and tertiary structures of DNA is somewhat arbitrary and it is thus often difficult to distinguish between different structures. In principle the secondary structure refers to the steric relationship of bases close to each other in the linear arrangement of nucleotides. There can be both interstrand and intrastrand steric relationships of a regular kind giving rise to periodic structures. The steric relationship between bases can be polar (hydrogen bonds) and apolar (stacking effect).

Hydrogen Bonds

One type of the steric relationships is the pairing of complementary bases, which is based on hydrogen bonds. The A–T and G-C base pairs are shown in Fig. 2.29.

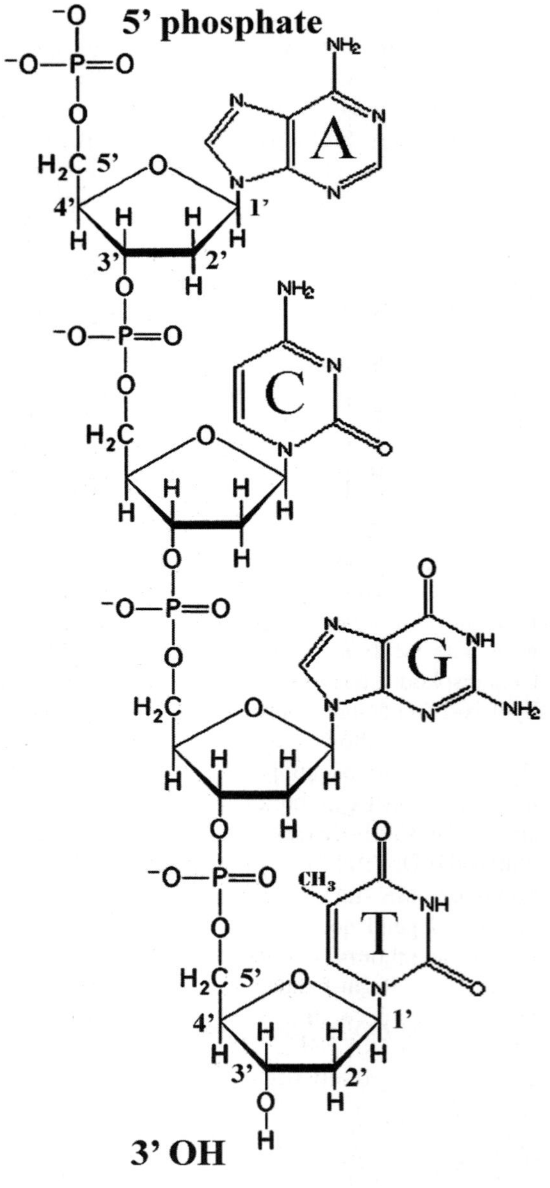

Fig. 2.27 Primary structure of DNA. Nucleotides are linked together by phosphodiester bonds to form a single stranded DNA. The building blocks of DNA are deoxyribonucleoside monophosphates. The chain has a polarity with the 5' phosphate in one end and the 3'hydroxy group at the other end. The DNA single strand can be represented in a simple way by its base sequence from 5' to 3' direction, which is in our example: ACGT

There exist several distinct forms of secondary DNA structures being either left- or right-handed forms. The full palette of double helical structures is observed for naturally occurring and synthetic DNA. Under certain conditions, different double helical structures represented as A, A', B, α-B', β-B', C, C', C'', D, E, T, Z etc. were found. The letters A to Z denote structural polymorphysm. Prefixes α and β refer to

DNA Structure

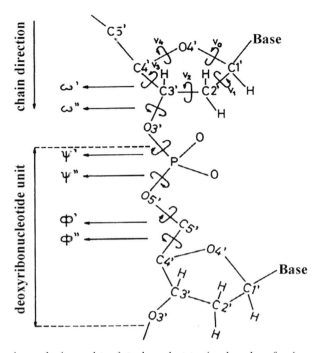

Fig. 2.28 Atomic numbering and ten interdependent torsional angles of a deoxyribonucleotide chain. Counting of nucleotides starts from the C5' → O3' direction. Torsional angles involve four atoms in sequence e.g. the torsional angle ω' is defined as the angle between projected bonds C5' – C4' and C3' – O3'. If C5' – C4' and C3' – O3' coincide in the projection, they are *cis*-planar and ω' is 0°. The angle is positive if the far bond C3' – O3' rotates clockwise with respect to the near bond C5' – C4'. An IUPAC-IUB comission worked out standard definitions for interatomic bond distances, bond angles, and torsion angles for the description of primary structures of polynucleotide chains (IUPAC-IUB, 1983)

packing differences, and additional primes indicate small variations within one type of structure. Right-handed forms are: A, B, C, D, T-DNAs as well as DNA-RNA hybrids, cruciform structures, etc. Z-DNA is a left handed form. In contrast to the polymorphysm of DNA, RNA displays some structural conservativism. The three best known, naturally occurring DNA models are A, B and Z forms (Fig. 2.30). C, D and T-forms are not discussed here: their structural data were reviewed by Zimmerman (1982). It is unlikely that all these forms realistically portray the structure of DNA since these models originate from proposals based on data of solid DNA samples instead of samples in aqueous solution.

B-DNA

The structure of B-DNA was elucidated by Watson and Crick (1953) based on the remarkable contribution of Rosalind Franklin's crystallographic imaging of B-DNA.

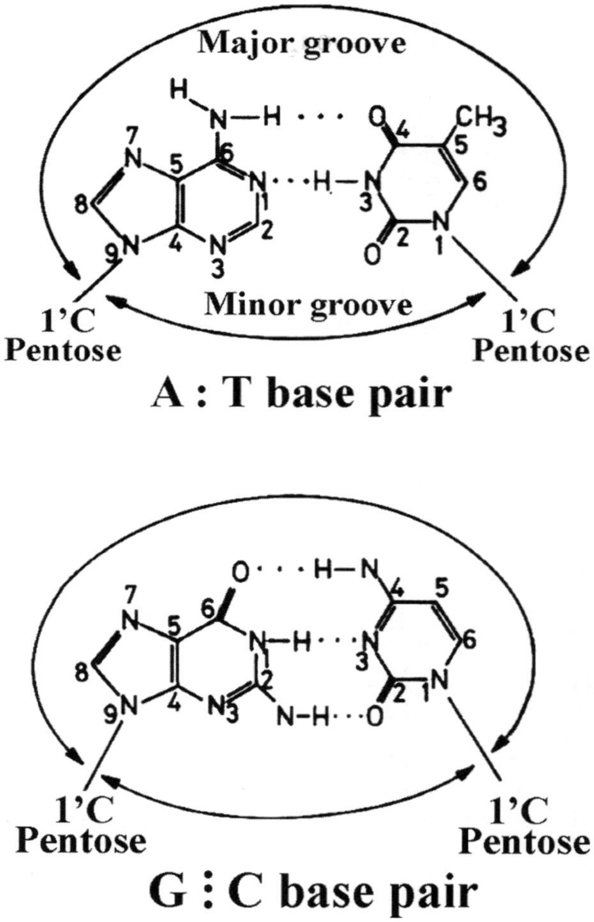

Fig. 2.29 Standard Watson-Crick base pairs. The A:T pair contains two, the G:C base pair three hydrogen bonds (indicated by *broken lines*). The distance between the two pentoses inside (*minor groove*) is smaller than outside (*major groove*)

The B-DNA is composed of two complementary strands running in opposite direction, held together by hydrogen bonds between base pairs and by the stacking effect of apolar interactions of bases. It is generally accepted that DNA in prokaryotic and eukaryotic cells exists in the B-form. B-DNA is less compact than the A-form containing 10 base pairs versus 11 per turn of helix. Both A- and B-forms are right-handed double helices. In contrast Z-DNA is a left handed duplex (Fig. 2.30). The base pairs of B-DNA are perpendicular to the helical axis. The radius of the helix is roughly 10Å. The base pairs have about 12° torsional twist relative to each other. Structural data of A-, B- and Z-DNA are given in Table 2.2. The phosphate groups which are on the outside of the B-DNA are highly charged and are presumed to interact with positive ions and water. When Watson and Crick discovered the structure of DNA they immediately recognized that each DNA strand holds the same genetic information and both strands can serve as templates for the reproduction of

DNA Structure

Fig. 2.30 Backbone models of A, B and Z-DNA. Three-dimensional structures were built using the Blackwell Molecular Models Kit following the instructions of Fletterick et al. (1985)

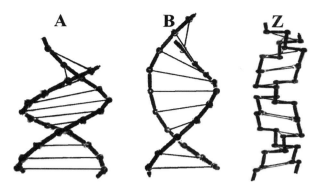

the opposite strand. The complementarity immediately suggested a possible mechanism for DNA replication (Watson and Crick, 1953) (Fig. 2.31).

Most DNA duplexes are generally considered to adopt the B-form in solution, whereas short streches of RNA duplexes and DNA-RNA hybrids are only found in A-forms (Arnott et al., 1966; Milman et al., 1967; O'Brien and McEwan, 1970). Model building experiments revealed that A and B-forms can adopt a *cis* conformation i.e. a bend in the direction of the axis of the double-helix, involving a single pair at the junction (Selsing et al., 1978). The B' variant containing 10.4 base pairs per helical turn is related to the nucleosomal organization in eukaryotic cells. Due to negative supercoiling the B-variant of bacterial DNA contains somewhat less, 10 base pairs per helical turn.

Smooth transition between A and B-forms has been demonstrated (Selsing et al., 1978). Based on own estimation the A-DNA – B-DNA transition would affect only one base pair (Fig. 2.32a). In the transition between Z- and B-DNA three base

Table 2.2 Structural data of A-, B- and Z-DNA

Characteristics	A-DNA	B-DNA	Z-DNA
Rotation per base pairs	32.7°	34.6°	30°
Base pair per helical turn	~11	~10.4	~12
Angle of incidence of bases relative to the helical axis	19°	1.2°	9°
Distance of base pairs (nm)	0.23	0.33	0.38
Helical pitch (nm)	2.46	3.4	4.56
Diameter (nm)	2.55	2.37	1.84
Conformation (glycosidic bond)	*anti*	*anti*	*anti* at C, *cis* at G
Sugar pucker conformation	3' *endo*	2' *endo*	2' *endo* at C, 3' *endo* at G
Direction of thread	right-handed	right-handed	left-handed
Repeating unit	base pair	base pair	two base pairs

Fig. 2.31 Double helix formation suggesting a copying mechanism

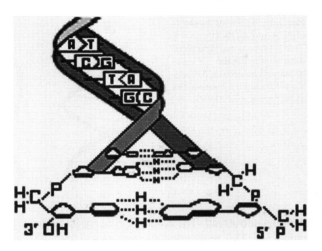

Fig. 2.32 Transition between A-, B- and Z-DNA. Imperfect base pairing at the transitions is indiacted by the arrows and the missing base pairs. Structures were built using the Blackwell Molecular Models Kit (Fletterick et al., 1985)

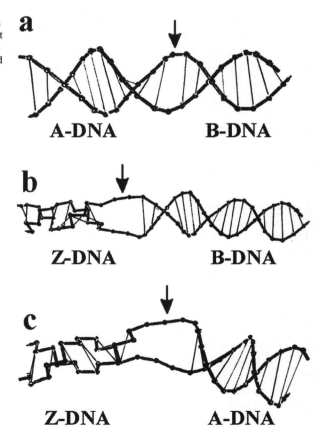

pairs would be missing (Fig. 2.32b) and in the Z- and A-DNA transition 4 imperfect base pairing would occur (Fig. 2.32c).

A-DNA

The physiological B-form can be converted into A-DNA by increasing the salt concentration of the DNA solution. Under appropriate conditions the A-form is more stable than B-DNA. The physiological salt concentration would not allow the existence of A-form. Its biological role is not clear but possibly may occur upon association with proteins specific to DNA.

A-DNA is structurally homologous to the double-stranded RNA, the major groove being almost flush with the surface of the molecule and the minor groove being deep. This conformation has eleven base pairs per turn and the base pairs are tilted by about 20° relative to the perpendicular base pairs of B-DNA. The pairs are not quite coplanar showing 15° of propeller twist (Fig. 2.33).

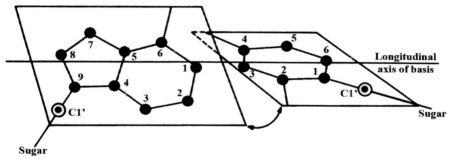

Fig. 2.33 Propeller twist of base pairs. The planes of the two bases do not coincide. The turn between the planes, called the propeller twist is indicated by the double headed arrow

Z-DNA

This conformation was discovered by Rich and his associates in concentrated salt solutions when the bases G and C were strictly alternating (Wang et al., 1979). The (G-C)$_n$ sequence allows a novel type of secondary structure which is left-handed. The backbone of Z-DNA is a zig-zag with a double nucleotide pair repeat, unlike the sinuous S-curve in the backbone of A- and B-DNA. The base pair propeller twist is minor: the base pairs are slightly tilted (10°) and virtually coplanar in contrast to A- and B-DNA. The minor groove of Z-DNA is deep while the major groove is shallow in this structure. The existence of Z-form in nature has been demonstrated. Antibodies raised against Z-DNA are bound to polytene chromosomes (Nordheim et al., 1981) pointing to its natural existence. A right-handed B-DNA structure can exist in close proximity to the left-handed Z-structure (Klysik et al., 1981). Weintraub proposed a dominant role for torsional strain and for the secondary structure in the formation of hypersensitive structures in chromatin (Weintraub, 1983).

In supercoiled plasmids, the junctions between Z-form and B-DNA are particularly susceptible to cleavage by single strand specific nucleases such as S1 (Wells et al., 1983) and under high salt concentration, by Bal-31 nuclease (Kilpatrick et al., 1983). The idea that the junction between Z-form and B-DNA is lacking hydrogen bonds is demonstrated by the model experiment shown in Fig. 2.32b.

B-Z transitions are easily accessible loci for the binding of specific proteins or ligands due to the differences in conformation. Z-DNA binding proteins occur in cells as diverse as those of *Drosophila*, human cancers, wheat germ, and *E. coli* (Kolata et al., 1983; Lipps et al., 1983). These proteins can flip some potential B-DNA sequences into the Z-form and stabilize it. The transition of synthetic poly (dG-dC) oligonucleotides from the B to Z-form at salt concentrations close to physiological can be facilitated by the methylation of cytosine residues (Behe and Felsenfeld, 1981; Moller et al., 1981). Methylation at the 5-C position of alternating CG sequences seems to stabilize Z-DNA (Fujii et al., 1982). Methylation may act as a "swith" in gene regulation. Generally speaking, the B–Z transition is facilitated by alternating GC tracts (Rich, 1983), whereas CCGG sequences favour transition from the B- to the A-form by steric inhibition of the B–Z conversion (Conner et al., 1982). The biological importance of radiation- or chemically-induced conformational changes has to do with the potential of B- Z-DNA conversion to control DNA transcription processes.

The word "palindrome" was created by the English writer Ben Jonson in the 1600s from Greek words *palin* (back) and *dromos* (way, direction).
Palindrome words: Madam, Adam
Palindrome names:
Ogopogo (hypothetical monster in Okanagan Lake, Canada).
Anuta Catuna (name of a Romanian runner)
Revilo Oliver (name of a Neo-Nazi philologist)
Palindrome sentences:
Rise to vote, Sir!
Was it a rat I saw?
Madam I'm Adam.
Madam, it is in Eden I sit—I'm Adam. (Nora Baron)
DNA hymn is in my hand. (Martin Clear)
Niagara, O roar again! (Leigh Mercer)
Too fat, so Lana lost a foot. (Bill A. O'Connor)

Cruciform DNA

Negative supercoiling (see below) is known to favour the conversion of interstrand base pairs to intrastrand pairs producing a set of hairpin structures also referred to as cruciform structures from palindromic DNA sequences. The first four-way junctions studied were cruciform structures formed by inverted repeats in supercoiled DNA

DNA Structure

Fig. 2.34 The cruciformation of palindromic DNA

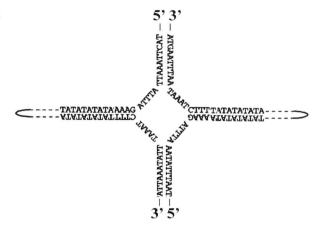

(Gellert et al., 1979). In *E. coli* it was demonstrated that cruciformation in the -35' promoter region of an operon has the potential to interfere with transcription by RNA polymerase (Horwitz, 1989). Proteins, such as the ribosomal S16 protein of *Escherichia coli*, displaying a nicking activity, may bind to cruciform DNA (Bonnefy, 1997). It is reasonable to assume that the junctions of the cruciform structure (Fig. 2.34) and the junction in the Holliday model (1965) postulated for genetic recombination are identical.

Secondary DNA structures are not easy to illustrate. When one would like to show the base pairs of the duplex (Fig. 2.35A), the helical structure can be omitted. When the helical structure is visualized the base pairing has to be sacrificed (Fig. 2.35B). The helical structure is most often viewed schematically by two intertwined ribbons (Fig. 2.35C) or simply by two parallel lines (Fig. 2.35D). Antiparallel nucleotide sequences are indicated by the letters of the nucleotide bases (Fig. 2.35E). Based on the complementarity, one sequence in itself is sufficient to

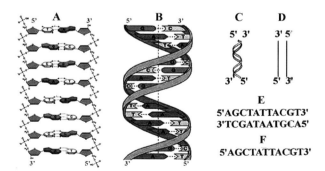

Fig. 2.35 Secondary structure of DNA. Structures show: (**A**) two dimensional ball and stick drawing of base pairing, (**B**) drawing of the three dimensional view of the double helix, (**C**) ribbon skeleton of helix, (**D**) linear double helix, (**E**) sequence of double stranded DNA, (**F**) the sequence of single stranded DNA in itself defines the sequence of the antiparallel strand

72   2 Structural Organization of DNA

read the sequece of the other chain (Fig. 2.35F). Due to the specificity of DNA poly-merases, in cells the DNA is always synthesized from 5' to 3' direction. Similarly the DNA sequence is read in this direction.

Bent DNA Structures

DNA bending plays a crtitical role in several biological functions e.g. DNA packing in viruses, eukaryotic DNA-protein complexes known as nucleosomes, transcription mediated DNA looping (Garcia et al., 2007). The physical forces associated with DNA bending generate torsional strain inside the molecule with significant struc-tural consequences. The bending of double-stranded DNA is regarded as a confor-mational variation, first described in the minicircles of the kinetoplast DNA (kDNA) network (Marini et al., 1982) and later in both prokaryotic and eukaryotic organisms. Sequence-directed bent DNA structures consisting of oligomeric adenine residues have been found in DNA molecules of various sources (Linial and Shlomai, 1988). The bent structure is likely to play some role in strand separation processes includ-ing transcription (Miyano et al., 2001), in the control of DNA replication (Zahn and Blattner, 1985; Ryder et al., 1986; Snyder et al., 1986) and in recombination (Ross et al., 1982). The bending parameters may change very rapidly within a few nanoseconds. The bending is sequence-dependent and has a significant effect on the overall helix bending in both B- and A-DNA (Dickerson, 1983). DNA-protein interactions are dependent upon the degree of the helix curvature and protein bind-ing decreases with the straightening of the binding site. The non-specific bending of double-stranded B-DNA is significant, especially if we take it into consideration that DNA in chromatin is packaged into nucleosomes, having about 1.8 circuits of helix around a small histone core.

Transition from Right-Handed to Left-Handed DNA

Local torsional strain. By fixing the circularized DNA molecule at a point and turning it at another point to the left, the revolution generates negative supercoils without appreciable distortion of the helical structure, since the structural constraint is distributed evenly along the molecule. There is no satifactory explanation so far as to why supercoils are formed generally in this direction and not to the right in nature. However, model experiments show that when DNA is twisted to the right the torsional constraint is not distributed evenly in the molecule. The local stress generated by right-handed twists decreases the number of helical turns in front of the twist and increases the helicity behind the twisting point. Thus it appears as if the torsional rigidity of the DNA backbone were different against right- and left-handed turns. This could explain why DNA is supercoiled preferentially to the left-hand direction and strand separation taking place by positive supercoiling (Fig. 2.36).

In strand separation processes such as DNA replication, transcription and recom-bination, the left-handed torsional strain generated by helicases (catalyzing strand separation) is located to only a relatively small portion of the genome. Helicases move along the double-stranded DNA by a wedge-like action to unwind DNA

DNA Structure

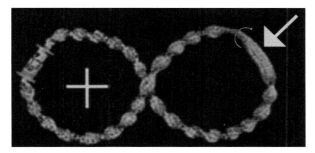

Fig. 2.36 Positive supercoiling generated by strand separation. Circular DNA was denatured at the region indicated by the arrow. Model experiment was carried out with Solomon's knot (macramé) DNA (Reproduced with permission of Bánfalvi, 1986b)

(Abdel-Monem et al., 1977). Forked DNA structures are formed by the concerted action of a set of enzymes. The local stress of melting could be releaved if the whole molecule were allowed to rotate. This is unlikely to happen because of the remarkable length of DNA in cells. Moreover, in bacteria DNA is usually attached to an infolding of cell membrane know as the mesosome. The unwinding of DNA clockwise in front of the fork leads to the formation of a positive supertwist (Fig. 2.36).

Formation of left-handed loops. Local torsional strain may not only melt duplex DNA but also initiate the formation of left-handed loops without strand separation or strand breaks (Fig. 2.37).

Right-handed (e.g. B-DNA) and left-handed DNAs have quite different conformations. In the right handed DNA the sugar-phosphate backbone is much more flexible and can exist in a very large number of conformations because each single bond in the backbone has a certain freedom of wobble. Although, the most stable conformation is the B-form, on account of the topology of the Solomon's knot DNA, the left-handed conformation can be "induced" by twisting a segment of DNA helix into a left handed configuration. Left-handed DNAs on the other hand, show one or more rigid kinks contributed by negative loops (Fig. 2.37a, b, c). This *cis* configuration of negative loops produces bands in the DNA backbone, whereas right-handed DNA corresponds to a *trans* configuration.

The formation of left-handed DNA needs strong base-pairing primarily at GC-rich sequences where C residues are methylated. Methylated bases arise by postreplicative modification at CG sequences of genomic DNA. In higher plants and animals the majority of repeating ..CG.. dinucleotides are modified by methylation. The methyl group is attached to the 5-carbon atom of cytosine in poly(dC-dG) which stabilizes the Z-DNA conformation over B-DNA (Behe and Felsenfeld, 1981). Methyl groups do not affect base-pairing and the hydrogen bonds of 5-methylcytosine with guanine are equivalent to those formed by cytosine. Methylation of guanine at C-6 has been shown to cause mispairing with uracil (Gerchman and Ludlum, 1973). Among non-radioactive nucleotides, pyrimidine (cytosine, thymine) analogues substituted at C-5 position proved to be versatile tools in nucleic acid research since they allowed the introduction of bulky groups without drastic reduction

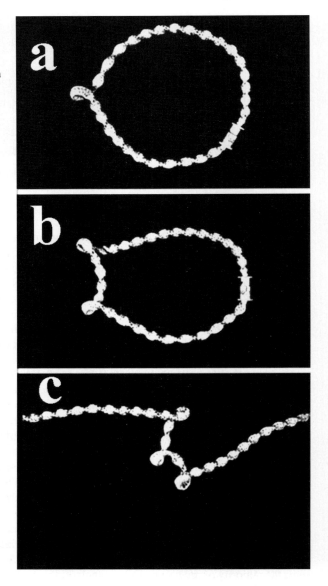

Fig. 2.37 Demonstration of left-handed loops. Circular DNA was denatured at a small region and turned into negative loop(s). Introduction of a. one, b. two and c. three negative loops (Reproduced with permission of Bánfalvi, 1986b)

of DNA polymerase activities (Dale et al., 1973; Langer et al., 1981; Bánfalvi and Sarkar, 1983; Bánfalvi et al., 1989).

Model experiments with Solomon's knot DNA indicate that at least 20 nucleotides are involved in the formation of a left-handed loop including two tetranucleotide junctions between right-handed and left-handed helical segments and 12 base pairs are forming the negative loop. The left-handed loop is slightly funnel-shaped, not quite symmetrical with 12 nucleotides at the inside circle and 13

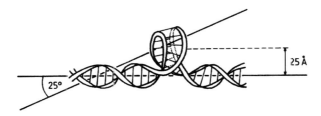

Fig. 2.38 Diagrammatic structure of left-handed DNA. Ribbons symbolise sugar-phosphate backbones held together by base pairs. The horizontal line indicates the fibre axis of right-handed DNA. The axis of left-handed DNA is marked by the slanting line (Reproduced with permission of Bánfalvi, 1986b)

nucleotides at the outer contour. Negative loops of less than 12 base pairs are probably not formed and would snap back to their right-handed counterparts. The angle of the helical axis of the left-handed DNA is twisted by at least 25 Å apart assuming a transition of 20 nucleotides (Fig. 2.38).

Tetraplex (Quadruplex) DNA

G-quadruplexes are four-stranded DNA structures formed by G-rich sequences. Such local DNA conformations have been suggested to be biologically important in processes such as DNA replication, gene expression and regulation, and the repair of DNA damage (Hackett et al., 2001; Rich and Zhang, 2003). Repeating sequences of pyrimidine-purine nucleotides $(d(TC)_n \cdot d(GA)_n)$ are readily transformed to new DNA structures on lowering the pH and increasing the ionic strength. The complex formed this way can be interpreted by triple-stranded (triplex) and four-stranded (tetraplex) polynucleotides (Johnson and Morgan, 1978). The idea of multistranded nucleic acids is an old one (Stent, 1958; Zubay, 1958). Clustered guanine residues in DNA readily generate hairpin or a variety of tetrahelical structures. Clusters of contiguous guanine residues in DNA can associate *in vitro* under physiological-like conditions to form four-stranded structures designated DNA tetraplexes or quadruplexes. The myogenic determination protein MyoD was reported to bind to a tetrahelical structure of guanine-rich enhancer sequence of muscle creatine kinase (Walsh and Gualberto, 1992), suggesting that tetraplex structures of regulatory sequences of muscle-specific genes could contribute to transcriptional regulation. Tetraplex DNA strands are expected to coil back on themselves with a fold of radians or *anti* folding, indicating opposite directions of duplexes and generating negative supercoils (Fig. 2.39a). The other possibility is the folding of two radians or *syn* folding leading to the formation of positive supercoils (Fig. 2.39b).

Differences in the Structure of DNA and RNA

Similarly to the DNA structure RNA consists of nucleoside monophosphates and has a polarity with 5'-phosphate and 3'-hydroxy ends (Fig. 2.40).

Fig. 2.39 Demonstration of tetraplex DNA structure using circular heliwire DNA model. (**A**) *Anti* folding of relaxed DNA generates negative supercoils. (**B**) *Syn* folding introduces positive supercoils. Arrows indicate the sites of tetraplex formation. (Reproduced with permission of Bánfalvi and Fieldhouse, 1988)

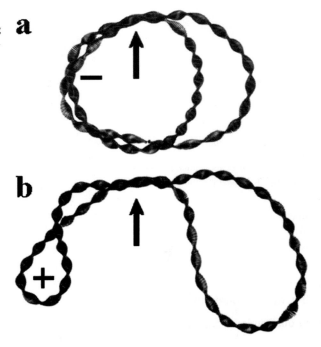

RNA differs in several structural details from DNA:

1. Sugar unit: RNA nucleotides contain ribose, DNA contains 2'-deoxyribose.
2. Base: RNA uses uracil instead of thymine present in DNA.
3. RNA is transcribed from DNA by enzymes called RNA polymerases, DNA is replicated from DNA by DNA polymerases.
4. RNA is a single stranded molecule containing short complementary streches for intramolecular base pairing with hydrogen bonds. The ratio of linear and helical streches is nearly the same. DNA can be single stranded in some viruses, but is double stranded in bacteria, animal and plant cells.
5. Complementary RNA chains are unable to form long double helical structures. The reason is that the 2'OH would not fit into a B-DNA-like structure since there is not enough space for it. The oxygen atom of the 2'-hydroxy goup is within van der Waals distance with four other atoms.
6. The RNA sequence folds back on itself and complementary bases pair by Watson-Crick base pairing (U–A, G–C). In addition, weaker G–U *wobble pairs* can be formed, where the bases bind in a skewed fashion inside the RNA molecule.
7. From the primary structure of RNA the pairing probabilities and computer predicted secondary structure folding can be calculated by energy minimization.
8. Numerous regulatory structural motifs of secondary RNA structures (stem-loops, hairpins, interior loops, bulges, multiloops, hammerhead and other

DNA Structure 77

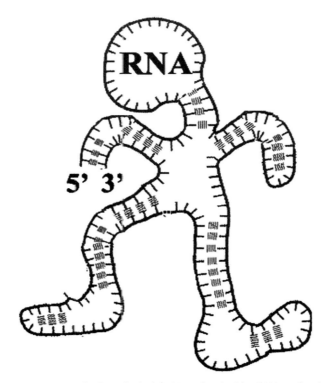

Fig. 2.40 Tertiary structure of a hypothetical RNA molecule. The RNA molecule consists of a single strand. Several segments may interact with each other as shown by the base pairing. Double helical RNA cannot assume a B-DNA-like conformation, due to the presence of the 2' hydroxy (absent in DNA) which would collide with the phophodiester backbone on the successive nucleotides. Distant regions of the single stranded RNA come together by Watson-Crick (A–U, G–C) base pairing and by the so called "tertiary base pairing". Other unusual structural features include interactions between bases and the phosphodiester backbone and the involvement of 2' hydroxy groups in hydrogen bond formation. 5' = 5' phosphate end; 3' = 3' hydroxy end of the RNA molecule

conformationally recurring motifs) have been identified as playing essential roles in gene expression. Such motifs can be recognized in the three types of RNA (rRNA, tRNA, mRNA) and in the self catalyzing RNA (ribosyme).

Owing in part to space limitations and in part to the scope of the book, less attention is paid to RNA structure.

Tertiary Structure of DNA (Topology)

Tertiary structure refers to the steric relationship of segments of DNA that are apart in the linear sequence. A circular double-stranded DNA without any folding is

78   2 Structural Organization of DNA

known as a relaxed DNA. This is the simplest and most stable tertiary structure of DNA. Viral DNAs can exist in a relaxed state in the cell. However, to fit any chromosomal DNA into a cell, the genome has to be completely folded. Therefore one can speak only about partially relaxed or relaxed parts of DNA in cells. Regarding the dinamic structure of DNA *in vivo*, the number of topoisomers must be innumerable and rapidly changing. In contrast the structure of isolated DNA is much more uniform and dependent on the conditions of isolation.

Molecular models

Model building proved to be an important step in the elucidation of the three dimensional structure of DNA (Watson and Crick, 1953). However, ball-and-stick models and space filling models fail to visualize tertiary changes in DNA structure as the introduction of a swivel is likely to destroy the model itself. It is possible to predict topological changes of DNA structures with the aids of computers. Calculations must take into account not only nucleotide sequence and minimum-free energy conformation of DNA based on covalent, electrostatic, van der Waals and steric interactions but also the free energy represented by the torsional strain of superhelicity. Because of the complexity of short and long-range factors, some of which may even be unknown, attempts toward this end remained limited.

Solomon's knot DNA is a simple version of the Watson-Crick model (Fieldhouse, 1981) which has the advantage from the educational point of view of being extremely inexpensive to construct at any part of the world and being very robust in the hands of students. Nevertheless it may be useful to illustrate some very important principles. The circularization of such linear macramé DNA allows a clear demonstration of superhelical structures (Bánfalvi, 1984). Topoisomers which are generated by the catalytic activities of topoisomerases (Gellert et al., 1976; Cozzarelli, 1980) can be distinguished by their linking numbers that specify the number of times two DNA strands are intertwined (Crick, 1976). „Molecular models can be made very cheaply from a variety of materials. Handling models helps students to appreciate three-dimensionality and stereochemistry and gives them different (and memorable) experiences than hearing or seeing" (Edward J. Woods, University of Leeds, personal communication).

Major groups of tertiary structures are: relaxed, knotted, catenated, looped, superhelical conformations and their combinations (Fig. 2.41). Linear, open DNA molecules are not regarded as topoisomers (Fig. 2.41a). A circular DNA without any folding or without superhelical turns is the simplest topoisomer known as relaxed DNA (Fig. 2.41b). The relaxed DNA can be supertwisted in two ways either by twisting the double helix to the left or in the direction of the helical turns to the right. The former will produce negative supercoil, the latter will generate a positive

DNA Structure

Fig. 2.41 Demonstration of topoisomers using Solomon's knot DNA. (**a**) linear DNA molecules, (**b**) relaxed circular DNA, (**c**) positively and negatively supertwisted DNA, (**d**) looped DNA, (**e**) topological knot, (**f**) catenated DNA, (**g**) catenated and supertwisted DNA, (**h**) catenated and knotted DNA (Reproduced with permission of Bánfalvi, 1984)

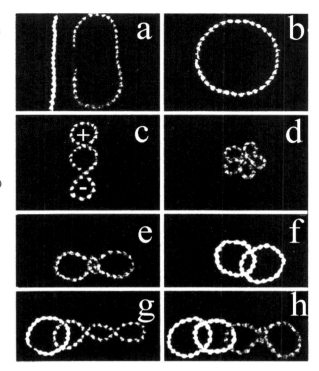

supercoil. However, in order to introduce both positive and negative supercoil into the same molecule, no net twisting of the double helix but only local uncoiling is necessary (Fig. 2.41c). Upon further rotation of one part of the molecule and fixing another part more and more loops are formed without strand breakage (Fig. 2.41d). On the other hand a topological knot can be obtained without supertwisting or local unwinding of the molecule (Fig. 2.41e). The formation of knotted DNA usually involves the danger of strand breakage. Two or more topoisomers may be linked together as catenanes in such a way that the duplexes are interlocked (Fig. 2.41f, g, h). The formation and separation of catenanes requires transient double-stranded breaks.

Thermodinamic Aspects of Supercoiling

If B-DNA is in relaxed state, there is no supercoiling. When the double helix is slightly turned before closing the circle, strain is introduced and winding up is compensated by the twisting of the circle into a superhelical form. Supercoiling is not only essential for the interaction with enzymes and for the folding of DNA but it also represents a kind of energy storage.

Under physiological conditions the parental double-helix must make one turn about its axis for every 10 base pairs in B-DNA. This is the so called winding of

DNA. In strand separation processes the double-helix has to be unwound to separate the DNA strands. For an unwound replication fork to move, the entire chromosome ahead of the fork would have to rotate rapidly requiring large amounts of energy. Alternatively, swiveling of the helix could solve this problem. The rotation of the entire genome in cell is unlikely to happen, not only because of the high energy consumption such a process would need, steric considerations and the membrane attachment of the chromosome also rule out this possibility. Swivels introduce tertiary changes in the DNA structure generating supercoiled topoisomers. With the discovery that the *E. coli* chromosome is a circular molecule, the development of a new method allowed John Cairns the visualization of DNA molecules by autoradiography after labeling with tritiated thymidine. Cairns has shown in his paper that DNA consists of a single molecule, that is replicated at a moving locus bidirectionelly (known today as the replicating forks) at which both new strands are beeing synthesized (Cairns, 1963a). Cairns was also the first to hypothesize swiveling in DNA replication (Cairns, 1963b). It turned out that virtually all duplex DNA in bacteria exists in a negatively supertwisted form. There is a difference between the topological state of DNA in bacterial and eukaryotic cells. The underwound state in bacteria is maintained by the catalytic action of gyrase (bacterial type II topoisomerase). In eukaryotes the underwound state of DNA is maintained by the left-handed coiling of DNA in nucleosomes. The lack of experimental approaches for manipulating the topological state of eukaryotic DNA is the reason for the relatively slow progress in this particular field, to be discussed in Chapter 3.

Negatively supertwisted DNA is generated by turning the circular double-helix in the opposite direction to the helical turns, i.e. to the left. Supertwisted DNA produced by right-handed twists has not been found in nature. A plausible explanation has already been given, namely the disruptive torsional strain generated by turning the double-helix to the right. To the contrary the torsional flexibility of DNA towards left-handed turns seems to be relatively higher (Bánfalvi, 1986a). The topological changes introduced by supercoiling apply not only to closed circular DNA but also to linear segments whose ends are fixed relative to each other. From the thermodynamic point of view, negatively supercoiled DNA is in a higher free-energy state than the relaxed one, thus any change that decreases the degree of supercoiling is energetically favoured (Wang, 1980). Examples for such changes are local denaturations, formation of cruciform structures, conversion of B-DNA to the Z-form, and intercalation of planar dye molecules (Bauer, 1978; Wells et al., 1980; Record et al., 1981; Gellert, 1981; Wang et al., 1983).

Topoisomerases

Supercoils and breaks are introduced into DNA by the catalytic activities of enzymes called topoisomerases. As their names suggest, topoisomerases change the tertiary structure of the DNA molecule either by relaxing supercoiled DNA through breaking and rejoining one strand at a time (type I) or by catalyzing catenation/

DNA Structure 81

decatenation, knotting/unknotting of DNA rings through breaking and rejoining DNA in a double-stranded fashion (type II enzymes) (Liu et al., 1980; Cozzarelli, 1980). In addition to these activities, bacterial topoisomerase II (gyrase) is responsible for keeping DNA in its underwound state by negatively supercoiling DNA (Gellert et al., 1976). The wide acceptance of the scheme that in bacteria topoisomerase I relaxes supercoiled DNA and gyrase, the bacterial topoisomerase II negatively supercoils DNA lies in its simplicity. Topoisomerase activity has been detected in diverse sources including bacteria (Gellert et al., 1976; Liu et al., 1980; Brown et al., 1979; Sugino and Eott, 1980), bacteriophages (Liu et al., 1979; Ghelardini et al., 1982) yeast (Goto and Wang, 1982), insects (Hsieh and Brutlag, 1980) amphibians (Baldi et al., 1980) and mammals (Liu et al., 1980; Cozzarelli, 1980).

The enzymology of supercoiling started with the discovery of *E. coli* ω potein by Wang (1971). This protein removes negative supercoils in DNA. The linking number is changed in the reaction referring to transient strand breaks. The ω protein is the archtype of DNA topoisomerase type I enzymes. Type I enzymes cleaving only one strand are further classified as type IA subfamily if the protein link is to the 5' phosphate (former type I-5') or type IB subfamily members when the enzyme generates 3' phosphate nicks and is attached to the 3' phosphate (earlier named type I-3').

The first type II enzyme, DNA gyrase, was isolated from *E. coli* by Gellert et al. (1976). Since the discovery of gyrase, a considerable amount of information has been accumulated regarding the properties, activities and functions of bacterial type II topoisomerases. The discovery of a novel type II topoisomerase from the thermophylic archebacterium *Sulfolobus shibatae* (Bergerat et al., 1997; Buchler et al., 1998) led to the division of type II topoisomerases into type IIA and type IIB subfamilies (Champoux, 2001). While DNA topoisomerases I and II (encoded by the *TOP1* and *TOP2* genes, respectively) are capable of relaxing both negatively and positively supercoiled DNA molecules, yeast DNA topoisomerase III (encoded by the *TOP3* gene) relaxes only weakly the negatively supercoiled DNA (Kim and Wang, 1992). Top3 seems to be involved in the processing of molecules generated during meiotic recombination (Gangloff et al., 1999). Further subfamilies are known which are based on structural alterations. Some of them are likely to be paralogues, such as eubacterial topoisomerase I and III, and different subfamilies of topoisomerse II and III. Regarding the several functions of topoisomerases one more very important aspect is emphasized, namely topoisomerases are essential for chromosome condensation (Adachi et al., 1991; Downes et al., 1994; Castano et al., 1996). The structural features common to all topoisomerases are:

a. the hinged clamps to open and close DNA,
b. binding cavities for temporary storage of DNA segments,
c. coupling conformational changes of topoisomerases to DNA rotation,
d. type I topoisomerases generally do not require ATP, while type II enzymes do,
e. for type II enzymes ATP hydrolysis results in further conformational changes (Champoux, 1978, 2001).

Models of Gyrase Action

DNA gyrase, the bacterial topoisomerase II is the only known enzyme that catalyses DNA supercoiling. The original meaning of supercoiling refers to DNA turned around itself generating plectonemic loops. As will be seen plectonemic DNA folding is of importance in chromatin condensation. Two models of gyrase actions are presented here. Gyrase is the prototype for type II enzymes interconverting topoisomers by way of transient double-stranded breaks. Negative supercoiling by gyrase influences all metabolic processes in which DNA is involved. Early hypothetical models of gyrase action assumed translocation of DNA past the enzymes and involved single-strand cleavage of DNA (Liu and Wang, 1978; Mizuuichi et al., 1978). The lack of evidence for translocation and the evidence of transient double-stranded scission during supercoiling gave rise to other models which are based on reversible double-standed breaks and require the transportation of a DNA segment through the double-chain cleavage. These models assume that the broken ends of DNA remain in close proximity and the enzyme complex holds the ends in a fixed orientation. This could be achived by reversible covalent linkage of the protein to each 5'-phosphate ends of DNA molecule (Morrison and Cozzarelli, 1979). The translocated segment is supposed to come either from a nearby (Mizuuichi et al., 1980) or from a distant segment of DNA oriented by its sequence with respect to the gating site (Brown and Cozzarelli, 1979). This latter mechanism is called the sign inversion. The sign inversion model of negative supercoiling is shown in Fig. 2.42.

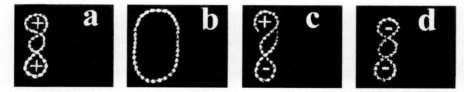

Fig. 2.42 Sign inversion model for DNA supercoiling. Positive supercoiled DNA (**a**) is relaxed by transient double-stranded DNA breaks and rejoining (**b**). Negative supercoils are generated by passing DNA segments through double-stranded breaks (**c**). The nagative supercoiled structure is stabilized by rejoining breaks (**d**). (Reproduced with permission of Bánfalvi, 1986a)

Sign Inversion Model of Negative Supercoiling

According to the sign inversion mechanism of supercoiling the positively supertwisted DNA which is normally formed in front of the separated strands during strand-separation processes (DNA replication, general recombination, transcription) is first relaxed through double breaks and rejoining. As high torsional energy is generated no ATP is needed for the relaxation. Gyrase then introduces negative supertwists into the closed circular DNA by transient strand breakage and by sealing the phophodiester bond at the distal side of the DNA backbone at the expense of ATP hydrolysis. The sign inversion was also thought to be responsible for catenation and

DNA Structure 83

decatenation of DNA rings. Each cycle of supercoiling is driven by conformational change induced by ATP. Fresh cycles are initiated by the hydrolysis of ATP. The inhibition of gyrase was observed by two classes of antibiotics reflecting a subunit associated composition of gyrase. The A subunit is associated with the breakage and rejoining of DNA. This step can be prevented by nalidixic acid. The B subunit mediates energy transduction. Novobiocin acts as a competitive inhibitor of ATP, interfering with energy coupling associated with supercoiling.

> Eukaryotic topoisomerase II enzymes are potential targets for clinically active anticancer drugs. Drugs targeting these enzymes may act by preventing the re-ligation of enzyme-DNA covalent complexes leading to protein-DNA adducts that include single- and double-stranded breaks. In mammalian cells, repair pathways are critical for repairing topoisomerase II–mediated DNA damage. Topoisomerase II–targeting agents, such as etoposide, can induce chromosomal translocations leading to secondary malignancies. The understanding of nonhomologous repair of topoisomerase II–mediated DNA damage thus may help to define strategies that reduce side effects on an important class of anticancer agents (Malik et al., 2006).

The sign inversion mechanism (Brown and Cozzarelli, 1979) consists of the following steps (Fig. 2.42):

1. Positive supertwists (Fig. 2.42a) are removed by nicking and closing phosphodiester bonds.
2. Gyrase binds to the relaxed DNA molecule (Fig. 2.42b) such that the two bound segments cross each other to form a right-handed (+) node and induces a counterposing left-handed (−) supercoil.
3. Gyrase induces a transient double-stranded break at the back of the right-handed node and the DNA passes through the break, inverting the handedness of the upper node from right to the left, changing the sign from (+) to (−) (Fig. 2.42c).
4. The phosphodiester bonds are resealed on the front side of the backbone. The net result is the reduction of linking number of DNA by 2 (Fig. 2.42d).

It is assumed that when sign inversion takes place DNA strands do not rotate relative to each other during the cycle based on the fact that the net result of a single cycle is the reduction of linking number of DNA by 2. However, the absence of odd-number changes in linking number does not exclude the possibility of rotation of free ends in the gyrase reaction. Energy is absorbed discontinuously; bioenergetic processes are known to be quantized. Discrete packets of ATP hydrolysis energy released at a time can cover the regeneration of doublets of negative supercoils as well in the gyrase reaction.

When sign-inversion experiments were carried out using Solomon's knot DNA (Bánfalvi, 1984) the linking number was expected to change by 2 in each cycle. These model experiments failed to confirm this theoretical value. We found odd-number changes in the linking number, suggesting that the reduction of the linking number does not depend only on the number of crossing of chains. The local torsional strain of strands rather than the crossing of chains turned out to be responsible primarily for the generation of supercoils which in turn distribute the strain evenly along the molecule. This observation led to the rotating model of negative supercoiling.

Rotating Model of Negative Supercoiling

The model is based on the rotation of broken strands, involving any odd-number change in linking-number and classical energetics for suprecoiling (Fig. 2.43).

The rotating mechanism of supercoiling expects neither transportation of DNA segments through transient breaks nor repeated strand-breaks and rejoining during the gyrase cycle:

1. Positively supercoiled DNA is relaxed by the introduction of a double-stranded break and by the rotation of broken ends relative to each other (Fig. 2.43a, b).
2. Strands are further rotated at their broken ends to the opposite direction of the helical turns (to the left) (Fig. 2.43c).
3. Continued rotation increases the number of supercoils without initiating fresh cycles (Fig. 2.43d, e).
4. Supercoiling is terminated by rejoining broken strands (Fig. 2.43f).

This continuous and economical model is consistent with the classical bioenergetics established by Lippmann (1941) and Calckar (1941, 1969) in which a high-energy intermediate is formed to drive the reaction. The sign inversion model cannot be

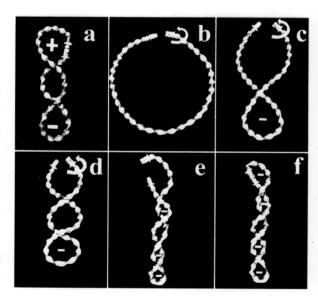

Fig. 2.43 Rotating model for supercoiling of DNA. Positive supercoils of Solomon's knot DNA (**a**) are removed by the introduction of double-stranded breaks (**b**). DNA is rotated around the breaking point to the left as indicated by the arrow (**c**). Further rotation to the left increases the number of negative supertwists (**d, e**). Negative supercoiled DNA is sealed (**f**). (Model experiment reproduced with permission of, Bánfalvi, 1986a)

DNA Structure   85

explained on the basis of classical bioenergetics and was supposed to operate by the conformational cupling theory also called "credit card" energetics ("buy now" = supercoiling, and "pay later" = ATP hydrolysis) (Cozzarelli, 1980).

That axial rotation of DNA during supercoiling takes place was confirmed in *E. coli* (Stupina and Wang, 2004). It is now generally accepted that the two lobes of the supercoiled DNA resembling figure eight can be rotated relative to one another.

Topology of Eukaryotic DNA

Nucleosome, the Supercoil Analogue of Eukaryotes

DNA strands interacting with polypeptide chains display additional levels of structural organization. The organization of DNA into loose associations of prokaryotic nucleoid and eukaryotic nucleosomal structures in cells plays a common role in packaging the DNA. The complex of DNA and of its associated proteins is known as chromatin.

The major difference between structural organization of prokaryotic and eukaryotic DNA comes from the fact that the eukaryotic DNA is not as bare as the prokaryotic DNA but is folded into chromatin structure. Eukaryotic DNA is closely associated with an equal mass of tightly bound proteins called histones. These histones form disc-like complexes around which the eukaryotic DNA is wound, creating a typical structure, the nucleosome. The basic chromatin structure contains nucleosomes. Nucleosomes consist of the double helical DNA and histones.

Histones:

- are basic proteins, about one in four residues being either arginine or lysine amino acids,
- five principal kinds of histones are known (H1, H2a, H2b, H3 and H4),

Nucleosomes:

- are the chromatin subunits,
- contain an octamer protein structure consisting of 2 subunits each of H2A, H2B, H3 and H4,
- alpha helical regions of histones contain the basic residues that contact the major groove of double-helix,
- H1 histone serves as a bridge between adjacent nucleosomes (Fig. 2.44),
- DNA is wrapped left-handed around a 3.2 nm radius core of histones,
- the core contains approximately 150 bp of DNA,
- there are 1.8 turns of DNA per nucleosome,
- linker DNA between two nucleosomes varies from 20 to more than 200 bp, with an average of 60 nucleotides long,
- the nucleosome core particle is representing the first level of DNA folding,

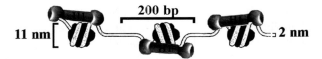

Fig. 2.44 Linear chain of nucleosomes known as "beads on string". DNA is wrapped around the histone octamer. The histone octamer is found at the center of a nucleosome core particle. The core particle consists of 2 copies of each of the four core histone proteins (H2A, H2B, H3 and H4). The double helical DNA has a cross-sectional diameter of approximately two nanometers, the diameter of the nucleosome is 11 nm. Nucleosomes are approximately 200 nucleotide base pairs apart from each other

- core structures would appear as "beads on string" if the DNA were pulled into a linear structure,
- folding of DNA in nucleosomes results in a compaction factor of about 7.

In the DNA structure there are several coiled structures which are essential in the compaction process. As alredy referred to, the first level of DNA condensation is the formation of the chromatin fibril, 11 nanometers in diameter, consisting of a linear chain of nucleosomes, protein core units around which DNA molecule is helically wrapped.

The slightly underwound state in the chromatin of eukaryotes arises from negative (i.e. left-handed) coiling of the DNA around the histone octamers to form nucleosome units. The ultimate motive force for negative supercoiling in chromatin is the spontaneous wrapping of DNA around the protein core. There is an interesting coincidence between the gyrase and the histone octamer, each protecting about 140 base pairs of DNA from digestion by staphylococcal nuclease and an estimated one-and-three-quarter supercoils from relaxation (Gellert et al., 1979). This similarity seems to come from evolutionary convergence rather than from homology to perform the same function, since the 400,000 dalton gyrase is larger than the 110,000 dalton histon assembly. Another distinguishing feature is the basicity of histones, while both gyrase subunits are acidic.

In eukaryotes the nucleosomal organization became the basis for the present sophisticated mechanisms controlling gene expression and cell differentiation. Negative superhelicity is limited and is unable to control promoter activity. Contrary to this, prokaryotic DNA is not as tightly associated with proteins and is highly supertwisted. Due to the many negative superstists ($\approx$ 80 in *E. coli*) the bacterial DNA strand separation processes (recombination, replication, transcription) driven by the torsional energy are much faster than those in eukaryotic cells. In eukaryotes the slightly underwound state is maintained by left-handed coiling of the DNA in nucleosomes. The nucleosomal structure localizes the torsional energy generated by turning the DNA around the histon core, but this type of supercoiling in not an additive property. Due to the relatively "lazy" DNA structure strand separation processes are slower. Further compaction of eukaryotic DNA results in heterochromatinization and temporary or permanent loss of gene expression. While these processes positive regulation dominates in eukaryotes, prokaryotes are mainly under negative control.

DNA Structure 87

The negative superhelicity became an essential tool for the precise control of overall gene expression in prokaryotes.

Compactness in the eukaryotic chromatin structure presents a significant barrier to DNA-dependent transactions, including DNA replication and gene transcription. Access to specific regions of chromatin is regulated at multiple levels, such as:

– chromatin remodeling of chromatin structure,
– local deposition of specific histone variants,
– post-translational modification of histones,
– recruitment of specific proteins to modified histones.

The detection, repair or degradation (apoptosis) of damaged double stranded DNA appears to be dependent on many of these same regulatory processes.

Supranucleosomal (Chromosomal) Organization of DNA

Nucleosomes are involved in the higher orders of structure that fold DNA into an extremely compact form found in the nucleus of eukaryotic cells. These levels involve thin (diameter 30 nm) fibrils, thin (300 nm) fibres, condensed fibres (600 nm) and chromosomes (1.4 µm). The thin fibril is the linear array of nucleosome core particles which are in contact with one another. Nucleosomes are arranged in thin filament with a diameter of 30 nm. Different regions of chromosomes can be folded somewhat differently. These differences are of functional significance for gene expression, but we do not know enough about structural orders of chromatin to reach conclusion at the functional levels. However, there is no doubt that increasing knowledge at structural levels leads to the understanding of function.

Early Chromatin Models

Despite the research in the field the supranucleosomal organization of the 11 nm chromatin string, the chromatin fibril remains poorly understood (Grechmann and Ramakrishnan, 1987; Hozak and Fakan, 2006), as indicated by the several models of the 30 nm fibril. Early models include the solenoid model (Finch and Klug, 1976), the helical-ribbon model (Worzel et al., 1981; Woodcock et al., 1984) and the crossed-linker double-helical model (Williams et al., 1986). Despite sharing certain similarities, these models differ considerably in detail, providing no evidence of transition from the 10 nm string to the 30 nm fibril (Grechmann and Ramakrishnan, 1987).

After extensive studies probing the structural and dynamic properties of chromatin (Bednar et al., 1998; Widom, 1998; Zlatanova et al., 1998; Luger, 2003), the question remained the same: how nucleosomes are arranged in the chain of a 30 nm chromatin fibril under physiological conditions (Widom, 1998; Luger, 2003).

Solenoid versus Zig-Zag Model

Early x-ray diffraction analysis favoured the solenoid model, where the nucleosomes are arranged in a helical fashion with a period of six nucleosomes per turn, the linker DNA bent, and nucleosomes forming a helical structure with a diameter of 30 nm (Finch and Klug, 1976). The condensation of DNA and solenoid formation before replication is schematically viewed in Fig. 2.45.

Twenty years later data on cryoelectron and scanning electronmicroscopy led to proposals compatible with a less regular arrangement of nucleosomes (van Holde and Zlatanova, 1996; Woodcock and Horowitz, 1995; Bordas et al., 1986; Kubista et al., 1990). These so called cross-linker models served as a basis for the zig-zag model (Woodcock et al., 1993). Additional support for the zig-zag model was provided by radiation induced DNA fragmentation of chromatin, showing a characteristic fragmentation of chromatin which could be correlated with the predictions of the zig-zag model, but deviated from the solenoid arrangement of nucleosomes (Rydberg et al., 1998). Current models favour the idea that chromatin condensation occurs via an accordion type compaction of nucleosome zig-zag chains (Grigoryev, 2004).

The interphase nucleus was regarded as a largely immobile and homogeneous organelle (Manuelidis, 1990; Berezney et al., 1995), but this attitude has changed

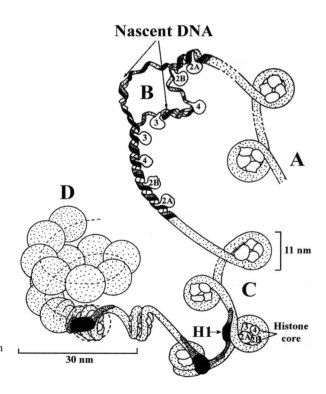

Fig. 2.45 Replication of eukaryotic DNA and hypothetic solenoid formation of nucleosomes. (**A**) Nucleosomes are decoiled before replication, (**B**) Replication bubble is formed and nascent DNA is synthesized at the two replication forks moving bidirectionally. (**C**) Active nucleosome string ("beads on string"). (**D**) Solenoid structure (30 nm)

DNA Structure   89

and chromatin as the most prominent constituent in the nucleus is believed to be an inhomogeneous and mobile structural component (Abney et al., 1997). When the salt-dependent rearrangement of nucleosomes was imitated by Monte Carlo simulation, the nucleosomal array of formation adopted an irregular 3D zig-zag conformation at high salt concentration and an extended beads on string conformation at low salt in agreement with hydrodynamic experiments (Sun et al., 2005). Contrary to the growing evidence that chromatin fibrils lack the regularity of the solenoid model and chromatin fibrils are flexible structures, the exact mechanism of chromatin folding remained unsolved.

Possible Fibril Arrangements

Since the solenoid model is seriously doubted, more attention has been paid to other models. There are at least four possibilities for fibril arrangements (Fig. 2.46), two of them are principal models known as the solenoid (Fig. 2.46A, B) and the zig-zag model (Fig. 2.46 C). The other two models include the leafed rosette (Fig. 2.46/D) and the hairpin structure (Fig. 2.45E).

In the leafed model the nucleosomes are arranged in a "double or triple beads on string" fashion without the spacer DNA strands crossing each other, giving rise to the 30 nm chromatin fibril (Fig. 2.47a). The zig-zag arrangement of chromatin would lead to two major topological structures, the parallel (Fig. 2.47b/1) and the perpendicular zig-zag chromatin structure (Fig. 2.47b/2). Further geometric variants of the zig-zag structure could be: zig-zag with stem, with stem and linker DNA and with the stem and the linker DNAs laying on planes perpendicular to the nucleosomal average plane.

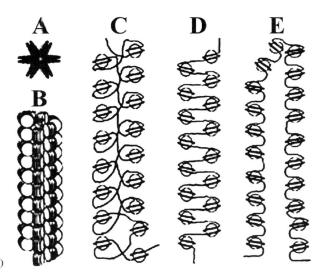

Fig. 2.46 Representation of geometric structures of possible nucleosome models. (A) and (B) two views of DNA solenoid structure, (C) extended zig-zag arrangement, (D) extended leafed rosette, (E) extended hairpin (Reproduced with permission of Bánfalvi, 2008)

Fig. 2.47 Models of nucleosome and linker DNA. In the simplified representation a short array of nucleosomes is connected with linker DNA. Transition from relaxed to condensed supranucleosomal structures include (**a**) leafed rosette, (**b**) zig-zag fibril, (**c**) plectonemic fibril resembles finger crossing, several plectonemic fibrils in a row resemble an accordion. (Reproduced with permission of Bánfalvi, 2008)

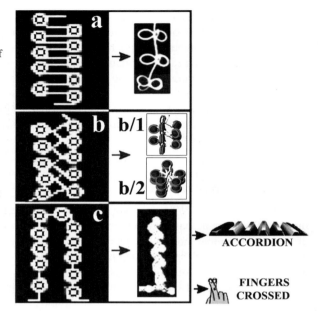

Hairpin Model

In the meandering, forming hairpin arrangement (Fig. 2.47 c) the strings could turn around themselves to form plectonemic loops. It should be mentioned at this point that no consensus has been reached whether or not the extended and condensed forms of the 30 nm chromatin fibrils are directly related to the euchromatin ↔ heterochromatin transition.

The condensation and decondensation pattern of the hairpin topology is similar to the movement of an accordion. Moreover, in the hairpin structure the nucleosome fibrils are turned around themselves keeping "fingers crossed" (Grigoryev, 2004) with the notable exception that during "finger crossing" the "fingertips" touch each other due to the continuity of the chromatin fibrils. It was suggested that nucleosome gaping ("the pushing together of the accordion") may not only compact the fibril, but may also be the driving mechanism for supercoiling the fibril loop in a condensed higher order structure (Mozziconacci et al., 2006).

The nucleosomal arrangements shown in Fig. 2.47 correspond to the characteristic properties of the 30 nm fibril, are consistent with experimental data of chromatin and meet theoretical considerations. However, the lack of definitive experimental evidence for these models, requires the refinement of existing techniques and the introduction of new experimental approaches to define the supranucleosomal organization and eliminate those theoretical forms which are not involved in structural transitions of the chromatin fibrils.

DNA Structure

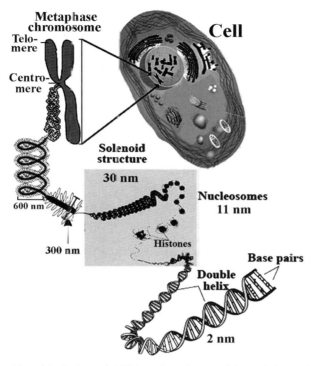

Fig. 2.48 Solenoid model of eukaryotic DNA condensation. Double standed DNA is turned around the histon core forming nucleosomes. According to this model it was assumed that nucleosomes (11 nm in diameter) coil further into a solenoid shape (30 nm), then form looped thinner fibers (300 nm) and compacted further to thick fibers (600 nm). These final coils are seen in a karyotyping spread as metaphase chromosomes. Metaphase chromosomes have chrarteristic sizes and banding patterns, regardless of tissue and cell type. The protein-DNA structure of chromatin is stabilized by attachment to a non-histone protein scaffold called the nuclear matrix. (Reproduced with permission of Bánfalvi, 2008)

The solenoid model was deduced from X-ray diffraction analysis and refers to a theoretical structure where nucleosomes are arranged in a helical fashion and the compaction corresponds to a 30 nm fibril of chromatin (Fig. 2.48).

Plectonemic Model of Chromatin Condensation

The analysis of fluorescent images of chromatin structures isolated from nuclei of *Drosophila* cells supported by computer image analysis resolved the fiber structure which was originally not visible for the human eye. Chromatin fibrils became detectable and could be magnified by computer image analysis resulting in the characterization of the nucleosomal arrangement in *Drosophila* nuclei. The nucleosomal string known as "beads on string" turned out to be folded in a meandrine fashion occasionally twisted around itself forming the 30 nm looped fibril and bringing about transitory plectonemic supercoiled structures. Solenoid formation was not observed in the supranucleosomal organization of *Drosophila* DNA (Bánfalvi et al., 2007).

The revaluation of experimental data of nucleosomal strings of *Drosophila* DNA (Bánfalvi, 2006b) led to a molecular model of chromatin condensation named plectonemic chromatin model. Figure 2.49 which is based on experimental data shows magnified nuclear substructures from the upper right corner to the lower left corner. The steps of magnification are indicated by numbers from 1 to 5 in Fig. 2.49. The condensation process is summarized from the opposite direction from the lower right corner to the upper left corner from a to e.

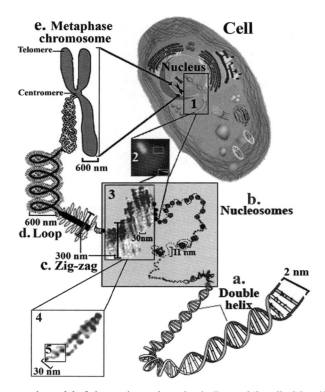

Fig. 2.49 Plectonemic model of chromatin condensation in *Drosophila* cells. Magnifications from the upper right corner to the lower left corner summarize our results in *Drosophila* cells after reversal of permeabilization (Bánfalvi, 2006a; Bánfalvi et al., 2007). 1. chain of chromosomes slipping out of the nucleus, 2. small unit of chromosomes (chromosome rodlet, the boxed area was converted to black and white negative image (right upper box), 3. computer image analysis of the boxed area showing the nucleosomal arrangement of chromatin fibrils, 4. beads on string forming a hairpin structure, 5. the formation of nucleosomes. Plectonemic model of chromosome formation from the lower right corner to the upper left corner: a. double stranded DNA (diameter 2 nm), b. nucleosomes (11 nm), c. nucleosomal arrangement including plectonemic fibrils (30 nm) and zig-zag loops of plectonemic fibers (average length 300 nm), d. 600 nm looped structure, e. metaphase chromosomes (Reproduced with permission of Bánfalvi, 2008)

DNA Structure   93

Experimental data of Fig. 2.49 include 1. the linear array of chromosomes slipping out of the nucleus, 2. small, visible chromosome rodlet (subunit of chromosome), with a boxed area containing chromatin fibrils which are invisible, but detectable by computer image analysis, 3. computer aided magnification of the boxed area shows meandering chromatin fibrils forming plectonemic loops, turned around themselves, 4. one uncoiled loop of the nucleosomal string was selected to magnify it further to the limit of microscopic and computer resolution leading to the clear visualization of individual nucleosomes, 5. formation of nucleosomes. The chromatin condensation pattern in Fig. 2.49 starts a. with double tranded DNA (2 nm), b. DNA turning around the histone octamer forming nucleosomes (11 nm), c. zig-zag arrangement of meandering nucleosomal strings, with a ~30 nm diameter and ~300 nm length plectonemic loops consisting of 12–15 nucleosomes, d. the compressed plectonemic loops (300 nm long) turn to chromatidal loops (600 nm), e. which are part of the metaphase chromosome. Condensed metaphase chromosomes are visible inside the nucleus of a cell. Chromosomes are highly ordered, organized packages designed for the storage of genetic material, for condensation during cell division and regulation of gene expression and for decondensation during DNA replication.

Chromosomes of Animal Cells

Eukaryotic Genome Size and DNA Compaction

The size of genomes of eukaryotes may vary ~80.000-fold (Hartl, 2000) and the size of individial chromosomes about 1200-fold from 0.2 Mbp to 245 Mbp. Chromosomes of yeast cells are much smaller (~0.2–2 Mbp) than those in humans (23–246 Mbp). Human chromosomes are listed in Table 2.3 and the size of each human chromosome is given in megabasepairs (Mbp, mega = million). The length of DNA in a single human cell is nearly 2m.

Table 2.3 The size of human chromosomes

Chr	Mega-base pair (Mbp)	Chr	Mbp
		13	113.0
1	246.1	14	106.3
2	243.6	15	100.2
3	199.3	16	90.0
4	191.7	17	81.8
5	181.0	18	76.0
6	170.9	19	63.8
7	158.5	20	63.7
8	146.3	21	46.9
9	136.3	22	49.3
10	135.0	X	153.6
11	134.4	Y	22.7
12	132.0		(Euchromatin)

A single human chromosome ranges in size from 49×10^6 nucleotide pairs in the smallest autosome (stretched full-length would be 1.7 cm) up to 246×10^6 nucleotide pairs in chromosome 1 (which would extend to 8.5 cm). The smallest chromosome is the Y, male chromosome (22.7×10^6 base pairs $\sim$0.8 cm DNA).

The total number of genes might not correspond with the phenotypic complexity in the anatomy and physiology of the organism. For example, the size of the human genome is not larger than the genomes of other mammals, not much larger than some invertebrates and plants, and may even be smaller than species of lower evolutionary complexity (Table 2.4). Nevertheless, more proteins are encoded in the human genome per gene than in other species (Rubin, 2001).

Chromosome Number

The number of chromosomes in the germ cell (sperm, egg) is half the diploid number; it is called the haploid chromosome number, represented as 1n. Somatic cells (cells other than germ cells) of sexually reproducing organisms are diploid, denoted by 2n. In humans, the haploid chromosome number is 23, the diploid chromosome number is 46. Most organisms are diploid, with two sets of chromosomes, and two copies (called alleles) of each gene.

Metaphase Chromosomes

Metaphase is the stage of the cell cycle when a chromosome is most condensed and easiest to distinguish and to study. The term metaphase comes from Strasburger (1884) (denotes the stage after, in Greek: meta means after), when he thought by mistake that nuclear division took place and chromosomes had divided into chromatids. It turned out later that chromosomes were already doubled by the time nuclear division begun. Due to the fact that metaphase chromosomes are too condensed to see more structural details, somewhat relaxed chromosomal forms (prometaphase, prophase) can be selected to study fine details. Moreover, the term metaphase chromosomes became obsolete, since it is based on the temporal division of the cell cycle. To update the terminology, the term condensed chromosome would be more appropriate. Nevertheless, fully condensed chromosomes are still referred to as metaphase chromosomes. Another objection against this oversimplification is that metaphase lasts for less than 1 h, while chromosomes of mammalian cells are in interphase for an average of 22 h. In contrast to those of condensed metaphase chromosomes, the shape, length, and architecture of decondensed chromosomes are not known (Lemke et al., 2002).

Centromere and Telomere Regions of Chromosomes

Centromeres. Mammalian chromosomes have constrictions in their rods. The primary constriction is called centromere. Centromeres are highly condensed chromosomal substructures. The kinetcochor protein complex attached to the centromere is responsible for directing chromosome movements in mitosis during the transition

DNA Structure

Table 2.4 Diploid chromosome number of different species

Species	Chromosome number
Bacteria	1
Ant (*Myrmecia pilosula*)	2
Parasitic roundworm (*Parascari equorum var. Univalens*)	2
Mold (*Penicillium species*)	4
Mosquito (*Culex pipiens*)	6
Indian muntjac (deer) (*Muntiacus muntjac*)	6
Fruit fly (*Drosophila melanogaster*)	8
Arabidopsis (plant, mustard family) (*Arabidopsis thaliana*)	10
Microscopic roundworm (*Caenorrhabditis elegans*)	12
House fly (*Musca domestica*)	12
Red clover (*Trifolium pratense*)	14
Honeybee (*Apis mellifera*)	16
Ginea pig (*Cavia porcellus*)	16
Green algae (*Alga mediterranea*)	20
Corn (maize) (*Zea mays*)	20
Bean (*Phaseolus vulgaris*)	22
Chinese muntjac (deer) (*Muntiacus reevesi*)	23
Tomato (*Lycopersicon esculentum* L)	24
Pepper (*Capsium annuum*)	24
Frog (*Rana pipiens*)	26
Siklworm (*Bombix mori*)	28
Budding yeast (*Saccharomyces cerevisiae*)	32
Fox (*Vulpes vulpes*)	34
Apple (*Malus silvestris*)	34
Tibetian fox (*Vulpes ferrilata*)	36
South African clawed frog (*Xenopus laevis*)	36
Domestic cat (*Felis catus*)	38
Domestic pig (*Sus scrofa*)	38
House mouse (*Mus musculus*)	40
Rat (*Rattus norvegicus*)	42
Rabbit (*Oryctolagus curriculus*)	44
Bat (*Eptesicus fuscus*) (Big Brown Bat)	44
Human (*Homo sapiens*)	46
Tobacco (*Nicotiana tabacum*)	48
Potato (*Solanum tuberosum*)	48
Gorilla (*Gorilla gorilla*), Chimpanzee (*Pan troglodytes*)	48
Domestic sheep (*Ovis aries*)	54
Alligator (*Alligator missississippiensis*)	60
Goat (*Capra hircus*)	60
Cow (*Bos taurus, Bos indicus*), Buffalo (*Bison bison*)	60
Donkey (*Equus asinus*)	62
Horse (*Equus caballus*)	64
Camel (*Camelus dromedarius*)	70
Chicken (*Gallus gallus*)	78
Domestic dog (Canis familiaris)	78
Crayfish (*Cambarus clarkii*)	200
Field horsetail (plant) (*Equisetum arvense*)	216
Fern (*Ophioglossum reticulatum*) (Polyploidy)	1260

96 2 Structural Organization of DNA

between metaphase and anaphase. The most frequently used classification besides chromosomal size is based on the position of the centromere, distingushing three morphological groups: acrocentric (constriction near the top), telocentric (constriction at the end), and metacentric (constriction in the middle).

Telomeres. Chromosomes have two ends called telomeres consisting of nucleoprotein complexes with the primary protective function and stabilization of the chromosome ends. Telomeres contain long streches of tandem repeats of DNA sequences rich in G and C bases. In vetebrates these terminal repeats are highly conserved sequences: $(TTAGGG)_n$. Telomer sequences are maintained by the catalytic function of telomerase, a reverse transcriptase responsible for the extension of the terminal repeats. Reduced or absent telomerase activity shortens the telomeres. When telomers become shorter than necessary, the telomer protecting protein complex is unable to bind and chromosomes become instable and free sticky ends emerge which may cause chromosome rearrangement and ultimately cancer. Gradual shortening of telomers is a natural process leading to the senescence of cells. When telomeric sequences in somatic cells shorten below the critical level, cell division ceases, contributing to cellular changes seen during ageing. The length of teleomeres in human blood cells is around 8,000 base pairs at birth and decreases to $\sim$1,500 in elderly people. Each cell division results in a telomere loss of 50–200 base pairs, indicating that cells can divide only 50–100 times due to the progressive shortening of telomers in a lifetime. Somatic cells with low telomerase activity are regarded as "mortal" cells, while germline cells (egg, sperm) with high telomerase activity are "immortal".

Telomer shortening is related to discontinuous DNA replication, namely to the lack of RNA primer formation in Okazaki fragments of the lagging strand at chromosome ends. The length of Okazaki fragments in eukaryotic cells is about 200 nucleotide long, corresponding to the nucleosomal organization of DNA. For the synthesis of Okazaki fragments a short piece of RNA (primer) is necessary to start DNA synthesis. After binding of RNA primer the Okazaki fragment can be synthesized. The primer cannot bind to the very end of the telomere, this Okazaki fragment remains single stranded and will be lost in the next cycle of replication. As cell division continues the telomere gets shorter and shorter.

The function of telomers is similar to plastic tips on shoelaces

By analogy telomeres have been compared with the plastic tips on shoelaces. Telomers similar to such plastic tips prevent chromosome ends from fraying and sticking to each other. Moreover, the shortening of these ends may affect vital streches of DNA and their fusion lead to cancer, other diseases and ultimately to death.

Telomeres have also been compared with a bomb fuse. The longer the fuse the later the bomb will explode.

DNA Structure 97

One would expect that there is a direct relationship between telomere length and lifespan. However, long-lived species like humans ($\sim$70–80 years) have much shorter telomers than other mammals like mice, which live only an average of two years. This evidence shows that telomeres alone do not determine lifespan. There are other factors which have to be taken into consideration regarding the lifespan namely the metabolic rate and body size. The calculation of the metabolic rate expressed in oxygen consumption per unit body mass (specific oxygen consumption) shows a striking picture of the relationship between body size and oxygen consumption. For example 1 g of shrew tissue consumes oxygen at a rate of some 100-fold as great as 1 g elephant tissue. This tremendous increase in oxygen consumption necessitates high oxygen supply and blood flow involving heart function, respiration and food intake must be similarly affected (Schmidt-Nielsen, 1997). Probably the major cause of aging is due to metabolic i.e. "oxidative" stress. It is the damage to DNA, proteins and lipids (fatty substances) caused by oxidants, which are highly reactive substances containing oxygen. These oxidants are produced normally during metabolism, and also result from inflammation, infection, consumption of alcohol, cigarettes, drugs, emotional stress. The calorie restriction diet has been developed to a strategy – a choice of lifestyle to extend healthy and maximum life span – experimentally confirmed by experiments in rodents and primates. Animal studies conducted over the past 20 years have shown up to a 40% increase in maximum life span by reducing the intake of calories to a level of 20–40%.

Caloric restriction with adequate nutrition

Although, there is much evidence that caloric restriction will extend the average lifespan of people, the extension of maximum lifespan seen in caloric restriction with experimental animals is not quite applicable to human beings. Other factors should also be taken into consideration. For example, death by cardiovascular disease shortens human life by an average of 13 years. It has long been known that cardiovascular mortality is lowest among those who are leanest. For this reason, caloric restriction may actually be of benefit to humans. The concept of caloric restriction is:

"Eat fast and die young or

Eat less and live longer".

Variation in DNA Content, C-Value Paradox

The constancy of DNA content of all cells in the organism of a given species was regarded as an evidence that genes are composed of DNA. This constancy is reflected by the term C-value (genome size). C-value is the haploid genome content per cell, 2C-value corresponds to the diploid DNA content. Paradoxically, there is no direct relationship between the DNA constancy and number of genes (Thomas, 1971). The amount of DNA in the haploid genome of a eukaryotic cell – known as C-value – appears to be greater than would be predicated from the evolutionary complexity in several species (Britten and Davidson, 1969). The C-value enigma was partially resolved when it turned out that eukaryotic cells contain split genes with non-coding regions, consequently the genome size does not reflect the gene number. In the eukaryotic genome only a very small proportion codes for proteins, ~10% in sea urchin, 5–10% in *Drosophila*, 1.5% in humans. The excess of the DNA also referred to as "selfish" DNA is indicated by the example of human genome, comprising only about 1.5% protein-coding genes and 85% non-coding sequences, such as introns, non-processed pseudogenes, transposable elements, short interspread elements (SINEs, such as the *Alu* I family in primates or the related B1 family in rodents), long interspread elements (LINEs, L1 family, similar to retroposons), retroposons known as processed R-genes (pseudogenes) mobilised via RNA form, e.g. retroviruses. SINEs and LINEs and processed R-genes moved around the genome over evolutionary time. Mobile genetic elements (transposons) can move around the genome within a single generation of an organism.

Excess DNA in genomes is one aspect of the C-value paradox. The other aspect of the C-value enigma is still not explained, namely why some species have a much higher amount of DNA than others of the same level of evolutionary complexity. Although, there is a basic relationship between the increase of evolutionary complexity and minimum C-values, but maximum values for the same group can be several orders of magnitude higher. The overall sigmoidal character of the evolutionary complexity shows an initial slow process with wide variation in C-values (Fig. 2.50). The complexity gradually speeded up during evolution, and slowed down at high complexity with a tendency of unifying C-values.

The C-value paradox indicates that non-coding DNA may have other (e.g. regulatory) functions. As already mentioned, a certain increase in genetic information is obviously accompanied by evolution. As a consequence of increasing complexity life has come to occupy places in which it did not exist before and the total mass of living matter increased. However, increase in evolutionary complexity does not rule out exception referred to as C-value paradox. Many species have much higher DNA content than the minimum in any class. Most classes of eukaryotes comprise a range of species with a variation in genome size of about 10-fold. Mammals, reptiles and birds have genomes which fall into a particularly small range of DNA contents; the C-value is usually 1–3 picograms per haploid genome ($1-3 \times 10^9$ base pairs). A 100-fold increase in C-values can be observed within the amphibia taxon, the DNA content varies widely, from less than 1 pg to almost 100 pg. Even closely related amphibian species have different contents of DNA in their nuclei, while their evolutionary complexity does not vary significantly. The DNA content of fishes similarly to amphibians, shows variations (1.5–50 pg). Flowering plants have the widest range of C-values. Bacterial genomes are compact and use effectively their genetic information with little excess DNA.

To explain the second aspect of the C-value paradox, the relationship between the osmolarity and evolutionary complexity is referred to. Through the maintenance of optimal conditions in the inner environment (primarily by ionic concentrations), adaptive changes in proteins, determined by changes in DNA sequences either coding for these proteins or regulating their function, may be minimized. Accordingly,

DNA Structure

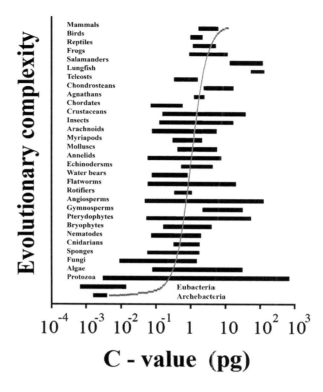

Fig. 2.50 C-value paradox. Eukaryotic genomes vary markedly in their DNA content (C-value) and all contain far more DNA than they would require to carry out all of their biological functions. The size of bacterial genomes is relatively uniform. 1 pg ≈ 1000 Mbp = 1 Gbp

species living in a relatively wide range of external and internal environmental stimuli and are unable to regulate their body temperature may have more specific DNA sequences coding for and regulating modulated enzymes. The existence of such isozymes permits the fine-tuning of metabolism to meet the particular needs of a given tissue or developmental stage at different internal and external environmental conditions. This would mean that species living under altered or extreme conditions have higher C-values than expected by their evolutionary complexity and correspondingly their plasma electrolytes expressed in osmolarity varies within wide range, e.g. in amphibians between 160 and 240 mOsm concentration. To the contrary, the C-value varies within narrow range in those species which live under constant osmotic conditions such as land vertebrates: mammals (~295 mOsm), reptiles (~286 mOsm), snakes (~300 Osm), lizards (~307 mOsm), turtles (~287 mOsm), birds (~317 mOsm) (Bánfalvi, 1991).

Variation in Chromosome Number

Euploidy, or euploid chromosome number is the normal diploid chromosome number listed in Table 2.4.

100   2 Structural Organization of DNA

Aneuploidy. This condition means that the cell or organism has additions or deletions of a small number of whole chromosomes from the normal diploid number of chromosomes. Aneuploidy is regarded as the most serious genetic imbalance of cancer cells and was considered to be the cause of cancer when it was discovered more than 100 years ago. After the discovery of the gene, the aneuploidy hypothesis has lost ground to the hypothesis that mutation of cellular genes causes cancer. According to this later hypothesis, tumour cells are diploid and aneuploidy is nonessential. However, it turned out that aneuploidy correlated 100% with chemical transformation in Chinese hamster cells. Due to the early association between aneuploidy, transformation, and tumourigenicity, it was concluded that aneuploidy is the cause rather than the consequence of malignant transformation (Li et al., 1997).

Among the genotoxic agents changing the chromosome number mevalonate deprivation is a good example which leads to aneuploidy. Chinese hamster (*Critecus griseus*) ovary (CHO) cells have the wild type chromosome number of 20–21, while mevalonate deprived cells have a broad distribution of chromosome number, often with a mean around 36–40 (Brown and Simoni, 1988).

Due to specific changes in chromosomes known as chromosome abnormalities the chromosome number for man was assumed to be 48 based on studies of sections of meiotic cells from human testis (Painter, 1923). The explanation for this higher chromosome number is that most karyotypes of human cells in the early 1950s were derived from abnormal cancer cells which were easier to grow than normal cells. As a general rule it is accepted, that the higher the aneuploidy, the wilder the growth of the cancer cells will be. Indeed, most cancer cells harbor chromosomal alterations such as abnormal chromosome number (aneuploidy, polyploidy) or structure (Jefford and Irminger-Finger, 2006). Aneuploidy is a major cause of human reproductive failure and may play an important role in the onset of cancer (Fenech, 2002).

Tjio (spell it: Cheeo) and Levan (1956) reported that the correct human chromosome number was 46. This observation was immediately confirmed independently by Ford and Hamerton (1956). An interesting episode in the chromosome research was the assumption that the chromosome number differed in whites (46) and Japanese (48) (Kodani, 1958), which turned out not to be the case (Tjio and Puck, 1958).

Rules for mouse genetic nomenclature were originally published by Dunn et al. (1940) and revised by the International Committee for Standardized Genetic Nomenclature in Mice (1963, 1973, 1981, 1989, 1993). In 2000 these guidelines were completely rewritten in order to explain more clearly the rules. Rules for rat genetic nomenclature were published by the Committee on Rat Nomenclature in 1992 and then by Levan et al. in 1995.

Karyotype, Chromosome Size

For the identification of chromosomes, they are normally arrested in late prophase or early metaphase of mitosis, when the chromosomes are duplicated and condensed, but the centromere has not yet divided. At this stage of the cell cycle individual

DNA Structure

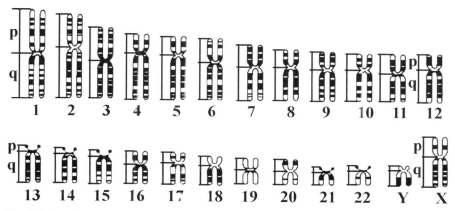

Fig. 2.51 Banding pattern of human chromosomes. Each chromosome has a short (p) and a long (q) arm. Each arm has 1–3 sections and further subsections within a section. Numbering of sections and subsections starts from the centromere. For example p 3.4 position in a chromosome refers to Sect. 3 and subsection 4, localized in the telomere region of the p arm (Reproduced with permission of Bánfalvi and Antoni, 1990)

chromosomes consist of two chromatids. Chromosomes are stained, photographed and arranged in order, from largest to smallest. The picture, or map of all chromosomes is called a karyotype. The karyotype can help to identify chromosome abnormalities that are evident in either the structure or the number of chromosomes. To help identify them, human chromosome pairs have been numbered from 1 to 22, the 23rd pair is labeled X and Y (Fig. 2.51). The size of human chromosomes is given in Table 2.3. Chromosomes can be grouped not only by size, but also by the location of the centromere (metacentric, submetacentric, acrocentric) which divides the chromosome into two arms (p, upper short and q, lower long arm). Each chromosome arm is defined further by numbering the bands that appear after staining; the higher the number, the more distant that area is from the centromere.

The genome size of individual chromosomes of eukaryotic cells varies:

- $\sim$80000-fold in size (Hartl, 2000), from
- $\sim$0.2–2 Mbp in budding yeast to
- 23–246 Mbp in humans.

Compaction:

- the linear compaction ratio of mitotic chromosomes relative to B-form DNA ranges from $\sim$160-fold to the $\sim$200-fold ratio in budding yeast (Dressler and Giroux, 1988; Guacci et al., 1994; Lavoie et al., 2000),
- $\sim$10,000–20,000-fold in mammalian chromosomes (Li et al., 1998),
- the chromosome compaction of mitotic versus interphase chromosomes is about two-fold higher in budding yeast (Lavoie et al., 2000),
- $\sim$4-fold at the euchromatin (Li et al., 1998) and about

102   2   Structural Organization of DNA

- 50-fold at heterochromatin regions of mammalian cells (Lawrence et al., 1990),
- phosphorylation of histone H3 is correlated spatially and temporally with condensation of interphase chromatin into metaphase chromosomes,
- phosphorylation is not required for the maintenance of the condensed state.

Genetic-Linkage

Genetic-linkage maps illustrate the order of genes on a chromosome and the relative distances between genes. The first maps were made by tracing the inheritance of multiple traits (e.g. hair and eye colour), through generations. Mapping of genetic-linkage is based on a normal biological process called crossing over, which occurs during meiosis – a type of cell division – producing sperm and egg cells. During meiosis in sex cells, chromosomes line up in pairs, "stick" to each other and exchange equivalent pieces of DNA. This sticking and exchanging is called crossing over. Recently, genetic-linkage maps are made by tracing variations of certain DNA sequences. Sequence alterations are found in many places in the genome and vary from person to person known as polymorphism.

Physical Mapping

The physical map is an ordered set of DNA pieces. In the late 1970s restriction endonucleases have been introduced to cut the genome into small pieces, which could then be analized individually and their sequences be determined. The overlapping DNA sequences were used to make the first physical maps and to localize them inside the genome. These overlapping DNA sequences were used later on to construct the entire DNA sequence of different species including man (Human Genome Project – HUGO).

Sex Chromosomes

Sex chromosomes of germ cells are normally designated X or Y in most animals and in some plants. These chromosomes determine the sex, XX resulting in female and XY in a male character. X chromosome is present: singly in males and doubly in females. The sex chromosome Y is carried exclusively in man. Healthy human females normally have two X chromosomes, and human males one X and one Y chromosome.

Most frequent genetic abnormalities are:

- XXX three X chromosomes in females,
- X0 Turner's syndrome,
- XXY Klinefelter's syndrome in man,
- XYY a man with two Y chromosomes.

Table 2.5 clearly shows that in man the male character is determined by the presence of Y chromosome, its absence results in female phenotype. Barr body is the

DNA Structure

Table 2.5 Composition of sex chromosomes in germ cells

Compostion of sex chromosomes	Phenotype of sex	Number of Barr bodies
XX (healthy women)	female	1
XY (healthy man)	male	0
XO (Turner syndrome)	female	0
XXY (Klinefelter syndrome)	male	1
XYY	male	0
XXX	female	2
XXXY	male	2
XXXX	female	3
XXXXY	male	3
XXXXX	female	4
XXXXXY	male	4

inactive X chromosome(s) in a female cell, or the inactive X chromosome(s) in a male (Lyon, 2003).

The normal female karyotype is: 46, XX. Triple X syndrome is characterized by the presence of an extra X chromosome in each cell of a human female with a 47, XXX aneuploidy. The extra X chromosome is due to *nondisjunction* during meiosis (sex cell division). The normal separation of chromosomes in meiosis is termed *disjunction*. The uneven distribution of the genetic material (*nondisjunction*) is a common mechanism for trisomy and monosomy formation. The condition also known as trisomy X or XXX syndrome, most often does not result in any physical irregularities or medical problems.

In Turner syndrome the karyotype is labeled 45, X (or 45, XO). Turner syndrome involves several abnormalities, of which X monosomy is the most frequent occurring in about every 2500th female births. The only sex chromosome X present is fully functional. Women with Turner syndrome are generally shorter than the average and sterile, unable to conceive a child due to the absence of ovarian function.

The normal male karyotype is: 46, XY. In Klinefelter syndrome the aneuploid karyotype is: 47, XXY. The effect of the extra X chromosome in Klinefelter syndrome is small testes development and reduced fertility. Affected patients are almost always sterile, often with some gynecomastia (increased breast tissue), low serum testosterone level but high follicle-stimulating hormone (FSH) and luteinizing hormone (LH) levels.

Patients with tetrasomy X, pentasomy X show joint and muscle abnormalities, hypotonia and joint looseness in the hips, skeletal problems including spinal deformities.

Pentasomy XXXXX syndrome affects mental, growth, and motor retardation.

Chromosomal Aberrations

After the chromosomal abnormalities due to higher or lower number of sex chromosomes the abnormalities of somatic chromosomes are discussed. Most of the abnormalities occur in sperm or egg cells. In such cases the mutation is present in each cell of the body. In abnormalities taking place after conception some cells are affected others are not, resulting in mosaicism.

There are two major types of irregularities:

– numerical and
– structural abnormalities.

The missing of a chromosome from a pair (monosomy) or more than two (three) chromosomes of a pair (trisomy) has alredy been mentioned in the previous subsection. The most frequently occuring numerical abnormality is trisomy 21, known as Down syndrome with three copies of chromosome 21, rather than two. The Turner syndrome is an example of chromosome X monosomy.

The most frequently occuring structural alterations are:

– deletion, when one nucleotide pair (A-T, or G-C) or a portion of chromosome is missing or deleted,
– duplication, a portion of duplicated chromosome is present in the genetic material,
– translocation, a portion of chromosome is transferred to another chromosome (it can also be reciprocal translocation e.g. the formation of the Philadelphia chromosome by translocation ($9q^+$, $22q^-$) is responsible for the development of chronic myeloid leukemia.
– inversions, the broken off portion of chromosome is reinserted with opposite polarity,
– ring formation, the broken off portion forms a ring.

Among the risk factors the maternal age and environmental factors are of particular concern. Girl babies are born with all the egg cells they will ever use as women for conception. Consequently, each egg cell is as old as the age of the woman. The older the women the higher the risk of chromosome abnormality of her baby. Paternal age is not increasing significantly the risk of chromosome abnormalities, as men produce new sperm cells throughout their life.

Altough, there is no conclusive evidence, but environmental catastrophies caused by exposure to ionic irradiation or carcinogens may contribute to an increased frequency of genetic errors.

Genetic Diversity

In population genetics there are several hypotheses regarding genetic diversity:

– neutral substitution theory of evolution is the diversity resulting in the accumulation of harmless substitutions,
– diversifying selection hypothesis proposes that two subpopulations of a species live in different environments selecting for different alleles at a particular locus,
– frequency-dependent selection: the more common an allele is, the less fit it will become.

DNA Structure   105

In principle, genetic diversity refers to any variation in the nucleotide sequence, genes, chromosomes, or whole genomes of organisms. Diploid cells have two alleles of each gene. Each allele codes for the production of a protein consisting of amino acids. Differences in the nucleotide sequences of alleles result in the production of slightly different variants of proteins. Structural variation has consequences on the development of the anatomical and physiological characteristics of the organism, and ultimately affects the behavior of the mutant organism.

Genetic diversity within a population is measured by:

- DNA sequence homology,
- DNA fingerprinting,
- microsatellite DNA loci,
- targeted analysis of fittness loci.

Genetic diversity is an indicator of ecosystem condition and sustainability.

Eukaryotic Cell Cycle

The sequence of events occuring repeatedly during the lifetime of a cell is named the cell cycle. The eukaryotic cell cycle consists of 4 major phases. Typical events take place in each period. The final step of the cell cycle is cytokinesis resulting in two identical daughter cells.

The 4 phases of a typical cell cycle are given below and depicted in Fig. 2.52.

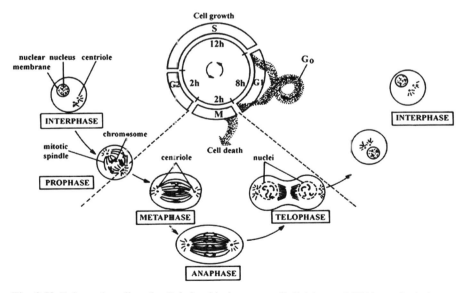

Fig. 2.52 Eukaryotic cell cycle: Relationship between cell division and DNA synthesis in eukaryotic cells. Major phases of the cell cycle: 1. interphase, involving all phases (G1/G0, S and G2) except the 2. mitotic phase (M), which involves mitosis (prophase, metaphase, anaphase and telophase) and cytokinesis (cell division). (Modified based on data from Rees and Sternberg, 1984)

2 Structural Organization of DNA

1. **G_1/G_0 phase.** The G_1 phase corresponds to the gap in the cell cycle that occurs following cytokinesis. During this phase cells either exit the cell cycle and become quiescent or terminally differentiated or start a fresh cycle. Terminally differentiated cells are in a non-dividing state. Quiescent and terminally differentiated cells are in G_0 phase. Cells in G_0 can rest in this state for extended periods of time. Specific stimulation induces them to leave the G_0 phase and re-enter the cell cycle through the G_1 phase or alternatively may undergo permanent terminal differentiation. In the G_1 phase cells start to synthesize cellular components needed for DNA synthesis and for cell division.
2. **S phase** is the phase of the cell cycle during which the DNA is replicated. This is the DNA synthesis phase. Additionally, some specialized proteins are synthesized during S phase, particularly the histones. The size and DNA content of cells increases synchronously during S phase.
3. **G_2 phase** follows the completion of DNA replication. The final stage of chromosome condensation takes place during G_2 phase. Nucleoli disappear and two microtubule organizing centers start to polymerize tubulins for spindle pole production.
4. **During M phase** cells prepare themselves for and undergo cytokinesis. During the **mitotic phase** or **mitosis** the chromosomes are paired, then divided. The subphases of this cell cycle stage leading to cell division are prophase, metaphase, anaphase and telophase.

The cell cycle of a typical eukaryotic cell is approximately 16–24 h under cell culture conditions. In multicellular organisms the cell cycles can be as short as 6–8 h and as long as 100 days or more. This variability depends on the G_1 phase of the cycle.

Cytogenetics

The structural investigation of the chromosomal structure is known as cytogenetics, including the analysis of metaphase chromosomes and structural chromosome abnormalities. Significant progression in the field took place in 1956, when the number of human chromosomes was determined (Tjio and Levan, 1956). Structural investigations of chromosomes use staining techniques, molecular approches, fluorescent *in situ* hybridization (FISH) and genomic hybridization. As this book deals primarily with the morphology of chromatin and apoptosis in cyclic phase (S-phase), only large-scale chromatin changes are discussed. Early interphase chromosomes do not show banding patterns, consequently Giemsa staining cannot be used. Based on the observation that the expanded length of lampbrush chromosomes of amphibian oocytes, relative to metaphase chromosomes, allows a higher resolution mapping of morphological chromosome features, our aim was to visualize decondensed interphase chromosomes with a higher resolution.

DNA Structure

107

Genes

The gene is a segment of DNA located on a chromosome, representing the fundamental physical and functional unit of inheritance, directing the synthesis of a protein, or associated with regulating function. It is generally accepted that genes fall into two basic categories:

– *Constitutive genes* that express themselves at a steady state level irrespective of the external or internal environment of the cell and
– *Inducible genes* which are subject to regulation (Massoud, 2003).

There are other categories of genes such as overlapping, independent, pseudogenes, tissue specific, spliced, cancer genes which will be briefly discussed. Further gene types are summarized in Table 2.6.

Constitutive (householding) genes. Housekeeping genes may be defined as those that are involved in routine cellular metabolism and always expressed at low level in all cells. Genes developed during the ancient reducing atmosphere such as the genes of glycolytic pathway, could be among the oldest constitutive ones. The most frequently used householding gene in the glycolytic pathway is phosphoglyceraldehyde dehydrogenase (PGADH), the enzyme of which is catalyzing the conversion of glyceraldehyde 3-phosphate to 1,3 bisphospho D-glycerate in the presence of NAD^+ and inorganic phosphate (Pi). During gene expression the normalization is based on householding genes, which are assumed to be equally expressed in different samples of interest. However, it is seriously doubted that the mRNA of this gene can be used as an internal standard, as it is not abundant, it is representing only a small portion of the total RNA and most importantly its level varies considerably from one tissue to another. This is not due to a direct effect on the PGADH gene, rather the reflexion of the amount of its mRNA relative to the expression of other genes (Ivell, 1998). Other householding genes with low frequencies may suffer similar fates. More common householding genes are the cell cycle gene cyclophilin, the gene of ubiquitin involved in the ATP-dependent degradation of proteins, or the gene of the structural protein of β-actin (Ivell, 1998).

Inducible genes. The expression of inducible genes is turned on by the presence of a substance named inducer. The induction can be explained by the strictly controlled energy metabolism of living cells. The machinery of RNA and protein synthesis needs a great amount of energy. Enzyme induction and repression serve both in prokaryotic and eukaryotic cells the rational energy need. The best known and thoroughly studied inducible system is the lactose operon in *Escherichia coli*. In the absence of glucose lactose induces the expression of the structure genes of the lac operon (lac z, y and a) encoding the information of three enzymes (β-galactosidase, transacetylase and lactose permease) necessary to utilize lactose as an energy source. Besides the structure genes the lac operon contains regulatory sites (operator and promoter region). In addition the regulator region encodes proteins that interact with the operator and promoter regions to stimulate or inhibit transcription. In the absence of lactose or other inducer the product of the regulator gene

108   2 Structural Organization of DNA

Table 2.6 Characterization of gene types

Type of gene	Brief description
Constitutive (householding) genes – Expressed constantly	
Inducible genes	Expression turned on by an inducer
Overlapping genes	Single stranded portion of DNA coding for two proteins
Independent genes	Alleles are not linked
Cancer genes	Overexpression of signaling factors or members of oncogenes leading to excessive cell divison, or mutation of tumour suppressor genes
Developmental genes	Reletad to developmental processes
Differentiated genes	Exhibiting differential patterns of gene expression
Epigenetic genes	Modifications resulting in changes in gene expression capacity without the change of DNA, e.g..inactivation of X chromosome
Essential genes	Their deletions (knockouts) are lethal to the organism
Extranuclear genes	In organelles (mitochondria, chloroplasts)
Hypothetical genes	Related to ORFan genes with putative open reading frames (ORF)
Early genes	Originally used for viral genes referring to their rapid and transient expression
Immunoglobulin genes	Encoding light and heavy chain segments of immunoglobulins
Interrupted genes	The split genes of eukaryotes consisting of exons and introns
Jumping genes	Transposons
Lethal genes	Dominant ones kill heterozygotes, recessive lethal genes kill only homozygotes
Marker genes	Inserted along with the gene for the new trait, used to flag the transformed cell
Mitochondrial (maternal) genes.	Coding for mitochondrial RNAs and 13 genes coding for polypeptides.
Nested genes	Their entire sequences are contained within other genes
Pangenes	Can govern stem cell's fate and block it until the proper time of transformation arrives
Plasmid genes	In microbes extrachromosomal small circular or linear self replicating DNA.
Pleiotropic gene	Affecting more than one charcateristic of the phenotype
Polygene (quantitative)	Acting together with other genes to determine quantitative traits (e.g. weight, size, colour).

(lac repressor protein) is binding to the operator region, inhibiting the transription of the structure genes of the lac operon. If glucose is absent and inducer (lactose) is added, its binding to the repressor, removes repressor from to the operator, RNA polymerase can move through and transcribe structure genes. Transcription takes places as long as lactose is present.

Another well known inducible system is represented by the electron transport system involved in biotransformation. This system is the cytochrome P_{450} (abbreviated: CYP or P450) superfamily of hemoproteins present in archea, bacteria and eukaryotes (Danielson, 2002). The P450 systems consists of NADPH-cytochrome P_{450} reductase, cytochrome P_{450}, NADH-cytochrome b_5 reductase and cytochrome

DNA Structure   109

b$_5$. The electrons are transferred either from NADPH or from NADH to the cytochrome P$_{450}$. The genes of these proteins accumulating in the smooth endoplasmic reticulum of cells of different organs (intestines, lung, skin and with the highest level in liver) are induced by xenobiotic compounds. Different components of the electrontransport chain are induced to different extent. The small vesicles generated during the homogenization of the endoplasmic reticulum are called microsomes isolated by differential centrifugation. Drug metabolism is investigated in the microsomal electrontrasnport system. Certain drugs (e.g. barbiturates) are strong inducers of the system. In the laboratory practice the N-demethylase activity of microsomes is measured. Microsomal fractions are isolated from the livers of untreated (negative control), phenobarbiturate (positive control) and xenobiotic treated rats.

Overlapping genes. In overlapping genes a single stranded portion of DNA is coding for two separate proteins. Overlapping genes may evolve as a result of the extension of an open reading frame caused by a switch to an upstream initiation codon, substitutions in initiation or termination codons, and deletions and frameshifts that eliminate initiation or termination codons (Rogozin et al., 2002). From the nucleotide where the translation starts, to the nucleotide where it stops, is called an **open reading frame.**

Overlapping is an unusual arrangement of the genetic code, as genes are normally sequentially linked, each gene coding one protein corresponding to the "one gene one enzyme (protein)" rule. Overlapping allows to produce more protein from a certain DNA sequence than if the genes were arranged linearly after each other. Such efficient packaging is unevoidable in circular single stranded viruses which would not have enough DNA to encode all their proteins if transcription took place linearly gene after gene (e.g. bacteriophage ΦX174). Overlapping genes were found not only in viruses but also in prokaryotic and eukaryotic cells. For instance in the human mitochondrial DNA, there are two overlapping genes. Some algae with a small genome content (*Guillaraia*, 5×10^5 base pairs) code for nearly 500 proteins, 44 of them produced from overlapping genes. The number of overlapping genes is particularly high in the genomes of mycoplasma species (Fukuda et al., 1999). The overlapping may serve as an efficient tool to reduce the tolerance for mutation (Clark et al., 2001). Overlapping genes are also thought to be the result of evolutionary pressure to streamline genome size. Alternatively, overlapping genes may be due to incidental elongation of the coding region (Fukuida et al., 1999)

The human genome was expected to encode more than 100,000 genes. By 2003 the Human Genome project came to an end, with a fairly complete listing of human genes which turned out to be less than 25,000. The so far unknown proteins may come from unknown overlapping genes and from alternative splicing of exons within a gene. Consequently, many new genes will be isolated. The utility of these new isolated genes remains a question.

Independent genes. Genes are genetically linked when their alleles are inherited jointly. The greater the distance between genes, the greater the chance that non-sister chromatids would cross over in the region between genes. During meiosis physically connected genetic loci on the same chromosome tend to segregate together and are genetically linked. Alleles on different chromosomes are generally not linked, their genes are independent genes. There is a higher probability of separation of alleles if they are far apart on the chromosome or are on different chromosomes.

Pseudogenes. Pseudogenes are relatives of known genes, but have lost their protein coding function. Pseudogenes and their associated functional genes share a common ancestor, but have diverged as separate genetic entities over millions of years and have become nonfunctional junk DNA. Nonfunctionality may be caused if any of the following steps fails in gene expression: transcription, pre-mRNA processing, translation, and protein folding.

Processed, retrotransposed pseudogenes. Repetitive nonfunctional, retroviral pseudogenes e.g. the human genome consists of 30–40% repetitive sequences. This group of pseudogenes can be further subdivided into "selfish" DNA and *de novo* generated repetitive sequences appearing in any location of the genome. Retroviral pseudogenes are produced by RNA containing retroviruses that convert the viral RNA genome to circular cDNA by the viral reverse transciptase. The DNA provirus then integrates randomly in the host genome. Short and long interspread elements (SINEs and LINES, respectively) disperse themselves throughout the genome by means of an RNA intermediate named retroposon or retrotransposon. Unlike LINEs, the SINE families present in the genomes of different organisms are mostly of independent origin.

Tandem repeats scattered throughout the genome can be as short as two nucleotides or as long as 20 kilobasepairs. Tandem repeats are classified according to the size of the individual repeat unit. The repeat units of 1–4 bases and locus sizes of less than 100 bp are called microsatellites. Microsatellites are used as genetic markers in mammalian genetics. Minisatellites with repeat units of 10–40 bp and locus sizes vary from several hundred base pairs to several kilobases. Midisatellites contain longer than 40 bp repeat units and macrosatellites has been proposed as the term to describe loci with large repeat units of 3–20 kb present in clusters.

Non-processed, duplicated pseudogenes. The functional gene may be copied as a result of gene duplication and after subsequent mutations may become nonfunctional. Such duplicated genes have the same characteristics and structural elements including intact introns and exons.

Disabled unique pseudogenes. These genes became nonfunctional by the same mechanism as the duplicated pseudogenes, but the gene became inactivated without duplication. The fixation of such an inactivation is unlikely to happen in a population, but may occur due to population effects such as genetic drift, natural selection, genetic bottleneck (when the population is reduced by at least 50% and often by several orders of magnitude). A typical example is the gene coding the enzyme L-gulono-γ-lactone oxidase in primates, responsible for the biosynthesis of ascorbic acid (vitamin C). This gene is functional in all mammals (except ginea pigs), but exists as a disabled gene in humans and other primates.

Tissue specific genes. Tissue specific gene expression is often used as a method to identify functionally relevant genes. Many of these changes are likely to represent polymorphism among individuals or populations and are not necessarily associated with disease. Those tissue-specific differences in gene expression which are unique to one population are unlikely to contribute to fundamental differences between

DNA Structure

tissue types (Whitehead and Crawford, 2005). The variation in gene expression among tissues, individuals, strains populations and species has a genetic basis (Brem et al., 2002; Cheung et al., 2003).

A significant source of biological polymorphism is related to the variability of gene expression among different tissues. Most of the genes (~75%) involved in central metabolic pathways such as fatty-acid metabolism, glycolysis, oxidative phosphorylation are differentially expressed in the major organs including brain, heart, liver. The major function of the heart is to act as a pump by contraction which is an energy requiring process (Weiss, 1983), consequently genes of the oxidative phosphorylation enzymes leading to the oxidative phosphorylation of ATP are more highly expressed in the heart. The brain which is comprising 2% of the human body weight produces about 16% of the total heat produced. Due to the high metabolic demand the oxygen consumption of brain is 8 times higher than the average oxidation process in the rest of the body. The metabolic rate in the brain is 7.5 times the average rate of the rest of the body (Guyton, 1991). Mitochondria are the principal cites for oxidative phosphorylation (ATP production) are abundant in heart, brain and skeletal muscle cells. Differences in gene expression among tissues were detected by measuring mRNA expression using microarrays (Whitehead and Crawford, 2005). The genes involved in the oxidative phosphorylation of liver represented a much smaller portion of gene expression, while genes involved in other metabolic functions such as in fatty acid and phospholipid biosynthesis were more highly expressed in liver than in other tissues.

Spliced genes. Eukaryotic genes are spliced genes organized into exons and introns. Genes contain introns that must be spliced out to form a mature mRNA. Exons are the expressed sequences of DNA forming mRNA. Introns as intervening sequences are removed from pre-mRNA. Introns are spliced out from pre-mRNA at splice junctions found at the ends of each intron. RNA splicing takes place at the surface of splicesosomes (snurps) the size of which is similar to ribosomes. Splicesosomes consist of small nuclear RNAs (U-RNAs), small ribonucleoproteins (snRNPs, snurps), other proteins and the small nuclear RNA (U-RNA) looped over, sourronded by the intron to be spliced out. The snurp recognizes the splice junction and cuts the pre-mRNA with great precision. Type I splicing removes introns from pre-tRNA, type II from rRNA and mitochondrial mRNA and type III splicing from pre-mRNA. The intron in type III splicing is removed as a lariat (lasso) structure, which is then degraded by the cellular machinery.

Alternative splicing. In eukaryotes an important mechanism for the regulation of gene expression is alternative pre-mRNA splicing. The importance of alternative slicing is indicated by an estimation that the primary transcripts of ~30% of human genes are subject to alternative splicing. Most of the functional categories of genes are alternatively spliced. Genes which are expressed in the nucleus or plasma membrane are generally alternatively spliced, those expressed extracellularly show lower levels of alternative splicing. There seems to be a correlation between the incidence of alternative splicing and intron number per gene. Studies of intron-rich genes with weak intron boundary consensus in ancestral organisms suggest that simple forms of alternative splicing could have been present in unicellular ancestors of plants, fungi and animals (Irimia et al., 2007).

Protein splicing discovered in 1990, is an unusual process by which the flow of information from a gene to its protein product is modulated post-translationally so as to yield two functionally unrelated proteins. It involves the precise, self-catalysed excision of an intervening polypeptide sequence, the intein, from an inactive precursor protein with the concomitant joining of the flanking sequences, the exteins, to produce a new functional protein (Paulus, 2000).

Genes involved in cancer. Cumulating mutations of different types of genes may cause cancer. Wild type genes normally control cellular functions. Among them are the proto-oncogenes which promote cell growth. Mutation may turn these protooncogenes to cancerous oncogenes. The overexpression of signaling factors or members of signaling pathways encoded by oncogenes may lead to excessive cell divison, ultimately to cancer. Tumour suppressor genes have an opposite effect, these genes prevent excessive cell growth. Rapid growth of cells is signaled by their neighbours which in turn triggers the production of inhibitory factors to prevent further growth. Due to mutations in the tumour suppressor genes, cancer cells ignore the inhibitory signals of their neighboring cells. The mutation may (a) knock out cell-surface receptors (antennas) of inhibiting factors which cannot pick up the external signals, (b) may affect the processing of the signal, blocking the signal pathway, or (c) disable proteins such as p53 which in healthy cells triggers cells to commit suicide (apoptosis). If the gene of p53 is damaged the suicide pathway is prevented, the singalling is out of control and an unlimited cell growth (cancer) follows.

Types of cancers:

- Carcinoma – epithelial origin
- Sarcoma – from connective tissue or muscle cell
- Leukemia – tumours from hemopoetic cells
- Neural tumours
- Adenoma – benign epithelial tumour (of glandular organization)
- Adenocarcinoma – malignant epithelial tumour
- Chondroma, chondrosarcoma – benign and malignant tumours of cartilage

Temporal and Spatial Order of Gene Replication

It has been recongized a long time ago that chromosomes are not replicated as a single element, but their different sections are synthesized at characteristic times during the S phase, indicating that replication of eukaryotic cells takes place at multiple replication units known as replicons (Taylor, 1963). These subchromosomal replication units have been visualized by autoradiography (Cairns, 1966). The chronological order of replication of specific DNA segments turned out to be invariant from cell generation to the next ones (Hubermann and Riggs, 1968). The size of the single replicon in prokaryotes corresponds to the genome size. Studies in eukaryotic cells showed that replication units (replicons) into which chromosomal

DNA is orgainzed range from 50 to about 300 kb. Replicons are initiated as synchronous clusters with 25–100 replicons per cluster spaced irregularly along the chromosome, but activated in a sequential fashion during S phase (Hand, 1978; Housman and Hubermar, 1975). Examinations in Chinese hamster and HeLa cells using a variety of synchronizaticn procedures showed that replication of specific genomic sequences follows a distinct temporal order (Amadi et al., 1969; Balaz and Schildkraut, 1971; Stambrook, 1974; Taljanidisz et al., 1989). A different order of ribosomal gene replication was reported for different eukaryotic cells (Balaz and Schildkraut, 1971; Giacomoni and Finkel, 1972; Gimmler and Schweizer, 1972; Epner et al., 1981). The comparison of replication times of genes and their levels of expression in Chinese hamster ovary cells showed that there was a general correlation between the transcriptional activity and replication, but the examination of specific genes also revealed several exceptions. These observed discrepancies were due either to their different temporal order of replication and transciption or to different methods used for synchronization and isolation of early and late repliacting DNA (Taljanidisz et al. 1989).

The genome of eukaryotic organisms is divided into a number of linear chromosomes. In the nuclei of higher eukaryotic organisms these chromosomes are distributed through the nucleus as individual chromosome territories. Chromosomes are likely to occupy specific territories with an internal spatial organisation but they are also organized within an interphase nucleus with gene rich chromosomes being found towards the nuclear interior and gene-poor chromosomes at the nuclear periphery. Within the functional space between chromosomes called the inter chromosome domain (ICD) could be located much of the transcription and splicing machinery, which begs the question whether gene expression transcription and RNA export occur in this domain (Bridger and Bickmore, 1998). Accumulating evidence is demonstrating that chromosome position affects translocation type frequency i.e. chromosomes that occupy similar nuclear addresses have higher translocation frequencies than chromosomes that do not (Bickmore and Teague, 2002).

DNA Is Replicated and Repaired in Several Subphases in S-Phase

A major response to DNA damage in eukaryotes is the activation of surveillance mechanisms called checkpoints that block the initiation of late processes of the cell cycle to allow time to repair damage. Whereas a large body of knowledge has been accumulated on control mechanisms regulating G1/S and G2/M transitions, little information is available on the control of DNA replication in response to DNA damage or replicational stress, the so called intra S-phase checkpoint(s). The number of checkpoint genes (rad) in fission yeast (Carr, 1995; Kaufmann. 1995) indicates that beside the three known cell cycle checkpoints, the S-phase itself may be regulated by more checkpoints. In view of the limited processivity and relatively slow rate of repair DNA polymerases, it is unlikely that the whole genome is tested for mutations only once during replication. The fact that the repair and replicative DNA

114   2 Structural Organization of DNA

polymerases are unable to work at the same time on the same template also suggests multiple surveillance and repair cycles. It was postulated that there is more than one pause site in S-phase. Eleven S-phase replication checkpoints in Chinese hamster ovary (CHO) cells (Bánfalvi et al., 1997b) and four replication checkpoints were detected in *Drosophila* cells (Rehak et al., 2000). It was also found that the rates of DNA repair and replication are inversely related and that there are at least as many repair checkpoints as replication peaks (Bánfalvi et al., 1997a). The coincidence of Chinese hamster ovary cells containing eleven chromosomes and *Drosophila* cells harboring four chromosomes and that these numbers correspond to the numbers of replication checkpoints indicate that chromosomes are likely to be repaired one by one before their replication.

References

Abdel-Monem, M., Lauppe, H.F., Kartenbeck, J., Durwald, H., and Hoffmann-Berling, H. (1977). Enzymatic unwinding of DNA. III. Mode of action of Escherichia coli DNA unwinding enzyme. J Mol Biol **110**, 667–685.

Abney, J.R., Cutler, B., Fillbach, M.L., Axelrod, D. and Scalettar, B.A. (1997). Chromatin dynamics in interphase nuclei and its implications for nuclear structure. J Cell Sci. **137**, 1459–1468.

Adachi, Y., Luke, M. and Laemmli, U.K. (1991). Chromosome assembly *in vitro*: topoisomerase II is required for condensation. Cell **64**, 137–148.

Altona, C. and Sundaralingam, M. (1972). Conformational analysis of the sugar ring in nucleosides and nucleotides. A new description using the concept pseudorotation. J Amer Chem Soc. **94**, 8205–8212.

Altona, C. (1975). Backbone conformation of several dinucleoside monophosphates in solution deduced from Fourier transform NMR spectroscopy at 270 MHz. In Structure and Conformation of Nucleic Acids and Protein-nucleic Acid Interactions. M. Sundaralingam and S.T. Rao, eds. Univ.Park Press, London.

Amadi, F., Giacomoni, D. and Zito-Bignami, R. (1969). Ont he duplication of ribosomal RNA cistrons in Chinese hamster ovary cells. Eur J Biochem. **11**, 419–423.

Arnott, S. and Hukins, W.L. (1972). The dimensions and shapes of the furanose rings in nucleic acids. Biochem J. **130**, 453–65.

Arnott, S., Hutchinson, F., Spencer, M., Wilkins, M.H., Fuller, W. and Langridge, R. (1966). X-ray diffraction studies of double helical ribonucleic acid. Nature. **211**, 227–232.

Balaz, J. and Schildkraut, C.L. (1971). DNA replication in synchronized cultured mammalian cells. II. Replication of ribosomal cistrons in thymidine-synchronized HeLa cells. J Mol Biol. **57**, 153–158.

Baldi, M.I., Benedetti, P., Mattoccia, E. and Tocchini-Valentini, G.P. *In vitro* catenation and decatenation of DNA and a novel eucaryotic ATP-dependent topoisomerase. Cell. **20**, 461–467.

Bánfalvi, G. (1984). Demonstration of topoisomers using Solomon's knot DNA. Biochem Educ. **12**, 155–156.

Bánfalvi, G. (1986a). Structural organization of DNA. Biochem Educ. **14**, 50–59.

Bánfalvi, G. (1986b). Transistion from right handed to left handed DNA. Biochem Educ. **14**, 7–10.

Bánfalvi, G. (1991). Evolution of osmolyte systems. Biochem Educ. **19**, 136–139.

Bánfalvi, G. (1994). The metabolic clockwork. Biochem Educ. **22**, 137–139.

Bánfalvi, G. (2006a). Why ribose was selected as the excusive sugar component of nucleic acids. DNA Cell Biol. **25**, 189–196.

Bánfalvi, G. (2006b). Structure of interphase chromosomes in the nuclei of *Drosophila* cells. DNA Cell Biol. **25**, 547–53. Bánfalvi, G. (2006c). Linear connection of condensing chromosomes in nuclei of synchronized CHO cells. DNA Cell Biol. **25**, 541–545.

References

115

Bánfalvi, G. (2008). Chromatin fiber structure and plectonemic model of chromosome condensation in *Drosophila* cells. DNA Cell Biol. **27**, 65–70.

Bánfalvi, G. and Antoni, F. DNA diagnostics. Orvosi Hetilap 131, 953–964 (in Hungarian).

Bánfalvi, G. and Fieldhouse, J. (1988). Heliwire DNA model to visualize *syn* and *anti* folding of tetraplex structures. Biochem Educ. **16**, 80–82.

Bánfalvi, G., Chou, W.M., Mikhailova M. and Poirier, A.L. (1997a) Relationship of repair and replicative DNA synthesis to cell cycle in Chinese hamster Ovary (CHO-K1) cells. DNA Cell Biol. **16**, 1155–1160.

Bánfalvi, G., Gacsi, M., Nagy, G., Kiss, B.Z. and Basnakian, A.G. (2005). Cadmium induced apoptotic changes in chromatin structure and subphases of nuclear growth during the cell cycle in CHO cells. Apoptosis **10**, 631–642.

Bánfalvi, G., Mikhailova, M., Poirier, L.A. and Chou, M.W. (1997b). Multiple subphases of DNA replication in CHO cells. DNA Cell Biol. **16**, 1493–1498.

Banfalavi, G. and Sarkar, N. (1983). Analysis of the 5′-termini of nascent DNA chains synthesized in permeable cells of *Bacillus subtilis*. J Mol Biol. **163**, 147–169.

Bánfalvi, G., Trencsenyi G., Ujvarosi, K., Nagy, G., Ombodi, T., Bedei, M., Somogyi, C. and Basnakian, A. (2007). Supranucleosomal organization of chromatin fibers in nuclei of *Drosophila S2* cells. DNA Cell Biol. **26**, 55–62.

Bánfalvi, G., Wiegant, J., Sarkar, N. and Van Duijn, P. (1989). Immunofluoresent visualization of DNA replication sites within nuclei of Chinese hamster ovary cells. Histochemistry. **93**, 81–86.

Bauer, W.R. (1978). Structure and reactions of closed duplex DNA. Annu Rev Biophys Bioeng. **7**, 287–313. Review.

Bednar, J., Horowitz, R.A., Grigoryev, S.A., Carruthers, L.M., Hansen, J C., Koster, A.J. and Woodcock, C.L. (1998). Nucleosomes, linker DNA, and linker histone form a unique structural motif that directs the higher-order folding and compaction of chromatin. Proc Natl Acad Sci USA. **95**, 14173–14178.

Behe, M. and Felsenfeld, G. (1981). Effects of methylation on a synthetic polynucleotide: the B–Z transition in poly(dG-m5dC).poly(dG-m5dC). Proc Natl Acad Sci USA. **78**, 1619–1623.

Belmont, A.S. (2006). Mitotic chromosome structure and condensation. Curr Opin Cell Biol. **18**, 6632–6638.

Berezney, R., Mortillaro, M.J., Ma, H., Wei, X. and Samarabandu, J. (1995). The nuclear matrix: A structural milieu for genomic function. Int J Cytol. **162A**, 1–65.

Bergerat, A., De Massy, B., Gadelle, D., Varoutas, P.C., Nicolas, A. and Forterre, P. (1997). An atypical topoisomerase II from Archaea with implications for meiotic recombination. Nature. **386**, 414–417.

Bickmore, W.A. and Teague, P. (2002). Influences of chromosome size, gene density and nuclear position on the frequency of constitutional translocations in the human population. Chromosome Res. **10**, 707–715.

Bolli, M., Micura, R. and Eschenmoser, A. (1977). Pyranosyl-RNA chiroselective self-assembly of base sequences by ligativa oligomerization of tetranucleotide-2′, 3′-cyclophosphates (with a comentary concerning origin of biomolecular homochirality) Chem Biol. **4**, 309–320.

Bonnefoy, E. (1997). The ribosomal S16 protein of *Escherichia coli* displaying a DNA-nicking activity binds to cruciform DNA. Eur J Biochem. **247**, 852–859.

Bordas, J., Perez-Grau, L., Koch, M.H.J., Vega, M.C. and Nave, C. (1986). The superstructure of chromatin and its condensation mechanism. I. Synchrotron radiation X-ray scattering results. Eur Biophys J Biophys Lett. **13**, 175–185.

Brem, R.B., Yvert, G., Clinton, R. and Kruglyak, L. (2002). Genetic dissection of transcriptional regulation in budding yeast. Science **296**, 752–755.

Bridger, J.M. and Bickmore, W.A. (1998). Putting the genome on the map. Trends Genet. **14**, 403–410.

Britten, R.J. and Davidson, E.H. (1969). Gene regulation for higher cells: A theory. Science **165**, 349–357.

Brown, D. and Simoni, R.D. (1983). Mevalonate deprivation leads to aneuploidy in Chinese hamster ovary cells. J Biol Chem. **263**, 13497–13499.

Brown, P.O. and Cozzarelli, N.R. (1979). A sign inversion mechanism for enzymatic supercoiling of DNA. Science. **206**, 1081–1083.

Brown, P.O., Peebles, C.L. and Cozzarelli, N.R. (1979). A topoisomerase from *Escherichia coli* related to DNA gyrase.Proc Natl Acad Sci U S A. **76**, 6110–6114.

Buhler, C., Gadelle, D., Forterre, P., Wang, J.C. and Bergerat, A. (1998). Reconstitution of DNA topoisomerase VI of the thermophilic archaeon *Sulfolobus shibatae* from subunits separately overexpressed in *Escherichia coli*. Nucleic Acids Res. **26**, 5157–5162.

Cairns, J. (1963a). The bacterial chromosome and its manner of replication as seen by autoradiography. J Mol Biol. **6**, 208–213.

Cairns, J. (1963b). The chromosome of *Escherichia coli*. Cold Spring Harbor Symp Quant Biol. **28**, 43–46.

Cairns, J. (1966). Autoradiography of HeLa cell DNA. J Mol Biol **15**, 372–373.

Calladine, C.R. (1982). Mechanism of sequence-dependent stacking of bases in B-DNA. J Mol Biol **161**, 343–352.

Carr, A.M. (1995). DNA structure checkpoints in fission yeast. Semin Cell Biol. **6**, 65–72,

Castano, I.B., Brzoska, P.M., Sadoff, B.U., Chen, H.Y. and Christman, M.F. (1996) Mitotic chromosome condensation in the rDNA requires TRF4 and DNA topoisomerase I in *Saccharomyces cerevisiae*. Genes Dev. **10**, 2564–2576.

Champoux, J.J. (1978). Mechanism of the reaction catalysed by the DNA untwisting enzyme: attachment of the enzyme to 3′-terminus of the nicked DNA. J Mol Biol. **118**, 441–446.

Champoux, J.J. (2001). DNA topoisomerases: Structure, function and mechanism. Annu Rev Biochem. **70**, 369–413.

Cheung, V.G., Conlin, L.K., Weber, T.M., Arcaro, M., Jen, K.Y., Morley, M. and Spielman, R.S. (2003). Natural variation in human gene expression assessed in lymphoblastoid cells. Nat Genet. **33**, 422–425.

Clark, M.A., Baumann, L., Thao, M.L., Moran, N.A. and Baumann, P. (2001). Degenerative minimalism in the genome of a psyllid endosymbiont. J Bacteriol. **183**, 1853–1861.

Conner, B.N., Takano, T., Tanaka, S., Itakura, K., and Dickerson, R.E. (1982). The molecular structure of d(ICpCpGpG), a fragment of right-handed double helical A-DNA. Nature. **295**, 294–299.

Committee on Rat Nomenclature, Cochairmen Gill T.J. III, Nomura T. 1992. Definition, Committee on Standardized Genetic Nomenclature for Mice. 1963. A revision of the standardized genetic nomenclature for mice. J. Hered. **54**, 159–162.

Committee on Standardized Genetic Nomenclature for Mice. 1973. Guidelines for nomenclature of genetically determined biochemical variants in the house mouse, Mus musculus-Biochem.Genet. **9**, 369–374.

Committee on Standardized Genetic Nomenclature for Mice, Chair: Lyon, M.F.: Rules and guidelines for gene nomenclature, pp. 1–7. In: Genetic Variants and Strains of the Laboratory Mouse, Green, M.C. (ed.), First Edition, Gustav Fischer Verlag, Stuttgart, 1981.

Committee on Standardized Genetic Nomenclature for Mice, Chair: Lyon, M.F.: Rules and guidelines for gene nomenclature, pp. 1–11. In: Genetic Variants and Strains of the Laboratory Mouse, Lyon, M.F., A.G. Searle (eds.), Second Edition, Oxford University Press, Oxford, 1989.

Committee on Standardized Genetic Nomenclature for Mice, Chairperson: Davisson, M.T. Rules and guidelines for gene nomenclature, pp. 1–16. In: Genetic Variants and Strains of the Laboratory Mouse, Lyon, M.F., Rastan, S., Brown, S.D.M. (eds.), Third Edition, Volume 1, Oxford University Press, Oxford, 1996.

Cozzarelli, N.R. (1980). DNA gyrase and the supercoiling of DNA. Science. **207**, 953–960.

Crick, F.H.C. (1976). Linking numbers and nucleosomes. Proc Natl Acad Sci USA. **73**, 2639–2643.

Dale, R.M.K., Livingston, D.C. and Ward, D.C. (1973). The synthesis and enzymatic polymerization of nucleotides containing mercury: potential tools for nucleic acid sequencing and structure analysis. Proc Nat Acad Sci USA. **70**, 2238–2242.

Danielson, P. (2002). The cytochrome P450 superfamily: biochemistry, evolution and drug metabolism in humans. Curr Drug Metab. **3**, 561–597.

References

117

Davies, B.D. (1978). Conformations of nucleosides and nucleotides. Prog NMR Spectrosc. **12**, 135–225.

Denisov, A.Y., Noronha, A.M., Wilds, C.J., Trempe, J.F., Pon, R.T., Gehring, K., and Damha, M.J. (2001). Solution structure o fan arabinonucleic acid (ANA)/RNA duplex in a chimeric hairpin: comparison with 2'-fluoro-ANA/RNA and DNA/RNA hybrids. Nucleic Acids Res. **29**, 4284–4293.

Dickerson, R.E. (1983). Base sequence and helix structure variation in B- and A-DNA. J Mol Biol **166**, 419–441.

Dickerson, R.E., Drew, H.R., Conner, B.N., Wing, R.M., Fratini, A.V., Kopka, M.L. (1982). The anatomy of A-, B-, and Z-DNA. Science **216**, 475–485.

Dickerson, R.E. (1983a). The DNA helix and how it is read. Sci Amer **249**, 94–111.

Downes, C.S., Clarke, D.J., Mullinger, A.M., Gimenez-Abian, J.F., Creighton, A.M. and Johnson, R.T. (1994). A topoisomerase II-dependent G_2 cycle checkpoint in mammalian cells. Nature **372**, 467–470.

Dresser, M.E. and Giroux, C.N. (1988). Meiotic chromosome behavior in spread preparations of yeast. J Cell Biol. **106**, 557–573.

Dunn, L.C., Gruneberg, H. and Snell, G.D. (1940). Report of the committee on mouse genetics nomenclature. J Hered. **31**, 505–506.

Epner, E., Rifkind, R.A. and Marks, P.A. (1981). Replication of α and β globin DNA sequences occurs during early S phase in murine erythroleukemia cells. Proc Natl Acad Sci USA. **78**, 3058–3062.

Eschenmoser A., and Dobler, M. (1992). Warum Pentose - und nicht Hexose- Nucleinsauren TeilI. Einleitung und Froblemstellung, Korformationanalyse für Oligonucleotid Ketten aus 2'3'-Dideoxygluvopyranosyl- Barsteinen ('Homo-DNA') sowie Betrachtungenzur Konformation von A- und B-DNA. Helv Chim Acta **75**,218–259.

Eschenmoser in Proc. Robert A. Welch Found. (1993). Conf. Chem. Res. **37**, (Robert A. Welch Foundation, Houston, TX) pp. 201.

Eschenmoser, A. (1994). The TNA-family of nucleic acid systems: Properties and prospects. Orig Life Evol Biosph. 2004. **34**, 277–306.

Fenech, M. (2002). Chromosomal biomarkers of genomic instability relevant to cancer. Drug Discov Today. **7**, 1128–1137.

Fieldhouse, J. (1981). A Solomon's knot DNA. Biochem Educ. **9**, 88.

Finch, J.T. and Klug, A. (1976). Solenoidal model for superstructure in chromatin. Proc Natl Acad Sci U S A. **73**, 1897–1901.

Fletterick, R.J., Schroer, T. and Matela, R.J. (1985). In Molecular Structure (ed. Staples, J.). Blackwell Scientific Publications, Oxford.

Ford, C.E. and Hamerton, J.L. (1956). The chromosomes of man. Nature. **178**, 1020–1023.

Fujii, S., Wang, A.H., Van Der Marel, G., Van Boom, J.H. and Rich, A. Molecular structure of (m5 dC-dG)3: The role of the methyl group on 5-methyl cytosine in stabilizing Z-DNA. Nucleic Acids Res. **10**, 7879–7892.

Fukuda, Y., Washio, T. and Tomita, M. (1999). Comparative study of overlapping genes int he genomes og Mycoplasma gentalium and Mycoplasma pneumoniae. Nucleic Acids Res. **27**, 1847–1853.

Gacsi, M., Nagy, G., Pirter, G., Basnakian, A.G., and Bánfalvi, G. (2005). Condensation of interphase chromatin in nuclei of Chinese hamster ovary (CHO-K1) cells. DNA Cell Biol. **24**, 43–53.

Gangloff, S., De Massy, B., Arthur, L., Rothstein, R. and Fabre, F. (1999). The essential role of yeast topoisomerase III in meiosis depends on recombination. Embo J. **18**, 1701–1711.

Garcia, H.G., Grayson, P., Han, L., Inamdar, M., Kondev, J., Nelson, P.C.. Phillips, R., Widom, J. and Wiggins, P.A. (2007). Biological consequences of tightly bent DNA: The other life of a macromolecular celebrity. Biopolymers. **85**, 115–130.

Gellert, M. (1981). DNA topoisomerases. Annu Rev Biochem. **50**, 879–910. Review.

Gellert, M., Mizuuchi, K., O'Dea, M.H. and Nash, H.A. (1976). DNA gyrase: An enzyme that introduces superhelical turns into DNA. Proc Natl Acad Sci U S A. **73**, 3872–3876.

118   2 Structural Organization of DNA

Gellert, M., Mizuuchi, K., O'Dea, M.H., Ohmori, H., and Tomizawa, J. (1979). DNA gyrase and DNA supercoiling. Cold Spring Harb Symp Quant Biol. **43**, 35–40.

Gerchman, L.L. and Ludlum, D.B. (1973). The properties of O 6 -methylguanine in templates for RNA polymerase. Biochim Biophys Acta. **308**, 310–316.

Ghelardini, P., Pedrini, A.M. and Paolozzi, L. (1982). The topoisomerase activity of T4 amG39 mutant is restored in Mu lysogens. FEBS Lett. **137**, 49–52.

Giacomoni, D. and Finkel, D. (1972). Time of duplication of ribosomal RNA cistrons in a cell line of *Potorus tridactilus* (Rat kangaroo). J Mol Biol. **70**, 725–728.

Gilbert, W. (1986). Evolution of antibodies. The road not taken. Nature. **320**, 485–486.

Gimmler, G.M. and Schweizer, E. (1972). rDNA replication in synchronized cultures of *Saccharomyces cerevisiae*. Biochem Biophys Res Commun. **46**, 143–149.

Goto, T. and Wang, J.C. (1982). Yeast DNA topoisomerase II. An ATP-dependent type II topoisomerase that catalyses the catenation, decatenation, unknotting, and relaxation of double-stranded DNA rings. J Biol Chem. **257**, 5866–5872.

Grechman, S.E. and Ramakrishnan, V. (1987). Chromatin higher-order structure studied by neutron-scattering and scanning transmission electron microscopy. Proc Natl Acad Sci USA. **84**, 7802–7806.

Grigoryev, S.A. (2004). Keeping fingers crossed: Heterochromatin spreading through interdigitation of nucleosome arrays. FEBS Lett. **254**, 4–8.

Groebke, K. (1998). Warum Pentose- und nicht Hexose- Nucleinsauren? Teil V. (Purin-Purin)- Basenpaarung in der homo-DNA-Reiche: Guanin, Isoguanin, 2,6- Diaminopurin und Xanthin. Helv. Chim Acta. **8**, 375–491.

Guacci, V., Hogan, E. and Koshland, D. Chromosome condensation and sister chromatid pairing in budding yeast. J Cell Biol. **125**, 517–530.

Guerrier-Takada, C., Gardiner, K., Marsch, T., Pace, N. and Altmann, S. (1983). The RNA moiety of ribonuclease P is the catalytic subunit of the enzyme. Cell. **35**, 849–857.

Guschlbauer, W. and Janlowski, K. (1980). Nucleoside conformation is determined by the electronegativity of the sugar substituent. Nucl Acids Res. **8**, 1421–1433.

Guyton, A.C. (1991). Textbook of Medical Physiology. 8th edition. Philadelphia: W.B. Saunders Company.

Hackett, J.A., Feldser, D.M. and Greider, C.W. (2001). Telomere dysfunction increases mutation rate and genomic instability. Cell. **106**, 275–286.

Hall, L.D., Steiner, P.R. and Pedersen, C. (1970). Studies of specifically fluorinated carbohydrates. Part VI. Some pentafuranosyl fluorides. Can J Chem. **48**, 1155–1165.

Hand, R. (1978). Eukaryotic DNA: Organization of the genome for replication. Cell. **15**, 317–325.

Hartl, D.L. (2000). Molecular melodies in high and low C. Nat Rev Genet. **1**, 145–149.

Hawkinson, S.W., Coulter, C.L. and Greaves, M.L. (1970). The structure of vitamin B_{12}. VIII. The crystal structure of vitamin B_{12}-5'-phosphate. Proc Roy Soc Ser. A. **318**, 143–167.

Holliday, R. (1965). Induced mitotic crossing-over in relation to genetic replication in synchronously dividing cells of ustilago maydis. Genet Res. **10**, 104–120.

Holmgreen, A. (1989). Thioredoxin and glutaredoxin systems. J Biol Chem. **264**, 13963–13966.

Horwitz, M.S. (1989). Transcription regulation *in vitro* by an E. coli promoter containing a DNA cruciform in the '-35' region. Nucleic Acids Res. **17**, 5537–5545.

Housman, D. and Huberman, J.A. (1975). Changes int he rate of DNA replication fork movement during S phase in mammalian cells. J Mol Biol. **94**, 173–181.

Hozak, P. and Fakan, S. (2006). Functional structure of the cell nucleus. Histochem Cell Biol. **125**, 1–2.

Hsieh, T. and Brutlag, D. (1980). ATP-dependent DNA topoisonmerase from D. melanogaster reversibly catenates duplex DNA rings. Cell. **21**, 115–125.

Huberman, J.A. and Riggs, A.D. (1968). Ont he mechanism of DNA replication in mammalian chromosomes. J Mol Biol. **32**, 327–341.

References

Hunziker, J. et al. (1993). Warum Pentose- und nicht Hexose- Nucleinsauren? Teil III. Oligo (2'3'-dideoxy-β-D-glucopyranosyl) nucleotide ('Homo-DNA') Paarungseigenschaften. Helv Chim Acta. **76**, 259–352.

Irimia, M., Rukov, J.L., Penny, D. and Roy, S.W. (2007). Functional and evolutionary analysis of alternatively spliced genes is consistent with an early eukaryotic origin of alternative splicing. BMC Evol Biol. **7**, 188.

IUPAC-IUB. (1983). Joint Comission on Biochemical Nomenclature. Abbreviations and symbols for the description of conformations of polynucleotide chains. Eur J Biochem. **131**, 9–15.

Ivell, R. (1998). A question of faith – or the philosophy of RNA controls. J Endocrinol. **159**, 197–200.

Jefford, C.E. and Irminger-Finger, I. (2006). Mechanisms of chromosome instability in cancers. Crit Rev Oncol Hematol. **59**, 1–14.

Johnson, D. and Morgan, A.R. (1978). Unique structures formed by pyrimidine-purine DNAs which may be four-stranded. Proc Natl Acad Sci U S A. **75**, 1637–1641.

Joyce, G.F. and Orgel, L.E. (1993). In The RNA World (eds. Rf. Gesteland and J.F. Atkins), pp.1. Cold Spring Harbor Laboratory Press, Cold Spring Harbor, N.Y.

Kalckar, H. (1941). The nature of energetic coupling in biological synthesis. Chem Rev. **28**, 71–178.

Kalckar, H. (1969). Biological Phosphorylations: Development of Concepts. Prentice-Hall, Englewood Cliffs, N.J.

Kaufmann, W.K. (1995). Cell cycle checkpoints and DNA repair preserve the stability of the human genome. Cancer Metast Rev. **14**, 31–41.

Kierzek, R., He, L. and Turner, D.H. (1982). Association of 2'-5' oligoribonucleotides. Nucl Acids Res. **20**, 1685.

Kilpatrick, J.E., Pitzer, K.S. and Pitzer, R. (1947). The thermodynamics and molecular structure of cyclopentane. J Amer Chem Soc. **69**, 2483–2488.

Kilpatrick, M.W., Wei, C.F., Gray, H.B. Jr. and Wells, R.D. (1983). BAL 31 nuclease as a probe in concentrated salt for the B-Z DNA junction. Nucleic Acids Res. **11**, 3811–3822.

Kim, R.A. and Wang, J.C. (1992). Identification of the yeast *TOP3* gene product as a single strand-specific DNA topoisomerase. J Biol Chem. **267**, 17178–17185.

Klug, A., Rhodes, D., Smith, J., Finch, J.T. and Thomas, J.O. (1980). A low resolution structure for the histone core of the nucleosome. Nature. **287**, 509–516.

Klysik, J., Stirdivant, S.M., Larson, J.E., Hart, P.A. and Wells, R.D. (1981). Left-handed DNA in restriction fragments and a recombinant plasmid. Nature. **290**, 672–677.

Kodani, M. (1958). Three chromosome numbers in whites and Japanese. Science. **127**, 1339–1340.

Koepsel, R.R. and Khan, S.A. (1987). Static and initiator protein-enhanced bending of DNA at a replication origin. Science. **233**, 1316–1318

Kolata, G. (1983). Z-DNA moves toward "real biology". Science. **222**, 495–496.

Kruger, K., Grabowski, P.J., Zaug, A.J., Sands, J., Gottschling, D.E. and Cech, T.R. (1982). Self-splicing RNA: autoexcision and autocyclization of the ribosomal RNA intervening sequence of Tetrahymena. Cell. **31**, 147–157.

Kubista, M., Hagmar, P., Nielsen, P.E. and Nordén, B. (1990). Reinterpretation of linear dichroism of chromatin supports a perpendicular linker orientation in the folded state. J Biomol Struct Dyn. **8**, 37–54

Lam, S.L., Ip, L.N., Cui, X. and Ho, C.N. (2002). Random coil proton chemical shifts of deoxyribonucleic acids. J Biomol NMR. **24**, 329–337.

Langer, P.R., Waldrop, A A. and Ward, D.C. (1981). Enzymatic synthesis of biotin-labeled polynucleotides: Novel nucleic acid affinity probes. Proc Nat Acad Sci USA. **78**, 6633–6637.

Larralde, R., Robertson, M.P. and Miller, S. (1995). Rates of decomposition of ribose and other sugars: Implications for chemical evolution. Proc Natl Acad Sci USA. **92**, 8158–8160.

Lavoie, B.D., Tuffo, K.M., Oh, S., Koshland, D. and Holm, C. (2000). Mitotic chromosome condensation requires Brn1p, the yeast homologue of Barren. Mol Biol Cell. **11**, 1293–1304.

Lawrence, J.B., Singer, R.H. and Mcneil, J.A. (1990). Interphase and metaphase resolution of different distances within the human dystrophin gene. Science. **249**, 928–932.

120   2 Structural Organization of DNA

Lemke, J., Claussen, J., Michel, S., Chuboda, I., Muhling, P., Sperling, K., Rubtsov, N., Grummt, U.W., Ullmann, P. and Kromeyer-Hauschild, K., et al. (2002). The DNA basedstructure of human chromosome 5 in interphase. Am J Hum Genet. **71**, 1051–1059.

Levan G., Hedrich, H.J., Remmers, E.F., Serikawa, T. and Yoshida, M.C. (1995). Standardized rat genetic nomenclature. Mamm Genome. **6**, 447–448.

Levitt, M. and Warshel, A. (1978). Extreme conformational flexibility of the furanose ring in DNA and RNA. J Amer Chem Soc. **100**, 2607–2613.

Li, G., Sudlow, G. and Belmont, A.S. (1998). Interphase cell cycle dynamics of a late- replicating, heterochromatic homogeneously staining region: precise choreography of condensation/decondensation and nuclear positioning. J Cell Biol. **140**, 975–989.

Li, R., Yerganian, G., Duesberg, P., Kraemer, A., Willer, A., Rausch, C. and Hehlmann, R. (1997). Aneuploidy correlated 100% with chemical transformation of Chinese hamster cells. Proc Natl Acad Sci USA. **94**, 14506–14511.

Linial, M. and Shlomai, J. (1987). Sequence-directed bent DNA helix is the specific binding site for Crithidia fasciculata nicking enzyme. Proc Natl Acad Sci U S A. **84**, 8205–8209.

Linial, M. and Shlomai, J. (1988). Bent DNA structures associated with several origins of replication are recognized by a unique enzyme from trypanosomatids. Nucleic Acids Res. **16**, 6477–6492.

Lippmann, F. (1941). Metabolic generation and utilization of phosphate bond energy. Adv Enzymol. **18**, 99–162.

Lipps, H.J., Nordheim, A., Lafer, E.M., Ammermann, D., Stollar, B.D. and Rich, A. (1983). Antibodies against Z DNA react with the macronucleus but not the micronucleus of the hypotrichous ciliate stylonychia mytilus. Cell. **32**, 435–441.

Liu, L.F., Liu, C-C. and Alberts, M.M. (1979). T4 DNA topoisomerase: a new ATP-dependent enzyme essential for initiation of T4 bacteriophage DNA replication. Nature. **281**, 456–461.

Liu, L.F., Liu, C-C. and Alberts, M.M. (1980). Type II DNA topoisomerases: Enzymes that can unknot a topologically knotted DNA molecule via a reversible double-strand break. Cell. **19**, 697–707.

Liu, L.F. and Wang, J.C. (1978). Micrococcus luteus DNA gyrase: Active components and a model for its supercoiling of DNA. Proc Natl Acad Sci U S A. **75**, 2098–102.

Luger, K. (2003). Structure and dynamic behavior of nucleosomes. Curr Opin Genet Dev. **13**, 127–135.

Lyon, M.F. (2003). The Lyon and the LINE hypothesis. Semin Cell Dev Biol. **14**, 313–318. Review.

Malik, M., Nitiss, K.C., Enriquez-Rios, V. and Nitiss, J.L. (2006). Roles of nonhomologous end-joining pathways in surviving topoisomerase II–mediated DNA damage. Mol Cancer Ther. **5**, 1405–1414.

Manuelidis, L. (1990). A view of interphase chromosomes. Science. **250**, 1533–1540.

Massoud S. (2003). Genetic and environmental interactions in psychiatric illnesses [letter]. J Neuropsychiatry Clin Neurosci. **15**, 386–387.

Milman, G., Chamberlain, M. and Langridge, R. (1967). The structure of a DNA-RNA hybrid. Proc Natl Acad Sci U S A. **57**, 1804–1810.

Miyano, M., Kawashima, T. and Ohyama, T. (2001). A common feature shared by bent DNA structures locating in the eukaryotic promoter region. Mol Biol Rep. **28**, 53–61.

Mizuuchi, K., Fisher, L.M., O'Dea, M.H. and Gellert, M. (1980). DNA gyrase action involves the introduction of transient double-strand breaks into DNA. Proc Natl Acad Sci U S A. **77**, 1847–1851.

Mizuuchi, K., O'Dea, M.H. and Gellert, M. (1978). DNA gyrase: subunit structure and ATPase activity of the purified enzyme. Proc Natl Acad Sci U S A. **75**, 5960–5963.

Moller, A., Nordheim, A., Nichols, S.R. and Rich, A. (1981). 7-Methylguanine in poly(dG-dC).poly(dG-dC) facilitates z-DNA formation. Proc Natl Acad Sci U S A. **78**, 4777–4781.

Morrison, A. and Cozzarelli, N.R. (1979). Site-specific cleavage of DNA by E. coli DNA gyrase. Cell. **17**, 175–184.

References

Mozziconacci, J., Lavelle, C., Barbi, B., Lesne, A. and Victor, J.-M. (2006). A physical model for the condensation and decondensation of eukaryotic chromosomes. FEBS Letts. **580**, 368–372.

Muller, D., Pitsch, S., Kittaka, A., Wagner, E., Wintner, C.E., Eschenmoser, A. and Ohlof, G. (1990). Chemie von α-Aminonitrilen. Aldomerisierung von Glycolaldehid-phosphat zu racemischen Hexose-2,4,6-triphosphat und rac-Ribose-2,4-diphosphat sind die Reaktionshamptprodukte. Helv Chim Acta. **73**, 1410–1469.

Murray, L.J., Arendal, W.B. 3rd, Richardson, D.C. and Richardson, J.S. (2003). RNA backbone is rotameric. Proc Natl Acad Sci USA. **100**, 13904–13909.

Nagy, G., Gacsi, M., Rehak, M., Basnakian, A.G., Klaisz, M. and Bánfalvi, G. (2004). Gamma irradiation-induced apoptosis in murine pre-B cells prevents the condensation of fibrillar chromatin in early S phase. Apoptosis. **9**, 765–776.

Nordheim, A., Pardue, M.L., Lafer, E.M., Möller, A., Stollar, B.D. and Rich, A. (1981). Antibodies to left-handed Z-DNA bind to interband regions of *Drosophila* polytene chromosomes. Nature. **294**, 417–422.

O'Brien, E.J. and Macewan, A.W. (1970) Molecular and crystal structure of the polynucleotide complex: Polyinosinic acid plus polydeoxycytidylic acid. J Mol Biol. **48**, 243–261.

Orgel, L.E. (1968). Evolution of the genetic apparatus. J Mol Biol. **38**, 381–393.

Ottig, G. et al. (1993). Warum Pentose- und nicht Hexose- Nucleinsauren? Teil IV. 'Homo-DNA': 1H, 13C,31P und 15N NMR-spektoskopische Untersuchung von ddGlc (A-A-A-A-A-T-T-T-T-T) in wassriger Lösung. Helv Chim Acta. **76**, 2701–2757.

Ozaki, H., Nakajima, K., Izumi, C. and Sawai, H. (2000). Convenient synthesis of arabinonucleoside containing oligodeoxyrobonucleotides. Nucleic Acids Symp Ser. **44**, 37–38.

Paulus, H. (2000). Protein splicing and related forms of protein autoprocessing. Annu Rev Biochem. **69**, 447–495.

Pitsch, S. et al. (1995). Pyranosyl- RNA ('p-RNA'): Base-Pairing selectivity and potential to replicate. Preliminary communication. Helv Chim Acta. **78**, 1621–1636.

Pitzer, K.S. and Donath, E. (1959). Conformations and strain energy of cyclopentane and its derivatives. J Amer Chem Soc. **81**, 3213–3218.

Prakash, T.P., Roberts, C. and Switzer, C. (1997). Angew Chem Int Ed Engl. **36**, 1522.

Prusiner, P. and Sundaralingam, M. (1972). Stereochemistry of nucleic acids and their substituents. XXV. Crystal and molecular structure of allopurinol, a potent inhibitor of xanthine oxidase. Acta Crystallogr. **B28**, 2148–2152.

Record, M.T. Jr., Mazur, S.J., Melancon, P., Roe, J.H., Shaner, S.L. and Unger, L. (1981). Double helical DNA: Conformations, physical properties, and interactions with ligands. Annu Rev Biochem. **50**, 997–1024. Review.

Rees, A.R. and Sternberg, M.J.E. (1984). From Cells to Atoms. Blackwell Scientific Publications, Oxford.

Rehak, M., Csuka, I., Szepessy, E. and Bánfalvi, G. (2000) Subphases of DNA replication in *Drosophila* cells. DNA Cell Biol. **19**, 607–612.

Reichard, P. (1993). The anaerobic ribonucleotide reductase from *Escherichia coli.* J Biol Chem. **268**, 8383–8386.

Rich, A. (1983). Right-handed and left-handed DNA: Conformational information in genetic material. Cold Spring Harb Symp Quant Biol. **47**, 1–12.

Rich, A. and Zhang, S. (2003). Timeline: Z-DNA: The long road to biological function. Nat Rev Genet. **4**, 566–572.

Richmond, T.J., Finch, J.T. and Klug, A. (1983). Studies of nucleosome structure. Cold Spring Harb Symp Quant Biol. **47**, 493–501.

Rogozin, I.B., Spiridonov, A.N., Sorokin, A.V., Wolf, I., Jordan, I.K., Tatusov, R.L. and Koonin, E.V. (2002). Purifying and directional selection in overlapping prokaryotic genes. Trends Genet. **18**, 228–232.

Ross, M., Shulman, M. and Landy, A. (1982). Biochemical analysis of att-defective mutants of the phage lambda site-specific recombination system. J Mol Biol. **156**, 505–522.

Rubin, G.M. (2001). The draft sequences: comparing species. Nature. **409**, 820–821.

Rydberg, B., Holley, W.R, Mian, I S. and Chatterjee, A. (1998). Chromatin conformation in living cells: Support for a zig-zag model of the 30 nm chromatin fibril. J Mol Biol. **284**, 71–84.

122   2 Structural Organization of DNA

Ryder, K., Silver, S., Delucia, A.L., Fanning, E. and Tegtmeyer, P. (1986). An altered DNA conformation in origin region I is a determinant for the binding of SV40 large T antigen. Cell. **44**, 719–25.

Saenger, W. (1984). Principles of Nucleic Acid Structure. (ed. C.R. Cantor), pp. 17–21. Springer Verlag, New York.

Sasisekharan, V. and Pattabiraman, N. (1978). Structure of DNA predicted from stereochemistry of nucleoside derivatives. Nature **275**, 159–162.

Sawai, H., Seki, J., and Ozaki, H. (1996). Comparative studies of duplex and triplex formation of 2′,5′ and 3′,5′ linked oligoribonucleotides. J Biomol Struct Dyn. **13**, 1043–1051.

Schmidt-Nielsen, K. (1997). Animal Physiology. Cambridge University Press, Cambridge, pp. 192–193.

Schweyer, S., Hemmerlein, B., Radzun, H.J. and Fayyazi, A. (2004) Continuous recruitment, co-expression of tumour necrosis factor-a and matrix metalloproteinases, and apoptosis of macrophages in gout tophi. Virchows Archiv. **437**, 534–539.

Selsing, E., Wells, R.D., Early, T.A. and Kearns, D.R. (1978). Two contiguous conformations in a nucleic acid duplex. Nature. **21**, 249–250.

Seto, H., Otake, N. and Yonehara, H. (1972). The structures of pentopyranine A and C, two cytosine nucleotides with α-α configuration. Tetrahedron Lett. **35**, 3991–3994.

Sharp, P.A. (1985). On the origin of RNA splicing and introns. Cell. **42**, 397–400. Review.

Snyder, M., Buchman, A.R. and Davis, R.W. (1986). Bent DNA at a yeast autonomously replicating sequence. Nature. **324**, 87–89.

Stambrook, P.J. (1974). The temporal relication of ribosomal genes in synchronized Chinese hamster cells. J Mol Biol. 82, 303–313.

Stent, G.S. (1958). Mating in the reproduction of bacterial viruses. Adv Virus Res. **5**, 95–149.

Stryer, L. (1995). Biochemistry (Fourth Edition), pp. 746. Freeman and Company, New York.

Stupina, V.A. and Wang, J.C. (2004). Proc Natl Acad Sci USA. **101**, 8608–8613.

Sugino, A. and Bott, K.F. (1980). Bacillus subtilis deoxyribonucleic acid gyrase. J Bacteriol. **141**, 1331–1339.

Sun, J., Zhang, Q. and Schlick, T. (2005). Electrostatic mechanism of nucleosomal array folding revealed by computer simulation. Proc Natl Acad Sci U S A. **102**, 8180–8185.

Sundaralingam, M, (1971). Stereochemistry of Nucleic acids and their constituents. XVIII. Conformational analysis of α-nucleotides by X-ray crystallography. J Amer Soc. **93**, 6644–6647.

Suzuki, K., Nakano, H. and Sizuki, S. (1967). Natural occurrence of and enzymatic synthesis of α-nicotamide adenine dinucleotide phosphate. J Biol Chem. **242**, 3319–3325.

Taljanidisz, J., Pppowski, J. and Sarkar, N. Temporal order of gene replication in Chinese hamster ovary cells. Mol Cell Biol. **9**, 2881–2889.

Taylor, J.H. (1963). DNA synthesis in relation to chromosome reproduction and reunion of breaks. J Cell Comp Physiol. **62**, 73–85.

Thelander, L. and Reichard, P. (1979). Reduction of ribonucleotides. Annu Rev Biochem. **48**, 133–158.

Thomas, C.A. (1971). The genetic organization of chromosomes. Annu Rev Genet. **5**, 237–256.

Thorp, H.H. (2000). The importance of being r: Greater oxidation stability of RNA compared with DNA. Chem Biol. **7**, 33–36.

Tjio, J.H. and Levan, A. (1956). The chromosome number of man. Heraditas **42**, 1–6.

Tjio, J.H. and Puck, T.T. (1958). The somatic chromosomes of man. Proc Natl Acas Sci USA. **44**, 1229–1237.

Uesugi, S., Miki, H., Ikehara, M., Iwahashi, H. and Kyogoku, Y. (1979). A linear relationship between electronegativity of 2′-substituents and conformation of adenosine nucleosides. Tetrahedron Lett. **42**, 4073–4076.

Usher, D.A. (1972). RNA double helix and the evolution of the 3′, 5′-linkage. Nat New Biol. **235**, 207–208.

Usher, D.A. and McHabe, A.H. (1976). Hydrolytic stability of helical RNA a sensitive advantage for the natural 3′, 5′-bond. Proc Natl Acad Sci USA. **73**, 1149–1153.

Van Holde, K. and Zlatanova, J. (1996). What determines the folding of the chromatin fiber? Proc Natl Acad Sci USA. **93**, 10548–10555.

References

Walsh, K. and Gualberto, A. (1992). MyoD binds to the guanine tetrad nucleic acid structure. J Biol Chem. **267**, 13714–13718.

Wang, J.C. Interaction between DNA and an *Escherichia coli* protein omega. J Mol Biol. **55**, 523–533.

Wang, J.C. (1980). Superhelical DNA. Trends Biochem Sci. **5**, 219–221.

Wang, J.C., Peck, L.J. and Becherer, K. (1983). DNA supercoiling and its effects on DNA structure and function. Cold Spring Harb Symp Quant Biol. **47**, 85–91.

Wang, A.H., Quigley, G.J., Kolpak, F.J., Crawford, J.L., Van Boom, J.H., Van Der Marel, G. and Rich, A. (1979). Molecular structure of a left-handed double helical DNA fragment at atomic resolution. Nature. **282**, 680–686.

Watson, J.D. and Crick, F.H.C. (1963). A structure for deoxyribose nucleic acid. Nature. **171**, 737–738.

Weintraub, H. (1983). A dominant role for DNA secondary structure in forming hypersensitive structures in chromatin. Cell. **32**, 1191–1203.

Weiss, L. (Ed.). Cell and Tissue Biology: A Textbook of Histology. 6th edition. Baltimore, MD: Urban and Schwarzenberg; 1983.

Wells, R.D., Brennan, R., Chapman, K.A., Goodman, T.C., Hart, P.A., Hillen, W., Kellogg, D.R., Kilpatrick, M.W., Klein, R.D., Klysik, J., Lambert, P.F., Larson, J.E., Miglietta, J.J., Neuendorf, S.K., O'Connor, T.R., Singleton, C.K., Stirdivant, S.M., Veneziale, C.M., Wartell, R.M. and Zacharias, W. (1983). Left-handed DNA helices, supercoiling, and the B-Z junction. Cold Spring Harb Symp Quant Biol. **47**, 77–84.

Wells, R.D., Goodman, T.C., Hillen, W., Horn, G.T., Klein, R.D., Larson, J.E., Muller, U.R., Neuendorf, S.K., Panayotatos, N. and Stirdivant, S.M. (1980). DNA structure and gene regulation. Prog Nucleic Acid Res Mol Biol. **24**, 167–267.

Whitehead, A. and Crawford, D.L. (2005). Variation in tissue-specific gene expression among natural populations. Genome Biol. **6**, R13.

Widom, J. (1998). Structure, dynamics, and function of chromatin *in vitro*. Annu Rev Biophys Biomol Struct. **27**, 285–327.

Williams, S.P., Athey, B.D., Muglia, L.J., Schappe, R.S., Gough, A.H. and Langmore, J.P. (1986). Chromatin fibers are left-handed double helices with diameter and mass per unit length that depend on linker length. Biophys J. **49**, 233–248.

Woodcock, C.L., Frado, L.L. and Rattner, J.B. (1984). The higher-order structure of chromatin: Evidence for a helical ribbon arrangement. J Cell Biol. **99**, 42–52.

Woodcock, C.L., Grigoryev, S.A., Horowitz, R.A. and Whitaker, N. (1993). A chromatin folding model that incorporates linker variability generates fibers resembling the native structures. Proc Natl Acad Sci USA. **90**, 9021–9025.

Woodcock, C.L. and Horowitz, R.A (1995). Chromatin organization re-viewed. Trends Cell Biol. **5**, 272–277.

Worzel, A., Strogatz, A. and Riley, D. (1981). Structure of chromatin and the linking number of DNA. Proc Natl Acad Sci U S A. **78**, 1461–1465.

Zahn, K. and Blattner, F.R. (1985). Sequence-induced DNA curvature at the bacteriophage lambda origin of replication. Nature. **317**, 451–453.

Zimmerman, S.B. (1982). The three-dimensional structure of DNA. Ann Rev Biochem. **51**, 395–427.

Zlatanova, J., Leuba, S.H. and Van Holde, K. (1998). Chromatin fiber structure: Morphology, molecular determinants, structural transitions. Biophys J. **74**, 2554–2566.

Zubay, G. (1958). A template model for the synthesis of ribonucleic acid from deoxyribonucleic acid. Nature. **182**, 1290–1292.

Zubay, G. (1998). Orig Life Evol Biosph. **28**, 13.

Chapter 3
Chromatin Condensation

(How Rope is Made from String)

Summary

The controversy of chromatin condensation resulted in several models for chromosome condensation. However, microscopic methods failed to distinguish among predicted structural motifs. To briefly summarize these developments, the structural organization of chromatin below the metaphase chromosome and beyond the nucleosomal fibril remained unknown. Nucleosomes condense into higher order structures, the nature of which is not elucidated. Due to the stickiness of the nuclear material, decondensed chromatin could not be separated in the interphase nucleus; consequently, the fibrilar chromatin structure could not be analysed. To avoid the stickiness reversible permeabilization was introduced which not only allows the survival of cells, but also permits to open the nucleus any time during the cell cycle. This chapter summarizes general mechanisms that modulate chromatin function, covalent modification of nucleosomal histones, and ATP-dependent remodeling and describes the analysis of chromatin structures in a cell cycle-dependent manner in mammalian and *Drosophila* cells using high resolution of centrifugal elutriation.

Importance of Chromatin Condensation

Before going into the structural details of chromatin, the importance of chromatin condensation has to be defined based on some of the significant observations and conclusions of reasearchers and teams dealing with nuclear structure.

- The term chromatin was used for the first time by Walter Flemming (1882). Flemming assumed some kind of nuclear-scaffold in the nucleus.
- 1910 Nobel Prize in Physiology or Medicine (Albrecht Cossel) "in recognition of the contributions to our knowledge of cell chemistry made through his work on proteins, including the nucleic substances".
- 1933 Nobel Prize in Physiology or Medicine (Thomas Hunt Morgan) "for his discoveries concerning the role played by the chromosome in heredity".

126 3 Chromatin Condensation

- 1953 Nobel Prize in Physiology or Medicine (Francis Crick, James Watson and Maurice Wilkins) "for their discoveries concerning the molecular structure of nucleic acids and its significance for information transfer in living material".
- 1974 Nobel Prize in Chemistry (Roger Kornberg) "for his studies of the molecular basis of eukaryotic transcription".
- 1982 Nobel Prize in Chemistry (Aaron Klug) "for his development of crystallographic electron microscopy and his structural elucidation of biologically important nucleic acid-protein complexes".
- 1993 Nobel Prize in Physiology or Medecine (Philip A. Scharp and Richard J. Roberts) "for their independent discoveries of split genes".
- The extent and distribution of condensing euchromatin are concomitant with differentiation and thus tissue-specific in animals and plants (Nagl, 1985).
- The interaction of genes and chromatin structure plays an important role in gene regulation (Aasland et al., 1995).
- Chromatin condensation is accompanied by significant transcriptional activity (De Campos-Vidal et al., 1998; Juan et al., 1998).
- The nuclear matrix is involved in the formation of metaphase chromosomes (Lewis et al., 1984; Wan et al., 1999).
- The torsional dinamics of DNA contributes to the structural organization of chromatin (Xue et al., 1998; Saha et al., 2002).
- The chromodomains of nuclear proteins make various interactions with histones and ribonucleic acids (Brehm et al., 2004).

Active and Inactive Chromatin

Gene expression is influenced by the complex process of chromatin condensation and decondensation and involves acetylation and deacetylation of histones, DNA methylation and demethylation and *cis*-acting genetic elements. The transcription initiation complex is composed of promoter sequences and DNA binding proteins. These two components of transcription are normally described as *cis*-acting elements and *trans*-acting factors. The effect of chromatin condensation on several genes known as silencing is preventing multiple genes from being transcribed, which process is not to be confused with gene repression which is affecting only single genes.

Chromatin in its inactive (non-trasncribed) state:

- is morphologically condensed (heterochromatin),
- contains deacetylated histones, with low content of acetylated histones,
- contains methylated DNA,
- contains DNA resistant to DNase I, restriction endonucleases, and DNA methylases (other proteins are accessible),
- is transcriptionally inactive,
- contains regularly spaced nucleosomes,

Conversely, the active (transcribed) chromatin:

- is euchromatic,
- with particular histone amino groups acetylated,
- is less methylated than inactive chromatin,
- has hypersensitive sites to DNase I digestion.

Euchromatin and Heterochromatin

Chromatin is a DNA/protein complex present in the nucleus. The human cell nucleus is only 5 μm in diameter. In interphase nuclei one can distinguish euchromatin and heterochromatin regions. The space which is not occupied by chromatin or the nuclear matrix is called the nucleoplasm containing soluble factors involved in the metabolism within the nucleus.

Euchromatin is dispersed, decondensed chromatin, a threadlike structure most abundant in transcriptionally active DNA. The open euchromatin structure allows the recognition and activity on DNA by the transcriptional machinery.

Heterochromatin is the highly aggregated, condensed form of chromatin, seen as dense patches inside the nucleus. Heterochromatin has three major locations in nuclei:

- near the nuclear envelope (marginal chromatin),
- chromatin as discrete bodies throughout the nucleus (karyosomes or chromatin bodies; it is recommended to reserve the term chromatin bodies to condensing interphase chromosomes),
- chromatin associated with nucleolus (nucleolar chromatin).

The "cartwheel" pattern of heterochromatin which is seen occasionally in the nucleus is related to the linear connection of condensing chromosomes (see later in Fig. 3.15, Elutriation Fraction 7A). Heterochromatin is considered transcriptionally inactive. There are several parameters and protein factors that have been associated with heterochromatin formation and condensation of mitotic chromosomes, but which of these apply to chromatin condensation during apoptosis remains to be determined.

Histone Code Hypothesis

Histones are evolutionarily conserved proteins playing a critical role in the proper packaging of DNA within the eukaryotic nucleus. DNA (147 bp) associated with histones (two each of histones H2A, H2B, H3 and H4) form the fundamental repeating subunit of chromatin, known as the nucleosome. In the seventies of the last century many studies have shown the involvement of different forms of histones in regulation on transcription, while recent interest is marked by studies of chromatin in the regulation of gene activity. Taking the size of the human genome composed of around 3×10^9 (3 billion) base pairs, the estimated number of nucleosomes within a human nucleus is several tens of millions. The higher-order chromatin structure of nucleosomes being the substrate for DNA replication, recombination, transcription and repair is also indicating conflicting results. The primary structure of major histone forms demonstrated the extreme conservativism of these small proteins. When these proteins are organized into the basic unit of chromatin known as nucleosomes, another highly conserved structure is formed. This conception changed recently after the recognition that histones may be modified posttranscriptionally and histone variants are able to assemble into a number of different nucleosomal particles. With the beginning of the nucleosome era some 35 years ago our conception of chromatin structure underwent a metamorphosis. Moreover, participants of chromatin research developed different perspectives of the same pivotal event (Olins and Olins, 2003).

As a consequence of histon modification DNA packaging may result in active and inactive conformations of histone. All these observations can be integrated into the "histone code" hypothesis. The tight association of nucleosomes to DNA and the highly conserved sequences of amino acids in histones contradict the postulated roles of nucleosomes in different sometimes opposing nuclear processes such as transcriptional activation or inactivation, replication or DNA repair, recombination, etc. To resolve this controversy the histone code hypothesis suggested that the post-translational modifications of histones are responsible for the direction of specific and distinct DNA-templated processes. Among the modifications of histones the acetylation of specific lysine residues of histone H3 and H4 by enzymes known as histone acetyltransferases turned out to play a major role in transcriptional activation. To the contrary, deacetylation of histones by histone deacetylases is associated with the transcriptional inactivation of chromatin. Other observations demonstrated that other histone modifications (e.g. methylated H4) can prevent the enzymatic activity of histone acetyltransferases. Further evidence supported the histone code hypothesis, indicating that post-translational modification of histones serves as a selective binding site for specific proteins which can read the histone code to initiate or to prevent distinct nuclear processes. The histone code hypothesis implies that inactive regions may also contain modified histones. Such an example is the methylation of lysine residue 9 of histone H3 which causes the inactivation of chromatin resulting in heterochromatin, which then recruits and binds heterochromatin protein (HP1) through its evolutionary conserved chromodomain. There are at least eight different classes of covalent modifications involving more than 60 distinct modification sites within the major core histones characterized to date; these modifications include lysine acetylation, lysine and arginine methylation, serine and threonine phosphorylation, lysine ubiquination, glutamate poly-ADP ribosylation, lysine sumoylation, arginine deimination and proline isomerization (Kouzarides, 2007; Wang et al., 2007, review).

The compacted chromatin behaves then as a physical barrier for the transcriptional machinery (Jacobs and Khorasanizadeh, 2002). It was further proposed that a given nucleosome might not consist of heterotypic variants, but rather specific histone variants come together to form a homotypic nucleosome, a hypothesis refered to as the nucleosome code. Such nucleosomes might in turn participate in marking specific chromatin domains that may contribute to epigenetic inheritance (Bernstein and Hake, 2006). Consequently, the dynamics of chromatin structure is tightly regulated through multiple mechanisms including histone modification, chromatin remodeling, histone variant incorporation, and histone eviction (Li et al., 2007).

Chromosome Arrangement in the Nucleus

The genomes of somatic cells of a certain organism except for minor changes are identical, while different cell types possess different epigenomes including the variability of DNA methylation and histone modification patterns. Chromatin is the physiological carrier of the genetic and epigenetic information in eukaryotes. Epigenetic variability is responsible for the cell-type-specific gene expression and silencing patterns in multicellular organisms. The impact of higher-order nuclear architecture on the epigenetic variability is not yet known (Van Driel et al., 2003), not mentioning the opposite effect as of how the epigenetic factors influence the large-scale nuclear structure. Multiple attempts were made to map the large-scale organization and distribution of chromatin in resting (Go), cycling (S-phase) and mitotic (M-phase) cells (Cremer and Cremer, 2001; Parada et al., 2004; Pederson, 2004; Kosak and Groudine, 2004). Reliable topological maps, however, for the spatiotemporal arrangements of the chromosomes in the nucleus of the diploid somatic cell have been lacking so far.

Models of Chromosome Condensation

A long-standing problem of chromatin condensation is how chromatin intermediates cycle dinamically between decondensed interphase stages which support transcriptional, repair and replicational activities of cells and compact mitotic and meiotic forms required for segregation and faithful transport of genetic information to the next generation of cells. Although, chromosomal DNA is packaged into condensed nucleoprotein suprastructures, it has to be accessible during the cyclic condensation-decondensation process. To the question, how the cyclic chromosome condensation and decondensation takes place the answer is that the process is not understood.

The major obstacle to distinguish among different structural forms is that chromatin should be studied:

- under physiological conditions in intact cells,
- during the development of chromosomes in interphase and
- the imaging of chromatin structures should be of sufficient microscopic resolution.

As earlier models failed, the development of new imaging techniques is expected to shed light on this long standing problem in cell biology (Mora-Bermúdez and Ellenberg, 2007).

Earlier models of mitotic chromosome structure were based largely on the examination of fully condensed metaphase chromosomes. The structural analysis of such fully compact, native metaphase chromosomes led to three alternative conceptual classes of models for metaphase chromosome architecture:

a) radial loop,
b) hierarchical folding,
c) network models.

These three models are different in terms of the structural motifs postulated as giving rise to chromosome condensation (Swedlow and Hirano, 2003). Current microscopy methods cannot distinguish between these different predicted structural motifs in metaphase chromosomes (Belmont, 1998). The review of these models led to a fourth, so called d) unified model of chromosome structure in which hierarchical levels of chromatin folding are stabilized late in mitosis by an "axial glue" mechanism (Kireeva et al., 2004).

a) Radial loop model. The radial loop model postulates that chromatin folding beyond the level of the 30-nm chromatin fibril is directed by nonhistone scaffold protein interactions with specific DNA sequences. The prediction from this model is that mitotic chromosome condensation is dependent on scaffold proteins, and the temporal pattern of scaffold assembly will coincide with or precede the appearance of chromosome condensation above the level of the 30-nm chromatin fibril. Laemmli's group has removed histone and non-histone proteins from metaphase chromosomes using gentle extraction methods which led them to the radial loop model of chromosome organization (Paulson and Laemmli, 1977; Laemmli et al., 1978). In this model 30-nm chromatin fibrils are forming loops, specific regions of which are radially organized and anchored to an axial chromosome substructure termed chromosome "scaffold" (Marsden and Laemmli, 1979) (Fig. 3.1). Anchoring of loops involves special scaffold and matrix associated sequences of DNA (SARs/MARs) (Razin, 1996).

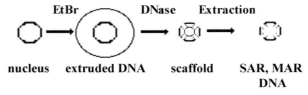

Fig. 3.1 Schematic view of isolating scaffold attached DNA regions. DNA sequences attached to the nuclear scaffold can be isolated by applying intercalating agents such as ethydium bromide (EtBr) to isolated nuclei. Intercalating agents cause partial unwinding of DNA extruding from and surrounding the nucleus. The extruded DNA can be removed with DNase treatment. The remaining DNA is about 10% of the total DNA and is thought to be associated with nuclear particles and is called scaffold attached regions (SARs) or matrix associated regions (MARs)

Models of Chromosome Condensation   131

A variation of the radial-loop model was the radial loop/helical coil model which suggested that the prophase chromatid which is already organized in a radial loop fashion is then helically folded to form the metaphase chromosome (Rattner and Lin, 1985; Boy De La Tour and Laemmli, 1988). This hypercondensed variant of metaphase chromosomes turned out to be present only in mitotically arrested cells (Maeshima and Laemmli, 2003).

b) Hierarchical folding. In this model the 10 nm chromatin strings and 30 nm chromatin fibrils are folded progressively into thicker fibers that coil to form the final metaphase chromosomes (Sedat and Manuelidis, 1978; Zatsepina et al., 1983; Belmont et al., 1987; Belmont and Bruce, 1994). The condensation process is not dependent on the formation of the core protein scaffold, consequently the temporal pattern of chromosome condensation will not necessarily coincide with scaffold assembly. Successive helical coiling and folded chromonema models are the two variants of this group of models (Sedat and Manuelidis, 1978; Belmont and Bruce, 1994). The description is missing between the folding levels of the 30 nm fibril and the postulated 250–300 nm coiling subunit.

c) Network model. Micromanipulation of metaphase chromosomes with brief nuclease treatment led to the loss of elasticity of chromosomes, which was taken as an evidence against a core protein scaffolding. Moreover, by mechanical streching of metaphase chromosomes the length could be increased several times of their normal length without significant change in diameter or sequential uncoiling as predicted by the hierarchical folding model (Poirier et al., 2000). These observation resulted in the proposal of the chromatin network model (Poirier and Marko, 2002) which suggests that chromosomes are stabilized by protein cross-links between chromatin fibers. This model paid less attention to possible condensation intermediates.

d) Axial glue mechanism. In this unified model of chromosome condensation the hierarchical folding is driving the condensation process in early mitosis, whereas cross-linking by condensins and other proteins stabilize the chromosome shape and compaction later in mitosis. The axial glue mechanism is based on Belmont's group observation that early prophase condensation takes place through the folding of large-scale chromatin fibers into condensed masses resolving 200–300 nm diameter chromatids that double by the end of prophase (Kireeva et al., 2004).

There is an overwelming experimental evidence that DNA is in its most decondensed state during strand separation processes (DNA replication, transcription, recombination). Chromatin domains are unfolded during interphase and compacted onto filaments and increasingly compact chromosomal structure. The condensation of eukaryotic chromosomal DNA starts in S phase, immediately after DNA synthesis when the newly replicated DNA is packaged into nucleoprotein structures. Eukaryotic DNA is closely associated with an equal mass of tightly bonded proteins called histones. At the first level of association of DNA with nucleoproteins, 1.8 superhelical turns of DNA are wrapped around an octamer of two copies each of four core histone proteins (H2A, H2B, H3 and H4) forming the nucleosome core particles. Recently significant progress has been made concerning how the nucleo-

132 3 Chromatin Condensation

somal subunits of chromatin fibers are disassembled and reassembled *in vitro* and *in vivo* (Luger and Hansen, 2005). Nucleosomes are repeating units of eukaryotic chromosomes containing superhelical turns of DNA. Improvement was achieved by adjusting the length of these turns of DNA from 146 to 147 base pairs (Davey et al., 2002). The nucleosomal core particles have been described in different organisms which delivered structural information in detail (Harp et al., 2000; White et al., 2001; Davey et al., 2002; Davey and Richmond, 2002).

The dynamic nature of nucleosomes is indicated by the several seemingly conflicting functions of nucleosomes:

– superhelix formation with histones causing a 7-fold compaction of double stranded DNA,
– more than 120 atomic interactions between DNA backbone and histones,
– sliding of histone octamers along DNA over long distances (Pennings et al., 1991), alleviated by ATP-dependent remodeling factors (Becker, 2002; Becker and Horz, 2002).

Nucleosomes connected by an average of 60 base pairs linker DNA are compacted further to a higher order, hierarchical architecture. The exact details of the three dimensional folding of the chromatin fibre are still controversial as the path between nucleosomes is unknown (reviewed by Horn, 1991; Hansen, 2002; Luger, 2003).

Since we know only the diameters of chromatin structures with increasing compaction, thus in the general scheme, guessing of folding possibilities have been neglected (Fig. 3.2). According to this general scheme of chromosome condensation the double helical DNA (Fig. 3.2A) is turned around the octameric histone core to create disc-shaped nucleosomes of 11 nm diameter. The nucleosomal attachment is known as "beads on string" or chromatin string (Fig. 3.2B). Neighbouring nucleosomes are packed together to form the 30 nm chromatin fibril, whose exact structure is not known (Fig. 3.2C). The fibril is thought to form chromatin loops of an estimated 300 nm diameter thick fiber (Fig. 3.2D). Chromatin loops turn around the non-histon proteinaceous component called scaffold (Jackson, 1991) bringing about the 600–700 nm condensed chromatin (Fig. 3.2E). Functional studies of many of the scaffold proteins indicated that they are essential for several aspects of chromosomal behaviour (Balczon, 1993). As far as the chromatin loops are concerned the existence of lampbrush chromosomes supports the concept that they are organized in a series of loops. Lampbrush chromosomes occur in a limited phase of meiosis during the oogenesis in amphibian oocytes.

The estimated diamater of the scaffold proteins is about 150 nm. The condensation of chromatin loops (thin fibers) and thick fibers takes place in unknown ways during the interphase and mitosis to form the most condensed metaphase chromosome (Fig. 3.2F). Lateral looping of the DNA from the scaffold core implies that the chromosome is regularly compacted along the long axis of the chromosome.

Eukaryotic cells must completely compact their chromosomes before faithfully segregating them during cell division. Failure to do so can lead to segregation defects with pathological consequences, such as aneuploidy and cancer. Duplicated interphase chromosomes are organized into tight rods before being separated and

Models of Chromosome Condensation

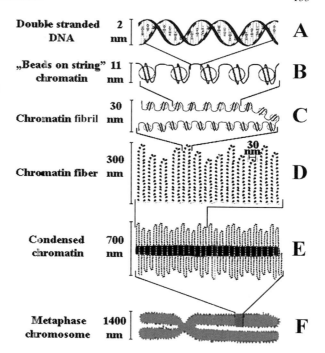

Fig. 3.2 General scheme of chromosome condensation. Structural levels from top to bottom: (A) double helical DNA, (B) string of nucleosomes, (C) 30 nm thin chromatin fibril, (D) 300 nm thick chromatin fiber, (E) chromatin loops turned around the central scaffold proteins (*black*), (F) metaphase chromosome

directed to the newly forming daughter cells. This vital reorganization of chromatin remains poorly understood (Mora-Bermúdez et al., 2007). The spatial organization of chromosomes and how chromosome organization changes during the cell cycle are incompletely known (Sarkar et al., 2002). It is known that:

1. during interphase different chromosomes are confined to different nuclear "territories" (Cremer et al., 1993),
2. individual chromosomes show random-walk correlations along their length (Yokota et al., 1995),
3. some chromosome loci are anchored to or near the nuclear envelope (Marschall et al., 1996), and to nuclear structures such as the "nuclear matrix" (Wolffe, 1995).

The questions of how chromosome structure is reorganized during mitosis and how large-scale chromosome structures during interphase and mitosis are related remain also unanswered. Even the large-scale structure of mitotic chromatids remains controversial (Koshland and Strunnikov, 1995; Strunnikov, 1998; Hirano, 2000), some groups favour the folding of chromatin loops onto a central protein axis (Earnshaw and Laemmli, 1983), others suggest successive levels of helical folding (Belmont et al., 1987).

Chromosome condensation in mammalian cells can be subdivided into several consecutive steps:

- The folding of DNA in nucleosomes results in a compaction factor of about 7.
- Winding of the 10 nm nucleosome fibril into 30 nm supernucleosomal arrays further compacts the genetic material. The folding of the nucleosome string into chromatin fibril accounts for an estimated 40-fold linear compaction.
- A 250 or more fold compaction is achieved by looping beyond the supranucleosomal level.
- Measurements by fluorescent *in situ* hybdridization (FISH) in mammalian nuclei yielded further compaction values ranging from hundreds to thousands, indicating additional levels of chromatin folding above the 30 nm chromatin fibril refered to as large-scale chromatin structure.
- Maximal chromosome compaction occurs by axial shortening in anaphase. Maximal compaction was not reached in metaphase, but in late anaphase, after sister chromatid segregation. Anaphase compaction proceeds by a mechanism of axial shortening of the chromatid arms from telomere to centromere (Mora-Bermúdez et al., 2007).
- A total of ≈10,000 fold compaction of double stranded DNA into metaphase chromosomes occurs in mammlian nuclei.

Chromosomes at mitosis are 2–10 μm in length. In somatic cells during mitosis the nucleus disappeares and chromosomes line up in the middle of the cell. The two

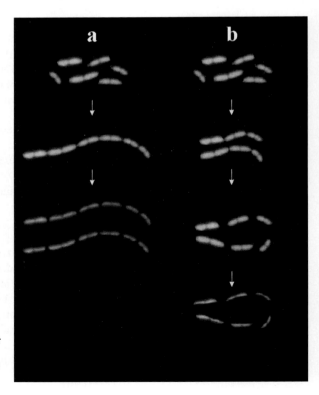

Fig. 3.3 Chromosome segragation in mitosis and meiosis. (**a**) In mitosis (*left*) chromosome condensation is followed by seggregation occuring prior to cell division. The fact that chromosomes line up in the middle of the cell indicates a defined chromosome order. (**b**) In meiosis (*right*), two cell divisions follow a single round of chromosome replication. In the first division (*meiosis I*), homologous chromosomes pair and segregate to daughter cells. In the, second division (*meiosis II*), sister chromatids segregate

sister chromatids are then evenly distributed to opposite halves of the cell undergoing division (mitosis) (Fig. 3.3a). In meiosis, two cell divisions follow a single round of chromosome replication (Fig. 3.3b).

Chromatin Folding in the Interphase Nucleus Is Poorly Understood

Chromatin in the mitotic chromosomes is packaged orderly and reproducibly, yet the supra-organization of the 10 nm chromatin string is still poorly characterized (Gassmann et al., 2004; Hozak and Fakan, 2006). This problem is indicated by the different models of the mitotic chromosome structure which have been suggested depending on experimental approaches. The four major models have been discussed in the previous subchapter.

Less attention has been paid to the structure of the interphase chromatin, which was deduced mainly by extrapolations from mitotic chromosome models. Although, chromatin is easily visualised by staining, hence its name (*coloured material*), yet the process and intermediates of chromatin condensation remained unknown. The scarcity of evidence regarding chromatin folding in interphase is due to the fact that individual chromosomes are invisible during the early stage of development in the eukaryotic nucleus. Among the limiting factors are mentioned: (1) low spatial resolution of light microscopy, (2) non-specific staining of DNA, (3) difficulty of visualizing 3-D structures, (4) sensitivity of chromatin to small environmental changes, and (5) stickiness of the nuclear material. As imaging techniques failed to show the topology of decondensed interphase chromosomes (Robinett et al., 1996; Bickmore, 1999; Kostriken and Weeden, 2001), there is little solid information concerning the architecture of interphase chromosomes (Lemke et al., 2002). By the time the typical structures of metaphase chromosomes develop, the nuclear membrane is fallen apart and only the end product of the condensation process can be seen. Direct approaches to visualize the topography of chromatin remained limited (Comings, 1980; Cremer et al., 1979; Sperling and Luedtke, 1981).

Chromatin remodeling, i.e. the dynamic modulation of chromatin structure, is a key component in the regulation of gene expression, apoptosis, DNA replication and repair, chromosome condensation and segregation. Chromatin remodeling complexes are known to facilitate transcriptional activation by opening chromatin structures (Xue et al., 1998). Nevertheless, reconstruction of the structure of interphase nuclei from electron micrographs of serially sectioned nuclei failed to distinguish among discrete chromosomes since most of the chromatin forms a continuous sticky network (Heslop-Harrison et al., 1988). It has been difficult to study the supraorganization of condensing chromosomes within the nuclear membrane as one cannot see how chromosomes develop behind the nuclear curtain. Consequently, only two typical structures are visible under microscope: the interphase nucleus and metaphase chromosomes.

Chromatin Models

Chromatin in the mitotic chromosomes is packaged orderly and reproducibly. Thomas Cremer's group defined the eukaryotic genome as a highly complex system, regulated at three major hierarchical levels: (1) the DNA sequence level, (2) the chromatin level and (3) the nuclear level, including the dynamic and three-dimensional spatial organization of the genome inside the cell nucleus. Some specific regions of the nuclear level chromatin are attached to the nuclear matrix. Their current research of chromosomal and chromatin organization focused mainly on: (1) the analysis of higher order nuclear architecture, (2) the dynamic interactions of chromatin with other nuclear components, (3) the functional organization of the cell nucleus, with the focus on chromosomal organization and (4) distribution of specific chromosomal areas within the cell nucleus (Cremer et al., 2006).

The group of Andrew S. Belmont described the reproducible and dynamic positioning of DNA within chromosomes during mitosis (Dietzel and Belmont, 2001). Belmont's group is interested in how the 10 nm strings and the 30 nm chromatin fibrils fold into interphase and mitotic chromosomes, how interphase chromosomes are positioned within nuclei, and what this means for DNA functions such as transcription and replication. It is acknowledged by this group that our current understanding of the higher levels of chromatin organization refered to as large-scale chromatin structure, is poor. To visualize folding dynamics of chromosome specific regions and general folding motifs, the combination of molecular biology, cell biology, genetic methods and imaging tools are being used.

Despite intensive research in the field the supranucleosomal organization of the chromatin fiber remains "terra incognita" (Grechmann and Ramakrishnan, 1987; Hozak and Fakan, 2006), as indicated by the several models of the 30 nm fibrils. Early models include the solenoid model (Finch and Klug, 1976), the helical-ribbon model (Worzel et al., 1981; Woodcock et al., 1984) and the crossed-linker double-helical model (Williams et al., 1986). Despite sharing certain similarities, these models differ considerably in detail, providing no evidence of transition from the 10 nm string to the 30 nm fibril (Grechmann and Ramakrishnan, 1987). After extensive studies probing the structural and dynamic properties of chromatin (Bednar et al., 1998; Widom, 1998; Zlatanova et al., 1998; Luger, 2003), the question remained the same: how nucleosomes are arranged in the chain of a 30 nm chromatin fibril under physiological conditions (Widom, 1998; Luger, 2003). The solenoid versus zig-zag model and the possible fibril arrangements of "beads on string" including the hairpin model and the plectonemic model of chromatin condensation have been discussed in Chapter 2.

Review of Methodologies to Follow Chromosome Dynamics

Light microscopy in intact cells. Chromatin dynamically cycles between the decompacted interphase state that supports transcription and replication to compact chromosomes required for their segregation, but the condensation-decondensation process is not understood. To address this problem since the first description of

Review of Methodologies to Follow Chromosome Dynamics

chromatin 130 years ago by Walther Flemming (1878), ideally the structure of chromatin should be studied in intact cells. Despite its immobile appearance, chromatin is in a constant dynamic exchange with the nucleoplasm (Phair et al., 2004; Beaudouin et al., 2006). Unfortunately, the perturbations are often a necessary part of the protocols used to study chromatin *in vitro* and are likely to disrupt the native structure, limiting the interpretation of the data. Moreover, the study of chromosomes within intact living cells suffers from the limited resolution of noninvasive methods, such as light microscopy (Mora-Bermúdez and Ellenberg, 2007).

Fluorescent microscopy in vivo. Fluorescence microscopy has been instrumental in advancing knowledge of chromosome dynamics (Belmont, 2001) and multidimensional *in vivo* studies of mitotic chromosomes (Hiraoka et al., 1989). Advances in imaging technology made these studies more sophisticated precise and accessible. We expect that by revealing major intermediates of chromatin condensation *in vitro*, live-cell imaging will become more meaningful and three dimensional visualization will become a routine technology to study chromatin structure. Recently fluorescence microscopy is a well-established and widely used technique for chromatin analyis. The high-resolution of electron microscopy has tremendous potential but technically still difficult to handle and is therefore less widely used. Regarding the physical properties of chromosomes three parameters can be studied: mechanical properties (elasticity and rigidity), structural properties such as folding and topology and dimensions such as length, width, diameter, volume. Our fluorescent studies focus mainly on the second and third categories, namely chromatin condensation, as well as size (length, diameter) and volume. Flexibility will be dealt with when *Drosophila* and mammalian chromosome condensation will be compared.

Analysis of chromosome dinamics by different methods. Electron microscopy (EM) studies have revealed that that the basal structure is the chromatin string, which is folded to higher levels of structures. The chromatin string consists of an array of core nucleosomes, known as "11 nm-beads-on-a-string", also referred to as the 10 nm string (Woodcock et al., 1993). *X-ray crystallography* provided evidence in atomic detail how these "beads" are formed by 147 bp of DNA wrapped 1.7 times around an octamer of core histones (Luger et al., 1997; Richmond and Davey, 2003).

The first level of chromatin folding is the 30 nm chromatin fibril structure (Bednar et al., 1998). Although, the arragment of nucleosomes in this thin fibril remained controversial, preference is given to the zig-zag over the solenoid arrangement based on electronmicroscopic studies (Woodcock et al., 1993). Results of electron tomography (Horowitz et al., 1998) and chromatin fragments generated by ionizating radiation (Rydberg et al., 1998) are compatible with the zig-zag arrangement, and chrystallographic observations have directly shown the zig-zag helical arrangement of nucleosomes (Dorigo et al , 2004; Schalch et al., 2005).

How the 30 nm fibrils are further organized into even higher order chromatin folds within chromosomes remains a mistery. Controversial conceptual models of chromatin condensation have been put forward based mainly on observations of fixed cells and isolated chromosomes (Swedlow and Hirano, 2003). It was suggested that higher orders of chromatin fibers, such as the 100 nm "chromonemas", could consist of "30 nm" fibrils successively folded into progressively thicker rodlets and

138 3 Chromatin Condensation

could be functional intermediates between the nucleosome array and the chromosome. This model has received support from engineered repetitive DNA sequences inserted into chromosomes and contradicted predictions of simple radial loop models while providing support for hierarchical models of chromosome architecture (Dietzel and Belmont, 2001; Strukov et al., 2003).

Isolation and staining of chromatin and chromosomes. Various types of tissues can be used to obtain chromosome preparations, such as peripheral blood, bone marrow, amniotic fluid, products of conception, etc.

Major steps involved in the isolation of chromosomes:

- Maintaning cell or tissue cultures
- Blocking cells in metaphase with a mitotic inhibitor
- Harvesting cells by centrifugation
- Exposing cells to hypotonic solution
- Exposing cells and suspending nuclei in fixative solution
- Washing nuclei in fixative
- Speading nuclei and chromosomes on slides
- Staining chromosomes
- Detection of chromosome preparations, numerical and structural changes under microscope.

For the isolation of chromatin structures the method developed for metaphase chromosomes to prepare nuclear spreads is used. Nuclei are centrifuged, washed several times in fixative solution and resuspended in fixative. Nuclei are spread over glass slides drop-wise from a height of approximately 30 cm. Dropping helps to distribute nuclei evenly on the slide surface (Henegariu et al., 2001). Slides are air dried, stored at room temperature overnight, rinsed with phosphate based saline (PBS) and dehydrated using increasing concentrations of ethanol (70, 90, 95 and 100%).

The Clark's fixative solution dates back to the 1850s and is used in cytogenetics, since it keeps cells in a "swollen" state, achieved after hypotonic treatment. The fixative is used to break the cell membrane of cells such as lymphocytes, fibroblasts, tumour cells. The fixative solution elutes some of the membrane lipids and proteins and makes the membrane more fragile and suitable for spreading flat on the slide when subjected to the drying techniques (Henegariu et al., 2001).

Giemsa (G) staining yields a series of lightly and darkly stained bands. The fluorescent quinacrine is used for Q-banding. The pattern of Q-bands is similar to that seen in G-banding. C-banding stains centromers.

Fluorescent staining of DNA takes place in the presence of an antifade compound, such as 1,4-diazobicyclo-(2,2,2)-octane (DABCO) and 2,6-diamino-2-phenylindole (DAPI) for blue fluorescent total staining of DNA (350 nm excitation). DAPI is known to form fluorescent complexes with natural double-stranded DNA showing fluorescence specificity at AT rich sequences in the minor groove of DNA (Parolin et al., 1995). Propidium iodide (PI, 488 nm excitation) staining of DNA with orange/red fluorescence emmission is also frequently used. Blue fluorescence of DAPI or red fluorescence of PI is monitored with a fluorescent microscope.

Fig. 3.4 Interphase nuclei of CHO cells (**A**) and metaphase chromosomes (**B**) from Chinese hamster ovary cells. H, heterochromatin; E, euchromatin

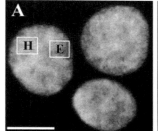
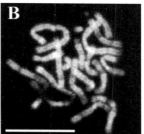

The two basic nuclear structures are seen as:

- interphase nuclei (Fig. 3.4A) and
- metaphase chromosomes (Fig. 3.4B)

visualized under the fluorescent microscope.

Methods to Monitor Chromosome Condensation During the Cell Cycle

Among the synchronization methods the most useful ones are based on two distinct strategies (Uzbekov, 2004). One principal approach is the chemical "arrest-and-release" strategy, in which cells are arrested at a certain stage of cell cycle, and collected at the time they enter the stage of interest. The other approach involves physical methods (mitotic detachment, density gradient centrifugation, centrifugal elutriation, cell sorting), to collect cells at specific stages of the cell cycle from an asynchronous culture.

Centrifugal Elutriation

Synchronized populations of cells can be obtained by centrifugal elutriation. Elutriation fractionates the cell population on the basis of sedimentation properties of small round particles, with minimal perturbation of cellular functions. Elutriation is regarded as an ideal method for the isolation of cell cycle phase specific subpopulations of synchronized cells. For cell synchronization by elutriation early log S phase cell populations are maintained for several generations prior to fractionation. Detailed protocol for cell cycle synchronization of mammalian cells and nuclei by centrifugal elutriation has been described (Bánfalvi, 2008).

The physical characteristics of cell size and sedimentation velocity are operative in the technique of centrifugal elutriation. Elutriator is an advanced device for increasing the sedimentation rate to yield enhanced resolution of cells separation. Separation is carried out in a specially designed centrifuge and rotor containing

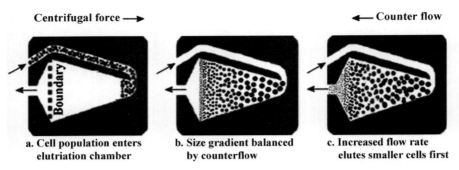

a. Cell population enters elutriation chamber **b. Size gradient balanced by counterflow** **c. Increased flow rate elutes smaller cells first**

Fig. 3.5 Schematic view of elutriation. (Reproduced with permission of Bánfalvi, 2008)

the elutriation chamber (Fig. 3.5). By increasing the flow rate of the elutriating fluid step-by-step, successive populations of relatively homogeneous cell size can be removed from the chamber and used as synchronized subpopulations of cells.

Synchronized cells were permeabilized reversibly, nascent DNA synthesized in the presence of dNTPs, including biotin-11-dUTP (as a dTTP analogue), dATP, dGTP and dCTP. Nascent biotinylated DNA does not perturb DNA replication (Hiriyanna et al., 1988; Blow and Watson, 1987), but seems to interfere with chromatin folding (Bánfalvi et al., 1989). Biotinylated-DNA by preventing chromatin folding caused the accumulation of early intermediates of chromatin folding. Reversal of permeabilization helped to restore membrane integrity (Bánfalvi et al., 1984). Chromatin condensation was blocked at its condensed state in metaphase by colchicin treatment. Normal and damaged chromosomal development were visualized by fluorescence microscopy.

Permeable Cells

A major obstacle in the attempt to follow DNA synthesis was the impermeability of the cell membrane to large molecules such as nucleotides, cyclic nucleotides and proteins. Thus many studies with intact cells used intermedier metabolites (e.g. thymidine, uridine) which undergo active transport and several steps before being incorporated into macromolecules (Castellot et al., 1978). It was originally described in bacteria that DNA synthesis in permeable cells depends on the presence of ATP, characteristic to replicative DNA synthesis (Moses and Richardson, 1970). The ATP-dependent incorporation of the four deoxyribonucleoside triphosphates (dNTPs) in permeable thymocytes reflected DNA replication and was hampered by inhibitors of DNA replication, while dNTP incorporation in the absence of ATP could be prevented by inhibitors of repair synthesis (Bánfalvi et al., 1984; Bánfalvi et al., 1997a, ibid. 1997b). These experiments served as a basis to follow simultaneously the two types of DNA synthesis.

Permeable cell systems have been used mainly for studying DNA synthesis and proved to be useful for analytical studies of DNA replication and repair synthesis

(Sheinin et al., 1978; Seki and Oda, 1986). It was observed that chromatin structure at the sites of DNA replicated in permeable cells is similar to that at the sites of DNA replicated in living cells, and that some structural change (possibly toward the maturation) of newly replicated chromatin occurs after the DNA replication in permeable cells (Seki et al., 1986). In permeable cells the reassembly of native chromatin structure upon completion of repair was not an active process (Kaufmann et al., 1983).

Technical Limitations to Visualize Large Scale Chromatin Structures

These included those limitations that have already been mentioned, namely the low spatial resolution of light microscopy, the nonspecific staining of DNA, the difficulty of visualizing 3-D structures even in thin sections of cells by electron microscopy, the stickiness of the nuclear materail and the sensitivity of chromatin conformation to even small environmental (e.g. ionic) changes. Improved visualization of chromatin folding patterns within diploid nuclei and polytene chromosomes was obtained using nonionic detergents in polyamine or divalent cation buffers to remove nucleoplasmic background staining (Bjorkroth et al., 1988). This allowed greatly improved visualization of individual chromatin fibers within a chromosome puff in polytene chromosomes. For diploid nuclei, buffer conditions were chosen which preserved *in vitro* chromosome morphology (Belmont et al., 1989). This approach of the Belmont laboratory revealed large-scale chromatin fibers, termed chromonema fibers appearing as short fiber segments or more continuous fiber lengths, roughly 100 nm in diameter (60–130 nm depending on cell cycle position and cell type).

Reversible Permeabilization

Reversibly permeabilized cells (Bánfalvi et al., 1984) have been used to visualize interphase chromatin structures in the presence and absence of biotinylated nucleotides (Bánfalv., 1993; Gacsi et al., 2005). By reversing permeabilization, it was possible to confirm the existence of a flexible chromatin folding pattern through a series of transient geometric forms such as supercoiled, circular forms, chromatin bodies, thin and thick fibers and elongated chromosomes. These results showed that the incorporation of biotin-11-dUTP interferes with chromatin condensation, leading to the accumulation of early intermediates of chromatin condensation being in decondensed state. To study the temporal order of chromatin condensation the incorporation of nucleotides was omitted and subpopulations of cells were synchronized by counterflow centrifugal elutriation. After reversal of permeabilization, chromatin structures were isolated from cells synchronized by centrifugal elutriation and visualized by fluorescent microscopy. Decondensed veil-like structures were observed in early S phase (between 2.0 and 2.25 C-value), supercoiled chromatin later in early S (2,25–2.5 C), fibrous structures in early mid S phase (2,5–2.75 C), ribboned structures in mid S phase (3.0–3.25 C), continuous chromatin strings later in mid S phase (3.25–3.5), elongated prechromosomes in late S phase (3,5–3.75 C), precondensed chromosomes at the end and after S phase ($\approx$4 C) (Bánfalvi et al., 2006). Fluorescent microscopy revealed that interphase chromosomes are not separate entities but form a linear array arranged in a semicircle. Linear arrangement was confirmed by computer image analysis.

Isolation of Interphase Chromosomes: Visualising "Babies" Before Birth

The main reason that higher order chromatin structure could not be followed by fluorescent microscopy was that from intact cells only interphase nuclei and metaphase chromosomes could be isolated (Fig. 3.4). To overcome this technical difficulty chromatin structures were isolated from nuclei after reversal of permeabilization of cells by maintaning their viability, yet able to open them without inducing apoptosis. This powerful method allows the analysis to monitor the kinetics of chromatin condensation in time-lapse sequences during the cell cycle by using synchronized cells. By using this new approach we have confirmed the existence of a flexible folding pattern by visualizing transient geometric forms of condensing chromosomes. As model systems we have used murine preB, rat HeDe (hepatocarcinoma, Debrecen), Chinese hamster ovary (CHO-K1, ATCC #CCL61), Indian muntjac, human erythroleukemia (K562) mammalian cell lines and *Drosophila* S2 cells.

Visualization of Intermediates of Chromatin Condensation in CHO Cells

Biotinylation Interpheres with Chromatin Folding

Nascent DNA synthesis was carried out in permeable cells in the presence of biotin-11-dUTP (Fig. 3.6). Cells were then allowed to recover from permeabilization in serum-enriched medium. Colcemid treatment was used to block the cell cycle in metaphase and prevent cells from proceeding to the next cell cycle. Replication sites in nuclei of exponentially growing cells were visualized after immunofluorescent amplification (Bánfalvi et al., 1989). Red propidium iodide and green FITC signals could be visualized simultaneously using blue excitation light, but the FITC signals were significantly masked. Because DAPI is spectrally well separated from FITC fluorescence and different shades of its blue colour fluorescence indicate the degree

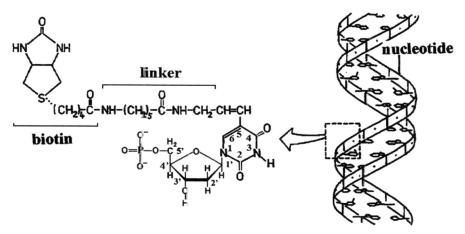

Fig. 3.6 Schematic representation of biotinylated DNA (*to the right*) containing biotin.11-dUMP (*to the left*)

of chromatin compactness, subsequent experiments employed DAPI for fluorescent staining of total DNA.

In control experiments the isolation of nuclei from permeable cells was attempted without reversing permeabilization. A sticky mass of nuclear material was obtained indicating that the nuclear membrane was affected by permeabilization. The stickiness could have caused problems for other scientists who attempted to use permeable cells. This is indicated by electron micrographs of nuclei which formed a continuous sticky network (Heslop-Harrison et al., 1988). After reversal of permeabilization stickiness of nuclear material was not observed.

Control experiment included the visualization of nuclei without permeabilization and after reversal of permeabilization (Fig. 3.7A and 7B). After reversal of permeabilization biotinylated-DNA was expelled from nuclei as decondensed fibres (Fig. 3.7C, D, E). Of the two nuclei visible in Fig. 3.7F, one is replicating (lower one) and immunofluorescent amplification with avidin-TITC and biotinylated goat-antiavidin shows the spatial distribution of nascent biotin-DNA. The upper nucleus is silent from the point of view of replication. As biotin-label could not be seen in metaphase chromosomes, pulse-chase experiments were attempted to drive the biotin label into metaphase chromosomes. However, pulse chase failed to process biotin-labeled DNA from interphase to metaphase chromosomes (Fig. 3.7E); instead, the formation of big chromatin clusters and separation and expulsion of biotinylated DNA was observed by immunofluorescent amplification of nascent DNA (Bánfalvi et al., 1989). Nevertheless, biotinylation turned out to be a potential tool to observe early intermediates of chromatin condensation and initiated the analysis of further intermediates in the presence and in the following experiments in the absence of biotinylated nucleotides.

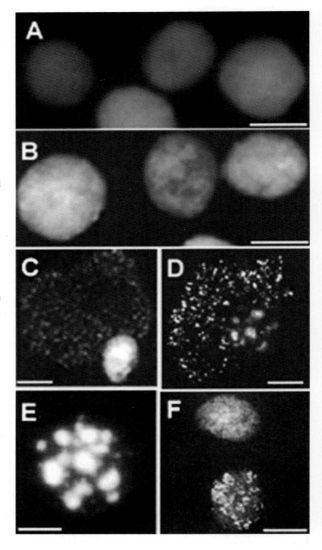

Fig. 3.7 Effect of reversal of permeabilization and biotinylation of nascent DNA on the organization of chromatin in CHO cells. (**A**) DAPI-stained nuclei of unpermeabilized CHO cells. (**B**) DAPI-stained nuclei of CHO cells after reversal of permeabilization. (**C**) CHO cells were permeabilized, labeled with biotin-11-dUTP, and permeabilization was reversed, nuclei were isolated and stained with DAPI. (**D**) and (**F**) are the same as (**C**), except that the biotin-labeled nuclei were subjected to immunofluorescence analysis. (**E**) Same as (**D**), except that DNA labeling with biotin-11-dUTP was followed by a 1 h chase in the presence of 1 mM dTTP. Bar, 5 μm. (Reproduced with permission of Gacsi et al., 2005)

Decondensed Chromatin Structures After Biotinylation

Permeabilization is blurring the two generally known chromatin structures visualized in exponentially growing cells, characteristic to interphase nuclei and metaphase chromosomes. When the isolation of nuclei from permeable cells was attempted without restoring cellular membranes by reversing permeabilization, a sticky mass of nuclear material was obtained. These experiments suggested that not only the cellular but also the nuclear membrane was affected by permeabilization. Restoration of cellular membrane function was confirmed by ^{3}H-thymidine incorporation and trypan blue exclusion (Bánfalvi et al., 1984). In those control experiments, where biotinylation was omitted, reversal of permeabilization did not lead to the accumulation of early chromatin condensation intermediates, indicating that biotin-DNA can be used as a tool for slowing down the condensation process, thereby allowing the accumulation of early intermediates of chromatin folding.

It was observed regularly that chromatin condensation in exponentially growing CHO cells starts with a polarization of the nuclear material. Figure 3.8 shows the early stage of chromatin decondensation. Although, in these experiments most of the cells were in S phase, they represented different subphases and stages of chromatin compaction. The polarization causes the extrusion of spherical chromatin (Fig. 3.8A, B, C, F, G, O). DNA is in its decondensed form and the disruption of the nuclear membrane revealed extruding chromatin as a fuzzy, veil-like structure (Fig. 3.8D, O, P, Q), leading to the formation of a nuclear plate (Fig. 3.8D, J, K, L, M, N, P). The nuclear material either maintained its round shape or unfolded into a supercoiled ribbon structure (Fig. 3.8Q). The polarization and supercoiling of chromatin strarts at a polarized end-plate region forming the head portion. Based on these regularly occuring extrusions very similar to micronuclei, one cannot escape the idea that micronucleus formation may be a regularly occuring early element of chromatin condensation.

Round shaped chromatin bodies are regarded as sperical forms of chromosomes, while supercoiled ribbon structure is representing the early elongated form of prechromosomes.

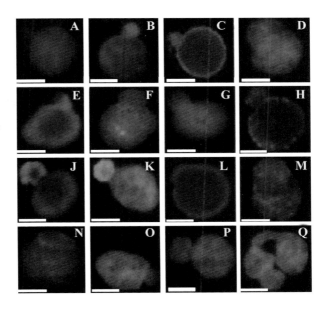

Fig. 3.8 Polarized condensation of biotinylated interphase chromatin. CHO cells were permeabilized, labeled with biotin-11-dUTP, and permeabilization was reversed by resuspension in growth medium containing 10% fetal bovine serum. Nuclei were isolated and stained with DAPI. The images were selected to illustrate polarization of the nuclear material and are at the same magnification. Bar, 5 μm. (Reproduced with permission of Gacsi et al., 2005)

Globular, Supercoiled, Fibrous, Ribboned Structures upon DNA Biotinylation

Condensing chromatin structures gradually changed their shape from globular and circular forms to elongated chromosomes as illustrated in Fig. 3.9. Spherical chromatin bodies, which represent less condensed structures (Fig. 3.9A–D), turn into more compact circular and fibrous chromatin (Fig. 3.9D–E). The formation of globular structures is probably related to the supercoiling of the chromatin veil (Fig. 3.9C). Judged by the number of supercoils, they are likely to correspond to chromosomes and to chromatin bodies (early decondensed chromosomes) in interphase.

Some regions of the nuclear material are still in a decondensed cotton-like state (Fig. 3.9D–F), such as the chromosomes seen at the upper right corner, at the bottom and in the center of Fig. 3.9F. Globular intermediates turn first into elongated ropes (Fig. 3.9D, E), then into elongated chromosomes of different length and thickness, which may represent the next stage of chromosome condensation (Fig. 3.9E–G). In addition, several transition forms were observed, in which elongated pre-chromosomes are thickened, probably by additional folding and protein binding. In several instances, chromosomes were found to be arranged in two lobes, which join at the upper stalk region (Fig. 3.9C, E, G). At the final stage of condensation chromosomes are aligned in an arc (Fig. 3.9E). In the lower portion of the nucleus shown in Fig. 3.9F the chromosomes form a condensed cluster while its upper region consists of a loose structure composed of decondensed fibrous chromatin and the elongated forms of condensing chromosomes. The side-by-side occurence of two types of chromosomal arrangement in the same nucleus suggests that these represent consecutive stages in chromosome condensation. The shortening and thickening of elongated precondensed forms (Fig. 3.9G) finally lead to metaphase chromosomes (Fig. 3.9H). Consecutive stages of chromosome condensation will be studied in the next subchapters using synchronized cells.

Summarizing biotinylation. Earlier observations (Bánfalvi et al., 1989; Bánfalvi, 1993) suggested that biotinylation of DNA by the incorporation of biotin-11-dUTP into nascent DNA interferes with chromatin condensation. Fully condensed metaphase chomosomes did not contain fluoresceine isothiocyanate (FITC) foci of staining, indicating that biotinylation prevents chromatin folding and leads to decondensed chromatin structures. The aim was to build on these observations by examining intermediates in chromatin condensation, taking advantage of the fact that permeabilization allows labeling with biotinylated nucleotides at any time during chromosome replication. Exponentially growing CHO cells were used in these experiments to assure that most of the cells are in S phase, where DNA is unfolded and chromatin structures are in decondensed state. The resulting decondensed chromatin structures could then be visualized upon reversal of permeabilization.

These results supported the earlier notion that the incorporation of biotin-labeled nucleotides into nascent DNA prevented chromatin folding and thus provided a means of visualizing intermediates in chromosome replication. Fluorescent images

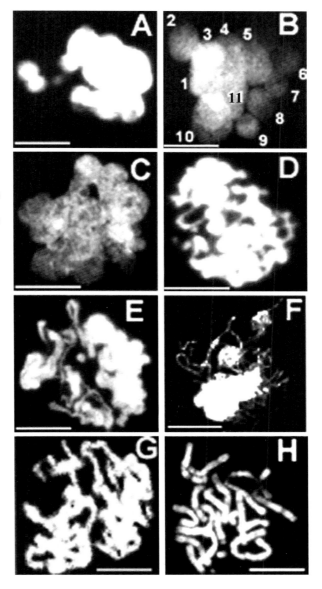

Fig. 3.9 Intermediates in the condensation of biotinylated chromatin. CHO cells were permeabilized, labeled with biotin-11-dUTP, and permeabilization was reversed by resuspension in growth medium containing 10% fetal bovine serum. Nuclei were isolated and stained with DAPI. The images were selected to illustrate frequently seen patterns, such as: (**A, B**) chromatin bodies, which have been numbered in (**B**); (**C**) supercoiled chromatin; (**D, E**) chromatin bodies turning to fibers; (**E–G**) transition to elongated chromosomes; (**G**) lobulate arrangement of elongated chromosomes; (**H**) metaphase chromosomes. Bar, 5 μm. (Reproduced with permission of Gacsi et al., 2005)

of chromatin structures confirm the existence of a flexible folding pattern including several transitional forms. Among these forms, it was possible to discern decondensed veil-like structures turning into ribboned, globular and fibrous structures. The existence of globular and fibrous structures had been shown in interphase chromatin by high resolution scanning electron microscopy (Iwano et al., 1997).

148   3 Chromatin Condensation

Chromatin Condensation of Non-biotinylated DNA Studied in Synchronized Cells

As biotinylation leads to the accumulation of early intermediates in chromatin condensation and interferes with the progression of the condensation process, we examined the intermediates of chromatin condensation in the absence of biotinylated nucleotides using cells synchronized at different stages of the cell cycle. In addition the omission of biotinylation in all further experiments contributed to a faster recovery of cells after reversal of permeabilization. Regular chromatin structures were identified as similar forms that recurred in the same synchronized population of cells. The analysis focused only on those chromatin structures observed under the fluorescent microscope, in which the chromatin network showed continuity and the structure was not deformed by physical forces. Centrifugal elutriation of an exponential CHO cell population yielded eight cell fractions, each of which was characterized by the flow rate at which cells were elutriated and by the average cellular and nuclear volume, nuclear diameter, and C-value, which define their cell cycle status (Table 3.1).

The cell cycle parameters were estimated by flow cytometry, which also provided an assessment of the degree of synchrony of cells in each fraction (Fig. 3.10 A–C). Flow cytometry was also used to assess whether cell permeabilization and its reversal led to apoptosis. This analysis showed less than 5% apoptotic cells (Ap) after permeabilization and reversal in elutriated fractions (Fig. 3.10D), which is similar to that seen in control cultures not subjected to elutriation and in cultures that were not permeabilized.

Table 3.1 Characterization of synchronized populations of CHO cells

| | | | Average | | | |
Fraction number	Flow rate of elutriation (ml/min)	Elutriated cells ($\times 10^6$)	Cell volume fl	Nuclear volume fl	Nuclear diameter μm	C-value
1	13.5	2.20	770	n.d.	n.d.	2.02
2	19	3.51	836	190	6.8	2.21
3	24	12.83	988	230	7.2	2.55
4	30	17.87	1119	250	7.6	2.76
5	36	23.57	1268	280	8.0	2.98
6	41	14.08	1498	335	8.5	3.28
7	47	9.12	2004	470	9.9	3.72
8	52	3.14	2618	530	10.3	3.99
Unfractionated cells		*9.15*	*1231*	*265*	*7.9*	

Cells subjected to elutriation 9.7×10^7 (100%). Elutriated cells: 8.6×10^7 (89%)
Loss of cells during manipulation: 1.1×10^7 (11%), n. d. = not determined.
Cellular and nuclear volumes and nuclear diameters were estimated with a Coulter Channelizer. C-values were estimated as described by Basnakian et al. (1989). (Reproduced with permission of Gacsi et al., 2005).

Fig. 3.10 Flow cytometric analysis of CHO cells synchronized by centrifugal elutriation at different stages in S phase. Elutriated fractions (E1 through E8) are described in Table 3.1. (**A**) Flow cytometric profiles of the elutriated fractions. (**B**) Nuclear size of the elutriated fractions. (**C**) Relationship between cell volume and diameter. (**D**) Elutriated fractions subjected to permeabilization, reversal and analysis by flow cytometry. Ap, apoptotic cells; PI, propidium iodide (Reproduced with permission of Gacsi et al., 2005)

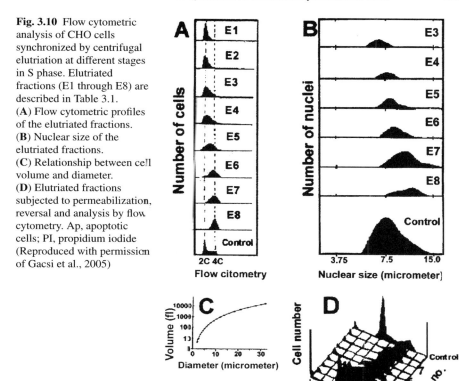

Decondensed Chromatin Structures in Cells Synchronized in Early S Phase (2.0 – 2.5 C)

The first fraction (Table 3.1), which contained primarily G$_1$ cells was not used for the study of chromatin condensation. Consecutive fractions contained unaggregated cells, which increased in size with each fraction elutriated. The cells in Elutriation Fraction 2 were in early S phase (2.21 C-value) and DNA was in a highly decondensed state. The chromatin in the fixed nuclei appeared as a fuzzy, veil-like structure (Fig. 3.11, left panels). In Elutriation Fraction 3 (2.55 C) the nuclear material either maintained its round shape with a polarized chromatin plate emerging from the nucleus (Fig. 3.11, right panels A–D), similar to biotinylated DNA seen in Fig. 3.8, or the chromatin began to unfold and turn into supercoiled structures (Fig. 3.11, right panels E–H). The ribboned chromatin was seen to be twisted in the middle stalk portion (e.g., Fig. 3.11 right panel E, G) and gradually turned into supercoiled chromatin structures (Fig. 3.11 right panel F, H,).

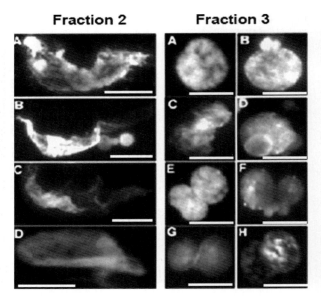

Fig. 3.11 Decondensed chromatin structures in cells synchronized at 2.0 – 2.5 C. CHO cells at different stages in S phase were obtained by centrifugal elutriation as described in Table 3.1. *Left panels*: Elutriation Fraction 2. *Right panels*: Elutriation Fraction 3. Bar, 5 μm. (Reproduced with permission of Gacsi et al., 2005)

Transition from Veiled to Ribboned Structures in Cells Synchronized in Early Mid S Phase (2.5–3.0 C)

Elutriation Fraction 4 represented cells in early mid-S phase (2.76 C). In the fixed nuclei from these cells, the veiled chromatin gradually developed into supercoiled loops (Fig. 3.12. left panels A–E), sometimes consisting of thick fibers (Fig. 3.12 left panels G–H) with a diameter of about 300 nm, which were similar to those of euchromatin fibers. Elutriation Fraction 5 (2.98 C) revealed further condensation of the fibrous structures, leading to continuous ribbon structures (Fig. 3.12, right panels A–F). Presumably due to further supercoiling, the thick chromatin fiber formed a thicker rope with a diameter, estimated at the right end of the chromatin in Fig. 3.12G (right panels) of about 700 nm. The chromatin fiber shown in Fig. 3.12H (left panels) is shown magnified for closer scrutiny in Fig. 3.13. The segment of the chromatin string shown in Fig. 3.13a includes supercoiled stretches that were more condensed than linear intervening stretches as indicated by fluorescent intensities. At the right end of the chromatin string (close to arrow 1), chromatin condensation resulted in a thicker precondensed chromosome, representing probably the most condensed part of the chromatin fiber. In Fig. 3.13b condensing supercoiled and less condensed regions appeared to alternate. In the region of Fig. 3.13c close to arrow 4, chromatin bodies were in a decondensed cotton-like state. Streches of chromatin (Fig. 2.13 a–c) show different stages of compaction, from arrow 1 with an almost completely condensed chromosome to arrow 4 with cotton-like decondensed chromatin structure.

Chromatin Condensation of Non-biotinylated DNA Studied in Synchronized Cells 151

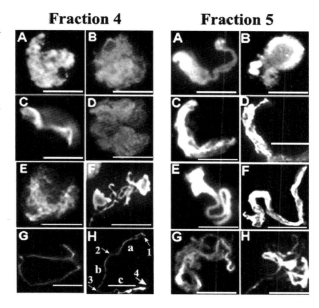

Fig. 3.12 Transition to ribboned chromatin structures in cells synchronized at 2.2–3.0C. CHO cells at different stages in S phase were obtained by centrifugal elutriation (Table 3.1). *Left panels*: Elutriation Fraction 4. Fraction 4H was subjected to further analysis shown in Fig. 3.13. *Right panels*: Elutriation Fraction 5. Bar, 5 μm. (Reproduced with permission of Gacsi et al., 2005)

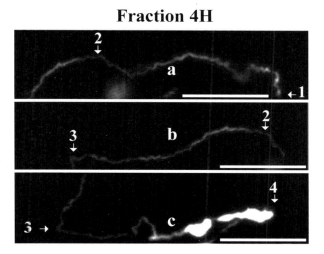

Fig. 3.13 Analysis of fibrous chromatin. The chromatin fiber from Elutriation Fraction 4, shown in Fig. 3.12H (*left panels*), at higher magnification. The fiber was broken into three segments, with the numbered arrows corresponding to those seen in Fig. 3.12H (*left panels*). Bar, 5 μm. (Reproduced with permission of Gacsi et al., 2005)

Linear Connection of Chromosomes in Condensing Interphase Chromatin in Cells Synchronized at Late Mid S Phase (3.0–3.5 C)

Further evidence for a continuous, linear chromatin structure was provided by the examination of Elutriation Fraction 6 (Fig. 3.14). Figure 3.14A and 3.14D show decondensed chromosomes that are wrapped around a long chromatin stem or are unfolding like new leaves. The ends of the chromatin structure appeared to

Fig. 3.14 Interconnected chromatin structures in cells synchronized at 3.28C. CHO cells at different stages in S phase were obtained by centrifugal elutriation (Table 3.1). The panels show examples of chromosomes seen in Elutriation Fraction 6. Panel (**C**) is identical to Panel (**B**) except that the backbone of the continuous chromosome fiber has been traced with a dotted *black line* and its ends are indicated by small *black circles*. Bar, 5 μm. (Reproduced with permission of Gacsi et al., 2005)

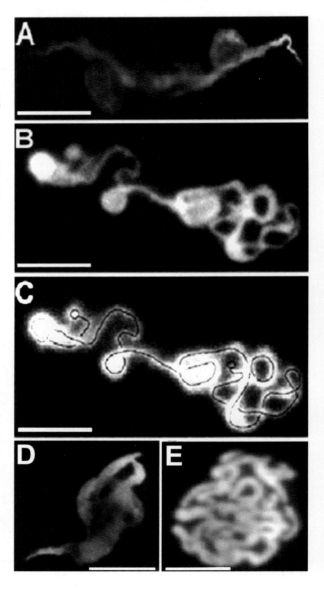

be supercoiled (Fig. 3.14A), suggesting that condensation from the thick fibers (300 nm) to chromatin ropes involves supercoiling. In Fig. 3.14B the chromatin is seen as a long fiber (300 nm) without distinguishable chromosomes, presumably corresponding to chromatin loops. The backbone of this chromatin fiber is traced with a black line and the two ends of the continuous structure, about 60 μm in

Chromatin Condensation of Non-biotinylated DNA Studied in Synchronized Cells 153

length, are indicated by two small circles (Fig. 3.14C). Continuity seems to be maintained during the formation of elongated pre-chromosomal forms visualized in Fig. 3.14D, E.

Distinct Forms of Early Chromosomes in Cells Synchronized at the End of S Phase (3.5–4.0 C)

In late S phase (Elutriation Fraction 7; average C-value of 3.72) chromosomes became visible and are arranged in arcs or circles (chromatin bodies) (Fig. 3.15, left panels). In some cases individual chromosomes were clearly distinguishable (Fig. 3.15, left panels E, G, H). In the last synchronized population (Elutriation Fraction 8, C-value of 3.98), which approached metaphase, chromosomes were compact but not yet completely condensed (Fig. 3.15, right panels), but in a few cases (Fig. 3.15, right panels E–G) began to resemble metaphase chromosomes (Fig. 3.15, right panel H). Linear connection seems to be maintained in condensing chromosomes (Fig. 3.15, right panels C–F). Their arrangement is strengthening the view that chromosomes are arranged head to tail.

Chromatin Image Analysis

The fluorescent image of a stained nucleus from Elutriation Fraction 7 (3.72 C) was subjected to further analysis (Fig. 3.16). A precalculated intensity mask was fitted to

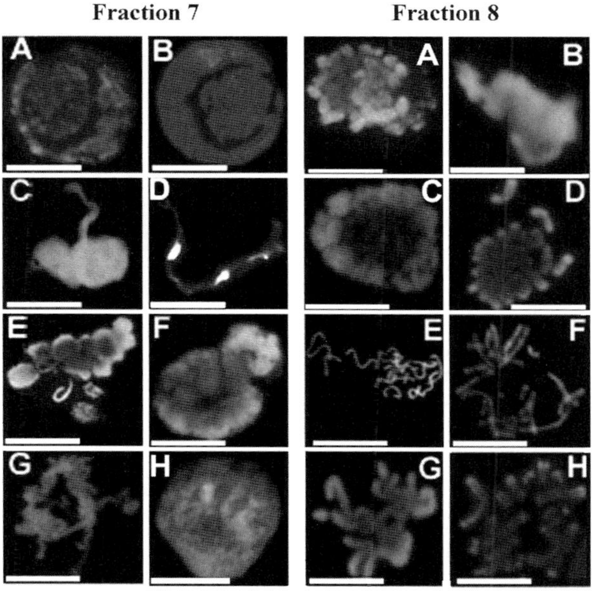

Fig. 3.15 Circular arrangement of pre-chromosomes in cells synchronized at 3.5–4.0 C. CHO cells at different stages in S phase were obtained by centrifugal elutriation as described in Table 3.1. *Left panels*: Elutriation Fraction 7. *Right panels*: Elutriation Fraction 8. Bar, 5 μm. (Reproduced with permission of Gacsi et al., 2005)

the chromatin image to yield a colour-coded intensity map, which is shown in black and white in Fig. 3.16B–F, and allowed the intensity assignments shown diagrammatically in Fig. 3.16G. From the estimated intensities of the 11 chromosomes, it was possible to tentatively correlate each of these with the 11 known chromosomes of CHO cells based on chromosome size, as indicated by the numbers in Fig. 3.16F. Although, the chromosome assignment is tentative at best, it serves to show that individual chromosomes can be visualized in late S-phase and are arranged in circular patterns, suggesting some sort of physical association of neighbouring chromosomes. The chromosomes are presented schematically in a linear array, the length of the loops corresponding to the intensity of the corresponding chromatin body in Panel F. The top row of numbers correspond to the numbers in Panel C, the bottom row to the numbers in Panel F, which order the chromosomes by size (Fig. 3.16G).

Summarizing synchronization of CHO cells. Biotinylation of DNA interfered with chromatin folding by blocking chromatin condensation at the fibrous stage. Thus it was expected that the omission of biotinylation would reveal further intermediates in the condensation process. Experiments carried out with synchronized populations of cells in the absence of biotinylation not only confirmed the existence of globular and fibrous structures but revealed further forms. These transitions involved the supercoiling of the chromatin veil, which seemed to separate decondensed chro-

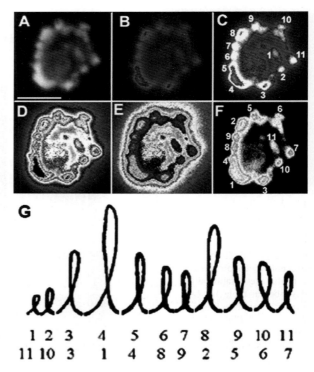

Fig. 3.16 Chromatin image analysis of precondensed chromosomes. (**A**) A nucleus from Elutriation Fraction 7 stained with DAPI. (**B–F**) Image analysis of the nucleus shown in panel A effected by fitting a precalculated intensity mask to the chromatin image. (**C**) Chromosome bodies are numbered consecutively by the white numbers on the outside of the circular arrangement and (**F**) in order of size. (**G**) Schematic view of chromosomes presented as loops. Bar, 5 μm. (Reproduced with permission of Gacsi et al., 2005)

mosomes as chromatin bodies and chromatin loops. The spherical bodies observed may correspond to the globular forms of interphase chromosomes. These gradually changed their shape to chromatin loops, then to semicircular chromatin structures resembling horseshoe-like arrays. Finally the precondensed folded-back structures of semicircles turned into elongated linear chromosome forms, which may be the immediate precursors of shorter metaphase chromosomes.

It was described earlier that there is a correlation between the number of sub-phases of DNA replication and the number of chromosomes in CHO and in *Drosophila* cells (Bánfalvi et al., 1997; Rehak et al., 2000). Although, the chromatin folding pattern is not completely understood, the linear arrangement of chromosomes suggests that chromosome replication is related to a defined temporal order of chromosome condensation.

Computer image analysis confirmed that condensed chromosomes seem to be organized in a circular arrangement, suggesting connectivity. The chemical basis of chromosome connectivity remains to be identified, whether the linkage of chromosomes plays a role in chromosome replication and whether it is related to the temporal order of chromosome replication. An interesting question concerns the role of telomers in the linkage between chromosomes and their replication.

Linear Connection of Condensing Chromosomes in Nuclei During Cell Cycle

Linear Connection of Isolated Chromosomes

Reversible permeabilization of CHO cells allowed the analysis of large-scale chromatin condensation in a cell cycle dependent manner. This subchapter confirms the existence of major intermediates of chromatin condensation and visualizes connectivity between different forms of chromosomes. Later in S phase the veil-like fibrillary chromatin is supercoiled to form chromatin bodies representing the earliest visible chromosomes Supercoiling results in chromatin fiber, which turns to rope (thick fiber) forming loops and chromatin ribbon. The elongated shape of prechromosomes indicates that they are arranged head to tail. Bent forms (loop, c and v shaped structures) of interphase chromosomes open at the end of S phase. Linear arrangement of chromosomes was observed till the end of the condensation process suggesting that connectivity of chromosomes is maintained throughout the cell cycle.

The two generally known chromatin structures which can be visualized in exponentially growing cells, namely interphase nuclei and metaphase chromosomes are not discussed here. Only large-scale chromatin changes are dealt with which take place in the interphase nucleus. Synchronized CHO cells were permeabilized reversibly and used to isolate chromatin structures. Figure 3.17 visualizes changes which reflect connectivity both in chromatin and in chromosomal structures.

Fig. 3.17 Linear connection of interphase chromosomes in condensing chromatin. Chinese hamster ovary cells were kept in suspension culture in spinner flasks. Exponentially growing cells were synchronized by elutriation and eight fractions were collected. Details of cell growth, synchronization, FACS analysis, reversal of permeabilization, isolation of nuclei, spreads of nuclear structures and visualization of chromatin structures were carried out as described (Gacsi et al., 2005). The characterization of elutriation fractions is described in Table 3.1. Bar, 5 μm. (Reproduced with permission of Bánfalvi, 2006a)

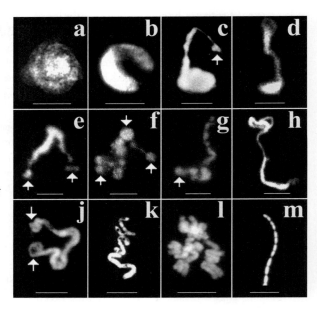

In the early S phase the fuzzy veil-like chromatin in fixed nuclei appeared as a helical structure ("cartwheel" pattern). The nuclear material maintained its round shape with a polarized end-plate emerging from the nucleus (Fig. 3.17a) which gradually came unwrapped (Fig. 3.17b). The extruded veiled chromatin gradually developed into supercoiled loops (Fig. 3.17c). Further coiling of fibrous chromatin formed continuous ribbon structure (Fig. 3.17d). Presumably supercoiling generated chromatin bodies at the ends of the fibrous chromatin (indicated by the arrows) (Fig. 3.17e–g). Introduction of more supercoils turns the whole chromatin structure into round shaped bodies (Fig. 3.17f), the supercoils correspond to chromatin bodies which are regarded as the earliest visible forms of chromosomes in the interphase nucleus. It is noted at this point that these chromatin bodies (to be distinguished from karyosomes) should be named chromatosomes (round shaped chromosomes). However, the name chromatosome is already occupied by a complex consisting of the nucleosome core particle and an additional 20 bp linker DNA associated to linker histone H1 or H5.

The next level of condensation of the fibrous chromatin bodies involves the turning of chromatin bodies around their hinge regions leading first to elongated chromatin bodies (Fig. 3.17g), then to thick chromatin fibers (Fig. 3.17h). Further evidence for a continuous, linear chromatin structure is provided by Fig. 3.17j where the thinner chromatin fiber is converted to a relatively loose but thick chromatin rope. Supercoiling tightens the rope and generates loops, which correspond to elongated prechromosomes arranged in a regular three dimensional zig-zag (Fig. 3.17k). This structural arrangement should not be confused with a lower level of folding, namely the so called 30 nm chromatin fibril, where there is a possibility that nucleosomes can be organized in a three-dimensional zig-zag, but irregularly. At the end of the folding process linearity seems to be maintained in folded prechromosomes (Fig. 3.17l). Linear connection is clearly demonstrated in metaphase chromosomes (Fig. 3.17m).

Linear Connection of Chromosomes Inside the Nucleus

Once we have seen the isolated chromatin structures outside the nucleus it is easier to distinguish among early chromosomes and recognize the linear arrangement of early chromosomes inside the nucleus (Fig. 3.18a). Spatial considerations dictate that chromosomes are folded into loops and other bent forms and arranged in such a way that their bulky parts point away from each other. Consequently, most of the neighbouring chromosomes are *trans* oriented relative to each other resulting in a zig-zag ladder (Fig. 3.18b).

Considering the round shape and the intense supercoiling taking place inside the nucleus, the string of chromosomes has to be of great flexibility. To alleviate the molecular strain generated by the coiling process chromosomes have to maintain their flexibility up to the highest level of compaction. Possible solutions to relax intramolecular strain can be (a) changing from *trans* to *cis* position of certain chromosomal loops in the zig-zag arrangement, (b) the presence of small chromosomes (rodlets) at sharp turning points, (c) hinge regions of chromosomes providing flexibility of condensing chromosomes. Microscopic images of individual chromosomes suggest all three solutions in interphase and show the existence of chromosome rodlets and hinge regions in nuclei of *Drosophila* cells (Bánfalvi, 2006b).

The finding that interphase chromosomes are clustered in domains, and that condensed interphase chromosomes are linearly arranged (Bánfalvi, 2006b) raise ques-

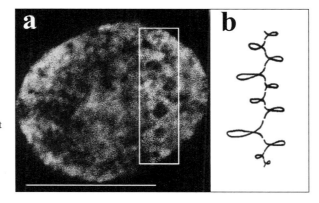

Fig. 3.18 Linear connection of chromosomes inside the nucleus of a CHO cell. The zig-zag arrangement of chromatin bodies in the nucleus of a cell from fraction 7 is shown inside the framed box (**a**). Schematical model of zig-zag arrangement of chromosomes, each loop representing a chromosome (**b**). Bar, 5 μm. (Reproduced with permission of Bánfalvi, 2006a)

158 3 Chromatin Condensation

tions about the nature of chromosome connectivity, mechanisms of its regulation, the involvement of telomeres in this phenomenon, the linear order of chromosomes and the resulting temporal order of gene replication. This notion has been confirmed by other studies in the female *Drosophila* S2 cell line, in which the chromosome order of 1-1-3-3-2-2-X-X has been determined (Bánfalvi, 2006c) (to be discussed later).

In conclusion, both *in vitro* (by isolation) and *in vivo* (in cells) it was possible to discern decondensed continuous veil-like structures turning to ribboned, globular and fibrous structures. The existence of globular and fibrous structures (Iwano et al., 1997) and the transition of loops onto fibrous chromatin has been suggested by others (Sarkar et al., 2002). Structural analysis not only confirms the existence of these forms but expresses their temporal order of condensation as C-values. The connectivity of chromosomes raises several questions related to the role of telomers and proteins in chromosome linkage, the spatial arrangement of chromosomes, the linear array of chromosomes in order of size etc. Unfortunately chromosome specific probes are not available for Chinese hamster ovary cells to show the spatial order of chromosomes. We have seen connectivity of chromosomes in other cells (human erytroleukemia, K-562, *Drosphila*, Indian muntjac) and came to the conclusion that large-scale chromatin condensation follows a common patway (Bánfalvi et al., 2006). As far as the conceptual classes of models for chromosome architecture are concerned recent structural analysis favours a hierarchical arrangement not necessarily identical with the one called hierachical folding. Further, fine structural analysis is necessary to come to a conclusion regarding the levels and intermediates of chromosome condensation.

Common Pathway of Chromatin Condensation in Mammalian Cell Lines

Although, the chromatin folding pattern is not quite understood, there is a growing evidence for the linear arrangement of chromosomes suggesting a defined temporal order of chromosome replication. That chromosomes are connected and arranged in a circular fashion has been demonstrated by computer image analysis (Fig. 3.16) in CHO cells. Moreover, the discontinuous nuclear growth revealed as stop sites coinciding with the path of chromosome replication referred to as subphases of DNA replication also pointed to a temporally and spatially ordered process in cadmium treated CHO cells (Bánfalvi et al., 2005). These observations raised the question whether the chromatin condensation is similar in mammalian cells or different cell types have their own characteristics.

To decide whether chromatin condensation patterns of mammalian cells are similar or different, the chromatin structures of four cell lines, namely Indian muntjac, CHO, murine preB and human erytroleukemia (K562) were compared in early S-phase between 2.0 and 3.0 C-values (Fig. 3.19) and in late S phase lasting from 3.0 to 4.0 C values (Fig. 3.20).

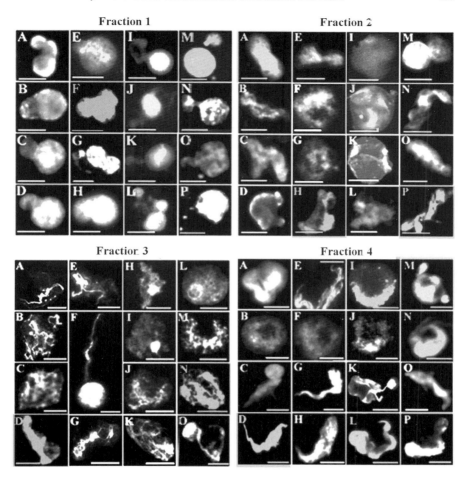

Fig. 3.19 Intermediates of chromatin condensation in early S phase. Indian muntjac (IM), Chinese hamster ovary (CHO), murine preB (preB) and human erythroleukemia (K562) cells at different stages of the cell cycle were obtained by centrifugal elutriation (8 fractions). Cells were permeabilized and permeabilization was reversed by resuspension in growth medium containing 10% fetal bovine serum. After 3 h at 37°C, nuclei were isolated, stained with DAPI and visualized by flurescent microscopy. Elutriation Fraction 1: Decondensed chromatin in IM, CHO, preB and K562 cells synchronized at 2.0–2.25 C-value. Typical images were selected to illustrate the polarization of the nuclear material. Polarized condensation of interphase chromatin in nuclei of IM (**A–D**), CHO (**E–H**), preB (**I–L**) and K562 (**M–P**) cells. Elutriation Fraction 2: Veil-like decondensed chromatin structures synchronized at 2.25–2.5 C. Elongated, twisted, veil-like chromatin in IM (**A–D**), CHO (**E–H**), preB (**I–L**) and K562 (**M–P**) cells. Elutriation Fraction 3: Fibrous chromatin at 2.5–2.75 C. Veiled chromatin turning to fibrous structures in nuclei of IM (**A–D**), CHO (**E–G**), preB (**H–K**) and K562 (**L–O**) cells. Elutriation Fraction 4: Transition to ribboned chromatin at 2.75–3.0 C. Ribboned chromatin in IM (**A–D**), CHO (**E–H**), preB (**I–L**) and in K562 cells (**M–P**). Bar, 5 μm each. (Reproduced with permission of Bánfalvi et al. 2006)

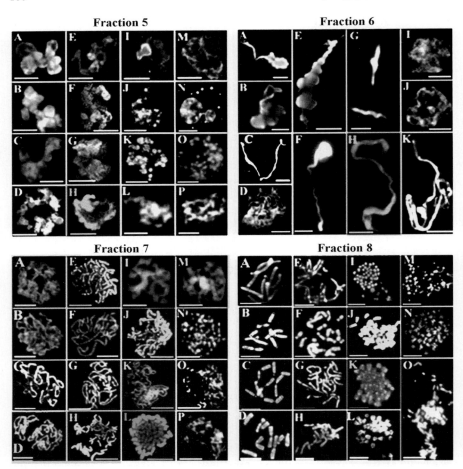

Fig. 3.20 Intermediates of chromosome condensation in late S phase. Elutriation Fraction 5: Formation of chromatin bodies at 3.0–3.25 C. Chromatin bodies in decondensed cotton-like stage in nuclei of IM (**A–D**), CHO (**E–H**), preB (**I–L**), K562 cells (**M–P**). Elutriation Fraction 6: Continuous linear connection of chromatin at 3.25–3.5 C. Linear chromatin structure in IM (**A–D**), CHO (**E–F**), preB (**G–H**) and in K562 (**I–K**) cells. Elutriation Fraction 7: Elongated and clustered forms of early chromosomes at 3.5–3.75 C. Elongated chromosomes in IM (**A–D**), CHO (**E–H**), preB (**J–L**), and K562 (**M–P**) cells. Clustered chromosome domains in IM (**A, D**), CHO (**H**), preB (**I**) and in K562 (**M, O, P**) cells. Elutriation Fraction 8: Final stage of chromosome condensation at 3.75–4.0 C. Chromosomes in IM (**A–D**), CHO (**E–H**), preB (**I–L**) and K562 (**M–O**) cells. Elongated chromosomes: **A, E, G**. Bent chromosomes: **F, G, K, L, M**. Linear attachment still visible in C. Bar, 5 μm each. (Reproduced with permission of Bánfalvi et al., 2006)

Common Pathway of Chromatin Condensation in Mammalian Cell Lines 161

Decondensed Chromatin Structures at the Unset of S Phase (2.0–2.25 C Value)

Exponentially growing cells were synchronized by centrifugal elutriation. Eight Elutriation Fractions were collected. Each fraction was characterized by flow cytometry based on its C-vaue (haploid genom content). In Figs 3.19 and 3.20 vertical columns consist of two to four fluorescent images, representing intermediates of the four different cell types including Indian muntjac (IM), Chinese hamster ovary (CHO), murine preB (preB), and human erytroleukemia (K562) cells.

Elutriation Fraction 1 contained G1 cells and cells belonging to the initial stage of S phase. The polarization of the nuclear material caused the extrusion of spherical chromatin. In Elutriation Fraction 1 the extruding chromatin is in its most decondensed state appearing as a fuzzy, veil-like structure in all four cell types (Fig. 3.19, Fraction 1).

Supercoiled Chromatin in Early S (2.25–2.5 C)

Elutriation Fraction 2 represented cells in early S phase. The nuclear material began to unfold and turn to supercoiled structures (Fig. 3.19, Fraction 2) twisted in the middle stalk portion. As a result, supercoiling gradually changes the shape of the decondensed nuclear material to elongated chromatin forms.

From Veiled to Fibrous Structures Later in Early S Phase (2.5–2.75 C)

In Elutriation Fraction 3 the veiled structure gradually turned to supercoiled loops consisting of chromatin fibers (Fig. 3.19, Fraction 3). The diameter of the thin fibers is at least by an order of magnitude thinner than that of the thicker fibers and may correspond to the known 30 nm chromatin fibril. The diameter of the thicker fibers is about 300 nm, similar to the euchromatin fibers (Fraction 3B, E, G, J, L, N).

Transition to Ribboned Chromatin Structures in Early Mid S Phase (2.75–3.0 C)

In Elutriation Fraction 4 the condensation of the fibrous structure by supercoiling is leading to a continuous, bent chromatin ribbon. These structures show a continuous linear arrangement of chromosomes within the ribboned chromatin (Fig. 3.19, Fraction 4). Chromosomes are not yet visible in these structures.

Summarizing the results of Fig. 3.19 one can conclude that the following chromatin forms were found between G1 and mid S phase: veil-like, supercoiled, fibrous, ribboned structures. These structures are similar in all four cell types.

Chromatin Bodies After the Mid S Phase Pause (3.0–3.25 C)

In Elutriation Fraction 5 the appearance of chromatin bodies represents the earliest forms of interphase chromosomes formed from the loops of the chromatin ribbon by supercoiling. Their arrangement provides further evidence for a continuous, linear chromatin structure. Decondensed chromosomes are seen either as chromatin loops or in their more compact forms as round-shaped chromatin bodies (Fig. 3.20, Fraction 5).

Early Chromosomes Later in Mid S Phase (3.25–3.5 C)

Elutriation Fraction 6 represented cells in late mid S phase. Chromatin structures isolated from nuclei of this fraction formed thick fibers presumably due to further supercoiling and linearization of chromatin loops and chromatin bodies. The diameter of fibers seems to be related to the degree of supercoiling and is probably a reflexion of a gradual shortening of fibers turning to chromatides (Fig. 3.20, Fraction 6). Individual chromosomes are often seen as supercoiled loops, chromatin bodies, thicker and thinner portions in the continuous chromatin structure. Thick segments in the linear arrangement indicate chromosomes, thinner portions are interchromosomal chromatin connections.

Early Elongated Forms of Chromosomes (3.5–3.75 C)

In late S phase chromosomes appear in their elongated forms (Fig. 3.20, Elutriation Fraction 7). Clusters of chromosomes were observed in Indian muntjac (Fraction 7 A), murine preB (Fraction 7I), human K562 erythroleukemia (Fraction 7M, O, P). Similar clusters of chromosomes were observed in resting rat hepatocytes (Trencsenyi et al., 2007a).

Final Stage of Chromosome Condensation (3.75–4.0 C)

Elutriation Fraction 8 contained synchronized cells belonging to the end of S phase, to G2 and M phase (Fig. 3.20, Fraction 8). This fraction consists of further subphases representing chromosomes of different compactness. The most frequently observed chromosomes were bent condensing forms, elongated forms (Fraction 8A, E), arcs or circles (Fraction. 8 I, K, L, M), not yet completely condensed chromosomes, approaching and occasionally reaching metaphase. Linear connection is maintained not only in the precondensed chromosomal forms (Fraction. 8E) but also clearly visible in metaphase chromosomes (Fraction. 8 C, H, L, O).

To summarize the pattern of chromatin condensation, the intermediates were compared in a cell cycle dependent manner in Indian muntjac, Chinese hamster ovary, murine preB and K562 human erythroleukemia cells. Synchronized cells were reversibly permeabilized and used to isolate interphase chromatin structures. The temporal order of chromatin condensation was indicated by the C-values of intermediates. Fluorescent microscopy showed that chromosome condensation follows a general pathway: a. Supercoiling of the veiled chromatin resulted in twisted fibrous nuclear material. turning to ribboned chromatin. b. The first visible forms of precondensed chromosomes appeared only after the mid S phase pause as chromatin loops or spherical bodies named chromatin bodies. Spherical chromatin bodies are regarded as early globular forms of condensing chromosomes in the interphase nucleus. The existence of globular and fibrous structures in interphase chromatin has been described in interphase chromatin by scanning electron microscopy (Iwano et al., 1997) and the transition of chromatin loops onto fibrous interphase chromatin has also been suggested by others (Sarkar et al., 2002). The existence of these intermediates is confirmed and their temporal order of condensation expressed in C-values. Chromatin bodies change their shape to semicircular structures resembling horseshoe-like arrays. These folded back structures gradually turn to linear extended forms of chromosomes, which are the immediate predecessors of the shorter metaphase chromosomes.

The higher order arrangement of the eukaryotic nuclei includes discrete substructures. One could ask whether chromatin bodies are related to other round shaped bodies such as coiled bodies or nuclear bodies.

Coiled bodies are a class of conserved nuclear bodies in animal and plant cells. Fractionation studies demonstrated that coiled bodies are a part of the underlying nuclear matrix. Coiled bodies contain splicing ribonucleoproteins (RNPs), the phosphoprotein p80 coilin and nucleolar antigens such are fibrillarin and snoRNPs (Sleeman and Lamond, 1999) Since coiled bodies vary in size, number, nuclear localization (associated with the nucleolus and free in the nucleoplasm) and composition, they do not seem to be related to chromatin bodies which are larger condensed chromatin structures.

Higher-eukaryotic nuclei contain numerous morphologically distinct substructures that are collectively called nuclear bodies, which have been implicated in a number of biological processes such as transcription and splicing. However, for most nuclear bodies, ranging in size from 0.1 to several micrometers, the details of involvement in these processes in relation to their three-dimensional distributions in the nucleus are still unclear (Wang et al., 2002). Each type of nuclear bodies can be identified immunologically with antibodies against their specific "marker proteins" or other factors such as the small nuclear (sn)RNAs (Lamond and Earnshaw, 1998; Misteli and Spector, 1999; Matera, 1999; Misteli, 2001). Among the known nuclear bodies are the Cajal body or coiled body (Matera, 1999 and references therein), the promyelocytic leukemia (PML) body or POD (Salomoni and Pandolfi, 2002 and references therein), splicing-related bodies including the SC35 speckles or interchromatin granule cluster (Misteli and Spector, 1999 and references therein), the GEM body (Liu and Dreyfuss, 2000), the matrix-associated deacetylase body (Downess et al., 2000), HAP body (Chiodi, 2000), and nucleoli-associated paraspeckles (Fox et al., 2002). The relationship of chromatin bodies to these nuclear bodies awaits examination.

Chromatin Condensation in Resting Cells and Chemically Induced Tumours of Rat Hepatocytes

In this subsection we turn from *in vitro* cell cultures to *in vivo* aspects of chromatin condensation to pinpoint the question whether the pattern of chromatin condensation changes in resting hepatocytes, in hepatocellular tumour, and in hepatocellular (HeDe) cell culture. To answer the question, we have isolated chromatin structures from rat liver nuclei of cells isolated from the normal quiescent organ, from hepatocellular tumour and from hepatocellular cell culture. Chemically induced hepatocellular tumour cells were obtained from rats, previously implanted with solid tumour pieces under the renal capsule (Paragh et al., 2005). From this solid hepatocarcinoma tumour named HeDe, we have started a primary culture which was further grown for more than 20 cell cycles resulting in the establisment of the hepatocarcinoma HeDe cell line (HeDe). Liver was chosen since despite of its large metabolic load, the liver is a quiescent organ in terms of cell proliferation with only 0.01–0.001% of hepatocytes undergoing mitosis at any given time (Starzl et al., 1977; Michalopoulos and De Francis, 1977; Fausto and Webber, 1994; Diehl and Rai, 1996). Consequently chromatin structures isolated from resting cells are representing the Go phase of the cell cycle.

Three questions related to chromatin condensation were addressed: 1. Is there any difference between the chromatin structure of fast growing tumour cells and slow growing resting cells? 2. Is the chromatin structure of tumour cells isolated from hepatocellular HeDe tumour different from that isolated from the HeDe cell line? 3. Are fast growing tumour cells prone to apoptosis and if this were the case could the apoptotic process be detected at the structural level of chromatin? It was found that most of the nuclear structures isolated from reversibly permeabilized resting liver cells maintained their round shape, containing highly decondensed chromatin referred to as chromatin veil, typical to G_1/Go cells. Nuclei isolated either from hepatocellular carcinoma HeDe cells or HeDe cell cultures revealed a variety of chromatin structures of high degree of supercoiling and occasionally forming apoptotic bodies.

Transmission of Hepatocellular Tumour with the HeDe Cell Line

With the HeDe cell line (10^6 cells/rat) injected under the liver or under the renal capsule the same tumour formation was observed as with the transplanted tumour cells (Trencsenyi et al., 2007a). The HeDe tumour cell line caused tumours in Fisher 344 rats, in hybrids of Fisher 344 and Long-Evans rats, but not in Long-Evans rats. These strains of rats are closely related. Strains of rats for research are derived almoust exclusively from the Norway rat (*Rattus norvegicus*). Almoust all the modern strains of rats in use today originated primarily from the Wistar Institute in Philadelphia. Long-Evans rats are believed to have originated from Wistar females crossed to a wild Norway rat.

Chromatin Condensation in Resting Cells and Chemically Induced Tumours 165

To estimate the malignancy of the tumour cell line the number of administered HeDe cells was gradually decreased. Malignancy was confirmed by placing gelatin discs conatining 10^6, 10^5, 10^4 and as few as 10^3 hepatocellular carcinoma cells (HeDe) under the liver capsule of Fisher 344 rats causing death in 2, 3, 4 and 6 weeks, respectively.

Decondensed Chromatin Structures in Resting Hepatocytes

The analysis of chromatin structures by fluorescent microscopy revealed that nuclei of resting cells being in Go phase tended to maintain their round shape even if the nuclear memrane was disrupted. The chromatin in resting hepatocytes was in a highly decondensed state as shown in Fig. 3.21. The chromatin in fixed nuclei appeared as a fuzzy, veil like structure (Fig. 3.21 panels a–d). The chromatin veil turned to loose chromatin ribbons (Fig. 3.21, panels e–h). The nuclear material either maintained its round shape (Fig. 3.21, panels a–h) or the chromatin began to unfold to elongated supercoiled structures (Fig. 3.21, panels k, l, m, p, q). Typical unfolded structures contained six chromatin clusters (Fig. 3.21 n. o).

The chromatin loops consisting of fibers (Fig. 3.21 q) are shown magnified for closer scrutiny in Fig. 3.22, where the chromatin is seen as a continuous supercoiled structure. The chromosomes are not distinguishable, yet. The backbone of the chromatin is traced with a white line showing the two ends of the continuous interphase chromatin structure (Fig. 3.22 b).

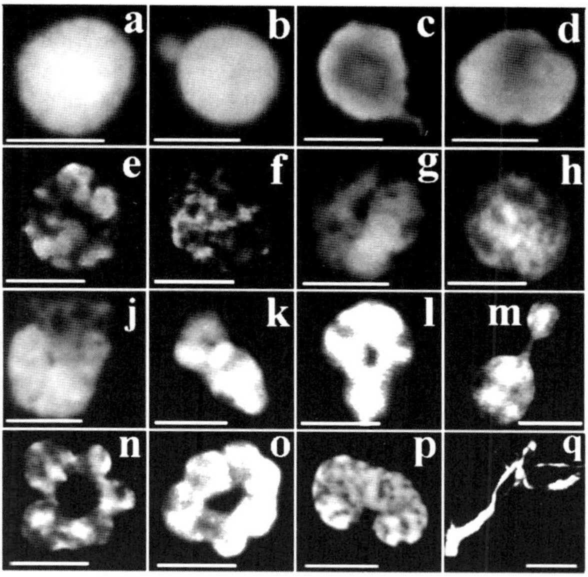

Fig. 3.21 Decondensed chromatin structures in nuclei of resting hepatocytes. Hepatocytes from rats were obtained after the perfusion with collagenase solution through the portal vein. Isolation of hepatocytes and nuclei were carried out as described (Reproduced with permission of Trencsenyi et al., 2007a, cover page illustration). Bar, 5 μm each

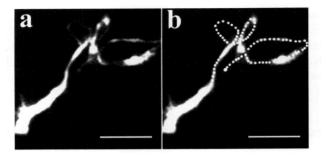

Fig. 3.22 Analysis of chromatin coil of continuous chromatin fiber isolated from resting hepatocytes. The chromatin bundle shown in Fig. 3.21 q at higher magnification. The backbone of the continuous chromatin structure has been traced with a dotted white line. Bar, 5 μm. (Reproduced with permission of Trencsenyi et al., 2007a)

Transition from Round Chromatin Bodies to Linear Chromosomes in Murine preB and CHO Cells

In our initial experiments the escape of chromosomes from reversibly permeabilized nuclei was regarded as artifacts. By using purified chromatin preparations, high resolution of fluorescent microscopy, and computer image analysis structures excluded from the nucleus of synchronized cells turned out to be real intermediates of chromosome condensation. Chromatin structures and individual interphase chromosomes escaping nuclei of reversibly permeabilized murine preB cells were analysed in a cell cycle dependent manner. Murine pre-B cells were preferred due to the fact that murine chromosome specific probes are vailable, which will be necessary later on for the determination of the linear order of chromosome arrangement. Cells were synchronized by counterflow centrifugal elutriation. Individual interphase chromosomes became visible as distinct fibrous chromatin bodies from middle S phase, turning to elongated chromosomes by the end of S phase. Major interphase chromosomal forms include: (a) mid-S-phase chromatin bodies at 3.0 C-value, (b) elongated chromatin bodies later in mid-S-phase (3.25 C-value), (c) chromatin bodies with head and leg portions later in S-phase (3.5 C), (d) supercoiled ribbons later in S-phase seen as twisted prechromosomes (3.7 C). (e) end-S-phase elongated, bent prechromosomal structures (3.9 C). The first karyotype analysis of the earliest forms of chromatin bodies referred to as chromatin bodies was performed.

Visible forms of interphase chromosomes appeared from the mid S-phase (Bánfalvi et al., 2006). Those interphase chromosomes could be sufficiently resolved which escaped the incompletely sealed nucleus after reversible permeabilization of cells. The aim of this subchapter is by opening the nucleus during the interphase: 1. to isolate intermediates of the chromatin condensation process, 2. to visualize individual chromosome structures excluded from the nucleus 3. to analyse these early chromosomal forms in a cell cycle-dependent manner. Early globular interphase chromosomes and elongated prechromosomes were used for compiling karyograms.

Table 3.2 Characterization of synchronized populations of murine preB cells

Fraction number	Flow rate of elutriation ml/min	Elutriated cells (×10⁶)	Cell diameter μm	Cell volume fl	Average C-value
1	n.d.	n.d.	n.d.	n.d	-
2	17.5	1.36	9.4	395	-
3	21.5	2.96	9.7	420	2.03
4	25.5	2.72	9.9	445	2.19
5	29.5	2.90	10.2	470	2.34
6	34.5	4.80	10.4	500	3.57
7	38	2.96	10.6	530	3.04
8	42	2.30	10.8	565	3.26
9	46	2.22	11.1	600	3.51
10	50	1.72	11.3	635	3.66
11	54	0.62	11.5	680	3.86

Cells subjected to elutriation: 2.8×10^7 (100%)
Number of elutriated cells 2.56×10^7 (91.4%)
Loss of cells during manipulation: 0.24×10^7 (8.6%)
n.d., not determined.

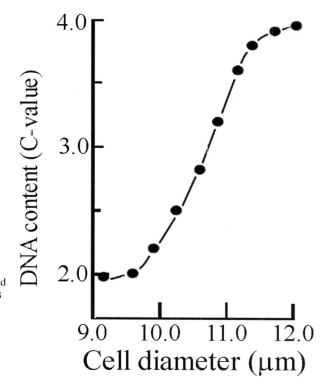

Fig. 3.23 Correlation between cellular diameter and DNA content in murine preB cells. Cell diameters in μm and DNA content expressed as C-values of individual elutriated fractions were determined and related to each other

Cellular volumes and diameters were estimated with the Coulter multisizer II. C-values were estimated as described by Basnakian et al. (1989).

Synchronization of murine preB cells was performed by centrifugal elutriation and eleven cell fractions were collected, which were characterized based on flow rate, cell growth, cell diameter and volume and average C-value (Table 3.2). The correlation between DNA content (C-value) and cell diameter in murine preB cells is shown in Fig. 3.23.

Cytometric Analysis of Synchronized preB Cell Populations

After permeabilization of murine preB cells without restoring cellular membranes, the sticky mass of the nuclear material prevented the isolation of chromatin structures. Kinetic experiments related to cell regeneration after permeabilization revealed that restoration of cellular membrane function takes place within a few hours (Bánfalvi et al., 1984).

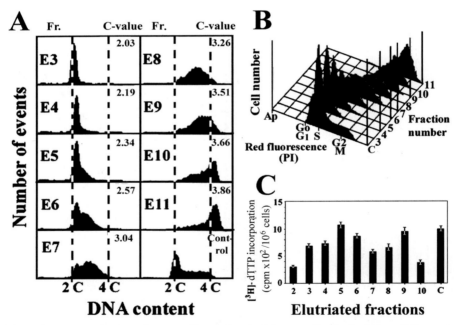

Fig. 3.24 Characterization of preB cells synchronized by centrifugal elutriation. Cells were grown, permeabilized, and after reversal of permeabilization elutriated as described by Trencsenyi et al. (2007b). (**A**) Cell cycle pattern of elutriated fractions (E3–E11). (**B**) After sealing cells were subjected to elutriation and analysed by FACS using propidium iodide staining. Ap, apoptotic cells; Go/G1 phase; S, replicating cells; G2/M, phase; C, control, unelutriated cells. (**C**) DNA synthesis was assayed by ^{3}[H]-thymidine incorporation using elutriation fractions of reversibly permeabilized cells. Columns represent the averages of three measurements; bars, standard deviation. (Reproduced with permission of Trencsenyi et al., 2007b)

Exponentially growing murine preB cells were subjected to elutriation and synchronized cellular fractions used for FACS analysis. The profile of the DNA content showed a gradual increase between 2C and 4C (Fig. 3.24A). Another cell population after reversal of permeabilization was subjected to elutriation and flow cytometric analysis. This experiment confirmed that cells reversed from permeabilization did not undergo apoptosis. Figure 3.24B shows that small apoptotic cells (Ap) are missing. In another experiment synchronized fractions of cells after reversal of permeabilization were used for DNA synthesis. Following the permeabilization-resealing protocol, synchronized cells were tested for their DNA synthesizing capability. The incorporation of ^{3}[H]-thymidine shows a biphasic replication profile typical for healthy non-permeable S phase cells (Fig. 3.24C). In the control experiment ^{3}[H]-thymidine incorporation was not detectable in permeable cells. To the contrary after a 3 h period of regeneration more than 70% of sealed cells restored their ability to incorporate ^{3}[H]-thymidine and excluded trypan blue (results not shown).

Chromatin Structures Excluded from Nuclei of preB Cells

Nuclei of sealed cells appeared to have normal physiology and are similar, though still distinguishable and differing sligtly form those of non-permeabilized cells (Fig. 3.25a), their shape being oval rather than round and can be at least partially

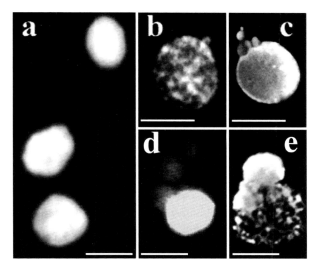

Fig. 3.25 Escape of interphase chromosomes from nuclei of preB cells. (a) Nuclei were isolated after reversal of permeabilization from exponentially growing preB cells and spread over glass slides from a hight of approximately 5 cm and visualized by fluorescent microscopy. (b–e) Extrusion of interphase chromatin structures from nuclei. The same population of nuclei as in panel (a) was spread over glass slides dropwise from a hight of approximately 30 cm. Chromatin structures were visualized by fluorescent microscopy. Bar, 5 μm each. (Reproduced with permission of Trencsenyi et al., 2007b)

opened when dropped to glass slides. Reversibly permeabilized cells provided a means of visualizing intermediates of chromatin condensation. Due to the incomplete regeneration intermediates of chromosome condensation were budding out from the nucleus (Fig. 3.25 b, c, d, e). Interphase chromosomes apparently slippped out of the nucleus (Fig. 3.25 b, c, d). Since reversible permeabilization allowed the nucleus to be opened any time during the cell cycle, fluorescence microscopy revealed not only chromatin structures such as fibrous chromatin (Fig. 3.25 d) but also more condensed chromatin structures (Fig. 3.25 e).

Condensation of Round Shaped Interphase Chromosomes in preB Cells

Major intermediates of interphase chromosomes are summarized in Fig. 3.26 showing the temporal development from mid S phase till the end of S phase. The earliest visible chromosal forms are nearly round or slightly oval shaped chromatin bodies. Individual globular forms of chromosomes appear in mid S phase as chromatin bodies visualized in Fig. 3.26 a–c (Elutriation fraction 7 corresponding to 3.04 C-value). In some of these early structures one can already see early head and leg formation (Fig. 3.26 d–f) (Elutriation fraction 8, 3.26 C). The appearance of head and leg portions is more evident in the next fraction (Elutriation fraction 9, at 3.51 C value) (Fig. 3.26 g–j). The fibrous chromatin gradually turns to ribboned chromatin which is twisted at the neck portion forming the head and the two legs (Elutriation fraction 10, at 3.66 C) (Fig. 3.26 k–m). The chromatin ribbon gradually opens at the leg portions, with the head portion still twisted (Fig. 3.26 n–o) (Elutriation fraction 10, at 3.86 C-value). Finally, the chromatin ribbon seems to be decoiled at its head portion and is likely to turn to a linear, but still bent chromosome with several other turning points which may serve as hinge regions (Fig. 3.26 p) providing higher rotational freedom for the coiling.

Karyotype of Early Interphase Chromosomes

A karyotype is an organisms picture of all chromosomes. In a karyotype, condensed chromosomes are arranged and numbered by size, from the largest to the smallest. This profile helps to identify quickly chromosomal alterations that may result in a genetic disorder. To make a map from the picture of chromosomes individual chromosomes are cut out and matched up using size, banding pattern and centromere position as guides. The karyogram is a graphic representation of a karyotype. Karyogram and karyotype are often used as synonyms.

Karyotype analysis was carried out using early interphase chromosomes. Nuclei of preB cells were dropped from about 30 cm to slides which allowed to open nuclei and increased the frequency of single chromosomes found on the slides. Mid- and late-S-phase chromosomes were used to carry out karyotype analysis. Although,

Transition from Round Chromatin Bodies to Linear Chromosomes

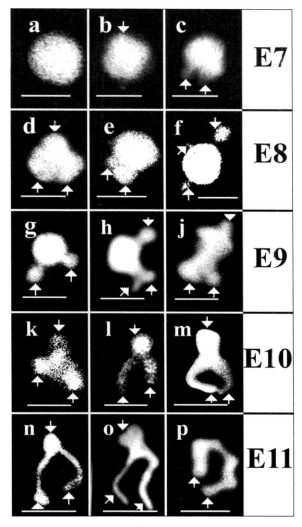

Fig. 3.26 Interphase chromosome structures in late-S-phase. Eleven synchronized fractions were obtained by centrifugal elutriation of preB cells (Table 3.2). Fractions collected during loading and cells containing continuous chromatin structures, but no visible forms of chromosomes (fractions 1–6) were not used for the analysis of interphase chromosome structures. Individual chromosomal structures were visualized by fluorescent microscopy. (**a–c**) Elutriation Fraction 7: fibrous chromatin bodies in mid S-phase (3.04 C-value). (**d–f**) Elutriation Fraction 8: round shaped chromatin bodies with emerging leg and/or head portions later in mid-S phase (3.26 C). (**g–j**) Elutriation Fraction 9: elongated fibrous chromatin bodies with distinct leg and head portions later in S phase (3.51 C). (**k–m**) Elutriation Fraction 10: chromatin bodies unfolded at leg portions forming chromatin ribbons in late-S phase (3.66 C). (**n–p**) Elutriation Fraction 11: Chromatin ribbons turned around their head portions (**n, o**) or opened to form a bent elongated prechromosome (**p**) at the end of S phase (3.86 C-value). Arrows pointing upward indicate protruding leg portions, arrows directed downward show the "heads" of interphase chromosomes. Bar, 1 μm each. (Reproduced with permission of Trencsenyi et al., 2007b)

karyotyping is traditionally performed by banding analysis, such pattern is not seen in early forms of chromosomes. Thus, DAPI staining due to its higher resolution was used to differentiate among chromosomes with different diameter and fluorescent light intensity. Mid-S-phase round shaped chromatin bodies have different spatial distribution than reduplicated, elongated and bent chromosomal forms. Newertheless, the diameter and fluorescence of chromatin bodies are reasonable selection criteria to distinguish among them. Figure 3.27 shows the karyogram of mid-S-phase chromatin bodies based on the size and fluorescent light intensity of early chromosomes.

That the karyotype analysis of mid S phase chromatin bodies corresponds to the chromosome number of murine cells was confirmed by another analyis carried out by using late S phase murine chromosomes (Fig. 3.28). The comparison showed that the chromosome numbers are matching and correspond to the real murine chromosome number.

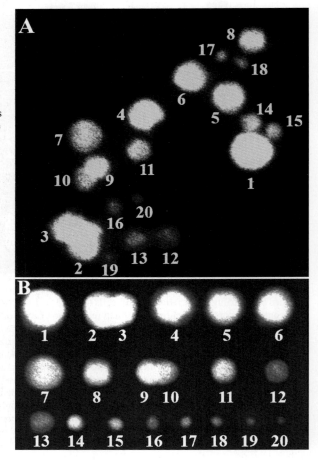

Fig. 3.27 Fluorescent karyotyping of interphase chromatin bodies. Early round shaped chromosomes (chromatin bodies) isolated from nuclei belonging to cells of Elutriation Fraction 7 were stained with DAPI and visualized under fluorescent microscope. Chromatin bodies are arranged and numbered according to decreasing size and fluorescent light intensity after DAPI staining. (**A**) Murine preB chromosomes appear as twenty chromatin bodies. (**B**) Chromatin bodies were selected and lined up according to size and fluorescence intensity from 1 to 20. At this early stage of karyotyping the pair of sex chromosomes (X and Y) could not be distinguished. Corresponding to decreasing length chromosome X would be the 3rd and chromosome Y the 19th chromosome (Dev et al., 1971; Schnedl, 1971). (Reproduced with permission of Trencsenyi et al., 2007b)

Fig. 3.28 Fluorescent karyotype of murine preB chromosomes stained with DAPI. Chromosomes isolated from cells of Elutriation fraction 11 were stained with DAPI and visualized. The twenty murine prechromosomes were selected and lined up based on size and fluorescent light intensity. The karyotype analysis is incomplete in the sense that X and Y chromosomes could not be distinguished at this stage of compaction. (Reproduced with permission of Trencsenyi et al., 2007b)

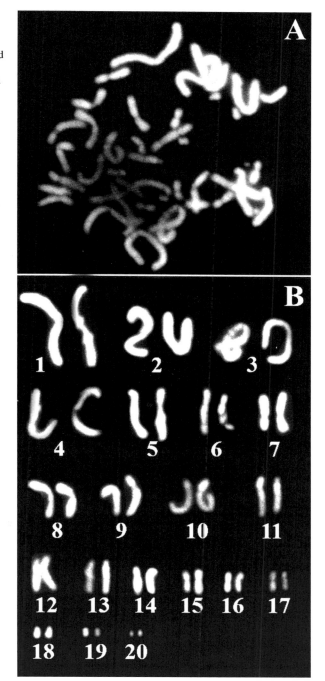

Cell cycle variations in chromatin structure detected by DNase I supported the concept of a chromatin condensation cycle during interphase and in mitosis. Prentice et al. (1985) subdivided the cell cycle into 11 stages, four for mitosis and seven of them belonging to interphase. The S phase started with the formation of small chromatin aggregates which appeared at stage II, mid-sized at stage III, and large ones at stage IV. Chromatin "bulges" appeared at stage III and were enlarged at stage IV, while heterochromatin disappeared (Leblond and El-Alfy, 1988). These observations are in conformity with the findings described in this suchapter regarding the existence of chromatin bodies in nuclei during interphase. Chromatin aggregates could be the places of replication factories since electron microscopy proved that these factories consist of nuclear bodies (Hozak et al., 1994). Replication factories were present in normal, untreated cells, therefore they could not be aggregation artifacts.

Control experiments with sealed cells after permeabilization have proved that 1. the reversal of permeabilization did not induce apoptosis, 2. reversibly permeable cells showed regular [3][H]-thymidine incorporation with a typical biphasic replication profile during S phase. Reversal of permeabilization was used to maintain the viability of murine thymus cells (Bánfalvi et al., 1984) and rat ventricular myocytes (Fawcett et al., 1998). After regeneration murine preB cells maintained their viability, but allowed chromosomes to slip out of the nucleus and to visualize and chraracterize them individually. The interphase chromosomes became visible as chromatin bodies in mid S phase. It is reasonable to assume that chromatin bodies could be related to chromatin aggregates, also referred to as chromatin factories or nuclear bodies since chromatin bodies are regarded as replicating chromosomes and DNA replication takes place in chromatin factories. This idea seems to be supported by the abrupt transition from early to late S and by the sudden switch of replication between different types of chromatin (Vogel et al., 1985). However, there is no evidence for the relationship between chromatin bodies and nuclear bodies.

The correlation between the number of subphases of DNA replication and the number of replication peaks in CHO and *Drosophila* cells is an indication that replication takes place chromosome by chromosome, consequently replication units could be identical with individual replicating chromosomes (Bánfalvi et al., 1997; Rehak et al., 2000). The fact that chromatin aggregates resembling chromatin bodies were also found by others using other approches and that chromatin structures in nuclei of untreated cells (Gacsi et al., 2005) differed significantly from those observed upon genotoxic effects such as gamma irradiation and cadmium treatment (Nagy et al., 2004; Bánfalvi et al., 2005) excluded the possibility that we were dealing with arteficial structures.

The temporal order of interphase chromosomal transitions was followed from mid- to late-S-phase involving round chromatin bodies, chromatin bodies with leg and head portions, supercoiled ribbons, elongated and bent interphase chromosomes. One could ask the obvious question what is the evidence that chromatin bodies correspond to individual chromosomes. The answer is given by the interphase karyogram of mid-S-phase chromatin bodies and late-S-phase chromosomes. The karyograms are based on the decreasing size and fluorescent light intensity of early chromatin bodies and elongated prechromosomes the number of which corresponds to the number of murine chromosomes. The 20 pairs of chromosomes of the mouse were identified by fluorescent or Giemsa staining (Buckland et al., 1971; Dev et al., 1971; Francke and Nesbitt, 1971; Schnedl, 1971). Judged by the visible number of globular forms termed chromatin bodies, eleven decondensed round shaped chromatin structures could be distinguished in CHO cells, which corresponds to the chromosome number of CHO cells. Moreover, from the estimated intensities of the eleven round or oval shaped chromosomes it was possible to tentatively correlate each of these with the 11 known chromosomes of CHO cells based on chromosome size. Although, the chromosome assignment was tentative, it showed that individual interphase chromosomes could be visualized in the late-S phase and arranged in circular patterns (Gacsi et al., 2005). The repeated coincidence between chromatin bodies and chromosome number seen both in CHO cells and in murine preB excludes the possibility that chromatin buds are chromatin clumps consisting of several chromosomes.

The linearization of bent chromosomes is likely to take place at the flexible hinge regions. The most flexible turning points seem to be located near the head portion of the interphase chromosome. Hinge regions connect smaller rigid units called rodlets at least in *Drosophila* chromosomes (Bánfalvi, 2006b).

Linear Connection of Chromosomes Excluded from Nuclei of CHO Cells

The escape of interphase chromosomes was utilized for the visualization of condensing interphase chromosomes in nuclei of Chinese hamster ovary cells (Bánfalvi, 2006a).

Figure 3.29 shows a comparative experiment with nuclei of CHO cells confirming the notion that interphase chromosomes escaping the incompletely sealed nucleus can be resolved by fluorescent microscopy. Visible chromosomes are seen inside the nucleus and an escaped chromosome is sitting at the top of the nucleus in Fig. 3.29a. One of the escaped chromatin bodies (Fig. 3.29b) was magnified for closer scrutiny (Fig. 3.29c, boxed) and showed at least three polar regions termed head and leg portions, tyical to early forms of interphase chromosomes (Bánfalvi, 2006c). Decondensed chromosomes appear in different forms, depending on the stage of compaction. Chromosomes escaping nuclei represent different stages of condensation (Fig. 3 29 c–g). The extruded chromatin ribbon seen in Fig. 3.26 d and g (boxed) clearly demonstrates that early chromosomal forms maintain connectivity. The extruded chromatin in Fig. 3.29g (boxed) selected for further magnification (Fig. 3.29f) consists of at least four elongated chromosomes (numbered from 1 to 4). These are U, V, S shaped chromosomal forms and one is twisted around its head portion (boxed and magnified 4). The extruded prechromosomes seen in nuclei of

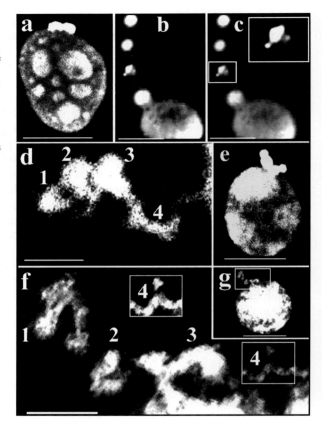

Fig. 3.29 Interphase chromosomes extruded from nuclei of CHO cells. (**a**) Chromosome islets inside the nucleus and an escaped chromosome outside the nucleus. (**b**) Early interphase chromosomes (chromatin bodies) escaping the nucleus. (**c**) Same as (**b**) with one of the escaped chromatin bodies (*in box*) enlarged showing some polarization. (**d**) Escaping chromatin ribbon supercoiled as looped chromosomes. (**e**) Escaped chromosomes sitting on the nucleus. (**g**) Extruded chromatin ribbon [boxed portion is magnified in (**f**) showing four precondensed chromosomes (1–4)]. Chromosome 4 (in box) is viewed separately. Bar, 5 μm each. (Reproduced with permission of Trencsenyi et al., 2007b)

CHO cells (Fig. 3.29) initiated the visualization of condensing chromosomal forms in other mammalian cells in a cell cycle dependent manner.

Condensation of Interphase Chromosomes in CHO Cells

The escape of individual interphase chromosomes from nuclei of reversibly permeabilized Chinese hamster ovary (CHO) cells was utilized for the visualization of condensing interphase chromosomes in a cell cycle-dependent manner in synchronized cells. It was observed that premature chromosomes were budding out or slipped out of the nucleus. The escape of an interphase chromosome seen in Fig. 3.30 with its head and two leg portions, resembles a starship between two planets. The diameter of this interphase chromosome is about 1 μm, and its length approximately 2 μm.

A similar early chromosomal form near the K562 nucleus can be seen in Fig. 3.31 resembling a meteorite approaching the surface of a planet. The next two figures show the early development of CHO chromosomes and give an answer to the ques-

Transition from Round Chromatin Bodies to Linear Chromosomes 177

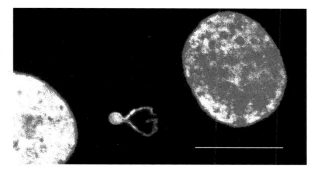

Fig. 3.30 The escape of an interphase chromosome from the nucleus of a CHO cell. This interphase schromosome is voyaging between two CHO nuclei. Bar, 5 μm. (Reproduced with permission of Bánfalvi, 2006c)

tion regarding the shape, length, and architecture of the decondensed interphase chromosome. Decondensed chromosomes appear in different forms, depending on the stage of their compaction.

The earliest visible chromosomal forms are round or oval-shaped chromatin bodies (Fig. 3.32a and c). The fibrous material gradually forms ribboned chromatin turned around its neck forming the head (Fig. 3.32 d and e) and leg portions (Fig. 3.32 f–l). The chromatin ribbon finally turns to a linear, but still bent chromosomal form (Fig. 3.32m). It is noteworthy to mention that chromosomes are linearly annealed in interphase (Fig. 3.33).

The folded chromosome that escaped (Fig. 3.30) and resembles a number eight with two open ends at its bottom, is similar to the one sticked to the nucleus

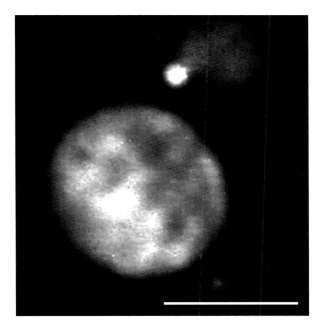

Fig. 3.31 Round shaped early chromosome escaping the nucleus of a human K562 cell. Bar, 5 μm

178   3 Chromatin Condensation

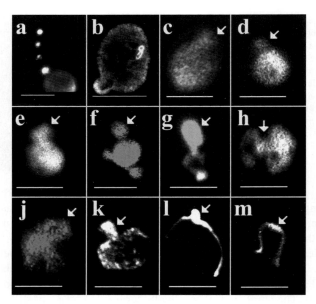

Fig. 3.32 Condensation of CHO interphase chromosomes in the late S-phase. Synchronized CHO cells were obtained by centrifugal elutriation. The fractions collected during loading, and cells containing continuous chromatin structures, but no visible forms of chromosomes (Table 3.1, fractions 1–4) were discarded. Fractions 5–8 collected at increasing flow rates (between 36 and 52 ml/min) were analyzed by FACS. (**a, b**) Chromatin from control non-fractionated cells. (**a**) Chromatin bodies excluded from the nucleus representing interphase chromosomes (same as Fig. 3.29b and c). (**b**) Precondensed chromosome sticking to the nucleus. Bar in a and b, 5 μm each. (**c–m**) Chromatin structures excluded from synchronized cells. (**c–e**) Chromatin bodies of Elutriation fraction 5 in the mid-S-phase (3.01 C). (**f–j**) fibrous chromatin bodies of fraction 6 with elongated shapes later in the mid-S-phase (3.27 C). Arrows indicate the head portion of the interphase chromosome. (**k–m**) Early elongated forms of chromosomes (3.71 C) in fraction 7. Unfolded chromatin bodies forming chromatin ribbons. Heads of chromosomes are indicated by arrows. Chromatin ribbons are turned around their head portions. Bar from c to m, 1 μm each. (Reproduced with permission of Bánfalvi, 2006c)

(Fig. 3.32b). These structures were initially regarded as artifacts, but now we know that these forms are real intermediates of the condensation pattern. Figure 3.32 contains the major intermediates of interphase CHO chromosomes showing their development from mid-S-phase (3.0 C-value) to the end of S-phase (4.0 C).

Elutriation Fraction 8 contained synchronized CHO cells belonging to the end of the S-phase, G2 and M phase, and to the final stage of chromosome condensation (4.0 C) (Fig. 3.34). This fraction is represented by precondensed CHO chromosomes, some of them still turned around their flexible heads (Fig. 3.34b, c, e and f); others are already linearized (Fig. 3.34d, f and g–m). The most frequently occuring chromosomal forms are the incompletely condensed arcs (Fig. 3.34b–f), hemicircles (Fig. 3.34 g), elongated linear forms (Fig. 3.34 h, j and k) approaching and occasionally reaching metaphase.

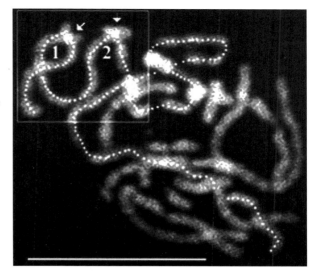

Fig. 3.33 Hinge regions in linearly connected, early elongated CHO chromosomes. Fraction 7 (3.71 C-value). Boxed: linear connection of two elongated chromosomes with their hinge regions indicated by the arrows. Bar, 5 μm. (Reproduced with permission of Bánfalvi, 2006c)

To summarize the results related to chromosome formation it is assumed that decondensed round-shaped bodies are the earliest forms of chromosomes which consist of a fibrous chromatin veil twisted around the head portion. In the round shaped chromatin bodies the head and the leg portions are inwardly folded, which gradually turn outward during their development at their flexible hinge regions.

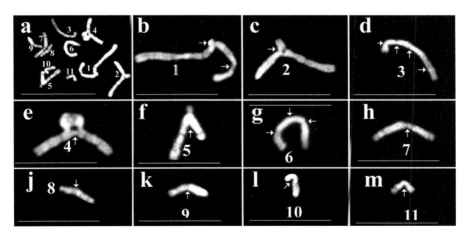

Fig. 3.34 Final stage of chromosome condensation. (**a**) Elongated CHO prechromosomes from Elutriation Fraction 8 (3.98 C-value) are numbered (1–11). (**b–m**) Chromosomes as seen under the fluorescent microscope after DAPI staining. Bar, 5 μm. Individual chromosomes with their flexible hinge regions indicated by the arrows. Bars, 2 μm each. (Reproduced with permission of Bánfalvi, 2006c)

These flexible parts of chromosomes are the turning points of further condensation by supercoiling.

The analysis of the temporal stages of chromatin condensation using synchronized cells revealed that the condensation process consists of two major phases, similarly to the observations in *C. elegans* embryos (Maddox et al., 2006). The initial phase corresponds to the organization of diffuse (veil-like) chromatin into distinct chromatin bodies in early S-phase. The second phase of chromatin condensation corresponds to the change from chromatin bodies to elongated highly curved, linear elongated distinct chromosomes to bar-shaped compact chromosomes. The condensation also resolves the two sister chromatids, which form morphologically distinct rods that remain connected along one side by sister chromatid cohesion.

The linearization of chromosomes takes place around the hind regions. The head portion of the interphase chromosome may correspond to the centromere region of the condensed chromosome. Consequently, mammalian chromosomes consist of two or more rod-like structures which can be folded at their flexible hinge regions. In *Drosophila* nuclei there are smaller condensed chromosomal units, so called rodlets, which are visible in the early stage of the condensation process (Bánfalvi, 2006b) to be discussed in the next subchapters.

Structure of Interphase Chromosomes in *Drosophila* Cells

Chromatin condensation was also visible in interphase nuclei of *Drosophila* cells. Our experiments show that the early fibrous chromatin structure is different in *Drosophila*, but the connectivity between interphase chromosomes resembles that of mammalian chromosomes (Bánfalvi, 2006b). A similar observation regarding the distribution of DNA replication proteins suggests that whilst the DNA replication machinary in *Drosophila* is organized into replication foci similar to most higher eukaryotes, no accumulation of replication factors could be detected. These observations point to fundamental differences between *Drosophila* and mammalian cells (Easwaran et al., 2007).

Synchronization of Drosophila Cells at Low Resolution of Elutriation

The aim of the investigation with *Drosophila* chromatin was to decide whether: (1) intermediates of chromatin condensation in lower eukaryotes follow the same pattern as mammalian cells, (2) chromosomes are separate entities or linearly annealed in nuclei of *Drosophila* cells. Emphasis is placed on the differences in chromosomal organization between *Drosophila* and mammalian cells.

Fluorescent images of interphase chromatin structures and chromosome structures isolated from reversibly permeable *Drosophila* cells were analyzed. By using

Structure of Interphase Chromosomes in *Drosophila* Cells   181

lower resolution of elutriation, collecting 8 fractions and high resolution of synchronization, collecting 36 elutriation fractions we confirmed that major intermediates of chromatin condensation include: decondensed veil-like chromatin at the unset of S phase (2.0–2.2 C-value), polarization of veiled chromatin (2.2–2.6 C), fibrous chromatin (2.6–3.0 C), chromatin bodies (3.0–3.3 C), early precondensed chromosomes (3.3–3.6). The compaction of *Drosophila* chromosomes did not reach that of the mammalian cells in the final stage of condensation (3.6–4.0 C). *Drosophila* chromosomes consist of smaller units called rodlets. Results demonstrate that nucleosomal chromatin ("beads on string") does not form solenoid structure, rather the topological arrangement consists of meandering and plectonemic loops. Early-S phase chromatin fibrils in nuclei of *Drosophila* cells are thinner than the veil like structures in mammalian cells. The connectivity of chromosomes shows linear arrangement [3(X), 1, 2, 4], with larger chromosomes (1 and 2) inside and smaller chromosomes [3(X), 4] at the two ends in the chromosomal chain (Bánfalvi, 2006b).

Cytometric Analysis of Drosophila Cells Synchronized at Low Resolution of Elutriation

For synchronization by centrifugal elutriation early exponential phase cell cultures were used. Centrifugal elutriation yielded 8 fractions, characterized by their flow rates at which cells were elutriated, by the average cellular diameter and volume and by their C-values (Table 3.3). The profile of cell growth is shown at increasing flow rates in Fig. 3.35A. Cell cycle parameters of elutriated fractions were estimated by flow cytometry, providing an assessment of the degree of synchrony of cells in each

Table 3.3 Characterization of synchronized populations of *Drosophila* cells at lower resolution of elutriation

Fraction number	Flow rate of elutriation (ml/min)	Elutriated cells ($\times 10^6$)	Cell diameter µm	Cell volume fl	Average C-value
1	27.0	0.904	8.1	290	n.d.
2	33.8	0.811	8.6	315	2.28
3	40.6	1.494	9.1	355	2.56
4	47.4	2.011	9.6	410	2.84
5	54.2	1.747	10.1	475	3.16
6	61.7	1.048	10.6	545	3.46
7	67.8	0.644	11.1	619	3.73
8	74.6	0.391	11.7	702	3.97

Cells subjected to elutriation: 10^7 (100%)
Number of elutriated cells: 9.05×10^6 (90.5%)
Loss of cells during manipulation: 0.95×10^6 (9.5%)
n.d. = not determined
Cellular volumes and diameters were estimated with the Coulter multisizer II. C-values were estimated as described by Basnakian et al. (1989). (Reproduced with permission of Bánfalvi, 2006b).

Fig. 3.35 Characterization of *Drosophila* cell populations synchronized at lower resolution of centrifugal elutriation. (**A**) Profile of cell growth after centrifugal elutriation (●-●), flow rate (▲-▲). (**B**) Cell-cycle pattern of various fractions analyzed by FACS using propidium iodide staining. The DNA content was expressed in C-values. The nuclear content increases from 2C to 4C as cells progress through the S phase. The flow cytometric profile giving the distribution of DNA content was used to calculate the average DNA content of a sample. Cell-cycle analysis by Coulter Multisizer II is given on the abscissa and cell number on the ordinate. (Reproduced with permission of Bánfalvi, 2006b)

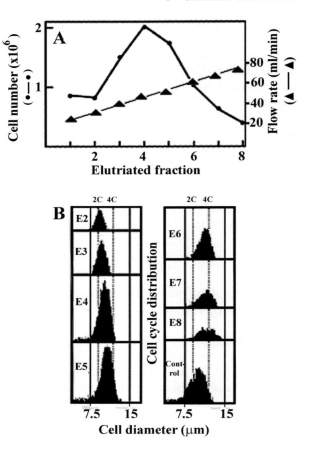

fraction (Fig. 3.35B). Flow cytometric profiles were used to express the C values (Basnakian et al., 1989).

Chromatin Condensation in Synchronized Drosophila Cells

Analysis was facilitated by earlier experience of studying chromatin structures in murine preB, CHO, Indian muntjac and human erythroleukemia (K562) cells (Nagy et al., 2004; Gacsi et al., 2005; Bánfalvi et al., 2005; Bánfalvi et al., 2006). Analysis focused on regularly occuring chromatin structures which were not disrupted by physical forces (Fig. 3.36).

Decondensed chromatin fibrils in early S phase (2.25–2.5 C value). The cells in Elutriation Fraction 2 were in early S phase and DNA was in its highly decondensed state. Since the fibers inside and around the nuclei are much finer than in mammalian cells, the chromatin structures were blurred especially at their peripheral regions (not shown).

Structure of Interphase Chromosomes in *Drosophila* Cells 183

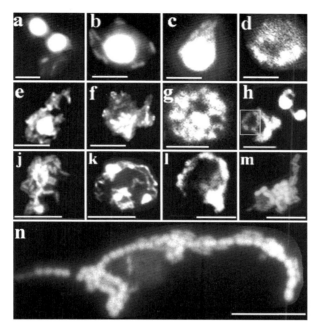

Fig. 3.36 Intermediates of chromatin condensation in *Drosophila* cells. Cells representing different stages of the cell cycle were obtained by centrifugal elutriation. Cells were permeabilized and permeabilization was reversed by resuspension in growth medium. Chromatin structures were isolated, stained with DAPI and visualized by fluorescent microscopy. (**a**, **b**) Elutriation fraction 3, decondensed chromatin. (**c**, **d**) Elutriation fraction 4, fibrous chromatin. (**e**, **f**) Elutriation fraction 5, chromatin ribbon. (**g**, **h**) Elutriation fraction 6, chromatin bodies. Box in h shows sister chromatids. (**j**, **k**) Elutriation fraction 7, elongated prechromosomes. (**l**, **m**, **n**) Elutriation fraction 8, condensed chromosomes. Bar, 5 μm each. (Reproduced with permission of Bánfalvi, 2006b)

From fibrils to chromatin rodlets (2.5–2.75 C). In nuclei isolated from cells belonging to Elutriation Fraction 3 thin chromatin fibrils formed rodlets ≈1–2 μm in length and 500–600 nm in diameter (Fig. 3.36a, b). The estimated diameter of thin fibrils around the rodlets is around 30 nm similar to the diameter of the zig-zag structure in mammalian chromatin.

Fibrous structures later in early S phase (2.75–3.0 C). In Elutriation Fraction 4 the veiled structure turned to looped chromatin fibers (Fig. 3.36c, d). These fibers are thinner than the 300 nm euchromatin fibers.

Transition to ribboned chromatin in mid S phase (3.0 – 3.25 C). Elutriation Fraction 5 represented cells in mid S phase. The condensation of the fibrous structures resulted in a bent chromatin ribbon pointing to a continuous linear arrangement of chromosomes within the ribboned chromatin (Fig. 3.36e, f). Individual chromosomes are not visible at this stage of the S-phase.

Chromatin bodies later in mid S phase (3.25–3.5 C). In Elutriation Fraction 6 round and elongated chromatin bodies (Fig. 3.36g, h) represent the first distinct forms of interphase chromosomes, formed from the loops of the chromatin ribbon by supercoiling. Chromatin bodies are linked to each other by thin and thick fibers. The number of rodlets inside the chromosomes varies depending on the size of the chromosome. The arch seen in the box of Fig. 3.36h is holding two sister chromatids together, probably by encircling them after replication. The ring is not complete, actually it is a hemicircle, the right sister chromatide being attached to the chromosome which is standing underneath.

Elongated prechromosomes in late S phase (3.5–3.75 C). Chromosomes appear in their elongated forms in late S phase (Fig. 3.36j, k, Elutriation Fraction 7). In elongated chromosomes and in the less condensed regions of chromosomes the rodlets are clearly visible.

Final stage of condensation (3.75–4.0 C). Elutriation Fraction 8 contained cells belonging to the end of S phase, to G2 and M phase (Fig. 3.36 l, m, n). Chromosomes are approching the final stage of condensation and occasionally reach metaphase. That chromosomes consist of rodlets is best seen in Fig. 3.36m. Linear connection of chromosomes is evident in Fig. 3.36n.

Chromatin Rodlets in Nuclei of Drosophila Cells

Chromatin rodlets similar to those seen in Fig. 3.36b are shown magnified for closer view in Fig. 3.37. The structure around these rodlets contains supercoiled fibrils of approximately 30 nm in diameter. Outside the fibrils is located an even finer network of fibrils seen as faint ghost. The diameter of the fibrils ranged between 10 and 12 nm. This structure is likely to correspond to the 11 nm nucleosomal organization of mammalian cells. The diameter of fibrils and fibers was estimated by analysing the content of the boxes by the drawing function of Adobe Photoshop 5.0 program. The negative image of the analysis is seen at the bottom of each panel in Fig. 3.37. The negative images in the boxes represent thin decondensed fibrils seen as lighter parts. The coiling of fibrils is clearly visible in Fig. 3.37a. Similarly, the outher thin

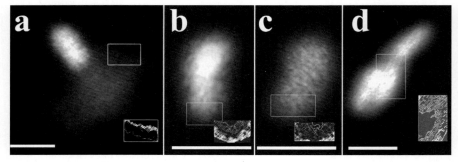

Fig. 3.37 Chromosome rodlets, the building blocks of condensing *Drosophila* chromosomes. Parts of these rodlets indicated by the boxes have been processed by the drawing function of Adobe Photoshop 5.0 program and are shown under their chromatin images. Bar, 1 μm each. (Reproduced with permission of Bánfalvi, 2006b)

fibrils could be visualized by this conversion, while the inner condensed parts of the rodlets were less visible (insets of Fig. 3.37b, c, d).

Linear Arrangement of Drosophila Chromosomes

The connection of interphase chromosomes seems to be maintained throughout the cell cycle and allows the determination of chromosome order in the linear arrangement at different stages of chromosome compaction (Fig. 3.38). Fluorescence intensities and sizes of chromatin bodies indicate the size of the chromosomes. Figure 3.38 a–d shows four sets of interphase chromosomes and the attachment of chromosome pairs. Following the rule that chromosome numbering should start at that end of the linear array which is closest to the largest chromosome (chromosome 1), the linear order of Drosophila chromosomes is 3 (X), 1, 2, 4. Figure 3.38c shows the connections between the chromosomes indicated by the black arrows.

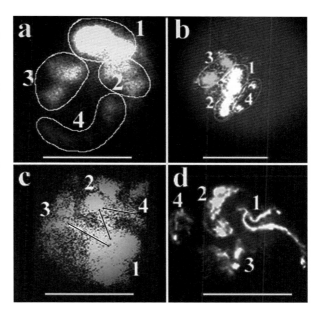

Fig. 3.38 Determination of the linear order of Drosophila chromosome attachment. Chromosomes are numbered based on their size. Four sets of chromosomes representing different stages of condensation have been visualized. Pairs of chromosomes are circled in (**a**) and (**b**) Bar, 5 μm each. (Reproduced with permission of Bánfalvi, 2006b)

Interphase Chromosomal Forms in Nuclei of Drosophila Cells

The first visible chromosomes appear as round shaped or elongated chromatin bodies in mid S phase (Fig. 3.39a–d, Elutriation Fraction 5). The supercoiling of chromatin inside the chromosomes (Fig. 3.39e–h) results in rodlets of condensed chromatin. Rodlets are the subunits of chromosomes (Fig. 3.39j–m, Elutriation Fraction 6). Bent elongated chromosomes with the rodlets inside the chromosomes

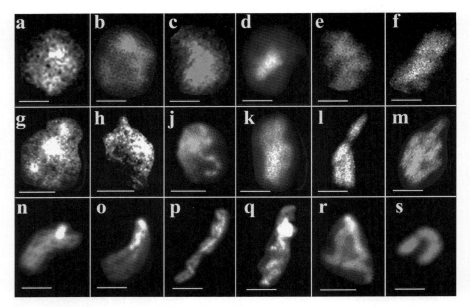

Fig. 3.39 Development of interphase chromosomes in *Drosophila* cells. Elutriation Fraction 5 (a–h). Chromatin bodies (a–d), supercoiled chromatin (e–h). Elutriation fraction 6, rodlets forming chromosomes (j–m). Elutriation Fraction 7, bent elongated chromosomes (n–p). Elutriation Fraction 8, bent condensed chromosomes (q, r, s,). Bar, 1 μm each. (Reproduced with permission of Bánfalvi, 2006b)

are still recognizable in Fig. 3.39n–p (Elutriation Fraction 7) confirming the idea that chromosomes consist of several rodlets. At the final stage of the condensation (Fig. 3.39 q, r, s, Elutriation Fraction 8) chromosomes are bent arcs or circular forms.

The general mechanism of chromosome condensation in *Drosphila* shows evolutionary similarities to that of mammalian cells. Two major structures of intermediates could be distinguished: chromatin forms and chromosomal structures. Chromatin structures involve the early S-phase decondensed chromatin veil, the texture of which seems to be finer, less condensed than that of the mammalian chromatin. The fine structure of the extruding chromatin halo encircles not only the nucleus, early chromosomal structures are also blurred (e.g. Fig. 3.39a). The average diameter of chromatin fibrils is 10–12 nm. The chromatin in this stage could correspond to the 11 nm nucleosomal arrangement of the mammalian chromatin. The second stage of condensation can be charaterized by the estimated 30 nm thin chromatin fibrils. The diameter of thick fibers ranged between 180 and 260 nm. The rodlets (1–2 × 0.5–1 μm) are likely to correspond to supercoiled thick fibers. Noteworth to mention, that condensed mammalian chromosomes can also be regarded as rod-like structures, but they are thicker. The rodlets in *Drosophila* nuclei are small condensed chromosomal units, seen in the early stage of the condensation process as structural elements of interphase chromosomes. Several of these chromosomelets form the condensed chromosome. The association of rodlets is probably dictated by the

Structure of Interphase Chromosomes in *Drosophila* Cells 187

flexible folding process in the nucleus. The folding of chromosomes is likely to take place at their hinge regions. These flexible parts among the rodlets are probably the turning points for supercoiling. Hinge regions seem to be important in the folding process both in the linearization of chromosomes at the end of the condensation process and in decondensation before chromosome replication.

The magnification of microscopy (1000x) would have allowed the analysis of objects larger than 100 nm. That the number of rodlets could not be estimated is explaned by: 1. the linear arrangement of chromosomes suggesting that chromosome replication is related to a defined temporal order of condensation which would mean that some chromosomes are in more condensed state than others, 2. small chromosome units are rolled up in chromosomes and became less visible.

Drosophila Schneider 2 cells contain only four pairs of chromosomes the sizes of which differ significantly from each other. Based on the linear arrangement of chromosomes and on the estimated fluorescent intensities it was possible to tentatively correlate each other with the 4 known chromosomes of *Drosophila* cells. Although, the chromosome assignment is tentative, it serves to show the linear order of chromosomes: 3 (X chromosome), 1, 2, 4, indicating that larger chromosomes (1, 2) are inside, smaller ones (3 (X), 4) fall into line at the two ends of the chromosomal chain. More rigorous chromosome identification is expected with chromosome specific probes and *in situ* hybridization. These experiments will bring us closer to Boveri's (1907) century old hypothesis regarding the territorial localization of chromosomes inside the nucleus.

Intermediates of Chromatin Condensation in Drosophila Cells at High Resolution of Synchronization

Chromatin-remodeling factors have been isolated from a variety of organisms ranging from yeasts and *Drosophila* to humans. Chromatin remodeling complexes can be divided into several main groups characterized by their ATPase subunits and associated factors (Becker and Horz, 2002; Narlikar et al., 2002). In yeast and human cells, two subclasses have been recognized: one comprises yeast SWI/SNF and human BAF, and the other includes yeast RSC and human PBAF (Mohrmann et al., 2004). In contrast to yeast and mammalian cells, *Drosophila* cells seem to differ in their remodeling factors (Tamkun et al., 1992; Simon and Tamkun, 2002), consequently their chromatin structure may be different. This idea seems to be supported by polytene chromosomes of *Drosophila* salivary glands, which are the only high resolution interphase chromatin structures indicating significant variability and helicity (Hochstrasser et al., 1986; Mathog and Sedat, 1989). The fundamental difference between lower and higher eukaryotes has also been emphasized by Easwaran et al. (2007).

The aim was to determine whether the difference in chromatin-remodeling could have structural consequences in *Drosophila* Schneider 2 cells at the supranucleosomal level of chromatin organization. Recent studies on chromosome condensation in *Drosophila* cells (Bánfalvi, 2006b) indicate that this process differs from those of mammalian cells (Bánfalvi et al., 2006).

High Resolution of Synchrony of Elutriated Fractions and DNA Synthesis in Synchronized Cells

Exponentially growing Schneider 2 *Drosophila* cells were synchronized using the highest possible resolution of centrifugal elutriation by collecting 36 fractions during the cell cycle. Subphases of DNA synthesis in elutriated fractions were confirmed by ^{3}[H]-thymidine incorporation revealing four replication peaks (Fig. 3.40A). Synchronization was tested by flow cytometry and by measuring the cell size of each fraction up to fraction 29 (Fig. 3.40B). The C-values of elutriated fractions are given in Table 3.4.

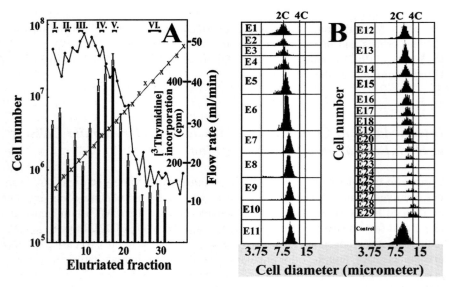

Fig. 3.40 Characterization of elutriated fractions of *Drosophila* cells. (**A**) Actively growing S2 cells (24 h culture) were separated into 36 fractions (●-●) of which 30 could be used due to low cell number at the end of elutriation. [^{3}H]-Thymidine incorporation calculated for 10^5 cells (columns). Each measurement of DNA synthesis contained 2.5×10^5 cells. Flow rate (x-x). (**B**) Cell-cycle patterns of the various fractions were analyzed by FACS using propidium iodide staining. The DNA content was expressed in C-values (haploid genome content). The nuclear content increases from 2C to 4C as cells progress through the S phase. The flow cytometric profile giving the distribution of DNA content was used to calculate the average DNA content of each fraction (Basnakian et al., 1989). Cell-cycle analysis by Coulter Multisizer II is given on the abscissa and cell number on the ordinate. (Reproduced with permission of Bánfalvi et al., 2007)

Structure of Interphase Chromosomes in *Drosophila* Cells 189

Table 3.4 C-values of elutriated fractions of *Drosophila* cells

Fraction number	Average C-value	Fraction number	Average C-value
1	2.00	16	3.41
2	2.07	17	3.45
3	2.14	18	3.55
4	2.20	19	3.62
5	2.28	20	3.66
6	2.38	21	3.69
7	2.58	22	3.72
8	2.75	23	3.76
9	2.86	24	3.77
10	2.89	25	3.79
11	2.96	26	3.83
12	3.06	27	3.87
13	3.17	28	3.92
14	3.28	29	3.97
15	3.34	30	n.d.
Control, nct elutriated (average)			2.76

C-values were calculated from the flow cytometric profiles of elutriated fractions as described earlier (Basnakian et al., 1989).
n.d., not detected.
(Reproduced with permission of Bánfalvi et al., 2007).

Major Steps and Structures of Chromatin Condensation in Drosophila Cells

Decondensed chromatin structures in early S phase. Chromatin was isolated from fractions (a) 2–4 (combined), (b) 5, (c) 8–9 (combined), (d) 14, e) 16–17 (combined) and (f) 26–30 (combined). Cells in early S phase contained chromatin being in its most decondensed state appearing as a slightly folded chromatin-veil (Fig. 3.41, Fractions 2–4).

Polarization of the nuclear material. It was observed first in CHO cells, then in other mammalian cells, that chromatin condensation in exponentially growing cells starts with the polarization of the nuclear material (Gacsi et al., 2005; Bánfalvi et al., 2006). The polarization of veil-like chromatin is chraracteristic to the early stage of chromatin condensation mainfested mainly by the extrusion of the chromatin (Fig. 3.41, Fraction 5).

Formation of fibrous chromatin. The condensing fibrillary material gradually changes its shape turning to a more compact fibrous chromatin. Some regions of the fibrous chromatin remain in a decondensed cotton-like state (Fig. 3.41, Fractions 8–9). Attention is called to the linear connection of interphase chromatin structures which is maintained throughout the S phase.

Globular structures. The condensing fibrous chromatin gradually turns to globular structures. The formation of the spherical or slightly elongated chromatin bodies is probably the result of the supercoiling of the fibrous chromatin-veil (Fig. 3.41, Fraction 14).

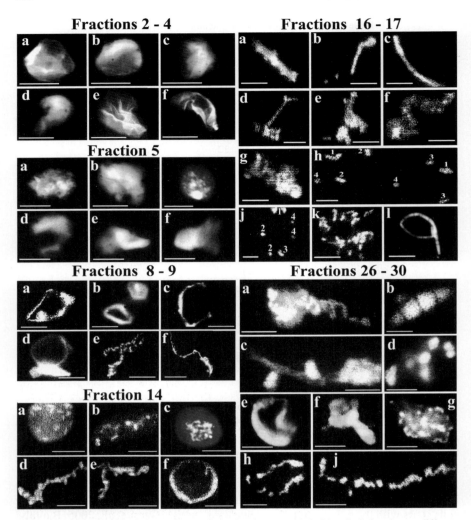

Fig. 3.41 Intermediates of chromatin condensation in *Drosophila* S2 cells belonging to different stages of the cell cycle obtained at high resolution of centrifugal elutriation. Cells were permeabilized and permeabilization was reversed by resuspension in growth medium conatining 10% fetal bovine serum. After 3 h at 37°C, nuclei were isolated, stained with DAPI and visualized by fluorescent microscopy. Elutriation Fractions 2–4: Decondensed chromatin at 2.07–2.20 C-value. Elutriation Fraction 5: Decondensed chromatin and typical images illustrating the polarization of the nuclear material at 2.28 C-value. Fractions 8–9: Fibrous chromatin at 2.75–2.86 C. Fraction 14: Chromatin bodies and continuous linear connection of interphase chromosomes at 3.28 C. Fractions 16–17: Early precondensed chromosomes at 3.41–3.45 C. Fractions 26–30: Final stage of chromosome condensation at 3.83–4.0 C. Bar, 5 μm each. (Reproduced with permission of Bánfalvi et al., 2007)

Structure of Interphase Chromosomes in *Drosophila* Cells   191

Early chromosomal forms. Distinct forms of chromosomes were observed in late S phase. Individual chromosomes can be clearly distinguished in Fig. 3.41, Fractions 16–17. The linear arrangement of chromosomes is clearly demonstrated in Fig. 3.41, Fractions 16–17/l.

Interphase chromosomes at the end of S phase. The end of the S phase is represented by Fractions 26–30 and shows chromosomes belonging to the final stage of chromosome condensation (Fig. 3.41, Fractions 26–30), which obviously did not reach the compaction of mammalian choromosomes.

Building Units of Drosophila Chromosomes

Drosophila chromosomes consist of smaller units called rodlets (Bánfalvi, 2006b). Condensing chromosomes and rodlets are seen in Fig. 3.42. Rodlets are extruded from the nucleus. The length of the rodlets is between 1 and 2 μm and their diameter ranges from 0.5 to 1.0 μm. The boxes in Fig. 3.42d and f are shown magnified

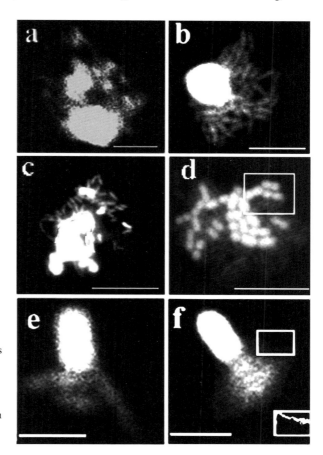

Fig. 3.42 Chromosome bodies (rodlets), the building blocks of condensing chromosomes. Parts of rodlets indicated by the boxes have been processed by the drawing function of Adobe Photoshop 5.0 program. Bar (**a–d**) 5 μm each. Bar (**e–f**) 2 μm each. (Reproduced with permission of Bánfalvi et al., 2007)

Fig. 3.43 Formation of chromosome rodlets in *Drosophila* cells. (**A**) Part of condensing chromosomes shown in the box of Fig. 3.42d at higher magnification. (**B**) Nucleosomal organization of *Drosophila* DNA. Decondensed part of a rodlet shown in Fig. 3.42f (*box*) was analyzed and magnified by Adobe Photoshop computer program. (**a**) Decondensed strings of nucleosomes. (**b**) condensed loops of strings. Arrows pointing upward indicate decondensed regions of nucleosomes. *Arrows* directed downward show coiled, plectonemic loops of nucleosomal strings. (**c**) Nucleosomal chromatin loop. Bar, 1 μm each. (Reproduced with permission of Bánfalvi et al., 2007)

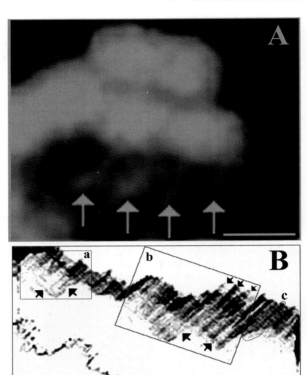

in Fig. 3.43A and B, respectively. The segment shown in Fig. 3.43A contains supercoiled stretches (indicated by arrows) under a chromosome which are less condensed as indicated by fluorescent light intensities. These chromatin streches are part of the thicker precondensed chromosome and are regarded as early forms of chromosome rodlets.

Folding of Nucleosomal Chromatin String in Drosophila Cells

Figure 3.42 e and f show two chromosome rodlets. The box in Fig. 3.42f has been subjected to computer image analysis and magnified to study the nucleosomal arrangement in chromatin rodlets (Fig. 3.43B).

Box a in Fig. 3.43B shows the meandering nucleosomal string, with the neighbouring nucleosomes 33 nm apart, their distance is higher than in loops seen in

Structure of Interphase Chromosomes in *Drosophila* Cells   193

Table 3.5 Transitions in the "beads on string" of *Drosophila* chromatin

Chromatin structure	Topology diameter nm	*Nucleosome distance nm*
Decondensed string	12	33
Condensed loop	31	25
Coiled condensed loop	24	23

Measurements are average values from Fig. 3.43B/a, b and c.
(Reproduced with permission of Bánfalvi et al., 2007).

Fig. 3.43B, box b. The average diameter of the decondensed nucleosomal string is 12 nm in Fig. 3.43B, box a). In Fig 3.43B, box b the nucleosomes and the loops of nucleosome strings are closer to each other, occasionally they are twisted around themselves. Solenoid formation could not be traced. Supranucleosomal chromatin fibers 33 nm in diameter predominate over nucleosomal fibers 12 nm in diameter. Transitions take place between 12 and 33 nm thicknesses, which may be important *in vivo*. Transitions are summarized in Table 3.5.

The loop of a nucleosomal string (Fig. 3.43B, circled c) was selected to magnify it further to the limit of resolution of the microscopic and computer analysis (Fig. 3.44a). This analysis beside showing the formation of nucleosomes (boxed), clearly demonstrates that strings of nucleosomes in chromosome rodlets follow a meander line. Meander folding is schematically viewed in Fig. 3.44b. Some chromatin models suggest that the supranucleosomal fiber results from the solenoidal winding of the nucleosomal fiber (Fig. 3.44 c). Based on Fig. 3.44a the formation of solenoid structure can be ruled out.

To the question how nucleosomes are spatially arranged, randomly or aligned precisely, that is, phased there have been earlier several reports which confirmed that nucleosomes are phased in chicken (Wittig and Wittig, 1979), in rat (Igo-Kemenes et al., 1980), in *Xenopus* (Gottesfeld and Bloomer, 1980) and in *Drosophila* (Louis et al., 1980; Levy and Noll, 1980). There are several other parameters beside the physical property of DNA, DNA binding proteins, ATP-dependent conversions, DNA and RNA polymerases, that can affect the chromatin structure. However, how these parameters determine nucleosome position and organization *in vivo* is poorly understood.

Plectonemic Model of Chromatin Condensation in Drosophila Cells

The revaluation of experimental data of nucleosomal strings (Bánfalvi et al., 2007) led to a molecular model of chromatin condensation called plectonemic chromatin model (Fig. 3.45) discussed in Chapter 2. Figure 3.45 is identical with Fig. 2.49.

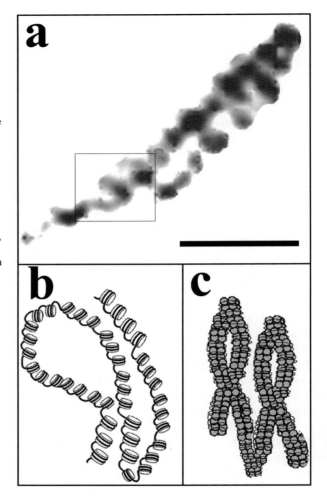

Fig. 3.44 Two models showing supranucleosomal organization of looped chromatin in *Drosophila* cells. (**a**) Nucleosomal chromatin loop circled in Fig. 3.43/B/c was magnified further to the limit of the resolution of computer image analysis. *Box* shows the formation of nucleosomes. Bar, 100 nm. Schematic representation of the two views of nucleosomal DNA: (**b**) Model of nucleosomes showing the DNA wrapped twice around a histone octamers and loop formation, (**c**) solenoid structure. (Reproduced with permission of Bánfalvi et al., 2007)

Conclusions. We studied the chromatin structure in *Drosophila* cells primarily during S phase when the DNA replication machinery is copying the genome with complete precision on the complex chromatin template to be duplicated. Exponentially growing *Drosophila* Schneider 2 cells, most of them in early S phase (average 2.76 C), were used to assure that DNA is unfolded and chromatin structures were in their decondensed state. The resulting decondensed chromatin structures were visualized after reversal of permeabilization. *Drosophila* chromosomes have been chosen since drawings of chromosomes of lower eukaryotes such as protozoans demonstrated coiling and supercoiling providing lower compaction, visible with the light microscope (Cleveland, 1949). When the chromatin structure was investigated in *Drosophila* line HS-2 using digestion with micrococcal nuclease, DNase I, and restriction enzymes, an extensive ordered array of nucleosomes was found. Moreover, a distinct 10-nucleosome-size DNA fragment could be observed in the micrococcal

Fig. 3.45 Plectonemic model of chromatin condensation in *Drosophila* cells. This figure is identical with Fig. 2.49

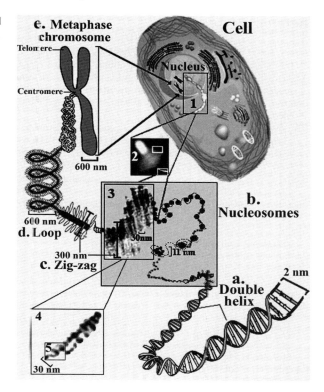

nuclease digestion pattern (Sun et al., 2001). Micrococcal nuclease mimics nuclease S1 by selectively cleaving inverted repeats (Dingwall et al., 1981), which have the potential to form hairpins or cruciforms (Lilley, 1980; Panayotatos and Wells, 1981). While we acknowledge the existence of these ordered arrays of nucleosomes consisting of 8–12 units with an average of 10 nucleosomes lined up one after the other we could not confirm the existence of a helical solenoid structure.

Finally, our study confirms the existence of a flexible chromatin folding pattern through transient geometric forms including (a) decondensed chromatin structures in early S phase being in its most decondensed state and appears as a slightly folded chromatin-veil, (b) polarized nuclear material seen as extrusions of the chromatin, (c) formation of fibrous structures and the linear connection of interphase chromatin without visible chromosomes, (d) fibrous chromatin supercoiled to globular structures as spherical or slightly elongated chromatin bodies, which are earliest visible chromosomes, (e) distinct forms of chromosomes observed in late S phase, (f) interphase chromsomes at the end of S phase. The lack of completely folded chromosomes may well be an indication that chromosomes of *Drosophila* cells do not reach the compaction of metaphase seen in mammalian cells (Bánfalvi et al., 2006). This idea seems to be supported by the salivary gland chromosomes which are known as relatively decondensed structures.

References

Abney, J.R., Cutler, B., Fillbach, M.L., Axelrod, D. and Scalettar, B.A. (1997). Chromatin dynamics in interphase nuclei and its implications for nuclear structure. J Cell Sci. **137**, 1459–1468.

Aasland, T., Gibson, T.J. and Stewart, A.F. (1995). The PHD-finger: implications for chromatin-mediated transcriptional regulation. Trends Biochem Sci. **20**, 56–59.

Balczon, R. (1993). Autoantibodies as probes in cell and molecular biology. Proc Soc Exp Biol Med. **204**, 138–154.

Bánfalvi, G. Fluorescent analysis of replication and intermediates of chromatin folding in nuclei of mammalian cells. (1993). In: Bach, P., Reynolds, C.H., Clark, J.M., Mottley, J., Poole, P.L. (eds). Biotechnology applications of microinjection, microscopic imaging, and fluorescence. New York and London, Plenum Press pp. 111–119.

Bánfalvi, G. (2006a). Linear connection of condensing chromosomes in nuclei of synchronized CHO cells. DNA Cell Biol. **25**, 541–545.

Bánfalvi, G. (2006b). Structure of interphase chromosomes in the nuclei of *Drosophila* cells. DNA Cell Biol. **25**, 547–553.

Bánfalvi, G. (2006c). Condensation of interphase chromosomes in nuclei of synchronized CHO cells. DNA Cell Biol. **25**, 641–645.

Bánfalvi, G. (2008). Cell cycle synchronization of animal cells and nuclei by centrifugal elutriation. Nature Protocols **3**, 663–673.

Bánfalvi, G, Mikhailova, M., Poirier, L-A. and Chou, M.W. (1997a). Multiple subphases of DNA replication in Chinese hamster ovary (CHO-K1) cells. DNA and Cell Biol. **16**, 1493–1498.

Bánfalvi, G., Nagy, G., Gacsi, M., Roszer, T. and Basnakian, A.G. (2006). Common pathway of chromosome condensation in mammalian cells. DNA Cell Biol. **25**, 295–301.

Bánfalvi, G., Poirier, L.A., Mikhailova, M. and Chou, M.W. (1997b). Relationship of repair and replicative DNA synthesis to cell cycle in Chinese hamster ovary (CHO-K1) cells. DNA Cell Biol. **16**, 1155–1160.

Bánfalvi, G., Sooki-Toth, A., Sarkar, N., Csuzi, S. and Antoni, F. (1984). Nascent DNA chains synthesized in reversibly permeable cells of mouse thymocytes. Eur J Biochem. **139**, 553–559.

Bánfalvi, G., Wiegant, J., Sarkar, N. and van Duijn, P. (1989). Immunofluorescent visualization of DNA replication sites within nuclei of Chinese hamster ovary cells. Histochemistry. **93**, 81–86.

Basnakian, A., Bánfalvi, G. and Sarkar, N. (1989). Contribution of DNA polymerase delta to DNA replication in permeable cells synchronized in S phase. Nucleic Acids Res. **17**, 4757–4767.

Brehm, A., Tufteland, K.R., Aasland, R. and Becker, P.B. (2004). The many colours of chromodomains. Bioessays. **26**, 133–40. Review.

Beaudouin, J., Mora-Bermudez, F., Klee, T., Daigle, N. and Ellenberg, J. (2006). Dissecting the contribution of diffusion and interactions to the mobility of nuclear proteins. Biophys J. **90**, 1878–1894.

Becker, P.B. (2002). Nucleosome sliding: facts and fiction. EMBO J. **21**, 4749–4753.

Becker, P.B. and Horz, W. (2002). ATP-dependent nucleosome remodeling. Annu Rev Biochem. **71**, 247–273.

Bednar, J., Horowitz, R.A., Grigoryev, S.A., Carruthers, L.M., Hansen, J.C., Koster, A.J. and Woodcock, C.L. (1998). Nucleosomes, linker DNA, and linker histone form a unique structural motif that directs the higher-order folding and compaction of chromatin. Proc Natl Acad Sci USA. **95**, 14173–14178.

Belmont, A.S. (1998). Nuclear ultrastructure: transmission electron microscopy and image analysis. Methods Cell Biol. **53**, 99–124.

Belmont, A.S. (2001). Visualizing chromosome dynamics with GFP. Trends Cell Biol. **11**, 250–257.

Belmont, A.S. and Bruce, K. (1994). Visualization of G1 chromosomes: a folded, twisted, supercoiled chromonema model of interphase chromatid structure. J Cell Biol. **127**, 287–302.

Belmont, A.S., Sedat, J.W. and Agard, D.A. (1987). A threedimensional approach to mitotic chromosome structure: Evidence for a complex hierarchical organization. J Cell Biol. **105**(1), 77–92.

References

Berezney, R., Mortillaro, M.J., Ma, H. Wei, X. and Samarabandu, J. (1995). The nuclear matrix: a structural milieu for genomic function. Int J Cytol. **162**A, 1–65.

Bernstein, E. and Hake, S.E. (2006). The nucleosome: a little variation goes a long way. Biochem Cell Biol. **84**, 505–517.

Bickmore, W. (1999). Fluorescence *in situ* hybridization analysis of chromosome and chromatin structure. Methods Enzymol. **304**, 650662.

Bjorkroth, B., Ericsson, C., Lamb, M.M. and Daneholt, B. (1988). Structure of the chromatin axis during transcription. Chromosoma (Bert.). **96**, 333–340.

Blow, J.J. and Watson, J.V. (1987). Nuclei act as independent units of replication in Xenopus cell-free DNA replication system. EMBO J. **6**, 1997–2002.

Bordas, J., Perez-Grau, L., Koch, M.H.J., Vega, M.C. and Nave, C. (1986). The superstructure of chromatin and its condensation mechanism. I. Synchrotron radiation X-ray scattering results. Eur Biophys J Biophys Lett. **13**, 175–185.

Boveri, T. (1907) Zellenstudien VI. Die Entwicklung dispermer Seeigelier. Ein Beitrag zur Befruchtungslehre und zur Theory des Kernes. Jena Zeit Natur. **43**, 1–292.

Boy De La Tour, E. and Laemmli, U.K. (1988). The metaphase scaffold is helically folded: sister chromatids have predominately opposite helical handedness. Cell. **55**, 927–944.

Chiodi, I., Biggiogera, M., Denegri, M., Corioni, M., Weighardt, F., Cobianchi, F., Riva, S. and Biamonti, G. (2000). Structure and dynamics of hnRNP-labelled nuclear bodies induced by stress treatments. J Cell Sci. **113**, 4043–4053.

Clark, J.M., Mottley, J. and Poole, P.L. (eds). Plenum. Press/New York, London. pp. 111–119.

Cleveland, L.R. (1949). The whole life cycle of chromosomes and their coiling systems. Trans Am Philosophical Soc. **39**, 1–97.

Cremer, T. and Cremer, C. (2001). Chromosome territories, nuclear architecture and gene regulation in mammalian cells. Nat Rev Genet. **2**, 292–301.

Cremer, T, Cremer, M., Dietzel, S., Müller, S., Solovei, I. and Fakan, S. (2006). Chromosome territories-a functional nuclear landscape. Curr Opin Cell Biol. **18**, 307–316.

Cremer, C., Cremer, T., Zorn, C. and Cioreanu, V. (1979). Partial irradiation of Chinese hamster cell nuclei and detection of unscheduled DNA synthesis in interphase and metaphase. A tool to investigate the arrangement of interphase chromosomes in mammalian cells. Hoppe Seyler's Z Physiol Chem. **360**, 244–245.

Cremer, T., Kurz, A., Zirbel, R., Dietzel, S., Rinke, B., Schnock, E., Speicher, M. R., Mathieu, U., Jauch, A., Emmerich, P., Scherthan, H., Ried, T., Cremer, C. and Lichter, P. (1993). Role of chromosome territories in the functional compartmentalization of the cell nucleus. Cold Spring Harbor Symp Quant Biol. **LVIII**, 777–792.

Comings, D.E. (1980). Arrangement of chromatin in the nucleus. Hum Genet. **53**, 131–143.

Davey, C.A. and Richmond, T.J. (2002). DNA-dependent divalent cation binding in the nucleosome core particle. Proc Natl Acad Sci USA. **99**, 11169–11174.

Davey, C.A., Sargent, D.F., Luger, K., Maeder, A.W. and Richmond, T.J. (2002). Solvent mediated interactions int he structure of the nucleosome core particles at 1.9 Å resolution. J Mol Biol. **319**, 1097–1113.

De Campos Vidal, B., Russo, J. and Mello, M.L. (1998). DNA content and chromatin texture of benzo[a]pyrene-transformed human breast epithelial cells as assessed by image analysis. Exp Cell Res. **244**, 77–82.

Diehl, A.M. and Rai, R. (1996). Review: regulation of liver regeneration by pro-inflammatory cytokines. J Gastroenterol Hepatol. **111**, 466–470.

Dietzel, S. and Belmont, A.S. (2001). Reproducible but dynamic positioning of DNA within chromosomes during mitosis. Nat Cell Biol. **3**, 767–770.

Dingwall, C., Lomonossoff, G.P. and Laskey, R.A. (1981). High sequence specificity of micrococcal nuclease. Nucl Acids Res. **9**, 2659–2673.

Dorigo, B., Schalch, T., Kulangara, A., Duda, S., Schroeder, R.R. and Richmond, T.J. (2004). Nucleosome arrays reveal the two-start organization of the chromatin fiber. Science. **306**, 1571–1573.

198   3 Chromatin Condensation

Downes, M., Ordentlich, P., Kao, H.Y, Alvarez-Jacqueline, G.A. and Evans, R.M. (2000). Identification of a nuclear domain with deacetylase activity. Proc Natl Acad Sci USA. **97**, 10330–10335.

Earnshaw, W.C. and Laemmli, U.K. (1983). Architecture of metaphase chromosomes and chromosome scaffolds. J Cell Biol. **96**, 84–93.

Easwaran, H.P., Leonhardt, H.A. and Cardoso, M.C. (2007). Distribution of DNA replication proteins in *Drosophila* cells. BMC Cell Biol. **8**, 42.

Fausto, N. and Webber, E.M. (1994). Liver regeneration. In Arias, I., Boyer, J. and Fausto, N., (eds). The Liver Biology and Pathology. Raven Press, New York. pp. 1059–1084.

Finch, J.T. and Klug, A. (1976). Solenoidal model for superstructure in chromatin. Proc Natl Acad Sci USA. **73**,1897–1901.

Flemming, W. (1879). Beitrage zur Kenntniss der Zelle und ihrer Lebenserscheinungen. Arch Mikroskop Anat. **16**, 302–436.

Flemming, W. (1882). Zellsubstanz, Kern und Zelltheilung. Verlag von F.C.W. Vogel. Leizig.

Fox, A.H, Lam, Y.W., Leung-Anthony, K.L., Lyon, C.E., Andersen, J., Mann, M. and Lamond, A.I. (2002). Paraspeckles: a novel nuclear domain. Curr Biol. **12**, 13–25.

Gacsi, M., Nagy, G., Pinter, G., Basnakian, A.G. and Bánfalvi, G. (2005). Condensation of interphase chromatin in nuclei of synchronized Chinese hamster ovary (CGO-K1) cells. DNA Cell Biol. **24**, 43–53.

Gassmann, R., Vagnarelli, P., Hudson, D. and Earnshaw, W.C. (2004). Mitotic chromosome formation and the condensin paradox. Exp Cell Res. **296**, 35–42.

Gottesfeld, J.M. and Bloomer, L.S. (1980). Nonrandom alignment of nucleosomes on 5S RNA genes of X. laevis. Cell **21**, 751–760.

Grechman, S.E. and Ramakrishnan, V. (1987). Chromatin higher-order structure studied by neutron-scattering and scanning transmission electron microscopy. Proc Natl Acad Sci USA. **84**, 7802–7806.

Grigoryev, S.A. (2004). Keeping fingers crossed: heterochromatin spreading through interdigitation of nucleosome arrays. FEBS Lett. **254**, 4–8.

Hansen, J.C. (2002). Conformational dynamics of the chromatin fiber in solution: determinants, mechanisms and functions. Annu Rev Biophys Biomol Struct. **31**, 361–392.

Harp, J.M., Hanson, B.L., Timm, D.E. and Bunick, G.J. (2000). Asymetries int he nucleosome core particle at 2.5 Å resolution. Acta Crystalogr D Biol Crystalogy. **56**, 1523–1534.

Henegariu, O., Heerema, N.A., Lowe Wright, L., Bray-Ward, P., Ward, D.C. and Vanve, G.H. (2001). Improvements in cytogenic slide preparation: Controlled chromosome spreading, chemical aging and gradual denaturing. Cytometry. **43**, 101–109.

Heslop-Harrison, J.S., Huelskamp, M., Wendroth, S., Atkinson, M.D., Leicht, A.R. and Benett, M.D. (1988). Chromatin and centromeric structures in interphase nuclei. In Kew chromosome conference III. Brandham, P.E. (ed.), Allan & Unwin/London, pp. 209–217.

Hirano, T. (2000). Chromosome cohesion, condensation, and separation. Annu Rev Biochem. **69**, 115–144.

Hiraoka, Y., Minden, J.S., Swedlow, J.R., Sedat, J.W. and Agard, D.A. (1989). Focal points for chromosome condensation and decondensation revealed by three-dimensional *in vivo* time-lapse microscopy. Nature. **342**, 293–296.

Hiriyanna, K.T., Varkey, J., Beer, M. and Benbow, R.M. (1988). Electron microscopic visualization of sites of nascent DNA synthesis by streptavidin-gold binding to biotinylated nucleotides incorporated *in vivo*. J Cell Biol. **107**, 33–44.

Hochstrasser, M., Mathog, D., Gruenbaum, Y., Schwaumweber, H. and Sedat, J.W. (1986). Spatial organization of chromosomes in the salivaty gland nuclei of *Drosophila melanogaster.* J Cell Biol. **102**, 112–123.

Horn, P.J. (2002). Chromatin higher order folding–wrapping up transcription. Science **297**, 1824–1827.

Horowitz, R.A., Agard, D.A., Sedat, J.W. and Woodcock, C.L. (1994). The three-dimensional architecture of chromatin *in situ*: electron tomography reveals fibers composed of a continuously variable zig-zag nucleosomal ribbon. J Cell Biol. **125**, 1–10.

References

Hozak, P. and Fakan, S. (2006) Functional structure of the cell nucleus. Histochem Cell Biol. **125**, 1–2.

Igo-Kemenes T., Omori, A. and Zachau, H.G. (1980). Different repeat lengths in rat satellite I DNA containing chromatin and bulk chromatin. Nucl Acids Res. **22**, 5377–5390.

Iwano, M., Fukui, K., Tkaichi, S. and Isogai, A. (1997). Globular and fibrous structure in barley chromosomes revealed by high-resolution scanning electron microscopy. Chromosome Res. **5**, 341–349.

Jackson, A.A. (1991) Structure±function relationships in eukaryotic nuclei. Bioessays **13**, 1–10.

Jacobs, S.A. and Khorasanizadeh, S. (2002). Structure of HP1 chromodomain bound to a lysine 9-methylated histone H3 tail. Science. **295**, 2080–2083.

Juan, G., Traganos, F., James, W.M., Ray, J.M., Roberge, M., Sauve, D.M., Anderson, H. and Darzynkiewicz, Z. (1998). Histone H3 phosphorylation and expression of cyclins A and B1 measured in individual cells during their progression through G2 and mitosis. Cytometry. **32**, 71–77.

Kaufmann, W.K., William J., Bodell, W.J. and Cleaver, J.E. (1983). DNA excision repair in permeable human fibroblasts. Carcinogenesis.**4**, 179–184.

Kireeva, N., Lakonishok, M., Kireev, I., Hirano, T. and Belmont, A.S. (2004). Visualization of early chromosome condensation: a hierarchical folding, axial glue model of chromosome structure. J Cell Biol. **166**, 775–785.

Kosak, S.T. and Groudine, M. (2004). Form follows function: The genomic organization of cellular differentiation. Genes Dev. **18**, 1371–1384.

Koshland, D. and Strunnikov, A.V. (1995). Mitotic chromosome condensation. Annu Rev Cell Dev Biol. **12**, 305–333.

Kostriken, R. and Weeden, C.J. (2001). Engineered interphase chromosome loops guide intrachromosomal recombination. EMBO J. **20**, 29072913.

Kouzarides, T. (2007). Chromatin modifications and their function. Cell. **128**, 693–705.

Kubista, M., Hagmar, P., Nielsen, P.E. and Nordén, B. (1990). Reinterpretation of linear dichroism of chromatin supports a perpendicular linker orientation in the folded state. J Biomol Struct Dyn. **8**, 37–54.

Laemmli, U.K., Cheng, S.M., Adolph, K.W., Paulson, J.R., Brown, J.A and Baumbach, W.R. (1978). Metaphase chromosome structure: the role of nonhistone proteins. Cold Spring Harb Symp Quant Biol. **42** 351–360.

Lamond, A.I. and Earnshaw, W.C. (1998). Structure and function in the nucleus. Science. **280**, 547–553.

Lemke, J., Claussen, J. Michel, S., Chudoba, I., Muhlig, P., Westermann, M., Sperling, K., Rubtsov, N., Grummt, U.W., Ullmann, P., Kromeyer-Hauschild, K., Liehr, T. and Claussen, U. (2002). The DNA-based structure of human chromosome 5 in interphase. Am J Hum Genet. **71**, 1051–1059.

Levy, A. and Noll, M. (1980). Multiple phases of nucleosomes in the hsp 70 genes of *Drosophila melanogaster*. Nucl Acids Res. **8**, 6059–6097.

Li, B., Carey, M. and Workman, J.L. (2007). The role of chromatin during transcription. Cell. **128**, 707–719.

Lewis, C.D., Lebkowski, J.S., Daly, A.K. and Laemmli, U.K. (1984). Interphase nuclear matrix and metaphase scaffolding structures. J Cell Sci Suppl. **1**, 103–122.

Lilley, D.M. (1980). The inverted repeat as a recognizable structural feature in supercoiled DNA molecules. Proc Natl Acad Sci USA. **77**, 6468–6472.

Liu, Q. and Dreyfuss, G. (2000). A novel nuclear structure containing the survival of motor neurons protein. EMBO J. **15**, 3555–3565.

Louis, C., Schedl, P., Samal, B. and Worzel, S. (1980). Chromatin structure of the 5S RNA genes of *D. melanogaster* Cell. **22**, 387–392.

Luger, K. (2003). Structure and dynamic behavior of nucleosomes. Curr Opin Genet Dev. **13**, 127–135.

Luger, K. and Hansen, J.C. (2005). Nucleosome and chromatin fiber dynamics. Curr Opin Struct Biol. **15**, 188–196.

Luger, K., Mader, A.W., Richmond, R.K., Sargent, D.F. and Richmond, T.J. (1997). Crystal structure of the nucleosome core particle at 2.8 Å resolution. Nature. **389**, 251–260.

Maddox, P.S., Portier, N., Desai, A. and Oegema, K. (2006). Molecular analysis of mitotic chromosome condensation using a quantitative time-resolved fluorescence microscopy assay. Proc Natl Acad Sci USA. **103**, 15097–15102.

Maeshima, K. and Laemmli, U.K. (2003). A two-step scaffolding model for mitotic chromosome assembly. Dev Cell. **4**, 467–480.

Manuelidis, L. (1990). A view of interphase chromosomes. Science. **250**, 1533–1540.

Marsden, M.P.F. and Laemmli, U.K. (1979). Metaphase chromosome structure: evidence for a radial loop model. Cell. **17**, 849–858.

Marshall, W.F., Dernburg, A.F., Harmon, B., Agard, D.A. and Sedat, J.W. (1996). Specific interactions of chromatin with the nuclear envelope: Positional determination within the nucleus in *Drosophila melanogaster*. Mol Biol Cell. **7**, 825–842.

Matera, A.G. (1999). Nuclear bodies: multifaceted subdomains of the interchromatin space. Trends Cell Biol. **9**, 302–309.

Mathog, D. and Sedat, J.W. (1989). The three-dimensional organization of polytene nuclei in male *Drosophila melanogaster* with compound XY or ring X chromosomes. Genetics. **121**, 293–311.

Mohrmann, L., Langenberg, K., Krijgsveld, J., Kal, A.J., Heck, A.J.R. and Verrijzer, C.P. (2004). Differential targeting of two distinct SWI/SNF-related *Drosophila* chromatin-remodeling complexes. Mol Cell Biol. **24**, 3077–3088.

Mora-Bermúdez, F. and Ellenberg, J. (2007). Measuring structural dynamics of chromosomes in living cells by fluorescence microscopy. Method Cell Cycle Res. **41**, 158–167.

Michalopoulos, G.K. and Defrances, M.C. (1997). Liver regeneration. Science. **276**, 60–66.

Misteli, T. (2001). Protein dynamics: implications for nuclear architecture and gene expression. Science. 2001; **291**, 843–847.

Misteli, T. and Spector D.L. (1998). The cellular organization of gene expression. Curr Opin Cell Biol. **10**, 323–331.

Mora-Bermúdez, F., Gerlich, D. and Ellenberg, J. (2007). Maximal chromosome compaction occurs by axial shortening in anaphase and depends on Aurora kinase. Nat Cell Biol. **9**, 822–831.

Moses, R.E. and Richardson, C.C. (1970). Replication and repair of DNA in cells of Escherichia coli treated with toluene. Proc Natl Acad Sci USA. **67**, 674–681.

Nagl, W. (1985). Chromatin organization and the control of gene activity. Int Rev Cytol. **94**, 21–56. Review.

Narlikar, G.J. and Kingston, R.-E. (2002). Cooperation between complexes that regulate chromatin structure and transcription. Cell. **108**, 475–487.

Olins, D.E. and Olins, A.L. (2003). Chromatin history: our view from the bridge. Nature Rev Mol Cell Biol. **4**, 809–814.

Panayotatos, N. and Wells, R.D. (1981). Cruciform structures in supercoiled DNA. Nature. **289**, 466–470.

Parada, L.A., Mcqueen, P. and Misteli, T. (2004). Tissue-specific spatial organization of genomes. Genome Biol. **5**, R44.

Paragh, G., Foris, G., Paragh, G. Jr., Seres, I., Karanyi, Z., Fulop, P., Balogh, Z., Kosztaczky, B., Teichmann, F. and Kertai, P. (2005). Different anticancer effects of fluvastatin on primary hepatocellular tumours and metastases in rats. Cancer Lett. **222**, 17–22.

Paulson, J.R. and Laemmli, U.K. (1977). The structure of histone depleted chromosomes. Cell. **12**, 817–828.

Pederson, T. (2004). The spatial organization of the genome in mammalian cells. Curr Opin Genet Dev. **14**, 203–209.

Pennings, S., Meersseman, G. and Bradbury, E.M. (1991). Mobility of nucleosomes on 5S rDNA. J Mol Biol. **220**, 101–110.

Phair, R.D., Gorski, S.A. and Misteli, T. (2004). Measurement of dynamic protein binding to chromatin *in vivo*, using photobleaching microscopy. Methods Enzymol. **375**, 393–414.

Poirier, M.G. and Marko, J.F. (2002). Mitotic chromosomes are chromatin networkswithout a mechanically contiguous protein scaffold. Proc Natl Acad Sci USA. **99**, 15393–15397.

References

Poirier, M., Eroglu, S., Chatenay, D. and Marko, J.F. (2000). Reversible and irreversible unfolding of mitotic new chromosomes by applied force. Mol Biol Cell. **11**, 269–276.

Rattner, J.B. and Lin, C.C. (1985). Radial loops and helical coils coexist in metaphase chromosomes. Cell. **42**, 291–296.

Razin, S.V. (1996). Functional architecture of chromosomal DNA domains. Crit Rev Eukaryot Gene Expr. **6**, 247–269.

Rehak, M., Csuka, I., Szepessy, E. and Bánfalvi, G. (2000). Subphases of DNA replication in *Drosophila* cells. DNA Cell Biol. **19**, 607–612.

Richmond, T.J. and Davey, C.A. (2003). The structure of DNA in the nucleosome core. Nature.**423**, 145–150.

Robinett, C.C., Straight, A., Li, G., Willhelm, C., Sudlow, G., Murray, A. and Belmont, A.S. (1996). *In vivo* localization of DNA sequences and visualization of large-scale chromatin organization using lac operator/repressor recognition. J Cell Biol. **135**, 1685–700.

Rydberg, B., Holley, W.R, Mian, I.S. and Chatterjee, A. (1998). Chromatin conformation in living cells: support for a zig-zag model of the 30 nm chromatin fibril. J Mol Biol. **284**, 71–84.

Saha, A., Wittmeyer, J., Cairns, B.R. (2002). Chromatin remodeling by RSC involves ATP-dependent DNA translocation. Genes Dev. **16**, 2120–2134.

Salomoni, P. and Pandolfi, P.P. (2002). The role of PML in tumour suppression. Cell. **108**, 165–170.

Sarkar, A., Eroglu, S., Poirier, M.G., Gupta, P., Nemani, A. and Marko, J.F. (2002). Dynamics of chromosome compaction during mitosis. Exp Cell Res. **277**, 48–56.

Schalch, T, Duda, S., Sargent, D.F. and Richmond, T.J. (2005). X-ray structure of a tetranucleosome and its implications for the chromatin fibre. Nature. **436**, 138–141

Sedat, J. and Manuelidis, L. (1978). A direct approach to the structure of eukaryotic chromosomes. Cold Spring Harb Symp Quant Biol. **42**, 331–350.

Seki, S., Mori, S. and Oda, T. (1986). Deoxyribonuclease I sensitivity of DNA replicated i permeable mouse sarcoma cells. Acta Med Okayama. **40**, 183–188.

Seki, S. and Oda, T. (1985). DNA repair synthesis in bleomycin-pretreated permeable HeLa cells. Carcinogenesis. **7**, 77–82.

Sheinin, R. and Humbert, J. (1978). Some aspects of eukaryotic DNA replication. Annu Rev Biochem. **47**, 277–316. Review.

Simon, J.A. and Tamkun, J.W. (2002). Programming off and on states in chromatin: mechanisms of Polycomb and trithorax group complexes. Curr Opin Genet Dev. **12**, 210–218.

Sleeman, J.E. and Lamond, A.I. (1999). Nuclear organization of pre-mRN splicing factors. Curr Opin Cell Biol. **11**, 372–377. Review.

Sperling, K. and Luedtke, E.-K. (1981). Arrangement of prematurely condensed chromosomes in cultured cells and lymphocytes of the Indian muntjac. Chromosoma. **83**, 541–553.

Starzl, T.E., Porter, K., Francavilla, J.A., Benichou, J. and Putnam, C.W. (1977). A hundred years of the hepatotrophic controversy. Ciba Found Symp. **55**, 111–129.

Strukov, Y.G., Wang, Y. and Belmont, A.S. (2003). Engineered chromosome regions with altered sequence composition demonstrate hierarchical large-scale folding within metaphase chromosomes. J Cell Biol. **162**, 23–35.

Strunnikov, A.V. (1998). SMC proteins and chromosome structure. Trends Cell Biol. **8**, 454–459.

Sun, F-L., Cuaycong, M.H. and Elgin, S.C.R. (2001). Long-range nucleosome ordering is associated with gene silencing in *Drosophila melanogaster* pericentric heterochromatin. Mol Cell Biol. **21**, 2867–2879.

Sun, J., Zhang, Q. and Schlick, T. (2005). Electrostatic mechanism of nucleosomal array folding revealed by computer simulation Proc Natl Acad Sci USA. **102**, 8180–8185.

Swedlow, J.R., and T. Hirano. (2003). The making of the mitotic chromosome: modern insights into classical questions. Mol Cell. **11**, 557–569.

Tamkun, J.W., Deuring, R., Scott, M.P., Kissinger, M., Pattatucci, A.M., Kaufman, T.C. and Kennison, J.A. (1992). Brahma: a regulator of *Drosophila* homeotic genes structurally related to the yeast transcriptional activator SNF2/SWI2. Cell. **68**, 561–572.

Trencsenyi, G., Kertai, P., Somogyi, C., Nagy, G., Dombradi, Z., Gacsi, M. and Bánfalvi, G. (2007a). Chemically induced carcinogenesis affecting chromatin structure in rat hepatocarcinoma cells. DNA Cell Biol. **26**, 649–655.

Trencsenyi, G., Ujvarosi, K., Nagy, G. and Bánfalvi, G. (2007b). Transition from chromatin bodies to linear chromosomes in nuclei of murine preB cells synchronized in S phase. DNA Cell Biol. **26**, 549–556.

Uzbekov, R.E. (2004). Analysis of the cell cycle and a method employing synchronized cells for study of protein expression at various stages of the cell cycle. Biochemistry. **69**, 485–496.

White, C.L., Suto, R.K. and Luger, K. (2001). Structure of the yeast nucleosome core particle reveals fundamental changes in nucleosome interactions. EMBO J. **20**, 5207–5218.

Van Driel, R., Fransz, P.F. and Verschure, P.J. (2003). The eukaryotic genome: A system regulated at different hierarchical levels. J Cell Sci. **116**, 4067–4075.

Van Holde, K. and Zlatanova, J. (1996). What determines the folding of the chromatin fiber? Proc Natl Acad Sci USA. **93**, 10548–10555.

Wan, K.M., Nickerson, J.A., Krockmalnic, G. and Penman, S. (1999). The nuclear matrix prepared by amine modification. Proc Natl Acad Sci U S A. **96**, 933–938.

Wang, G.G., Allis, C.D. and Chi, P. (2007). Chromatin remodeling and cancer, part I: covalent histone modifications. Trends Mol Med. 13, 363–372.

Wang, I.F., Reddy, N.M. and Shen, C.K. (2002). Higher order arrangement of the eukaryotic nuclear bodies. Proc Natl Acad Sci USA.**99**, 13583–13588.

Widom, J. (1998). Structure, dynamics, and function of chromatin *in vitro*. Annu Rev Biophys Biomol Struct. **27**, 285–327.

Williams, S.P., Athey, B.D., Muglia, L.J., Schappe, R.S., Gough, A.H. and Langmore, J.P. (1986). Chromatin fibers are left-handed double helices with diameter and mass per unit length that depend on linker length. Biophys J. **49**, 233–248.

Wittig, B. and Wittig, S. (1979). A phase relationship associates tRNA structural gene sequences with nucleosome cores. Cell. **18**, 1173–1183.

Wolffe, A.P. (1995). "Chromatin," 2nd ed. Sect. 2.4, Academic Press, New York.

Yokota, H., Van Den Engh, G., Hearst, J.E., Sachs, R. and Trask, B.J. (1995). Evidence for the organization of chromatin in megabase pair-sized loops arranged along a random walk path in the human G0/G1 interphase nucleus. J Cell Biol. **130**, 1239–1249.

Woodcock, C.L., Frado, L.L. and Rattner, J.B. (1984). The higher-order structure of chromatin: evidence for a helical ribbon arrangement. J Cell Biol. **99**, 42–52.

Woodcock, C.L., Grigoryev, S.A., Horowitz, R.A. and Whitaker, N. (1993). A chromatin folding model that incorporates linker variability generates fibers resembling the native structures. Proc Natl Acad Sci. USA. **90**, 9021–9025.

Woodcock, C.L. and Horowitz, R.A. (1995). Chromatin organization re-viewed. Trends Cell Biol. **5**, 272–277.

Worzel, A., Strogatz, A. and Riley, D. (1981). Structure of chromatin and the linking number of DNA. Pro Natl Acad Sci USA. **78**, 1461–1465.

Xue, Y., Wong, J., Moreno, G.T., Young, M.K., Côté, J. and Wang, W. (1998). NURD, a novel complex with both ATP-dependent chromatin-remodeling and histone deacetylase activities. Mol Cell. **2**, 851–861.

Chapter 4
Apoptosis

Summary

In this chapter the historical events of apoptotic research are followed by (a) the relationship of apoptosis to the genetic communication, (b) characterization of apoptosis and necrosis, (c) induction and chemical inducers of apoptosis, (d) apoptotic pathways, (e) antiapoptotic pathways and (f) apoptosis protocols.

History

1842	Vogt reported cell death in the notochord and adjacent cartilage of metamorphic toads (Vogt, 1842).
1883	Ilya Ilyich Mechnikov recognized phagocytosis associated with cell death in the muscles of metamorphic toads (Clarke and Clarke, 1996). Mechnikov received Nobel Prize with Paul Erlich in physiology and medicine in recognition of their work on immunity.
1889	Naturally occuring cell death was uncovered by John Beard (Clarke and Clarke, 1996).
1951	Cell death in normal vertebrate onthogeny (Glucksmann, 1951).
1961	Cell death in chick embryos as studied by electromicroscopy (Bellairs, 1961).
1964	The cell death in invertebrate development is "programmed". The term programmed cell death has been proposed (Lockshir and Williams, 1964)
1965	Kerr's first paper on cell death ("shrinkage nephrosis") (Kerr, 1965). John Foxton Ross Kerr from the University of Queensland received the Paul Ehrlich Prize Prize for his description of apoptosis, which he shared with Boston biologist Robert Horvitz. Horvitz and colleagues found that the *ced-3* gene was required for the cell death during the development of the nematode *Caenorrhabditis elegans*. The Paul Ehrlich award is considered second only to the Nobel Prize.
1966	Inhibitors of RNA and protein synthesis blocked cell death during amphibian metamorphosis. Cell death is an active process (Tata, 1966).

1969	Comparison of insect and vertebrate developmental cell death (Whitten, 1969).
1972	The term apoptosis for programmed cell death has been introduced by Kerr et al. (1972).
1973	Digestion of chromatin DNA at regularly spaced sites (nucleosome ladder) (Hewish and Burgoyne, 1973).
1974	Linkage between cell death and chromatin digestion (Williams et al., 1974).
1976	Programmed cell death in *Caenorrhabditis elegans* (Sulston, 1976).
1980	Biochemical changes accompany apoptotic morphology: DNA is degraded into a nucleosome ladder, apoptosis requires the activation of death genes (Wyllie, 1980).
1982	Genetic pathway in the nematode *Caenorrhabditis elegans* dedicated to programmed cell death (Horvitz et al., 1982).
1983	First cell death mutants (ced-1, ced-2) of *C. elegans* (Hedgecock et al., 1983).
1983	p53 can be used for the detection of primary mouse tumour cells (Rotter, 1983).
1986	Death from growth factor withdrawal by apoptosis (Duke and Cohen, 1986).
1984	p53 expression is not quantitatively related to metastatic behavior in lymphoma cells (Rotter et al., 1984).
1986	Death from growth factor withdrawal by apoptosis (Duke and Cohen, 1986).
1986	Ced-3, ced-4 mutants of *Caenorrhabditis elegans* (Ellis and Horvitz, 1986).
1986	Cloning of Bcl-2 (Tsujimoto and Croce, 1986; Cleary et al., 1986).
1988	Bcl-2 inhibits cell death (Vaux et al., 1988).
1989	Antibodies made to Fas/APO-1 (Yonehara et al., 1989; Trauth et al., 1989).
1991	Substrate specificity of granzyme (Poe et al., 1991; Odake et al., 1991), inhibition of cell death may lead to autoimmune disease (Strasser et al., 1991), p53 can cause apoptosis (Yonish et al., 1991), cloning of Fas (Itoh et al., 1991), p35 was identified as being responsible for blocking the apoptosis (Clem et al. 1991).
1992	Identification of *ced-9* (Hengartner et al., 1992), cloning of *ced-4* (Yuan and Horvitz, 1992), expression of phosphatidyl serine on apoptotic cells (Fadok et al., 1992), inhibition of *C. elegans* cell death by human Bcl-2 (Vaux et al., 1992).
1993	Cloning of bax (Oltvai et al., 1993) and *ced-3* (similarity to ICE) (Miura et al., 1993; Yuan et al., 1993). The role of p53 in apoptosis was proposed. Fas ligand was identified, ICE was found to be the mammalian homologue of *ced-3*. IAPS (inhibitors of apoptosis) were identified.
1994	Cloning of *ced-9* (similarity to Bcl-2) (White et al., 1994), structure of caspases (Walker et al., 1994; Wilson et al., 1994), unified pathway of CTL killing (granzyme) and apoptosis via caspases (Vaux et al., 1994).

History	205

1995   p53 is a caspase inhibitor (Xue and Horvitz, 1995), Fas and TNF inactivation is prevented by caspase inhibitor (CrmA) (Tewari and Dixit, 1995; Enari et al., 1995), cell death mediated by Fas-FasL (Alderson et al., 1995), cloning of Mort1/FADD (Boldin et al., 1995; Chinnaiyan et al., 1995), identification of cellular IAPs (Roy et al., 1995; Hay et al., 1995; Rothe et al., 1995; Uren et al., 1996; Liston et al., 1995; Duckett et al., 1996) and HID (Grether et al., 1995).

1996   Cytochrome c and dATP in a complex activate caspase 3 (Liu et al., 1996), cloning of caspase 8 and binding to FADD (Boldin et al., 1996; Muzio et al., 1996), structure of Bcl-x (Muchmore et al., 1996), establishment of Apotosis (journal, 1996), transglutaminase induction by various cell death and apoptosis pathways (Fesus et al., 1996).

1997   Isolation of CED-4 like protein, Apaf1 (Zou et al., 1997), interaction of CED-9, CED-4 and CED-3 (Spector et al., 1997; James et al., 1997; Wu et al., 1997; Chinnaiyan et al., 1997; Ottilie et al., 1997; Seshagiri and Miller, 1997; Irmler et al., 1997), caspase recruitment domain (Hofmann et al., 1997), BH3 interactions (Sattler et al., 1997), transglutaminase-dependent posttranslational modification of the retinoblastoma gene product in apoptosis (Oliverio et al., 1997), induction of apoptosis by retinoids (Szondy et al., 1997).

1998   Bcl-2 prevents caspase-independent cell death (Okuno et al., 1998), expression of bcl-2 and bax in TGF-beta 1-induced apoptosis (Motyl et al., 1998), cloning of human deoxyribonuclease II and its possible role in apoptosis (Krieser and Eastman, 1998), heat shock and apoptosis (Punyiczki and Fésüs, 1998), EG-1, a BH3 only protein essential for apoptosis in the worm (Conradt and Horvitz, 1998).

1999   Sperm chromatin structure undergoing sperm-specific apoptosis (McCarthy and Ward, 1999), necrosis and apoptosis contribute to HIV-1-induced killing of CD4 cells (Plymale et al., 1999), DFF40 caspase-activated endonuclease (Liu et al., 1999), activation of caspases and p53 by bovine herpesvirus 1 infection (Devireddy and Jones, 1999).

2000   Caspase-dependent and independent cell death in rat hepatoma (Pandey et al., 2000), stage-specific apoptotic patterns during *Drosophila* oogenesis (Nezis, 2000), apoptosis in the early bovine embryo (Matwee et al., 2000), apoptosis induced by ultrasound exposure (Ashush et al., 2000), the language of histone modification ("histone code") (Strahl and Allis, 2000), apoptosis in neuronal development and disease (Nijhawan et al., 2000).

2001   HIV induces lymphocyte apoptosis by a p53-initiated, mitochondrial-mediated mechanism (Genini et al., 2001), induction of histone degradation by Granzyme A (Zhang et al., 2001) BIR interaction motif in caspase 9 processing, p53 modulates base excision repair in a cell cycle-specific manner (Offer et al., 2001), translating the "histone code" (Jenuwein and Allis, 2001), MST1-JNK promotes apoptosis (Ura et al., 2001).

2002	Nobel Prize in Physiology or Medicine: Sydney Brenner, Robert, H. Horvitz, John E. Sulston "for their discoveries concerning genetic regulation of organ development and programmed cell death". Regulation of apoptosis through arachidonate cascade in mammalian cells (Nishimura et al., 2002), retinoblastoma activated transcription of the survival gene bcl-2 (Decary et al., 2002), IL-4 induced proteolytic processing of mast cell STAT6 (Sherman et al., 2002), Apaf-1 is a mediator of E2F-1-induced apoptosis (Furukawa et al., 2002).
2003	Characterization of histone deacetylases (HDACs) (de Ruijter et al., 2003), apoptosis as a source of soluble histones (Kanai, 2003), cytochrome C-dependent Fas-independent apoptotic pathway (Kim et al., 2003), involvement of cytochrome c oxidase inhibition in caspase-independent cell death (Yuyama et al.,2003).
2004	Inactivation and activation of CAD/DFF40 in the apoptotic pathway (Woo et al., 2004), nitric oxide induced apoptosis mediated through p53- and Bax-dependent pathways (Yung et al., 2004), caspase-like activity during the conjugation in *Tetrahymena* (Ejercito and Wolfe, 2004).
2005	Caspase-independent induction of apoptosis by proapoptotic Bcl-2-related protein Nbk / Bik (Oppermann et al., 2005), apoptin induced chromatin condensation in normal cells (He et al., 2005), lysosome is a primary organelle in B cell receptor-mediated apoptosis (He et al., 2005), antitumour drugs that alter chromatin structure (Rabbani et al., 2005), SUMO-1 protein represses apoptosis (Lee et al., 2005).
2006	Specific histone variants form a homotypic nucleosome (nucleosome code) (Bernstein and Hake, 2006), apoptotic endonuclease DFF40/CAD as a structural probe for chromatin (Widlak and Garrard, 2006), topoisomerase IIB and nuclease interaction to digest sperm DNA in an apoptotic-like manner (Shaman et al., 2006), protein methyltransferase 2 inhibits NF-kappaB function and promotes apoptosis (Ganesh et al., 2006).
2007	Autophagy is an ultrastructural marker of heavy metal toxicity (Di Gioacchino et al., 2008), RNAi silencing of the Wilms tumour (WT1) gene inhibits cell proliferation and induces apoptosis (Zamora-Avila et al., 2007), early-stage apoptosis associated with DNA-damage-independent *Ataxia Telangiectasia* Mutated (ATM) phosphorylation and chromatin decondensation (Schou et al., 2007), chromatin remodeling and cancer (Wang et al., 2007), chromatin modifications and their function (Kouzarides, 2007).

Increased Scientific Interest to Understand Apoptosis

The mistic sounding topic apoptosis (programmed cell death) gained recognition slowly as demonstrated by the number of references on apoptotic cell death between 1970–2007. The total number of publications on programmed cell death is

History

now over 140,000. Based on the increasing number of publications the research on cell death seems to be still in its early exponential growth phase (Fig. 4.1). The major challenge in apoptosis research is the involvement of protein factors in an increasingly complex web of signalling patways playing major role in cell proliferation and differentiation. The increased scientific interest is primarily due to the clinical implications of apoptosis involving its potential to treat cancer, autoimmune and neurodegenerative diseases (Jacobson et al., 1997). The discoveries related to apoptosis are of importance for cancer research to treat tumours by turning on "programmed cell death". If one could push the "self-destruct" button to kill off these abnormally dividing cells, cancer could be effectively cured.

Apoptosis

The word apoptosis can be found for the first time between the end of the 5th century BC and the early 4th century BC, in chapter 35 of *Mochlicon*—Hippocrates' treatise on dislocations of the bones, structural changes (bone erosion) related to tissue and cell death (André, 2003). Marcus Aurelius in political and social context used the word apotosis as a synonyme of failure and decline. This expression came from his physician Galen, who extended the medical meaning of apoptosis to wound healing and inflammation.

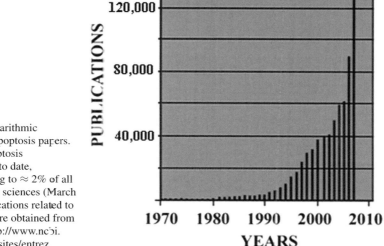

Fig. 4.1 Logarithmic increase of apoptosis papers. 143,400 apoptosis publications to date, corresponding to ≈ 2% of all papers in life sciences (March 2008). Publications related to apoptosis were obtained from PubMed. http://www.ncbi.nlm.nih.gov/sites/entrez

Relationship of Apoptosis to Genetic Communication

Destruction of Biological Information

Information is knowledge, but knowledge in itself is not information, yet. To become information knowledge (a) has to take its proper form, (b) needs a material carrier and (c) energy to reach its target. The flow of information is communication. Biological information refers to living organisms which are highly organized compared to the disorder of their environment. Thus, biological information can be defined as the constant struggle of living systems against molecular disorder (randomness). Genetic information is the result of continuous development. The information can be characterized by its: prospective character, measurability, and by the units of signal set. Informational systems in biology include behavioral forms, nervous and hormonal systems, signal transformation, genetic information. Genes are units of genetic information. Since information is neither material nor energy, the laws of material and energy constancy do not apply for information. Consequently information can be generated, manipulated and destroyed. Recombination may generate new information. Among the manipulations of biological information viral infections serve as an example. Apoptosis ruins unnecessary biological information. Although, in a negative sense, apoptosis as self destruction belongs to the processes of biological information.

Different stimuli initiate the self destruction program of cells such as the absence of survival factors, irreparable damages, conflicting signals attenuating the balance of the cell cycle. The fact that cell death is related to the cell cycle balance and is programmed reflects an instructive, genetically controlled process.

Intrinsic way of apoptosis is carried out as an orderly process that generally confers advantages during an organism's life cycle in the differentiation of a developing embryo. Tens of billions of cells die each day due to apoptosis in the average human adult. The yearly amount of self destruction amounts to a mass of cells which equals to an individual's body weight. Induction of extrinsic apoptosis is known to follow after high levels of DNA damage.

Apoptosis is a normal process, belonging to the protection of genetic information by the elimination of damaged cells which would need uneconomical excess energy for their repair (Fig. 4.2) and serves as a mechanism to destroy the genetic information which is not needed anymore after cellular differentiation. Defective apoptotic processes have been implicated in a variety of diseases (autoimmunity, myocardial infarction, stroke, diabetes, neurodegenerative diseases, Alzheimer's disease, infertility, hepatitis, sepsis, viral infections, etc.), whereas an insufficient amount of cell destruction results in uncontrolled cell proliferation leading to cancer. Epidemiological data provide evidence that it is possible to prevent cancer and other chronic diseases, some of which share common pathogenetic mechanisms, such as DNA damage, oxidative stress, and chronic inflammation. It is well established that mutations in somatic cells play a key role in cancer initiation and other stages of the carcinogenesis process. Diseases related to defects in apoptosis indicate that there are many genes implicated in the apoptotic process.

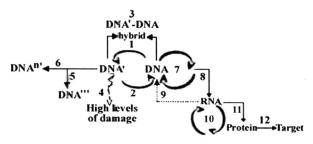

Fig. 4.2 Apoptosis involved in the processes related to the transfer of genetic information. Processes are numbered: 1, mutation; 2, DNA repair; 3, recombination; 4, apoptosis; 5, aging; 6, malignant transformation (cancer); 7, DNA replication; 8, transcription; 9, reverse transcription in retroviruses; 10, RNA replication in RNA viruses; 11, translation; 12, protein targeting. Figure 4.2 is identical with Fig. 1.10

The most important genes and products involved in apoptosis:

- Receptors and ligands (Fas, Trail, TNF),
- Bcl-2 family (more than a dozen),
- Proteases: Caspases (> 12), granzyme, calpain,
- Regulators of apoptosis:

  - eight IAPs (inhibitors of apoptosis proteins) in human cells (inhibit effector caspases and promote degradation), IAPs are also found in viruses.
  - Inhibitors of IAPs: Smac/DIABLO.
  - Two domain types: Bir domain and RING-Zn Finger. BIR domain is present in all IAP (Inhibitor of Apoptosis) family proteins. The domain binds Zn using 3 Cys and a His conserved across IAP proteins.
  - Survivin: expressed during development, not in differentiated cells. cIAP1, c-IAP2, XIAP. Survivin is expressed in most tumours, correlates to prognosis.
  - APAF1, a cytoplasmic protein that initiates apoptosis.

    Substrates of caspases: ICAD, PARP, laminA, Bcl-2, other caspases, BID etc.,
    Mitochondrial proteins: cytochrome c, VDAC, ANT AIF, SMAC,
    Kinases: MapK, PKA, Akt, CAP, RIP3,
    Transcription factors: p53 myc, Nur,
    Other genes: FADD, TRADD, DAD.

For abbreviations see the explanatory Glossary and abbreviation sections at the end of the book.

Have You Ever Seen Apoptosis?

The answer is yes, several times. To understand apoptosis the fall foliage of plants is a good example. The fall foliage season e.g. in the Northeastern States in New

England (Vermont, Maine, New Hampshire, US) is the busiest time of the year for visitors, who flock from all over the world to catch a glimpse of the phenomena what we regard as plant apoptosis. This apoptotic process starting in mid September is due to the fact that leaves are not lighted enough. In the fall, because of changes in the length of daylight and decrease in temperature, the leaves stop their food-making process. The chlorophyll breaks down, the colour of its green pigment disappears, and the yellow to orange xanthophyll pigments become visible and give the leaves part of their fall splendour (Fig. 4.3).

Along with the appearance of fall colours, other changes take place. At the point where the leaf is attached to the tree, a special layer of cells develops and gradually damages the tissues by an orchestrated series of biochemical events leading to characteristic cell morphology and death. When the leaf is shed and falls from the tree, it leaves behind a leaf scar. At the same time, the tree seals the cut. The shed leafs still contain many organic molecules, but their cells are dead. Most of broad leaf trees in zones with adverse winter conditions shed leaves in autumn. Succulent plants such as cacti and some euphorbias have no foliage leaves.

Plants eliminate cells, organs, and parts during responses to stress and express various developmental programs. Beside leaves non-pollinated flowers are fully thrown away. Ovaries with fertilised egg cells in ovules on the same plant are retained forming fruits while the other parts of flowers, e.g. petals and sepals fall off. Stigmas and pistils may also be eliminated. Hormons such as ethylene ($H_2C{=}CH_2$) produced in all higher plants are associated with fruit ripening, abscisic acid plays a major role in abscission of fruits, while gibberelins have antiapoptotic (among several other) effects delaying senescence in leaves and citrus fruits.

These elimination processes are highly controlled by internal factors and in some cases, such as fall foliage, are combined with external stimuli (light, temperature) leading to an array of cellular events including intracellular pathways. Fundamental elements of apoptosis, as characterized in animals in the next chapters, are conserved in plants. Although, plant apoptosis is out of the scope of this book, plant

Fig. 4.3 Fall foliage in Vermount (USA) – 1994 (with the author).

Fig. 4.4 Programmed cell death resulting in the loss of tadpole's tail during embryonic development. The four stages of embryonic development: (a) fertilized egg cells, (b) tadpole with long tail, (c) tadpole with shortening tail, (d) adult frog without tail

foliage is a good example to introduce apoptosis as a fundamental biologic process by which metazoan cells orchestrate their suicide. For futher details of plant apoptosis see Havel et al. (1999). Programmed cell death is an integral part of the life cycle of unicellular eukaryotes and even prokaryotes sometimes undergo regulated cell death (Ameisen, 2002). Programmed cell death is a widespread phenomenon, occuring in all kinds of metazoans including simple metazoans such as hydra (Cikala et al., 1999). Even bacterial colonies such as cyanobacteria or *E. coli*, the intestinal colony bacteria in many mammals, among them humans use apoptotic-like patways to control colony size (Berman-Frank et al., 2004; Kolter, 2007). During embrionic development of animals, cells are produced in excess which undergo programmed cell death and contribute to sculpturing many organs and tissues. Further implications of programmed cell death in animal development are the loss of cells in the interdigital tissue, the death of neurons in later stages of adult brain development or the development of reproductive organs. Cells of adult organisms also undergo constant and balanced physiological cell death to maintain homeostasis and constant cell numbers. A typical example of *intrinsic* animal apoptosis is the loss of tadpole's tail during its embryonic development (Fig. 4.4) or in mammals the interdigital cell death during the embryonic development which removes the inter-digital mesoderm initially formed between fingers and toes so that the fingers and toes can separate.

Sunburn
Similar apoptotic mechanism in humans is the peeling of sunburned skin. The so called "healthy tan" does not exist. Unprotected skin exposed to "healthy tan" undergoes premature aging, higher doses of ultraviolet doses cause apoptosis (sunburn).

What Triggers Apoptosis?

The fate of cells of multicellular organisms is determined by the balance between two types of cellular messages: die or survive. The cells of multicellular organisms communicate both by signals (messenger molecules moving through blood, lymph or other liquids) and by cues (messengers attached to the extracellular matrix or to intracellular receptors). Examples of plant apoptosis indicated that if the communication between cooperating cells of multicellular systems is modified by local environmental changes cells become detached from their proper place in the body and self-destruction takes place.

Apoptosis (Type I Programmed Cell Death) and Necrosis

Cell death can occur by two distinct mechanisms, necrosis or by programmed cell death (apoptosis). Necrosis ("accidental" cell death) is a pathological process occuring in cells exposed to physical or chemical insults. Apoptosis is a physiological and controlled process by which unwanted or useless cells are eliminated during development and other normal biological processes.

Apoptosis, or type I programmed cell death, is fundamental to both the development and homeostasis of multicellular organisms (Kerr et al., 1972; Vaux and Korsmeyer, 1999). These two cell death processes (necrosis, apoptosis) often occur simultaneously in a wide variety of pathological conditions as well as in cultured cells exposed to physiologic activators, physical trauma, environmental toxins and chemicals (Wyllie et al., 1980; Thompson, 1995).

Different types of cell death should be more precisely distinguished
A more precise distinction should be made among the various forms of cell death such as apoptosis, oncosis, necrosis and programmed cell death (Van Cruchten and Van Den Broeck, 2002). Apoptosis is characterized by cellular shrinking, condensation and margination of the chromatin and ruffling of the plasma membrane, breaking up of the cell in apoptotic bodies, followed by lysis. Cell death marked by cellular swelling should be called oncosis, whereas the term necrosis should refer to the morphological alterations appearing after cell death. Apoptosis and oncosis are regarded as pre-mortal processes, while necrosis is a post-mortal condition. Programmed cell death refers to one of the specific pathways irrespectieve of the characteristic morphology of apoptosis (Van Cruchten and Van Den Broeck, 2002).

Characterization of Apoptosis

Based on the analogy of leaves falling off trees, the word apoptosis (of Greek origin), means "falling off", emphasizes that death is an integral part of the life cycle of plants and animals. Apoptosis was first observed by Carl Vogt in 1842. The body's elaborate mechanism for programmed cell death is a complex and highly regulated process that allows a cell to self-degrade in order for the body to eliminate unwanted or dysfunctional cells. The term programmed cell death was used first by Lockshin and Williams (1964), who proposed that cell death during development is a locally

Apoptosis (Type I Programmed Cell Death) and Necrosis 213

and temporally defined self-destruction process. The term apoptosis has been introduced by Kerr et al. (1972) to describe typical morphological changes leading to controlled self-destruction of cells. The first demonstrated biochemical feature of this type of cell death was internucleosomal fragmentation (Wyllie, 1980), which was occasionally preceded by the generation of large DNA fragments (Pandey et al., 1994). Meanwhile it turned out that cytoplasmic changes occur in cells lacking nuclei, indicating that apoptosis may occur without endonuclease mediated DNA degradation. Furthermore reproducible proteolytic cleavage was observed pointing to the importance of proteases in the induction and completion of apoptotic cell death (Kaufmann et al., 1993). Genetic studies in *Caenorrhabditis elegans* provided evidence that the CED-3 gene, one of the three genes that are essential for regulating the cell deletion, is turned on during the normal development of this nematode. Ced-3 protein was shown to share extensive homologies with the human protease interleukin-1b converting enzyme (ICE) (Yuan et al., 1993). Soon at least ten homologues of the ICE/CED-3 protease have been identified. These aspartate-specific cysteine proteases are now called "caspases" (Alnemri et al., 1996). Beside caspases other proteases such as calpains, granzymes and the proteasome/ubiquitin pathway of protein degradation turned out to play a role in the process.

Selective degradation of proteins in eukaryotes primarily requires the ubiquitin system that functions to mark proteins for degradation by the multicatalytic protease, the proteasome. Successive ubiquitinations result in the formation of multiubiquitin chains attached to proteolytic substrate proteins. Multiubiquitinated proteins are then recognized and degraded by the proteasome. Proteasomes are multimeric proteinase complexes containing 2 copies each of 14 different polypeptides. Proteasomes play a central role in protein degradation in animal cells and are involved in many basic cellular processes including the cell cycle, apoptosis, stress response, and the regulation of immune and inflammatory responses (Coux et al., 1996; Hilt and Wolf, 1996; Voges et al., 1999; Glickman and Ciechanover, 2002).

Types of Programmed Cell Death

Since almost all programmed cell death is apoptotic, the terms apoptosis and programmed cell death are often used interchangeably. When the phrase programmed cell death is used without qualification, it usually refers to type I cell death (apoptosis). However, accumulating reports demonstrate that there are several other non-apoptotic forms of programmed cell death beside apoptosis or type I cell death. These irregular forms include: a. autophagy (Type II cell death), b. paraptosis, c. mitotic catastrophe, d. dark cell death and e. necrosis-like programmed cell death which will be discussed only briefly.

Type I Cell Death (Apoptosis)

Among the significant morphological changes taking place inside the eukaryotic cell we find processes such as cell motility, endocytosis, exocytosis and the most dramatic changes accompanying the process of cell division. In these processes the cytoskeleton, particularly the actin skeleton plays a key role. In another process the cell is subjected to physiological cell death known as apoptosis with the most striking morphological changes taking place during the active or execution phase (Wyllie, 1981; Arends and Wyllie, 1991; Fesus et al., 1991; Earnshaw, 1995; Lazebnik et al., 1995) in a sequence of events leading to the complete fragmentation of the cell into small rounded structures containing fragments of the preexisting cell, called apoptotic bodies.

Biochemical Changes in Apoptosis

Apoptotic changes in cells are distinguished according to their biochemical and morphological characters. The two best known biochemical changes are the general increase in protease and nuclease activity. Serine proteases play an important role in many models of cell death. The nuclease activity degdares chromatin into oligonucleosomes forming the well-known apoptotic ladder and ~30–50 kb fragments.

Other significant metabolic changes in the cytoplasm and nucleus include:

- loss of balance between inducer and repressor factors, belonging to the Bcl-2 family (e.g. Bcl 2 pro-survival, Bax pro-apoptotic factor),
- induction of p53 tumour suppressor gene,
- inreased level of phosphatidylserine,
- caspase-3 activation,
- the breakdown of cytoskeleton by caspases,
- cleavage of poly-ADP-ribose polymerase (PARP),
- mitochondrial release of cytochrome c,
- cleavege and degradation of several proteins,
- formation of reactive oxygen species (ROS),
- enhanced expression of glutathione-S-transferase (an antioxidant enzyme),
- high lipid peroxide content of cells (Tarachand and D'souza, 1999).

Morphological Changes in Apoptosis

Apoptosis is an *intrinsic* property of cells and shows the following characteristic morphology under the microscope:

1. Cytoskeleton disruption is one of the most important features in apoptosis leading to shrinking and rounding of cells and loosing contact with neighbouring cells.
2. Rapid movement of agonizing cells, seen as "apoptotic dance" under the microscope (Trencsenyi, G., Nagy, G., Pinter, G., Rozsa, D. and Bánfalvi, G. see in the attached DVD).
3. Displaying intracellular proteins on cell surface.
4. Chromatin is undergoing condensation forming compact patches inside the nucleus, known as pyknosis, the hallmark of apoptosis (Susin et al., 2000; Kihlmark et al., 2001).

Apoptosis (Type I Programmed Cell Death) and Necrosis 215

5. Discontinuities in the nuclear envelop and DNA cleavage producing first large fragments, referred to as karyorrhexis.
6. The chromatin (DNA and protein) in the nucleus is degraded. Further breakdown of nuclear material to small nucleosomal units (180 base pairs) is seen as laddering during electrophoresis, differentiating apoptosis from toxic (ischemic) cell death.
7. Irregular buds in the cell membrane known as blebs.
8. The Golgi apparatus disperses during apoptosis, without obvious degradation, in a manner similar to that occurring in mitosis (Philpott et al., 1996).
9. Cells that commit suicide undergo mitochondrial breakdown with the release of cytochrome c.
10. The phospholipid phosphatidylserine, which is normally hidden within the plasma membrane, is transferred from the cytoplasmic surface of the cell membrane to the outer leaflet.
11. The phospholipid phosphatidylserine is bound by receptors on phagocytic cells like macrophages and dendritic cells which then engulf the cell fragments.
12. The phagocytic cells secrete cytokines that inhibit inflammation e.g., interleukin-10 (IL-10) and tumour growth factor beta (TGF-β).
13. The nuclear lamina is not completely solubilized during apoptosis, as occurs in mitosis (Philpott et al., 1996).
14. The breakdown of cells into small, spherical membrane-bounded vesicles, called apoptotic bodies. Nearby phagocytic cells like macrophages and dendritic cells engulf apoptotic cell fragments. Phagocytosis removes dead cells with a minimum of risk or damage to neigbouring cells.

Among these morphological changes the two most characteristic ones are apoptotic chromatin condensation and the rapid uptake of dead cells by their neighbouring cells in a highly ordered and morphologically distinct process. Apoptotic cells undergoing the above set of morphological changes revealed the activation of a complex machinery leading to the disruption of cells into small, spherical, membrane-bounded fragments called apoptotic bodies. Morphological changes of apoptosis vary between cell types. This heterogeneity reflects the wide range of cellular proteins and enzymes involved in apoptotic pathways and points to genotoxic specific changes to be characterized later. Chromosomes could be seen earlier only after their condensation, consequently different types of morphological changes of apoptosis could be hardly detected or visualized since the condensation process takes place behind the nuclear curtain.

Apoptosis Is a Selection Process

1. As a consequence of self destruction cells that have served their purpose during fetal development are removed by apoptosis.
2. Killing off those cells takes place which have become damaged and threaten to grow out of control to form tumours, consequently cell suicide helps to prevent cancer.

216   4   Apoptosis

3. Apoptosis is selecting against certain immune cells after a brief infection-fighting active life to prevent them from turning their deadly weapons on body's normal cells.

Apoptosis Is a Controlled Process

Necrosis is uncontrolled, not a programmed cell death. There are many apoptotic signals and pathways leading to a general mechanism causing cells to undergo organized degradation of cellular organelles by the activation of a family of cysteine proteases called caspases (Thornberry and Lazebnik, 1998). The interaction of several "die" or "survive" signals from inside and outside the cell ultimately determine whether cells continue to reproduce or are terminated. Proteins carry these signals and the relative balance of the survival and apoptotic ("death") signals determines the cell's fate. Regulation of apoptosis is critical for tissue development and for the homeostasis of the organism. The loss of homeostasis has severe consequences. Apoptotic hyperactivity can impair development or cause neurodegenerative diseases, hypofunction in apoptosis may aggravate viral infections and lead to cancer.

During the process of apoptosis, chromatin condenses, initially in a circumferential pattern, but thereafter the nucleus becomes shrunken and often separates to several spherical bodies, called apoptotic bodies. Concomitantly with these appearances, DNA is cleaved, first to large chromatin domains of around 50–300 kbp, then to many oligonucleosomes, but not in all cell types (Susin et al., 2000).

Pathways of Type I Cell Death

Three mechanisms are known to be involved in the apoptotic process: (a) receptor-ligand mediated mechanism, (b) mitochondrial pathway and (c) mechanism in which the endoplasmic reticulum plays a central role. All three mechanisms activate caspases which are responsible for the characteristic morphological changes observed during apoptosis (Van Cruchten and Van Den Broeck, 2002).

Alternative Pathways of Programmed Cell Death

Under specific conditions, particularly in the absence of functional caspase-dependent pathway, activation of several alternative cell death models has been proposed, including autophagy, paraptosis, mitotic catastrophe and the descriptive model of apoptosis-like and necrosis-like programmed cell deaths (Broker et al., 2005).

Type II Programmed Cell Death

This type is also known as autophagy or *cytoplasmic* cell death, meaning the lysosomal digestion of a cell's own cytoplasmic material. Type II programmed cell death is characterized by the formation of large vacuoles which digest organelles in a specific sequence prior to the nucleus is degraded.

Apoptosis (Type I Programmed Cell Death) and Necrosis 217

In many tissues, dying cells display similar changes in morphological and chromosomal DNA organization. The process was such a widely accepted phenomenon that many authors assumed that all programmed cell deaths occur by apoptosis. However, the comparison of the patterns of cell death displayed by T cells and muscles proved that they differ in terms of cell-surface morphology, nuclear ultrastructure, DNA fragmentation, and polyubiquitin gene expression. These observations raised the question: Do all programmed cell deaths occur via apoptosis? The significant differences suggested that more than one cell death mechanism is used during development and researchers arrived at the autophagic or Type II cell-death mechanism (Schwartz et al., 1993). Similar question was raised and similar answer was given by Bursch et al. (2000).

Paraptosis

This process results in a non-apoptotic morphology indicating that the biochemical pathways of programmed cell death might be much more diverse than expected. Paraptosis was described by Sperandio et al. (2000). The features of paraptosis differ from those of apoptosis and involve cytoplasmic vacuolation, mitochondrial swelling and absence of caspase activation without condensation or fragmentation of the nuclei. These morphological and metabolic changes point to an internally programmed cell death, which could run side-by-side with the apoptosis of other cells, but are subject to different control mechanisms (Sperandio et al., 2000; Wyllie and Golstein, 2001).

Mitotic Catastrophe

This is a type of cell death that occurs during mitosis. In response to DNA damage, checkpoints are activated to delay cell cycle progression and to coordinate repair. The mitotic DNA damage checkpoint in mammalian cells delays mitotic exit and prevents cytokinesis and, thereby, is responsible for mitotic catastrophe (Huang et al., 2005). Mitotic catastrophe is the response of mammalian cells to mitotic DNA damage. Mitotic catastrophe seems to be the result of a combined deficiency of cell-cycle checkpoints, particularly the DNA structure checkpoints and the spindle assembly checkpoint in association with cellular damage. Unarrested cell cycle before or at mitosis triggers aberrant chromosome segregation, which culminates in the activation of the apoptotic default pathway and cellular demise (Castedo et al., 2004). The morphological aspect of apoptosis may be incomplete, although these alterations constitute the biochemical hallmarks of apoptosis. Cells that fail to execute an apoptotic program in response to mitotic failure are likely to divide asymmetrically in the next round of cell division, with the consequence of aneuploid cells. This implies that disabling of the apoptotic program may actually favour chromosomal instability, through the suppression of mitotic catastrophe. Mitotic catastrophe thus may be conceived as a molecular device that prevents aneuploidization, which may participate in oncogenesis.

218   4 Apoptosis

Necrosis-Like Programmed Cell Death

Besides the three types of programmed cell death mentioned, other pathways called "non-apoptotic programmed cell-death" ("caspase-independent programmed cell-death" or "necrosis-like programmed cell-death") have been discovered. These represent alternative and efficient backup mechanisms leading to death but can also function as main type of programmed cell death (Kroemer and Martin, 2005). Apoptosis and necrosis may share common initiating pathways and the final outcome may be dependent on the availability of downstream caspases. Apoptosis and necrosis are the major types of cell death, with major focus of this book on apoptotic chromatin changes.

Dark Cell Death

Dark cell death was observed during the slow neuronal death in Huntington's disease, characterized by cytoplasmic condensation, chromatin clumping, wavy cell membrane, without blebbing of nuclear or plasma membrane (Leist and Jaattela, 2001).

Necrosis

The word necrosis comes from the Greek Νεκρός meaning dead and refers to accidental cell or tissue death. Since there is no cell signaling system involved, the cleanup of the necrotic cellular debris by phagocytic immune cells is more difficult than in programmed cell death. Due to the lack of signalling sytem, dead cells of necrosis cannot be easily located and recycled, the lysis of cells following membrane damage causes inflammation.

Necrosis is characterized by:

– swelling of cells,
– nuclear swelling,
– mitochondrial damage and rapid loss of energy,
– loosing homeostatic control,
– loss of integrity of cell membrane by lysis is an early event,
– disruption of organelles,
– release of cellular content,
– inflammatory responses (oedema, damage to surrounding cells),
– rupture of necrotic cells.

If apoptosis does not occur fast enough the plasma membrane and intracellular organelles can undergo lysis in a process called secondary necrosis.

The type of cell death is dependent on the intensity of the damage, the length of exposure, the spatial orientation, stage of cell cycle toward the damaging agent. Short exposures or low concentrations of genotoxic agents cause the cells to undergo delayed apoptotic death. In contrast, cells exposed to high concentrations of the same agent rapidly undergo necrotic cell death. One can conclude from these characteristics that apoptosis is the "default" mode of death, while high level injury or severe ATP depletion does not initiate apoptosis, rather necrotic cell death ensues.

The major distinctive features of apoptosis were originally described by Kerr et al. (1972):

- DNA fragmentation
- Chromatin condensation
- Cell shrinkage
- Preserved cell membranes
- Preserved organelles
- Engulfment of apoptotic cells by neighbouring cells to avoid inflammation

A further morphological characteristic is the apoptotic dance, visualized in the attached DVD.

Morphological Differences Between Necrosis and Apoptosis

Depending on the cell type and the stimulus, a cell may die either by apoptosis or necrosis (Fig. 4.5). Upon death stimulus the necrotic cell is swelling, blebbing and lysed (Fig. 4.5 upper part). Apoptosis can be recognized by chromatin condensation, internucleosomal degradation of the DNA, cell shrinkage and disassembly into membrane-enclosed vesicles (Fig. 4.5 lower part) as a consequence of caspase activation (Rathmell and Thompson, 1999). The cells of the immune system undergo two different kinds of apoptotic processes: activation-induced cell death, and damage-induced cell death, a more generalized phenomenon in response to a variety of cellular insults, mainly oxidative metabolism by-products (Ginaldi et al., 2005).

Apoptotic and necrotic cells are both recognized by phagocytes, but only apoptotic cells are eliminated without the release of cytosolic components to the

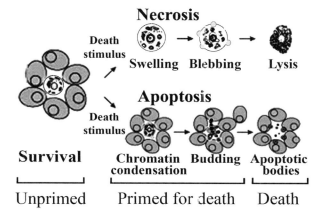

Fig. 4.5 Major morphological differences between necrosis and apoptosis. Unprimed cell prone to cell death is surrounded by six healthy cells. Upon death stimulus the cell in the middle may undergo either necrosis (*upper row*) or apoptosis (*lower part*)

environment, thereby preventing an inflammatory response (Fadok et al., 2000; Sauter et al., 2000). Apoptotic cells and bodies are swiftly recognized by macrophages and engulfed.

The milk fat globule epidermal growth factor (EGF) 8 (MFG-E8 protein) is involved in the engulfment of apoptotic cells. If apoptotic cells are not efficiently engulfed, they undergo secondary necrosis, leading to systemic lupus erythematosus-type autoimmune diseases at least in mice (Hanayama et al., 2004). In contrast to apoptosis, necrosis is characterized by the swelling of cells and organelles, resulting ultimately in disruption of the cell membrane and cell lysis (Majno and Joris, 1995) (Table 4.1).

Table 4.1 Pathological distinction between apoptosis and necrosis

	Apoptosis	Necrosis
Stimuli	Physiological or pathological	Pathological
Occurence	Single cells	Groups (neighbouring cells)
Cellular damage	Phagocytosis, without inflammation	Inflammation, presence of macrophages
Cell size	Shrinkage, fragmentation	Swelling
Cellular membrane	Continuity maintained, blebbed, phosphatidylserine turned outside	Smooth, early lysis
Adhesion of cells	Early loss	Late loss
Cytoplasmic swelling	Late stage	Early stage
Damage to organelles	Late	Early
Swelling of organelles	Contracted, "apoptotic bodies"	Swelling, disruption
Release of lysosomal enzymes	Absent	Present
Mitochondria	Surface relatively intact, increased permeability, release of cytochrome c and Apaf-1	Swelling, chaotic structure
Nucleus	Convoluted, breakdown, (caryorhexis)	Disappearance, (caryolysis)
Chromatin	Compaction in dense masses	Clumping, undefined forms
DNA breakdown	Internucleosomal cleavage, DNA ladder formation	Diffuse, random smearing

Induction of Apoptosis

There are a large variety of apoptotic stimuli which can trigger apoptosis affecting either the cytoskeleton, the replication process or genome expression. In contrast with the diversity of stimuli of apoptosis, signalling and execution mechanisms are conserved (Earnshaw, 1995). Apoptosis is triggered by a broad range of stimuli, including death ligands, DNA damage, and cellular stresses. Apoptosis as an active process can be induced from outside or inside the cell, e.g. by the binding of cell surface receptors, by large DNA damage causing defects the DNA repair mechanisms cannot cope with, viral infection, inefficient repair, treatment with cytotoxic drugs or irradiation, by the lack of survival signals, contradictory cell cycle signalling or by developmental death signals. Death signals of diverse origin appear to initiate similar cell death pathways leading to the characteristic features of apoptotic cell death, characterized by (a) cell shrinkage, detachment, (b) membrane blebbing, chromatin condensation, DNA fragmentation, (c) apoptotic body formation and (d) lysis (Fig. 4.6).

This type of cell death program is regulated by a proteolytic process mediated by caspases (Salvesen and Dixit, 1997; Cyrns and Yuan, 1998; Thornberry and Lazebnik, 1998). Caspases are cysteine proteases and are originally expressed as inactive precursors, but an apoptotic stimulus induces the autoproteolytic activation of initiator caspases such as caspase-8 and -9. Activated initiator caspases cleave and activate downstream effector caspases, such as caspase-3 and -6, which in turn cleave various cellular substrates that may be activated or inactivated (Nicholson, 1999). Among the substrates one of the most critical one is the cleavage of the inhibitor of caspase-activated DNase (ICAD). Cleavage of ICAD by effector caspase-3 releases caspase-activated DNase (CAD), which is responsible for nucleosomal DNA fragmentation during apoptosis (Enari et al., 1998; Liu et al., 1997, 1998). Since the molecular mechanisms underlying the other characteristic morphologies of apoptotic cells have yet to be elucidated, the question arises whether or not apoptotic processes carrying out apoptotic cascades (caspase activation,

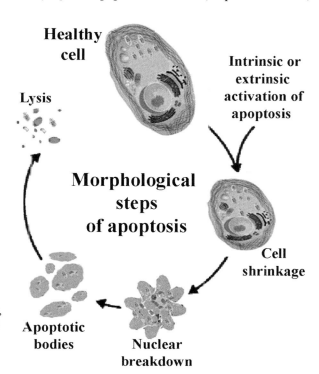

Fig. 4.6 The four major stages in apoptotic cell death. (**a**) Activation of apoptosis leading to cell shrinkage and chromatin condensation, (**b**) nuclear breakdown (budding), (**c**) the formation of apoptotic bodies, (**d**) lysis and clearance by engulfement, autophagy or other

proteolytic activity, degradation of cytosolic and nuclear targets) are the same or different irrespective of the types of the trigger in the commitment phase and the following execution phase. The characterization of different types of cell death, including the intensity, length, spatial and temporal factors suggest that multiple patways are involved. The diverging and expanding web of the constantly increasing apoptotic pathways seems to be far from being established. As morphological changes depend also on apoptotic stimuli, apoptosis can be characterized by at least one more criterion, namely by the type of genotoxic agents. This would allow a further classification of apoptotic changes based on the specificity of genotoxic effects. The multiple effects could explain the recently recognized diversity of cell death. This would also mean that not only the organisms die in thousand ways but also the cells as written in the Latin distich of Vesalius, the great physician of the XVI. century:

Mille modis morimur mortales nascitur uno
Sunt hominum morbis mille sed una salus.

The first row of the couplet states that:

We mortals die in thousand ways, but are born in one way
There are thousand kinds of human diseases, but only one health.

Similarly to humans, cells can be affected by several means but healthy only by one way as the second row of the chronogram continues.

Elegant Experiments in *Caenorhabditis elegans* Model Organism

C. elegans has played a key role in the understanding of apoptosis. There are several reasons why this parasitic worm was chosen for experiments related to apoptosis: it is a simple organism consisting of 1090 cells, the fates of all cells are known in the embrionic development, it is rapidly reproduced with 3 weeks of life span, easy to maintain worm cultures, easy to mutate worms with ethyl methylsulphonate (EMS), capable to form hermaphrodites. Of the 1090 cells 131 undergo apoptosis, without influencing the viability of the worm i.e. these cells are probably required only in the early stage of development (Fig. 4.7).

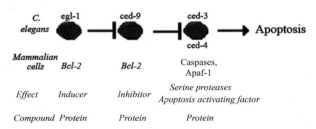

Fig. 4.7 Inducers and inhibitors of apoptosis in the parasitic worm *Caenorhabditis elegans* and in mammalian cells. The major activity of Bcl-2 oncogene is the inhibition of apoptosis. In *Caenorhabditis elegans* (*C. elegans*), three main cell death regulators have been isolated including *ced-3*, *ced-4* and *ced-9*. *Ced-3* encodes a *C. elegans* caspase and is activated by the adaptor molecule Ced-4. Ced-9 shares homology with the mammalian Bcl-2 protein and binds Ced-4, inhibiting its ability to activate Ced-3

Regulation of Apoptosis

The primary targets of regulation of apoptosis are the proteins involved. There are many proteins which have been shown to act on the apoptotic pathway. Among them, the cascade of members of the cysteine protease family, caspases, is thought to be indispensable for apoptosis to occur (Nicholson et al., 1995; Dubrez et al., 1996; Jacobson et al., 1996; Enari et al., 1996). Caspase-3, a member of this family, is the most downstream in the protease cascade, and the inhibition of caspase-3 by synthetic peptide inhibitors often prevents apoptosis induced by various stimuli. On the other hand, Bcl-2 and Bcl-xL are apoptosis inhibitory proteins. They function as suppressors of mitochondrial dysfunction related to apoptosis (Cossarizza et al., 1994; Zamzami et al., 1995; Petit et al., 1995), and inhibit the release of cytochrome c (Yang et al., 1997; Kluck et al., 1997) and apoptosis-inducing factor, from mitochondria (Susin et al., 1996). Healthy cells are balanced by the presence of pro- and anti-death Bcl-2-family proteins.

The Bcl-2 protein family plays a central role in the positive and negative regulation of cell survival and cell death. Bcl-2 is the prototype of inhibitors preventing the induction of apoptosis. Most important proteins involved in the regulation of apoptosis are: p53, p21, WAF1, Myc, Bcl-2, Bax, Bak (Fig. 4.8).

The proapoptotic proteins of the Bcl family promoting cell death: Bcl-X_5, Bcl-xL, Bax, Bad, Bak, Puma, Noxa, Bid, Bik, Bim, Hrk, *C. elegans* Egl-1 cause the depolarization of the mitochondrial membrane. Other components of the proapoptotic extrinsic pathway are the apoptotic protease activating factor-1 (APAF-1), a major component of the apoptosome; Fas/CD95, death receptor 4 (DR4) and DR5.

Natural-born killers, such as the peptide subunits of cell-signaling "BH3" protein domains either sensitize or activate mitochondrial apoptosis, and out-maneuver opposing "anti-death" proteins and trigger the suicide process.

BH-domains
Bcl-2 homology (BH) domains consist of BH1, BH2, BH3 and BH4.

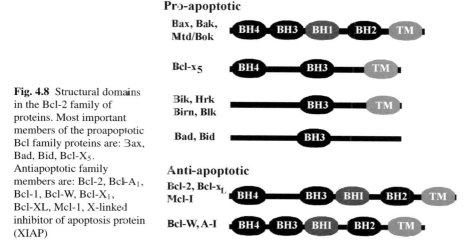

Fig. 4.8 Structural domains in the Bcl-2 family of proteins. Most important members of the proapoptotic Bcl family proteins are: Bax, Bad, Bid, Bcl-X_5. Antiapoptotic family members are: Bcl-2, Bcl-A_1, Bcl-1, Bcl-W, Bcl-X_1, Bcl-XL, Mcl-1, X-linked inhibitor of apoptosis protein (XIAP)

224   4 Apoptosis

- Functional pro-apoptotic members require the presence of BH3 domain.
- BH domains lack enzymatic activity, function by dimerization.
- BH1 and BH2 are death antagonists, allow heterodimerization with Bax to repress apoptosis.
- BH3 is a death agonists. In Bax, Bak heterodimerizes with Bcl-XL and Bcl-2 to promote apoptosis.
- BH4 is conserved in antagonist members (e.g. Bcl-XL), allowing interaction with death regulatory proteins such as Raf-1, Bad, and Ced-4.
- Bid contains one BH3. Caspase 8 cleavage of Bid results in translocation of cleaved Bid to the mitochondria inducing Cyt. C release. Bid links the death receptor pathway with the mitochondrial cell death pathway and amplifies caspase activation.

The proapoptotic activity of BH3 protein is regulated by transcriptional and post-transcriptional mechanisms to prevent undesirable cell death during development. To the contrary in healthy cells the balanced presence of anti-death Bcl-2-family proteins and their direct binding to p53 prevents apoptosis. Antiapoptotic proteins of the Bcl family inhibit the induction of cell death including: Bcl-2, Bcl-A$_1$, Bcl-1, Bcl-W, Bcl-X$_1$, Bcl-XL, Mcl-1 (Kelekar and Thompson, 1998). Similarly to normal cells, cancer cells also stay alive, but in these cells the overexpression of anti-death proteins dominates (Letai et al., 2002). Certain cells live in a state that can be distinguished by their dependence on antiapoptotic proteins for survival. These cells are desribed as being "primed for death", as death signaling has caused their anti-apoptotic family members to sequester significant quantities of proapoptotic BH3 proteins. Inhibition of the antiapoptotic proteins in these cells, but not unprimed cells, results in BAX/BAK oligomerization. It is suggested that there are three functionally distinguishable states with respect to programmed cell death: unprimed, primed for death, and dead. The state of being primed for death is a continuum, as the magnitude of BH3 proteins priming the mitochondrion can vary continuously until the antiapoptotic reserve is overwhelmed and the cell commits programmed cell death (Certo et al., 2006).

Extrinsic and Intrinsic Pathways of Apoptosis

Apoptosis and related forms of cell death have central importance in development, homeostasis, tumour surveillance, and in the function of the immune system. Based on the origin of the stimulus apoptosis can be categorized and divided into two principal pathways referred to as *extrinsic* and *intrinsic* pathways. The extrinsic pathway is activated by the ligation of death receptors, whereas the *intrinsic* pathway emerges from mitochondrial stress.

Extrinsic (Caspase 8) Pathway

The *extrinsic* pathway begins with the binding of a ligand (e.g. FasL, CD95L) to the death receptors (e.g. CD95) and results in the clustering of receptors and initiates the

Apoptosis (Type I Programmed Cell Death) and Necrosis 225

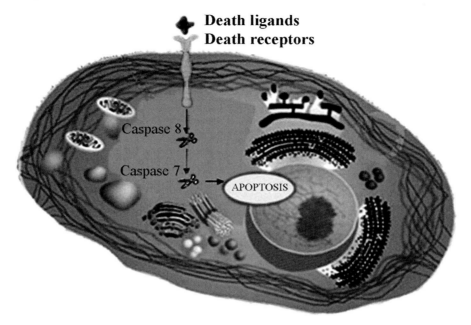

Fig. 4.9 General *extrinsic* pathway of apoptosis. Death ligand binds to the death receptor inducing first initiator then executioner caspases, leading to apoptosis. Caspase cleavage is indicated by the scissors

extrinsic pathway. Fas ligand belongs to the "death receptor" family, which includes the TNF receptor, to be discussed later. The general extrinsic pathway of apoptosis is schematically shown in Fig. 4.9.

Caspases can be activated by death receptors (FAS, TRAIL, TNF) activating caspase 8 and 10, or activated by granzyme B released by cytotoxic T cells and NK cells and activate caspase 3 and 7 and by apoptosomes (regulated by cytochrome c and the Bcl-2 family) which activate caspase 9. These pathways will be discussed later.

Mitochondrial involvement is probably not required for receptor mediated extrinsic cell death (Fig. 4.9). However, mitochondria may act as an amplification step, when the signal generated by the death receptor is weak. The relative contribution of mitochondria may depend on the cell type (Scaffidi et al., 1998).

Intrinsic (Mitochondrial) Pathway

Mitochondria are essential to cellular aging. Free radical production by mitochondria is increased with aging. The rate of oxidant production by mitochondria correlates inversely with the maximal life span of species. In many species, females live longer than males. It was reported that mitochondrial oxidant production by females is significantly lower than that of males (Sastre et al., 2002). Mitochondria play also an important role in the regulation of cell death. The signal activation of the *intrinsic* pathway is known as mitochondrial apoptosis (Fig. 4.10).

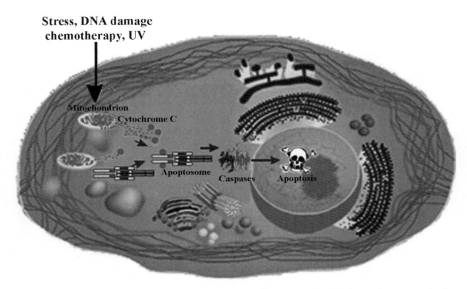

Fig. 4.10 *Intrinsic* (mitochondrial) pathway of apoptosis. External DNA damaging agents (radiation, chemotherapy) or internal stress (e.g. reactive oxygen species = ROS) causing mitochondrial damage initiate the intrinsic pathway, by releasing cytochrome c from damaged mitochondria. Cytochrome c released from mitochondria binds the cytosolic Apaf-1 protein and dATP to form the apoptosome. Apoptosome particles are attracting and activating pro-caspase-9. The *intrinsic* pathway activated by DNA-damaging agents is then coupled to the *extrinsic* pathway through the caspases

The *intrinsic* pathway is primarily activated by DNA-damaging agents and is coupled to the *extrinsic* pathway of caspase 8 or 10 induced cleavage of Bid (a proapoptotic Bcl2 family member). The intrinsic apoptotic pathway begins when DNA injury occurs within the cell. The injury does not result in necrosis and does not produce an inflammatory response since the apoptotic machinery is in place and ensures that the damaged cell is degraded cleanly, to prevent inflammation. The increased presence of p53 protein induces the transcription of pro-apoptotic BAX and APAF1 genes the gene products of which initiate the release of cytochrome c and Apaf1 proteins. Mitochondrial damage can initiate the intrinsic pathway by overcoming the protecting activity of the pro-survival protein Bcl2 and by the release of cytochrome c from damaged mitochondria. Mitochondria are involved in the activation of caspases by releasing cytochrome *c* into the cytosol where cytochrome c binds to the adaptor known as apoptotic protease activating factor 1 (Apaf-1) and causes its oligomerization. This renders Apaf-1 competent to recruit and activate the cell death initiator, pro-caspase-9 (Fig. 4.11).

Mitochondria were originally not thought to be involved in cell death, as cells without mitochondrial genome died by apoptosis and Bcl-2, a cell death inhibitor residing in the mitochondrial outer membrane, prevented this cell death (Jacobson et al., 1993). Subsequent, studies with cell-free systems of apoptosis proved that mitochondria are part of the central apoptotic pathway and several reports

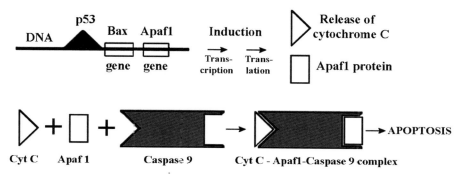

Fig. 4.11 The role of p53 in apoptosis. (**a**) p53 may activate the expression of Bax and Apaf1 genes. Bax protein can then stimulate the release of cytochrome c from mitochondria. (**b**) Apoptosis can be induced by the binding of cytochrome c and Apaf1 to Caspase 9

demonstrated that cytochrome *c* is released from mitochondria by various pro-apoptotic stimuli in intact cells (Liu et al., 1996; Adachi et al., 1997; Kharbanda et al., 1997; Kim et al., 1997; Kluck et al., 1997; Yang et al., 1997; Bossy-Wetzel et al., 1998). The role of cytochrome *c* in apoptosis was further supported by the finding that micro-injection of cytochrome *c* into the cytoplasm causes dose-dependent cell death in some cell types (Li et al., 1997; Zhivotovsky et al., 1998). To the contrary, micro-injection of neutralizing anti-cytochrome *c* antibodies can prevent cell death (Neame et al., 1998).

At low doses of apoptotic stimulus not all mitochondria dump their cytochrome *c*. Some mitochondria may escape the death signal, proliferate and rescue the cell. Inhibitor of apoptosis proteins (IAPs) and caspase inhibitors may limit small amounts of mitochondrial damage (Deveraux et al., 1997, 1998; Tamm et al., 1998). That cytochrome *c* release from mitochondria might be a reversible event has been evidenced by studies on neuronal cultures. When NGF was re-added after nerve growth factor (NGF) withdrawal to caspase-inhibited sympathetic neurons, cytochrome *c* loss from mitochondria became restored and cells survived (Martinou et al., 1999).

Once caspase 9 is activated, it cleaves and activates other downstream cell death effector caspases, the first of which is the effector caspase 3. Bcl-2, an apoptosis inhibitor localized to the mitochondrial outer membranes, prevents cytochrome *c* release, caspase activation and cell death. In living cells caspases are present as inactive zymogens and become activated in response to pro-apoptotic stimuli. Other proteins released from damaged mitochondria are Smac/DIABLO and Omi/HtrA2, inactivate IAPs (inhibitor of apoptosis proteins), which normally bind and prevent the activation of caspase 3. The correlation among Bcl family members, IAPs, Smac/DIABLO and Omi/HtrA2 is central to the intrinsic apoptosis pathway. In *C. elegans*, *ced-3* and *ced-4* were identified as genes essential for programmed cell death, while *ced-9* was found to be a regulator of cell death by preventing apoptosis. The first mammalian homolog for ced-3 was described as Bcl-2 which was found in B-cell lymphomas (Vaux et al., 1988). Noteworth to be mentioned that caspases

228   4 Apoptosis

have also critical nondeath regulatory functions, notably they are involved in the maturation of a wide variaty of cell types.

Other observations related to the decrease of the mitochondrial inner membrane potential also indicated that mitochondria are involved in apoptosis (Petit et al., 1996; Zamzami et al., 1995). In addition, the electron transport chain and the coupled oxidative phosphorylation generate reactive oxygen species (ROS) initiating apoptosis in damaged or in those mitochondria which do not function properly (Hockenbery et al., 1993; Kane et al., 1993; Sarafin and Bredesen, 1994; Greenlund et al., 1995; Zamzami et al., 1995; Adachi et al., 1997).

Glutamate Downregulates the Intrinsic Pathway of Apoptosis

In many cell types, limiting glutamate (Glu) supply or inhibiting glutamine (Gln) metabolism via glutaminase inhibits the supply of intracellular glutamate available for glutathione (GSH) biosynthesis. To the contrary, *in vivo* studies have shown that excess Gln reduces mitochondrial GSH (mtGSH) levels in some cancer cells. This controversy seems to be the result of enhanced ROS generation by oxidative Gln metabolism, which reduces mtGSH. Depressed mtGSH in turn induces the cancer cells to be more susceptible to TNF-alpha-induced apoptosis via NADPH action. In addition, glutamine is a substrate of oxidative metabolism and a precursor of ATP synthesis that can lead to superoxide production. Conversely, catalase is able to inhibit c-myc-induced apoptosis and superoxide dismutase also depletes ROS-induced cell death.

Cytochrome *c*: With or Without It

That cytochrome *c* is a double agent for life and death was discovered by its regulatory function in apoptosis besides being involved in bioenergetics (Reed, 1997; Bossy-Wetzel and Green, 1999, review). Cytochrome *c* is a mitochondrial protein encoded by the nuclear apo-cytochrome *c gene* and imported across the mitochondrial outer membrane. In the mitochondrial intermembrane space apo-cytochrome *c* acquires a heme group, resulting in holo-cytochrome *c* in the mitochondria. In living cells, holo-cytochrome *c* shuttles electrons from respiratory complex III to IV, supporting intracellular ATP production by oxidative phosphorylation. For caspase activation also the holo-cytochrome c is needed, not the apo-cytochrome *c* (Yang et al., 1997). Upon release from mitochondria both oxidized and reduced cytochrome *c* induce caspase activity with similar efficiency (Hampton et al., 1998). Even if cytochrome c is in its redox-inactive state (Cu- and Zn-substitution) it still maintains its pro-apoptotic efficiency (Kluck et al., 1997). The release of cytochrome c is regulated by the anti- and proapoptotic members of the Bcl-2 family. Once in the cytoplasm, cytochrome c binds Apaf-1 to pro-caspase-9, leading to the activation of caspase-9 and to the initiation of the caspase cascade (Fig. 4.11B). Then caspase-9 cleaves and activates the executioner caspase-3, which in turn cleaves poly (ADP-ribose) polymerase (PARP) and activates caspase activated DNase (CAD) to fragment DNA (Green and Reed, 1998).

Cytochrome *c* release may have short- and long-term effects:

1. the immediate effect of cytochrome *c* is that it activates caspases along with Apaf-1 leading to rapid caspase-dependent cell death,
2. the long-term effects of cytochrome *c* loss from mitochondria include:

   a. the disruption of electron transport chain,
   b. the decrease in intracellular ATP level,
   c. the accumulation of reactive oxygen species (ROS), which may result in necrotic cell death (Cai and Jones, 1998).

Apoptosome Formation

In response to *extrinsic* or *intrinsic* death stimuli cytochrome c is released from mitochondria and binds the cytosolic Apaf-1 protein and dATP to form the apoptosome. After the formation of apoptosome, a wheel-like particle with 7-fold symmetry, it is attracting and activating pro-caspase-9. Active caspase-9 is inducing other caspases triggering the final events of the apoptotic cascade (Fig. 4.10).

Apoptosis Inducing Signals

Among the inducers of apoptosis six groups have been distinguished:

- Chemicals (will be discussed separately),
- Physical insults (radiation, heat shock),
- Viruses (Sindbis, Baculo, HIV-1),
- Cells (Cytotoxic T cells),
- Cytokines (TNFα, TGFβ),
- Loss or absence of trophic factors (glucose, hormones, growth factors),
- Other factors (p53, c-myc, Ced-2,3,4, Ced 9 gene mutations, Fas/Apo-1, IL-1β convering enzyme).

Chemical Inducers of Apoptosis

The most frequently used chemical agents known to induce apoptosis are named (in alphabetical order):

- actinomycin D: an inhibitor of transcription, by binding to DNA at the transcription initiation complex prevents the elongation by RNA polymerase,
- aphidicolin: tetracyclic diterpene antibiotic, inhibitor of eukaryotic nuclear DNA polymerase α, δ,
- A23187: mobile divalent cation ionophore (Mn^{2+}, Ca^{2+}, Mg^{2+}) with antiboitic properties,

230 4 Apoptosis

- caffeine: xanthine analogue, central nervous system stimulant,
- camptothecin: quinoline-based alkaloid, naturally-occurring DNA topoisomerase I inhibitor,
- ceramide: lipid composed of sphingosine and a fatty acid which can be released from the cell membrane by enzymes and act as a signaling molecule,
- cycloheximide: inhibitor of protein synthesis in eukaryotic cells,
- dexamethasone: member of the glucocorticoid steroid hormones with anti-inflammatory and immunosuppressant properties,
- doxorubicin (adriamycin): anthracycline antibiotic, DNA intercalating agent commonly used in the treatment of cancer,
- etoposide: a chemotherapy drug used to treat lung, ovarian or testicular cancer,
- 5-fluorouracil: one of the oldest chemotherapy drugs, especially for the treatment of colorectal cancer,
- Fas: type II transmembrane protein belonging to the tumour necrosis factor (TNF) family,
- free radicals: reactive oxygen species (ROS) such as $OH^{\bullet}$, H_2O_2, superoxide,
- glucocorticoids: a class of steroid hormones regulating cardiovascular, metabolic, immunologic, and homeostatic functions,
- glutamate: downregulates the intrinsic pathway of apoptosis as a ROS scavenger,
- hydroxyurea (hydroxycarbamide): antineoplastic agent used in hematological oncology,
- methylating agents: a class of compounds that react with nucleophilic centers of organic macromolecules, including nucleic acids (Friedberg et al., 2005). The reaction of methylating agents with DNA is a unimolecular ($S_N 1$) nucleophilic type substitution characterized by high affinity for oxygen (Grover, 1979). The cytotoxicity of methylating agents, such as cell cycle arrest and apoptosis, are mostly linked to methylation of the O^6 position of guanine to form O^6–meG (Karran and Bignami, 1999). The cytotoxic effects of methylating agents are exploited in chemotherapy of several malignancies. Antitumor methylating agents are among the oldest and most widely used class of anticancer drugs (DeVita et al., 2001),
- plaxitaxel: an antineoplastic drug interfering with the growth of cancer cells used in the treatment of breast, ovary and lung cancer and AIDS-related Kaposi's sarcoma,
- staurosporine: isolated from the bacterium *Streptomyces staurosporeus,* with biological activities ranging from anti-fungal to anti-hypertensive, to anti-cancer effects,
- thymidine and nucleoside derivatives: involved in cellular metabolism and used as enzyme inhibitors, antiviral, and anticancer agents,
- tumour necrosis factor (TNF): if not indicated otherwise it refers to TNF-α, a member of cytokines causing cell death, cellular proliferation, differentiation, inflammation, tumorigenesis, and viral replication,
- vinblastin: the alkaloide of *Cataranthus roseus* (Vincaceae) with cytostatic properties, inhibiting the formation of the mytotic spindle, DNA repair mechanisms, and RNA polymerases.

Apoptosis Inducing Signals   231

These drugs are normally used between the nanomolar (nM, 10^{-9} M) and millimolar (mM, 10^{-3} M) concentration range. Other physiological apoptotic agents such as lactosylceramide, 15d-PGJ$_2$ increase prostaglandin (PG) release in parallel with the induction of apoptosis (Moore et al., 2003). Noteworth to mention that not all of these agents will induce apoptosis in every cell type. Consequently, the apoptotic effect depends on the agent, the cell type selected and the concentration used. The time course of induction may vary, maximal induction of a particular protein may take 8 to 72 hours post-treatment. The time course of induction of apoptosis is an indication that it takes several hours till morphological changes are manisfested. Necrotic changes reflect faster and more severe changes than apoptotic changes. Moreover, not all proteins are affected by each agent in a particular cell line. Even proteins involved in preventing apoptosis, such as Bcl-2, may induce apoptosis by chemical treatments, pointing to a delicate homeostatic balance.

Chromatin Breakdown

Unicellular organisms respond to the presence of DNA lesions by activating cell cycle checkpoint and repair mechanisms, while multicellular animals have acquired a further option of eliminating damaged cells by triggering apoptosis. Deficient DNA damage-induced apoptosis contributes to tumorigenesis and to the resistance of cancer cells to a variety of therapeutic agents (Norbury and Zhivotovsky, 2004). Chromatin degradation is an irreversible step in apoptosis, consequently the understanding of this process is critical to elucidate how the cell becomes irreversibly committed to the death process. Apoptotic chromatin changes will be discussed in Chapter 5.

The apoptotic process was originally described as morphological changes. Chromatin condensation is the most recognizable nuclear hallmark of apoptosis. These changes start in the nucleus during which the cells round up, the chromatin condenses, crescent-shaped masses aggregate at the membrane, the nucleolus dissolves (Kerr, 1971), indicating that chromatin condensation progresses in two distinct steps: stage I involves chromatin compaction while stage II involves DNA fragmentation (Susin et al., 2000).

Stage I is followed by cytoplasmic and nuclear condensation, fragmentation and cell death. The nuclear fragments formed during the process are surrounded by the nuclear membrane, now known as apoptotic bodies (Kerr, 1971; Oberhammer et al., 1993). A common marker for apoptosis is the activation of endonucleases (Wyllie, 1980; Cohen and Duke, 1984). Endonuclease activation was thought to be the initial event of chromatin condensation (Cohen and Duke, 1984), but this attitude changed when it turned out that phosphorylation of various protein substrates such as lamins and histon H1 are likely to precede the endonucleolytic breakdown (Gerace and Blobel, 1980; Peter et al., 1990; Nurse, 1990). Events of chromatin condensation in the interphase nucleus are still hardly known, as morphological changes could not be visualized and early apoptotic changes could not be distinguished from normal changes taking place during the normal condensation process. Due to these limitations in chromatin condensation the structural changes have been

232   4 Apoptosis

confirmed by other measurements such as the drastic decrease in the ratio of DNA replication/repair upon apoptotic treatment (Bánfalvi et al., 2000), by flow cytometry to show the presence of small apoptotic cells, by the increased level of p53 as an apoptotic indicator using anti-p53 antibodies, electrophoretic mobility shift assay etc. (Offer et al., 2001).

Repair and Apoptosis

Impairment of DNA repair capacity may impact on individual health in such areas as aging and susceptibility to certain diseases. Chromosomal instabilities and the resulting neoplasia are closely linked to abnormalities of DNA metabolism, DNA repair, cell cycle regulation and control of apoptosis. It has been shown that defects in some DNA repair genes increased cancer risk, accelerated aging and impaired neurological functions (Ruttan and Glickman, 2002). Some examples of such disorders are diseases such as *ataxia telangiectasia* (primary immune deficiency), Bloom syndrome (congenital telangiectatic erythema), Fanconi anaemia (inherited bone marrow failure), Werner syndrome (premature aging disease).

DNA damage is greater in actively proliferating cells than in differentiated cell compartments (Hong et al., 1999). DNA repair efficiency is maximized in those stages of the cell cycle when the rate of DNA replication is minimal or reduced such as G1 before the S phase, during the mid S phase pause and after the S phase in G2. One can predict from the duration of the S phase and from the lack of protective histones that unwound parental and newly synthesised DNA strands are more vulnerable to mutations than in any other phase of the cell cycle.

Repair Enzymes

In certain situations damaged areas, called lesions, can bring the replication machinery to a complete halt. Besides the cell cycle checkpoints the genetically most vulnerable cell cycle stage (S phase) is protected by an intra-S phase checkpoint (Bartek et al., 2004; Bartek and Lukas, 2001). The activation of the intra-S phase checkpoint in response to DNA damage and replication block depends on the combined actions of ATM and ATR and requires an intact MRN (Mre11-Rad50-Nbs1) complex (Lee and Paull, 2004, 2005). The protein coded by the *ataxia telangiectasia* mutated (ATM) gene functions to control the rate at which cells grow. The ATR (*ataxia-telangiectasia-* and Rad3-related) kinase is essential for the maintenance of genomic integrity. The main role of the S-phase checkpoint could be the coordination of the cellular DNA damage with DNA synthesis. Consequently the intra-S phase checkpoint would only delay, but not arrest permanently the cell cycle until the increased repair process restores genome integrity.

DNA polymerases are enzymes responsible not only for DNA copying, editing, but also repairing small damages of DNA with high fidelity. Taking the limited processivity of repair DNA polymerases it is unlikely that the whole genome is tested for mutations uninterruptedly only once during replication. Due to spatial

Apoptosis Inducing Signals 233

limitations, repair and repliactive DNA polymerases would be unable to work at the same time on the same template. It is logical to assume that there is more than one pause site during S phase in which repair activity is elevated. Consistent with this idea, replicative and repair DNA polymerases and other repair enzymes work intermittently on replicating DNA, with repair enzymes acting first on damaged DNA before duplication is followed by replicative enzymes. Consequently, considerable fluctuation in replicative and repair synthesis is expected during S phase in intact, undamaged cells. Indeed, by increasing the resolution power of cell synchronization 11 replication peaks, termed replication checkpoints were identified in Chinese hamster ovary (CHO) cells (Bánfalvi et al., 1997). In *Drosophila* S2 cells four replication checkpoints were found (Rehak et al., 2000). CHO cells have 11 chromosomes, *Drosophila* cells contain four chromosomes. Corresponding to replication checkpoints, multiple subphases of DNA repair also exist (Szepessy et al., 2003). According to this hypothesis unwound units of DNA, logically chromosomes, are first made error free by repair enzymes immediately prior to their reduplication. In agreement with this hypothesis it was descried that apoptosis is found predominantly in the region of actively cycling cells, where repair enzyme expression is lowest (Hong et al., 1999). Consistent with this idea the reduced repair capacity or mutation of repair enzymes will increase the apoptotic tendency.

Inactivation of the ubiquitin-conjugating DNA repair enzyme (HR6B) in mice caused male sterility associated with chromatin modification, leading to the suggestion that ubiquitination of histones modulates chromatin structure as an essential part of DNA synthetic processes (repair, replication) in which it is implicated (Roest et al., 1996). The HR6 repair enzyme is localized in the euchromatin regions of the nucleus, an indication that its function is related to an active chromatin conformation (Koken et al. 1996). The defect in spermatogenesis is consistent with the impairment of postmeiotic chromatin remodelling process and provides evidence of the ubiquitin pathway in chromatin dynamics. Moreover, it is assumed that disturbance of chromatin remodelling in spermatids of HR6B-deficient mice is the primary cause of the infertility (Roest et al., 1996). Experiments with transgenic mice revealed that the expression of protamine induces disruption of the normal dense chromatin structure of spermatozoa and results in infertility (Rhim et al., 1995). Interestingly, in yeast the ubiquitin-conjugating DNA repair enzyme (UBC1) was found to be required for the recovery of growth after germination of ascospores and may thus accomplish the reverse reaction catalysed by RAD6, namely the decondensation of chromatin (Jentsch, 1992).

The human apurinic/apyrimidinic endonuclease/redox factor-1 (hAPE/Ref-1) is a multifunctional protein. This factor is involved in the repair of DNA damaged by oxidative or alkylating compounds as well as in the regulation of stress inducible transcription factors such as AP-1, NF-kappaB, HIF-1 and p53. With respect to transcriptional regulation. both redox dependent and independent mechanisms have been described. Due to its involvement in DNA repair and apoptosis-related signaling mechanisms, APE/Ref-1 is also regarded as a novel target for tumour-therapeutic approaches (Fritz et al., 2003).

Breadown of Proteins

Proteolysis is the biochemical degaradation of proteins through the hydrolysis of peptide bonds. Targets of protein degradation are abnormally folded proteins, short-lived proteins and other cell contents, such as nuclear proteins.

The two major intracellular proteolytic systems are the:

- lysosomal proteolysis (uptake and enzymatic degradation of secretory vesicles, cytoplasm and organelles). Lysosomes are responsible for the degradation of long-lived proteins and for the enhanced protein degradation observed under starvation,
- non-lysosomal (proteins tagged e.g. by ubiquitination, recognition by proteolytic systems) degradation by proteasome.

Proteases

Proteases play an important role in signaling apoptosis by catalyzing the limited cleavage of enzymes. Major players are caspases, which are both activated and catalyse limited proteolysis. Other proteolytic enzymes may also be involved in the pathway leading to cell death.

Major classes of proteases:

- Cysteine proteases:

  - Papain, a well-known plant protease,
  - Cathepsins a large family of lysosomal cysteine proteases, digesting proteins engulfed into lysosomes,
  - Calpains (non-lysosomal) are Ca^{++}-activated neutral cysteine proteases, cleaving intracellular proteins involved in cell motility and adhesion,
  - Caspases are cysteine proteases involved in the activation and implementation of apoptosis.

- Serine proteases include the digestive enzymes trypsin, chymotrypsin, and elastase.
- Threonine proteases (the proteasome hydrolases constitute a unique family of threonine proteases).
- Zinc proteases (metalloproteases, e.g. carboxypeptidases).
- Aspartate proteases (the digestive enzyme pepsin, some proteases found in lysosomes, the kidney enzyme renin, the HIV-protease).
- Ubiquitin-proteasome.
- Mitochondrial proteases.

It has become clear that proteolysis of cellular proteins is a highly complex, temporally controlled and tightly regulated process which plays important roles in a broad array of basic cellular processes. Proteolyis is carried out by a complex cascade of enzymes and displays a high degree of specificity towards its numerous

Apoptosis Inducing Signals   235

substrates. With the multiple cellular targets, the proteolytic system is involved in the regulation of many basic cellular processes including cell cycle and division, differentiation and development the response to stress and extracellular modulators, morphogenesis of neuronal networks, modulation of cell surface receptors, ion channels and the secretory pathway, DNA repair, regulation of the immune and inflammatory responses, biogenesis of organelles and apoptosis.

Protease inhibitors are mainly proteins with domains that enter or block the active site of a protease to prevent substrate access. Inhibitors of apoptosis proteins (IAPs) are proteins that block apoptosis by binding to and inhibiting caspases. The apoptosis-stimulating protein Smac antagonizes the effect of IAPs on caspases. Only caspases belonging to cysteine proteases and involved in apoptosis will be discussed further.

Caspases

Cysteine Aspatate Specific ProteASEs belonging to cystein proteases use the thiol moiety of cysteine to catalyse peptide bond hydrolysis. Caspases catalyse the cleavage of proteins after the amino acid aspartic acid (Asp). The proteolysis of caspases mediates signal pathways downstream of FAS receptor, triggering apoptosis.

The activation of a protease cascade involving the caspase family, earlier referred to as the ICE/CED-3 protease family (Alnemri et al., 1996), is central to the execution of apoptosis (Nicholson, 1996). The observation that a variety of nuclear proteins, including poly(ADP-ribose)polymerase (PARP), lamin B, topoisomerase I, topoisomerase II, and histone H1, are degraded concomitant with the DNA fragmentation calls into question the selectivity of the degradative process for DNA during apoptosis (Kaufmann, 1989; Kaufmann et al., 1993). That caspases degrade topoisomerase I in cells undergoing apoptosis or necrosis was confirmed by Casiano et al. (1998). Many of the substrates cleaved during apoptosis are protein autoantigens targeted by antinuclear autoantibodies present in the sera of patients with systemic autoimmune diseases (e.g. systemic *lupus erythematosus scleroderma*). This observation has been exploited to define apoptotic events associated with proteolysis by using antinuclear antibodies (Casiano and Tan, 1996).

Paracaspases (such as human: MALT1) are related to caspases present in animals and slime mold and induce apoptosis in an analogous manner to caspases. *Metacaspases* are present in plants, fungi, and protists (Uren et al., 2000).

Fragmentation of DNA

It remains to be clarified whether proteases play additional roles in DNA degradation such as preparing chromatin for nucleolytic attack. Based on experimental evidence it was suggested that non-caspase proteases may be required for chromatin degradation during apoptosis which perhaps function by altering chromatin substructure and sensitize it to the nucleolytic attack (Hughes et al., 1998). Despite the clear involvement of proteases and nucleases in apoptosis and the well known upstream nuclease activity, the exact relationship between proteolytic activity and DNA degradation remains to be clarified.

Among the endonucleases responsible for internucleosomal DNA fragmentation during apoptosis, several candidates have been proposed.

- DNase I is a Ca^{2+}/Mg^{2+}-dependent endonuclease capable of degrading double stranded DNA at neutral pH, found in nuclear extracts (rat kidney, liver, spleen, bovine, chiken, human ovaries), implicated as the enzyme responsible for apoptotic DNA degradation (Kreuder et al., 1984; Peitsch et al., 1993; Peitsch et al., 1994; Bacher et al., 1993; Polzar et al., 1994).
- DNase II belongs to the cation-independent endonucleases, functioning optimally at acidic pH in the absence of divalent cations (Eastman, 1994). Its lysosomal localization and ubiquitous tissue distribution suggested that this enzyme played a role in the degradation of exogenous DNA encountered by phagocytosis (Evans and Aguilera, 2003). Three acidic DNases II are so far known: DNase II alpha, DNase II beta and L-DNase II (Leukocyte derived).
- NUC-1. This *C. elegans* nuclease is related in sequence and activity to mammalian DNase II (Lyon et al., 2000; Wu et al., 2000).
- NUC18/cyclophilin (Montague et al., 1997).
- NUC40 and NUC58 optimal activity of both DNases requires neutral pH and is inhibited by zinc ions. The physicochemical characteristics of these nucleases indicate that they are novel DNases associated with apoptosis in cytotoxyc T lymphocytes (CTLL2 cells) (Deng and Podack, 1995).
- DNase gamma (DNase G), a neutral endonuclease activated in the presence of both Ca^{2+} and Mg^{2+} and inhibited by Co^{2+}, Ni^{2+}, Cu^{2+}, and Zn^{2+} (Shiokawa et al., 1994). DNase γ mRNA is found predominantly in the spleen, lymph nodes, thymus, and kidney, and to a lesser extent in the brain, heart, and pancreas. Activity of DNase γ is found during γ- ray irradiation (Higami et al., 2003). The enzyme cooperates with histone H1.
- Endonuclease G, a mitochondrial protein released in apoptosis and involved in caspase-independent DNA degradation (Van Loo et al., 2001; Liu et al., 1997; Parrish et al., 2001).
- Caspase activated DNase (CAD) also known as DNA fragmentation factor (DFF) (Li et al., 2001) is a neutral Mg-dependent endonuclease. The DNA Fragmentation Factor (DFF) is cleaved and activated by upstream proteases to trigger downstream DNA degradation (Liu et al., 1997). The murine homolog of the DNA Fragmentation Factor has been cloned and is known as the inhibitor of caspase activated DNase (ICAD) that binds directly to the nuclease (CAD) (Enari et al., 1998). Cleavage of ICAD releases and activates CAD which translocates to the nucleus and is capable of internucleosomal DNA fragmentation. DFF40/CAD exists in living cell nuclei as an inactive complex with its inhibitor DFF45/ICAD. After the induction of apoptosis, DFF40/CAD is released from the complex by the caspase-3-mediated cleavage of DFF45/ICAD and catalyses nucleosomal DNA fragmentation.

Apoptotic Pathways 237

- LDFF, with large molecular weight DNA fragmentation activity is Mg^{2+}-dependent and Ca^{2+}-independent, can occur in both acidic and neutral pH conditions and can tolerate 45°C treatment. Causes large molecular weight DNA fragmentation occuring in *Xenopus* egg extracts in apoptosis.
- NM23-H1, a DNA nicking DNAse, is a tumour suppressor gene product with reduced expression in transformed, metastatic cells. The nicking DNAse activity plays an important role in immune surveillance. The "nickase" is activated by granzyme B.
- Inducible lymphocyte Ca^{2+}/Mg^{2+}-dependent endonuclease, or ILCME is an endonuclease that is up-regulated during apoptotic T cell death. This endonuclease is located in the nucleus, and its activity is increased up to eightfold by a variety of stimuli or conditions that induce apoptosis (Khodarev and Ashwell, 1996).
- 25 kDa activity (Schwartzman and Cidlowski, 1993), a 97 kDa (kilodalton) molecule (Pandey et al., 1997).

There are several apoptotic models in which internucleosomal DNA degradation does not occur (Hughes and Cidlowski, 1997). The existence of such models led to the discovery that DNA is degraded into large 30–50 Kb fragments during apoptosis (Oberhammer et al., 1993). This is the so called higher order chromatin degradation i.e. the excision of chromatin loops and their oligomers from chromosomes. Higher order chromatin degradation has been described as an integral part of programmed cell death that marks the commitment of cells to death (Brown et al., 1993; Sun et al., 1993; Oberhammer et al., 1993; Cohen et al., 1994; Zhivotovsky et al., 1994; Beere et al., 1995; Lagarkova et al., 1995). The DNA backbone digestion is thought to occur at matrix attachment regions (MARs), and proceed in distinct, spatio-temporal stages (Konat, 2003).

Summarizing DNases implicated in apoptosis: they may be classified into three major groups: the Ca^{2+}/Mg^{2+} endonucleases, the Mg^{2+-} endonucleases, and the cation-independent endonucleases. Although a number of candidate enzymes have been proposed, further investigations will prove which endonucleases are involved in the apoptotic process.

Apoptotic Pathways

The intranuclear mechanisms signaling apoptosis after DNA damage overlap with those that initiate cell cycle arrest and DNA repair. Probably these are evolutionary related processes with the early events of these pathways highly conserved, while downstream elements are increasingly diverging. Moreover, multiple independent routes have been traced which signal the nuclear damage to the mitochondria, upsetting the balance and kipping it in favour of cell death rather than repair and survival. The signal activation of the *intrinsic* pathway is known as mitochondrial apoptosis. The intrinsic pathway is primarily activated by DNA-damaging agents and is

238   4 Apoptosis

coupled to the *extrinsic* caspase pathway. The general mechanism of the *extrinsic* and the *intrinsic* patways of apoptosis has been shown in Figs. 4.9 and 4.10.

Irradiation and Stress-Induced Apoptosis

To explain the cytotoxic effects of radiation damage, reactive oxygen species (HO˙ and H_2O_2) have been repeatedly invoked as important intermediates, because hydroxyl radical (HO˙) and iron-catalysed homolytic cleavage of hydrogen peroxyde (H_2O_2) can greatly increase radiation-induced cell damage and death. Irradiation- and stress-induced apoptosis of tumour cells is thought to be triggered by signaling through the mitogen-activated protein kinase Jun NH_2-terminal kinase (JNK) (Adler et al., 1995; Chen et al., 1996; Johnson et al., 1996; Verheij et al., 1996), which phosphorylates the transactivation domain of c-Jun and increases its transactivation potential. Recent evidence suggests that JNK activation is also involved in the induction of apoptosis observed when growth factors are removed from growth factor-dependent cells. The transcription factor AP1 has been implicated in the induction of apoptosis in cells in response to stress factors and growth factor withdrawal (Jacobs-Helber et al., 1998).

Irradiation-Induced Apoptosis

Ionizing radiation induces apoptosis through a pathway that is largely dependent on p53 (Fig. 4.12). The exact mechanism of p53 induced apoptosis is unclear. One mechanism by which p53 mediates apoptosis is through its ability to transactivate members of the TNF receptor family of "Death Receptors" to be discussed later.

Explanation to Fig. 4.12: The ATM (*ataxia telangiectasia*) protein coordinates DNA repair by activating enzymes that fix the DNA damage. p53 (also known as TP53) is a transcription factor that regulates the cell cycle and hence functions as a tumour suppressor. Bax, is a Bcl-2 associated portein with proapoptotic property similar to Bad and Bid. Bcl-2 family proteins govern mitochondrial outher membrane permeabilization. In mitochondria the disruption of the inner transmembrane potential can be detected upon apoptosis-inducing stimulus such as dexamethasone, irradiation, etoposide, Fas/TNF ligation, ceramide. Cytochrome c released from mitochondria activate caspases. The Akt (Protein kinase B) pathway functions as a transmitter, sending signals calling into action other players in the apoptotic process. Phosphorylation of the eukaryotic initiation factor-2 (eIF2) is an important mechanism mitigating cellular injury in response to diverse stresses. STAT1 is a protein necessary for mounting an appropriate immune response to baterial and viral pathogens. Poly(ADP-ribose) polymerase (PARP) is involved in a number of cellular processes involving mainly DNA repair and programmed cell death. Tallin is an inhibitor of apoptosis. Paxillin is a multi-domain protein serving as docking site for other proteins to perturb or bypass normal signaling cascades necessary for controlled cell proliferation. Focal adhesion kinase (FAK) is playing a central role

Apoptotic Pathways

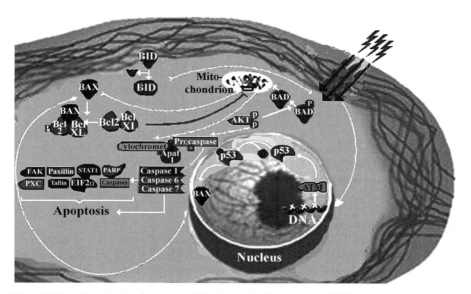

Fig. 4.12 Irradiation induced apoptosis. The signal of radiation caused DNA damage is transmitted to further downstream elements including the following major steps: ATM → p53 → Bax → mitochondria → cytochrome c release → caspase activation → apoptosis

in cell spreading, differentiation, migration, cell death and acceleration of the G1 to S phase transition of the cell cycle. Caspases as cysteine proteases have already been discussed as main effectors of apoptosis.

Free Radical Induced Apoptosis

Hydrogen peroxide (H_2O_2) is a potent mutagen and the major mediator of oxidative stress. Large amount of H_2O_2 is generated endogenously during cellular respiration and in other metabolic pathways in target cells, and exogenously by activated inflammatory cells via oxidative burst. Although H_2O_2 is in itself a mild oxidant, it can be converted in the presence of reduced transition metals, such as ferrous and cuprous ions, to highly reactive hydroxyl radicals that are believed to mediate the genotoxicity of H_2O_2 (Halliwell and Gutteridge, 1989).

In endothelial cells peroxide can be converted in the presence of ferrous ions to highly reactive hydroxyl radicals in the Fenton reaction:

$$Fe^{2+} + H_2O_2 \rightarrow Fe^{3+} + OH^{\cdot} + OH^-$$

The Fenton reaction is a ferrous ion-dependent decomposition of hydrogen peroxide, generating the most reactive hydroxyl radical. Reducing agents, such as ascorbate and glutathione in the asorbate-glutathione cycle of mitochondria and

240   4 Apoptosis

peroxisomes represent an important antioxidant protection system against hydrogen peroxide.

The Haber-Weiss cycle consists of the following two reactions:

$$H_2O_2 + OH^{\cdot} \rightarrow H_2O + O^{2-} + H^+ \quad \text{and}$$
$$H_2O_2 + O_2^- \rightarrow O_2 + OH^- + OH^{\cdot}$$

Iron(III) complexes catalyse this reaction. Fe(III) is reduced by superoxide, followed by oxidation by dihydrogen peroxide in the Fenton reaction.

Peroxides and hydroxyl radicals activate NF-kB and stimulate the expression of inflammatory genes (adhesion molecules, TNF and IL-8). (See below Figs. 4.16, 4.17 and 4.18). The apoptotic response of endothelial cells to oxidative stress seems to be involved in the development and progression of atherosclerosis.

Oxidative stress can trigger programmed cell death. Activated neutrophils respond to inflammatory stimulation by producing reactive oxygen species to kill invading bacteria. However, reactive oxygen species can also damage endothelial cells lining the vascular wall and trigger apoptosis. Reactive oxygen species produced by endothelial cells following ischemia aggravate the oxidative stress and lead to apoptosis. Although, superoxide dismutase (SOD) converts the reactive and damaging superoxide free radicals to peroxides that are less reactive than superoxide, nevertheless stimulate apoptosis. The glutathione (GSH) tripeptide is a reducing agent which removes the toxic metabolites generated by reactive oxygen species.

Induction of Nucleases

Intracellular acidification leads to the activation of a low pH-dependent nuclease caused by agents such as UV(C), etoposide or ceramide. It is suggested that cellular acidosis may set favourable conditions for a dormant, low pH-dependent (acidic) nuclease, which could be involved in intranucleosomal genome degradation, a hallmark of programmed cell death (Famulski et al., 1999). Figure 4.13 shows the schematic view of endonuclease G activation by cytochrome c released from mitochondria. EndoG is encoded in the nuclear genome but localized in mitochondria in normal cells. Endonuclease G (endoG) is a mitochondrial protein and one of the major endonucleases implicated in DNA fragmentation during apoptosis in cell types ranging from fungi to mammals. EndoG localized in mitochondria has been shown to translocate from mitochondria to the nucleus upon death stimulus when cell death is triggered and showed intrinsic nuclease activity. Overexpression of endoG strongly promoted apoptotic cell death under oxidant or differentiation-related stress. Loss of endoG expression conferred robust resistance to oxidant-induced cell death (Gannavaram et al., 2008), confirming the observation that depletion of endonuclease G leads to necrotic cell death (Büttner et al., 2007).

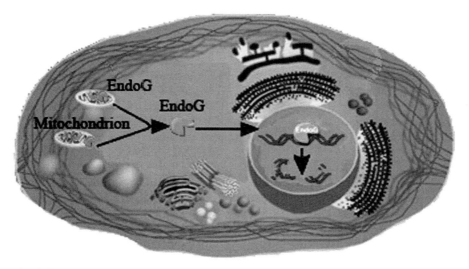

Fig. 4.13 Involvement of endonuclease G in apoptosis

Developmentally Induced Cell Death

Bcl-2 family members that have only a single Bcl-2 homology domain, BH3, are potent inducers of apoptosis, and some appear to play a critical role in developmentally programmed cell death (Puthalakath et al., 1999). Developmentally programmed cell death was illustrated by the loss of tadpole's tail in Fig. 4.4.

Granzyme Mediated Apoptosis

Cytotoxic T lymphocytes are the major effectors of the immune system for the elimination of virus-infected cells. Two main pathways serve the T cell-mediated cytotoxicity: 1. the perforin-granzyme-based mechanism (Kagi et al., 1994) is Ca^{2+}-dependent and can be suppressed with EGTA-Mg^{2+}, 2. the Ca^{2+}- independent Fas-based cytotoxic mechanism remains active in the presence of EGTA-Mg^{2+}. Among the intracellular apoptotic agents the serine-dependent and aspartate-specific protease granzyme B is of interest.

Cytotoxic T cells use a mechanism to kill tumour cells and virus-infected cells based on the release of perforin and granzyme proteins (Fig. 4.14). Perforins are proteins which form pores in the membranes of the attacked cell, allowing the entry of Granzyme A and Granzyme B. Granzymes are exogenous serine proteases released by cytoplasmic granules within cytotoxic T cells and natural killer cells. Granzyme B induced caspase activation results in the cleavage of factors like ICAD, the release of DFF40 to fragment DNA, which is one of the hallmarks of apoptotic cell death.

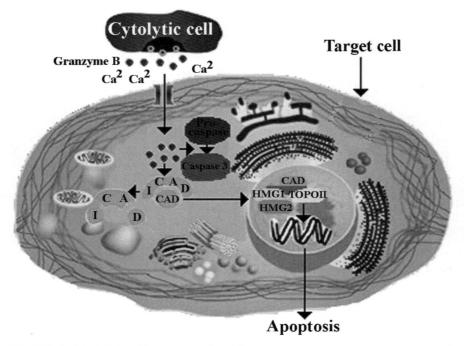

Fig. 4.14 Apoptosis induced by granzyme B protein

Targets of the serine protease Granzyme A include nuclear lamins responsible for maintaining nuclear structure and histones, the basic building blocks of chromatin structure. Granzyme A is also important in cytotoxic T cell induced apoptosis, activating caspase independent pathways. Granzyme A activates a DNA nicking DNAse (NM23-H1), a tumour suppressor gene product with reduced expression in transformed, metastatic cells. The nicking DNAse activity plays an important role in immune surveillance to prevent cancer through the induction of tumour cell apoptosis. The activation of nicking DNase takes place indirectly by the cleavage of proteins inhibiting the nicking activity of the SET complex including SET, Ape1, pp32 and HMG2.

In addition to the nicking inhibition, SET has nucleosome assembly activity and may help the interaction of transcriptional regulation with chromatin structure by interacting with the transcriptional coactivator. In the SET complex Ape1 repairs oxidative DNA damage, reduces transcription factors involved in immediate early responses. The cleavage of Ape1 by Granzyme A may contribute to DNA degradation and apoptosis. HMG2 is an acidic chromatin-associated protein responsible for DNA bending, altering chromatin structure and accessibility of genes for transcription. In addition to the nucleosome assembly activity and inhibiting the NM23-H1 nicking DNase, the nucleosome assembly protein SET inhibits DNA and histone methylation by the transcriptional coactivator (CBP) (Chakravarti and Hong, 2003; Fan et al., 2003). Exeption to the rule is the tumour suppressor pp32, which is not cleaved by Granzyme A, nevertheless part of the SET complex. The involve-

Apoptotic Pathways 243

ment of the SET complex in chromatin structure and DNA repair suggest that their cooperation protects the chromatin and DNA structure. The inactivation of the SET complex means the loss of its protective effect which maintained the integrity of DNA and chromatin structure.

Proteins (granzyme B, perforin) released by effector (cytotoxic T) cells may induce apoptosis in target cells by forming pores on their cell membranes and by cleavages of caspases, particularly caspase-3. Alternatively, caspase-independent granzyme activation of apoptosis may also take place. Caspase-3 activated DNAse (CAD) is stimulated by the cleavage of its inhibitor (ICAD). CAD interacting with topoisomerase II (Topo II) contributes to the decondensation of chromatin, leading to DNA fragmentation and apoptosis (Fig. 4.14).

Induction of Apoptosis Through Death Receptors

Death receptors have been mentioned among the members of the tumour necrosis factor receptor (TNFR) gene family. Members of this superfamily differ in their primary structure, but have similar extracellular subdomains with individual structural details allowing to recognize their specific ligands. The death receptors contain a homologous death domain at the cytosolic side of the plasma membrane. Adapter-molecules (FADD, TRADD, DAXX) contain death domains allowing them to interact with the death domain of the receptor and transmit the apoptotic signal to the death-machinery. As a consequence, many cellular proteins are degraded, leading to cell death.

Apoptosis is induced through a family of receptors known collectively as "death receptors". Members of the tumour necrosis factor (TNF) family of cytokines and receptors are of particular importance in the apoptotic process (Cosman, 1994; Smith et al., 1994; Nagata and Golstein, 1995; Lynch et al., 1995; Amakawa et al., 1996; van Parijs and Abbas, 1996). Among the major mediators of apoptosis within this superfamily are TNF, lymphotoxin α (LTa), Fas ligand (FasL) (Smith et al., 1994), Apo3L (Marsters et al., 1998), and TNF-related apoptosis-inducing ligand (TRAIL) (Wiley et al., 1995; Pitti et al., 1996; Marsters et al., 1996a) (Table 4.2). Other TNF-related cytokines and their receptors include the LTα1-LTβ2 complex. Another cytotoxic TNF-like ligand named Apo3 ligand (Apo3L) (Marsters et al., 1998) is a 249-amino-acid type II transmembrane protein that binds Apo3 (also known as DR3, WSL-1, TRAMP, or LARD) a member of the TNF family of death receptors (Marsters et al., 1996b; Chinnaiyan et al., 1996; Kitson et al., 1996). The extracellular domain of Apo3L shows highest identity to that of TNF. Like TNF, soluble Apo3L induces apoptosis and nuclear factor kappa B

Table 4.2 Tumour necrosis factor receptors (TNFR)

Death receptor	Ligands
Fas (CD95, ApoI)	CD95L (FasL) and DAXX
TNFR1 (p55, CD120a)	TNF, lymphotoxin alpha
DR3 (Apo3, WSL-1, TRAMP, LARD)	Apo3K (TWEAK)
DR4	TRAIL (Apo2L)

(NF-kB) activation in human cell lines. UnlikeTNF, however, Apo3L is expressed in a wide range of tissues (Marsters et al., 1998). The binding of ligands to death receptors initiates signaling by receptor oligomerization, recruitment of adaptor proteins and activation of caspases. Apo3L recruits initiator caspase 8 via the adapter protein FADD. Caspase 8 then oligomerizes and is activated via autocatalysis. Activated caspase 8 stimulates apoptosis via two parallel cascades: it directly cleaves and activates caspase-3, and it cleaves Bid (a Bcl-2 family protein) (Li et al., 1998). Truncated Bid (tBid) translocates to mitochondria, inducing cytochrome c release, which sequentially activates caspases 9 and 3. Death receptor 3L (DR-3L) can deliver pro- or anti-apoptotic signals. DR-3 promotes apoptosis via the adaptor proteins TRADD/FADD and the activation of caspase 8. Alternatively, apoptosis inhibited via an adaptor protein complex including RIP which activates NF-kB and induces survival genes including IAP. Induction of apoptosis via Apo2L requires caspase activity, but the adaptor requirement is unclear.

FAS Signalling Pathway

Apoptosis is triggered by a variety of stimuli, including cell surface receptors like FAS (also known as CD95 and APO-1) (Fig. 4.15), mitochondrial response to stress, and cytotoxic T cells.

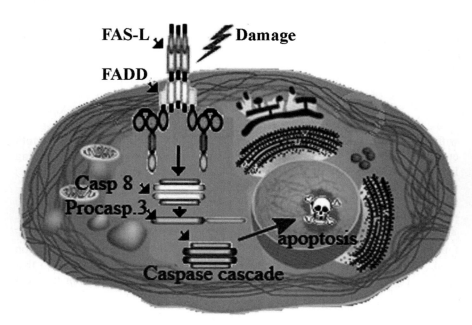

Fig. 4.15 Schematic view of FAS ligand induced apoptosis. Ligand (either FAS-L or TNFR-α or TRAIL) binds to its cell surface death receptor (FAS, TNFR-1, DR5, respectively), which attracts adaptors (FADD or TRADD), procaspase 8 and procaspase 10 forming the death signalling complex (DISC). After activation caspase 8 cleaves procaspase 3 which after autocatalysis activates downstream effector caspases 6, 7 which in turn cleave their protein substrates

Apoptotic Pathways

The Fas receptor (CD95) conveys the apoptotic signal of the Fas-ligand expressed on the surface of other cells. Binding of FAS to FasL on another cell activates the apoptotic patway through the cytoplasmic death domain that interacts with adaptor molecules (FAS, FADD and DAX) to activate the caspase proteolytic cascade. Caspase-8 and caspase-10 are activated first and cleave and activate downstream caspases, which degrade then a variety of cellular substrates that lead to cell death. The activated caspase 8 activates caspase 3 through two pathways. The less complex way is that active caspase cleaves procaspase 3 which after autocatalysis activates further downstream caspase 6 and 7 (Fig. 4.15). The more complex one is that caspase 8 cleaves Bcl-2 interacting protein (Bid) and its COOH-terminal part translocates to mitochondria where it triggers cytochrome c release. The released cytochrome c binds to apoptotic protease activating factor-1 (Apaf-1) together with dATP and procaspase 9 and activates caspase 9. The caspase 9 cleaves procaspase 3 and activates caspase 3. Caspase 3 cleaves DNA fragmentation factor (DFF) 45 in a heterodimeric factor of DFF40 and DFF45. Cleaved DFF45 dissociates from DFF40, inducing oligomerization of DFF40 that has DNase activity. The active DFF40 oligomer causes internucleosomal DNA fragmentation, which is an apoptotic hallmark indicative of chromatin condensation. Caspases also digest nuclear lamins, leading to the breakdown of the nucleus, loosing its regular structure. Further caspase substrates belong to the cytoskeletal structure, to the regulation of the cell cycle regulation and to signaling pathways (not shown).

Activation of kinases and the production of ceramide are also involved in Fas-mediated apoptosis. The FAS signaling patway is opposed by the inhibitor of FADD homologous ICE/CED3-like protease (I-FLICE) and FAS associated phosphatase (FAP) metabolism. Some viruses and tumours by using similar strategies may escape immune surveillance partly by the suppression of FAS-mediated apoptosis.

The Fas-FasL interaction plays an important role in the immune system. The lack of the immune system leads to autoimmunity. Inefficient apoptotosis due to the lack of Fas ligand or limited number of Fas receptors causes lymphoproliferation, leading to autoimmune diseases. In contrast to autoimmunity excessive apoptosis by Fas ligand causes tissue destruction such as hepatitis. Fas-mediated apoptosis removes self-reactive lymphocytes, transformed cells and virus infected cells.

Tumour Necrosis Factor (TNF) Induced Apoptosis

Based on sequence homology, a large family of molecules collectively called the nerve growth factor/TNF receptor family of apoptosis inducing signaling proteins has been identified. Receptors in the tumour necrosis factor (TNF) receptor family are associated with the induction of apoptosis, and inflammatory signaling. The tumour necrosis factor (TNF) receptor superfamily contains several members with homologous cytoplasmic domains known as death domains. The intracellular death domains are important in initiating apoptosis and other signaling pathways following ligand binding by the receptors.

Out of the nearly 20 members of the TNF superfamily, TNF is probably the most potent inducer of apoptosis. TNF is a pleiotropic cytokine eliciting a wide range of cellular responses. It is a major mediator of apoptosis as well as of inflammation and immunity. TNF interacts with its two receptors TNFR1 and TNFR2, that initiate several intracellular signal transduction pathways. The apoptotic pathway induces caspases and finally leads to programmed cell death. The NF-kB pathway activated by TNF counteracts apoptosis by inducing anti-apoptotic molecules. Although, many components of the TNF signaling pathways are indicated in Fig. 4.16, the complex network of interacting signals leading to the decision between cellular life and death is still poorly understood and not discussed.

The death domain (DD) of Fas (FADD) shows high-sequence homology to the *Drosophila melanogaster* protein reaper. The death domain is also found in tumour necrosis factor receptor 1 (TNF-R1), TNF-R1-associated death domain (TRADD), Fas-associated death domain (FADD), and Fas receptor-interaction protein (RIP). Fas and FADD associate through their homologous DD while TNF-R1 and TRADD associate through their homologous DD (Fig. 4.17).

In the absence of ligand, death domain-containing receptors are maintained in an inactive state and cells growth prevails. Alternatively, the binding of silencer of

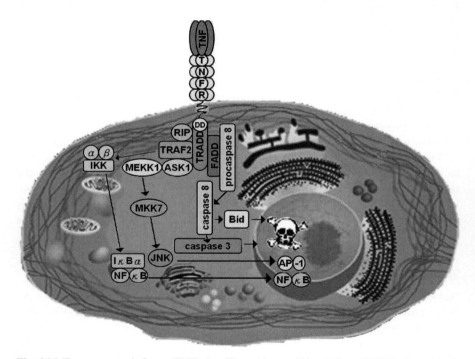

Fig. 4.16 Tumour necrosis factor (TNF) signalling pathway. Abbreviations: TNF-tumour necrosis factor; TNFR, tumour necrosis factor receptor; DD, death domain; RIP, Receptor Interacting Protein; TRADD TNF-R Associated Death Domain; FADD Fas Associated Death Domain protein; TRAF2 TNF-receptor associated factor 2; ASK1 Apoptosis signal-regulating kinase 1; MEKK1 MAPK/ERK kinase1; MKK7 Mitogen-activated protein kinase kinase7; JNK c-Jun N-terminal kinase; IKK IκB kinase; I kappa Bα NF-kappa B inhibitor alfa; NF-κB Nuclear Factor kappa B; AP-1 Activator protein 1; Bid Bcl2 Interacting Domain

Apoptotic Pathways

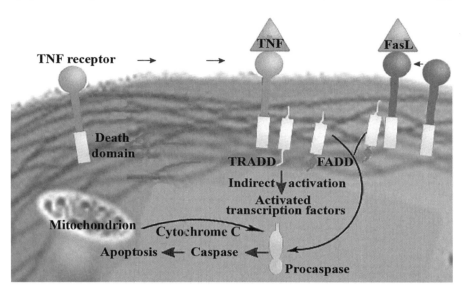

Fig. 4.17 The analogous mechanisms of apoptosis triggered by the external signals of FAS and TNF, with the mitochondrial caspase 8 pathway joining in

death domain (SODD) especially its overexpression suppresses TNF-induced cell death and NF-kB activation demonstrating its role as a negative regulatory protein for these signaling pathways.

The interaction between Fas and FasL results in the formation of the *death-inducing signaling complex* (DISC), which contains the FADD, caspase-8 and caspase-10 (Fig. 4.15). The mitochondrial pathway begins or joins in with the release of cytochrome c from mitochondria interacting with Apaf-1, causing self-cleavage and activation of caspase-9 (Fig. 4.17 not detailed).

Initially, only FasL was thought to trigger cell death by apoptosis. However, it turned out that inhibition of Fas-induced apoptosis in L929 fibrosarcoma cells by a caspase inhibitor lead to necrosis mediated by oxygen radicals (Vercammen et al., 1998a, b). Also, primary activated T cells can be efficiently killed by FasL, TNF-α and TRAIL in the absence of active caspases (Holler et al., 2000). These results suggested that Fas, like TNFR-1 (Laster et al., 1988), triggers apoptotic or necrotic death (Figs. 4.15 and 4.16). Indeed caspase-dependent initiation of apoptosis and necrosis by Fas receptor was found in lymphoid cells, indicating that two different pathways emerge from the Fas-receptor, one is leading to caspase-3-dependent apoptosis. The other caspase-dependent process is favouring necrosis in a manner dependent upon activation of caspase-8, but not execution caspases, like caspase-3 (Hetz et al., 2002). In these type II of cells, the *Fas*-DISC starts a feedback loop that spirals into increasing release of pro-apoptotic factors from mitochondria and the amplified activation of caspase-8 (Fig. 4.17).

To summarize the two types of cell lines differing in their kinetics of caspase activation: 1. upon Fas-ligation, type I cells (e.g. B lymphoblastoid cell line SKW6.4

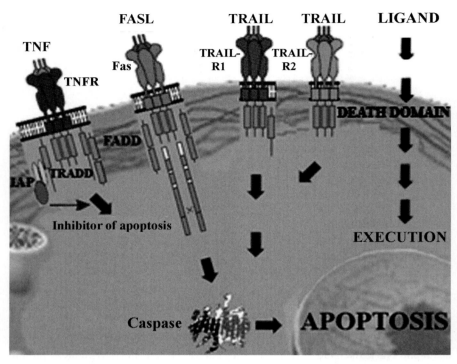

Fig. 4.18 Tumour necrosis factor-α-related apoptosis-inducing ligand (TRAIL). TRAIL is a member of the tumour necrosis factor-α (TNF-α) family of cytokines which is known to induce apoptosis upon binding to its death domain-containing receptors

and T lymphoma cell line H9) show early Caspase-8 and Caspase-3 activation, 2. type II cells such as human peripheral blood leukemia T (e.g. Jurkat) and human acute lymphoblastic T-cell leukemia (e.g. CEM) cells show delayed Caspase-8 and -3 activation (Scaffidi et al., 1998). The general mechanism of FAS, TNF and TNF-related (TRAIL) apoptosis is summarized in Fig. 4.18.

Living or dying is controlled through a class of proteins. Some proteins keep cell living (anti-apoptosis protein, e.g. Bcl-2), other proteins initiate apoptosis (pro-apoptosis proteins, e.g. Bax). After *TNF-R1* and *Fas* activation the delicate balance between pro-apoptotic (BAX, BID, BAK, or BAD) and anti-apoptotic (*Bcl-Xl* and *Bcl-2*) members of the *Bcl-2* family is lost and apoptosis prevails.

Caspase-Dependent Initiation of Apoptosis and Necrosis

Caspases have already been mentioned briefly among proteases. In mammals, more than a dozen caspases have been identified which cleave their substrates after aspartic acid (Asp) (Cyrns and Yuan, 1998). Activation of pro-caspases requires two

Apoptotic Pathways

Table 4.3 Human caspase family

Caspase	Other name	Family
Caspase-1	ICE	Caspase homologous to ICE
Caspase-2	ICH-1	Caspase homologous to ICH-1
Caspase-3	CPP32/Yama/Apopain	Caspases homologous to CED-3
Caspase-4	TX/ICH-2/ICEREL-II	Caspase homologous to ICE
Caspase-5	TY/ICEREL-III	Caspase homologous to ICE
Caspase-6	Mch2	Caspase homologous to ICH-1
Caspase-7	Mch3/ICE-LAP3/CMH1	Caspase homologous to ICH-1
Caspase-8	Mch5/MACH/FLICE	Caspases homologous toMACH/FLICE
Caspase-9	Mch6/ICE-LAP6	Caspases homologous to ICH-1
Caspase-10	Mch4	Caspases homologous to MACH/FLICE

caspase cleavages at Asp residues. These cleavages remove the amino-terminal pro-domain and separate the large and small catalytic subunits. Once activated, caspases can process and activate their own and other pro-caspases. Major human caspases are summarized in Table 4.3.

Homologies are based on phylogenetic relationships.

Caspases can be divided into three functional groups:

- initiator (activator) caspases are activated first. Activators cleave and activate effector caspases,
- effector (executioner) caspases cleave and activate cellular substrates,

In addition to caspases there are cytokine (inflammatory) processors.

The three groups contain 14 types of caspases based on sequence of activation in apoptosis (Table 4.4).

Caspases are synthesized as inactive zymogens, the so called procaspases. Initiator caspases are responsible for processing and activating the executioner caspases. The executioner caspases cleave a subset of proteins leading to characteristic morphological changes of apoptosis and DNA fragmentation. Cytokine caspases are involved in inflammatory processes.

Initiator caspases include procaspases-2, -8, -9 and -10, possessing long domains. Executioner caspases are procaspases-3, -6, and -7 possessing short domains. The group number of caspases is not identical with the group number of procaspases.

Among the long prodomains of initiator caspases are death effector domains (DED) in the case of procaspases-8 and -10 or caspase recruitment domains (CARD) as in the case of procaspase-9 and procaspase-2.

Activation of Initiator Caspases

- In death receptor-mediated apoptosis FADD/MORT1 serves as an adaptor for pro-caspase-8 (Muzio et al., 1996; Boldin et al., 1996).

Table 4.4 Classification of caspases based on sequence of activation in apoptosis

Type	Name	Regulatory unit	Adapter molecule
Activators:	Caspase 2	CARD	RAIDD
	Caspase 8	DED	FADD
	Caspase 9	CARD	Apaf 1
	Caspase 10	DED	
Executioners:	Caspase 3		
	Caspase 6		
	Caspase 7		
Cytokine processors:	Caspase 1	CARD	CARDIAK
	Caspase 4	CARD	
	Caspase 5		
	mCaspase 11		
	mCaspase 12		
	Caspase 13		
	mCaspase 14		

Abbreviations: CARD, caspase recruitment domain; DED, death effector domain; FADD, Fas-associated death domain containing protein; Apaf-1, apoptotic peptidase activating factor 1. RAIDD, RIP-associated ICH-1/CED-3-homologous protein with DD; CARDIAK, CARD containing interleukin (IL)-1 beta converting enzyme (ICE) associated kinase (RIP2, RICK, CCK).

– In drug or radiation-induced apoptosis Apaf-1 acts as an adaptor molecule to bind pro-caspase-9, which is regulated by cytochrome c and dATP (Zhou et al., 1997).

Among the different factors and numerous signaling pathways, some are dependent on caspase proteases, while others are caspase independent. Caspases transmit the apoptotic signal in a proteolytic cascade, with caspases cleaving and activating other caspases that then degrade other cellular targets that lead to cell death. Upstream caspases include caspase-8 and caspase-9. Caspase-8 is the initial caspase involved in response to receptors with a death domain like FAS. Fas clustering recruits FADD, caspase 8 and 10 and FLIP (FLICE inhibitory protein) to the death receptor forming the death inducing signalling complex (DISC). Concentration of activated caspase 8 cleaves pro-caspase 3, which in turn undergoes autocatalysis to form active caspase 3 and 7, the effector caspase of apoptosis. Briefly, in some cells the activation of caspase 8 is sufficient for the subsequent activation of effector caspases 3 and 7, which cleave their protein substrates to execute apoptosis.

The mitochondrial pathway begins with the release of cytochrome c from mitochondria, which then interacts with Apaf-1, causing self-cleavage and activation of caspase-9. Down stream caspases are caspase-3, -6 and -7 that are activated by the upstream proteases and act on themselves to cleave finally cellular targets. Granzyme B and perforin proteins are released by cytotoxic T cells inducing apoptosis in target cells, forming transmembrane pores, triggering apoptosis, probably through the cleavage of caspases. Caspase-independent mechanisms of Granzyme B mediated apoptosis have also been suggested.

The implication of caspases in apoptosis was discovered in *Caenorhabditis elegans* as the product of ced-3 gene (Ellis and Horvitz, 1986). Since then, a large

Apoptotic Pathways   251

family of these caspases has been described in a wide variety of organisms (Table 4.4). Caspase was first identified as an interleukin 1β converting (activating) enzyme (ICE), a cystein protease cleaving after the amino acid aspartic acid (Asp). In some types of cells (type I), processed caspase-8 directly activates other members of the caspase family, and triggers the execution of apoptosis. The apoptotic protein substrates of caspases are more than 40. Caspases are expressed as proenzymes and are activated during apoptosis either by autocatalytic cleavage or via other caspases (Salvesen and Dixit, 1997).

Engagement of the Fas receptor promotes apoptosis by activation of caspases. FasL binding stabilizes the trimeric form of the Fas receptor, thereby allowing recruitment of the Fas-associated death domain containing protein (FADD). FADD then binds to and activates caspase-8, an initiator caspase, which in turn activates downstream effector caspases, including caspase-3 (Martins and Earnshaw, 1997; Nagata and Golstein, 1995; Suda et al., 1993).

p53 in Normal and Deregulated Cancer Cells

General Structure and Levels of p53 in Cells

The p53 gene is located on human chromosome seventeen (17p13.1) and coding the 53 kilodalton phosphoprotein (393 amino acids), also known as TP53 or tumour protein (p53) (EC:2.7 1.37) regulating the cell cycle and functioning as a tumour suppressor. p53 has been described as "the guardian of the genome", referring to its role in conserving stability by preventing genome mutation. The molecular guardian p53 is the sensor of damage and disorder in cells and has a very powerful role in cancer prevention. p53 is faulty in most human tumours.

p53 is a tetramer consisting of four similar polypeptide chains and four structural domains including the:

- domain responsible for the tetramerization of the protein,
- domain activating transcription factors,
- domain that recognizes specific DNA sequences (core domain),
- domain recognizing damaged DNA (mismatched base pairs, single-stranded DNA).

p53 was discovered in 1979 as a cellular protein tightly bound to the large T oncoprotein of the SV40 DNA tumour virus (Linzer and Levin, 1979; Lane, 1984). The first subunit of p53 binds a consensus site on the DNA duplex and forms contacts to the bases and backbone. The second subunit binds 11 bp away in a region of modest homology, weakly interacting with the first subunit in a head-to-tail dimer. The third subunit does not bind DNA. However, it makes protein-protein contacts that stabilize crystal packing. DNA binding does not induce structural changes to the protein. It is thought that crystal-packing forces inhibit the fourth subunit from binding (Cho et al., 1994).

In normal cells the p53 protein level is low. Under normal unstressed conditions the low level of p53 is maintained by ubiquitine mediated proteasomal degradation (Woods and Vousden, 2001). p53 may harness cell damage in two ways, either by resting cells long enough to be repaired or by killing them. Although, it is not exactly known which of these functions are more important, it is likely that the mode of action tackling damaged cells depends on the degree of damage. Under stress conditions, p53 expression increases to induce apoptosis or to arrest the cell cycle allowing ample time for DNA repair (Bálint et al., 2001).

The history of p53 function since its discovery in 1979 shows an interesting change:

- 1980s: p53 is a proto-oncogene involved in cell cycle and development,
- 1990s: p53 is a tumour suppressor gene (transcription), guardian of genomic integrity (DNA repair, homologous recombination, chromosome segregation),
- 2000s: p53 is involved in senescence and apoptosis.

A more detailed historical perspective covering the first twenty years of p53 is given by Harris (1996), Oren and Rotter (1999). Further discussion summarizes the regulation of p53 level and the function of p53 in DNA repair, cell cycle and apoptosis.

Function of p53 in Cell Cycle

p53 plays an important role in cell cycle control and apoptosis. Defective p53 can allow abnormal cells to proliferate, resulting in cancer. For ten years the p53 gene was believed to be an oncogene when it turned out that it was a mutant gene. In 1989 Bert Vogelstein and his colleagues (Baker et al., 1989) showed that p53 gene is actually a tumour suppressor, much like the retinoblastoma gene. Cancer formation is regarded as an autonomous process of cells driven by mutations in genes that increase their proliferation and survival. In fact the p53 oncosuppressor is the gene which has been found to be most frequently altered in human cancers and mutant forms of p53 protein are present in more than 60% of human primary tumours. Since mutant forms of p53 have been found in many tumours and aroused so much interest *Science* has chosen the p53 protein as the "Molecule of the Year" in 1993.

Restoring missense point mutations in the p53 gene would be a major step in curing many cancers (Vogelstein et al., 2000). Various strategies have been proposed to restore proper tumour suppressor activity of p53 function in cancer cells (Blagosklonny, 2002). So far, no biologically active agent has been found. One possible candidate as anti-cancer drug could be the molecular chaperone Hsp90, which interacts with p53 *in vivo*.

DNA damage and other stress signals may trigger the increase of p53 proteins, which have the following major functions: DNA repair, growth arrest, apoptosis, and cellular senescence, which are essential for normal cellular homeostasis and maintaining genome integrity. The growth arrest stops the progression of cell cycle, preventing replication of damaged DNA to avoid proliferation of cells containing

Apoptotic Pathways 253

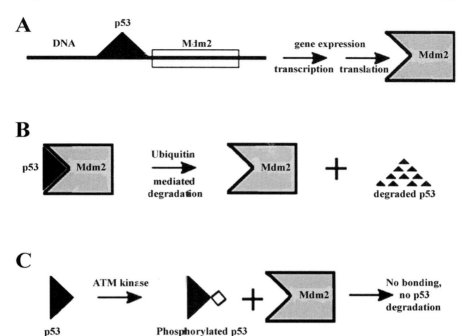

Fig. 4.19 Regulation of p53 level by Mdm2. (A) Gene expression of Mdm2: p53 binds to DNA and promotes the expression of Mdm2 gene. (B) Mdm2 ligase binds p53 and sensitizes the tumour suppressor for ubiquitin-dependent degradation by the proteasome. (C) Upon DNA damage ATM protein kinase (or DNA-PK, or CHK2) is activated and phosphorylates p53. Phosphorylated p53 is unable to bind to Mdm2. Consequently, p53 and Mdm2 levels increase. After DNA repair ATM kinase is inactivated, p53 binds to Mdm2 and p53 level decreases. In normal cells p53 is not phosphorylated and Mdm2 binding is keeping p53 at low level

abnormal DNA and may activate the production of proteins involved in DNA repair. The cellular p53 concentration is strictly regulated. Major regulator protein of p53 is the ubiquitine ligase Mdm2, which can trigger the degradation of p53 in the ubiquitin system by apoptosomes. Mdm2 serves its own regulation, p53 as a transcriptional activator regulates the expression of Mdm2 in a feed-back loop (Fig. 4.19).

p53 regulates also other genes involved in growth arrest, DNA repair and apoptosis. Details of the basic functions of p53 are reviewed by Almog and Rotter (1998). The major functions of p53 are summarized in Fig. 4.20.

Although, p53 protein functions mainly as a transcription factor (Polyak et al., 1997; Yu et al., 1999), the induction of apoptosis has also been described in the absence of transcription (Caelles et al., 1994; Haupt et al., 1995). Responses upon stress are channeled through a coordinated network, involving several negative and positive feedback loops (Harris and Levine, 2005). p53 is best known for its role in cell survival in response to DNA damage and in maintaining genomic integrity. The importance of p53 gene mutations is reflected by the most commonly diagnosed genetic disorders, responsible for the high percentage of cancerous cases. Most of

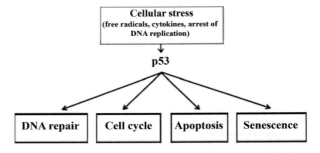

Fig. 4.20 Involvement of p53 protein in cellular homeostasis and genome integrity. The cellular stress may be caused by external exposure or endogenously

the malignancies including breast and lung cancers are associated with missence mutations or deletions of P53. Malignancies caused by p53 lesions are often resistant to the most commonly used antimetabolite treatments.

Role of p53 in DNA Repair

Deleterious intra- or extracellular effects of DNA lesions against the genetic integrity of cells are counteracted by DNA repair and other responses such as apoptosis, activation of cell cycle checkpoints or tolerance to DNA damage (Fig. 4.21). p53 protein is central to the cellular response of higher multicellular organisms to a variety of potentially damaging extracellular stimuli, including UV light, γ-irradiation, chemical carcinogens and chemotherapeutic agents. The guardian role of the p53 gene in maintaining the integrity of the genome is evident in that p53 is the most commonly altered gene in human cancer, with a mutation frequency exceeding 50% (Harris and Hollstein, 1993). Missense mutations occur most frequently within the evolutionarily conserved DNA binding domain (Hollstein et al., 1994). It is believed that stress signals like DNA damage induce the expression of p53, which in turn arrests the cell cycle, presumably to allow DNA repair (Lane, 1992), or drives cells to apoptosis if the damage is too extensive (Liebermann et al., 1995).

Most observations regarding the role of p53 in DNA repair seem to involve nucleotide excision repair (NER) (Ford et al., 1998; Lloyd and Hanawalt, 2000). It is believed that p53 plays an indirect role in NER, probably by transactivating the p48 gene (and some others), which is essential for efficient repair of certain

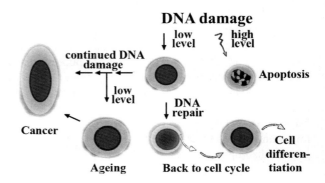

Fig. 4.21 The effect of low and high DNA damage on cellular processes

DNA lesions (Hwang et al., 1999; Tang et al., 2000). Experimental evidence for p53's direct involvement in the excision repair comes from its interaction with other proteins, since p53 with deletion at the C-terminal of more than 95 amino acids no longer supports the nucleotide excision activity (Chang et al., 2002).

Furthermore, base excision repair (BER) activity in human and murine cell extracts closely parallels their levels of endogenous p53, and BER activity is much reduced in cell extracts immunodepleted of p53 suggesting the role for p53 in DNA repair, which could contribute to its function as a key tumour suppressor (Zhou et al., 2001).

One can conclude from these observations that p53 is involved in multiple types of DNA damage repair mechanisms such as nucleotide excision repair, base excision repair and correction of double stranded breaks (Bálint et al., 2001).

Cell Cycle Arrest at Increased Level of p53

In proliferating cells, the cell cycle consists of four phases: 1. Gap 1 (G1), the interval between mitosis and DNA replication is characterized by cell growth. Transition occurs at the restriction point (R) in G_1 which commits the cell to the proliferative cycle. If the conditions signalling this transition are insufficient, the cell exits the cell cycle and enters G_0, a nonproliferative, resting phase. 2. Replication of DNA occurs during the synthesis (S) phase, which is followed by 3. a second gap phase (G2) during which growth and preparation for cell division occurs. 4. Mitosis and the production of two daughter cells occur in M phase.

The p53 protein by sensing DNA damage can halt progression of the cell cycle in G_1, by blocking the activity of cyclin-dependent kinase 2 (CDK2). Both copies of the p53 gene must be mutated for this to fail so mutations in p53 are recessive, and p53 qualifies as a tumour suppressor gene. p53 as a tumour suppressor gene prevents tumour formation under normal conditions. The gene product p53 protein binds to DNA as a tetramer, and activates or represses transcription of a large and still increasing number of genes including the expression of p21/WAF1/Cip1. The p21 protein interacts with cyclin-dependent protein kinase 2 (cdk2), a protein that stimulates cell division. The binding of p21 to cdk2 stops the progression of the cell cycle and cell does not divide (Fig. 4.22). Mutation of p53 gene may lead to

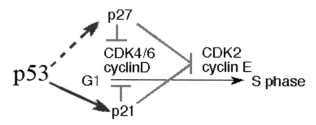

Fig. 4.22 Regulation of cell cycle by p53 protein. p53 binds to DNA, and stimulates the expression of genes to produce inhibitory proteins such as p21 and p27 proteins. The cell cycle progression from G1/G0 into the S phase requires cyclin-dependent protein kinase 2 (cdk2), that stimulates cell division. p21, p27, GADD45 or 14-3-3σ can interact with Cdk2 which arrest cell cycle. p53 regulates the expression of these inhibitory proteins to induce growth arrest

non-functional p53 protein while the lack of p53 gene prevents the formation of p53 protein and cell division may go on uncontrolled, forming tumours.

The CIP/KIP family (p21, p27) induced by p53, may mediate normal control and cell cycle response to damage and arrest the cell cycle between G1 and S phase (Hirao et al., 2000). p21 is exception to the rule, since it may act both positively and negatively, depending upon how much is present.

Role of p53 in Apoptosis

The p53 gene has been implicated in the regulation of programmed cell death. When the apoptosis machinery is disturbed, damaged cells that would normally die continue to grow, in some cases divide to cancerous masses. p53 mutations are associated with genomic instability and an increased susceptibility to cancer. It is still debated whether the mutant p53 protein – by itself – is functional in cells. When mutant p53 was inserted into cancer cells that were null for p53, they became resistant to apoptosis and made cancer cells more resistant to chemotherapy (Sigal and Rotter, 2000).

The physiological role of p53 in apoptosis:

1. The tumour suppressor activity summarized by Yee and Vousden (2005) has been confirmed in p53 null mice, tumour progression is correlated with the loss of apoptosis (Parant and Lozano, 2003).
2. p53 transcriptionally transactivates and transexpresses several genes to trigger apoptotic responses including *extrinsic* and intrinsic pathways (Fridman and Lowe, 2003).
3. Among other factors p53 balances between pro- and antiapoptotic signals to keep cells below or above the apoptotic threshold. One mechanism by which p53 promotes apoptosis is through the activation of Cdc42 and inactivation of Bcl-2. Cdc42 is a member of the small G protein family the activation of which leads to p53 mediated Bcl-2 phosphorylation and apoptosis.
4. Oxidative stress is transmitted by p53 leading to redox-related transactivation referred to as p53-inducible genes (PIGs) resulting in the upregulation of antioxidant enzymes including superoxide dismutase (SOD).
5. Transcription-independent p53-mediated apoptosis (Haupt et al., 1995; Yee and Vousden, 2005).
6. p53 has a direct apoptotic role in mitochondria. p53 can act as a BH3-only functional homologue by binding to and inhibiting Bcl-XL and Bcl-2 antiapoptotic proteins resulting in the release of cytochrome c from mitochondria and in caspase activation (Mihara et al., 2003).

Role of apoptosis under pathological conditions:

– down regulation resulting in cancer, autoimmune disorders, viral infections,
– up regulation: degenerative disorders e.g. Alzheimer's disease, ischemic heart injury (stroke), post-menopausal osteoporosis.

Apoptotic Pathways   257

The complexity and large number of molecular players, especially those of transcriptional and non-transcriptional downstream targets of p53, suggest that not only individual factors but one or more sets of components and their coordinated actions may be responsible for the p53 mediated apoptosis in different cell types (Yee and Vousden, 2005). All these possibilities confirm that apoptosis can be caused and prevented trough many interlinked pathways.

p53 Signaling Pathway

p53 is a transcription factor the activity of which is regulated by phosphorylation. The function of p53 is to prevent the cell from progressing through the cell cycle if DNA is damaged. This may be done in multiple ways such as blocking the cell cycle at a checkpoint until repair can be made or allow the cell to enter apoptosis if the damage cannot be repaired. The critical role of p53 is evidenced by the fact that it is mutated in a very large fraction of tumours from almost all sources.

Originally when p53 was tought to be a dominant oncogene. the cDNA clones isolated from tumour cells showed cooperativity with the RAS oncogene in transformation assays. However, it turned out that this was not the case since the cDNA clones used in these experiments were not wild-type but mutated forms of p53, while cDNAs from normal tissues were not capable of RAS co-trasformation. The analysis of several murine leukemia cell lines showed that the p53 locus was lost by either insertions or deletions on both alleles and mutant p53 proteins were shown to have modified conformation and increased stability. This increased stability of p53 is a characteristic feature of mutant forms of p53 proteins found in tumour cell lines. Mutation at the p53 locus occurs in cancers such as colon, breast, liver and lung carcinomas.

Complexing of p53 with proteins may result in a change of p53 mediated transactivation of gene expression. Another possibility to regulate the activity of p53 protein is by phosphorylation. During S phase the protein becomes phosphorylated by the M-phase cyclin-CDK complex of the cell cycle and also by casein kinase II (CKII). The level of p53 is lower after mitosis but increases again during G_1.

Hypoxia and p53

DNA damage induces p53 protein accumulation and p53-dependent apoptosis in cells which have undergone malignant transformation. p53 activity is probably differentially regulated by DNA damage and hypoxia. Unlike DNA damage, hypoxia does not induce p53-dependent cell cycle arrest. Although, hypoxia also induces p53 protein accumulation, but does not induce the production of endogenous downstream p53 effector mRNAs and proteins (e.g. p21, Bax, CIP1, WAF1, etc). Hypoxia does not prevent the induction of p53 target genes by ionizing radiation. This probably means that under hypoxic conditions p53 transactivation does not need a DNA damage-inducible signal. The Akt pathway inhibits p53-mediated transcription and apoptosis. The ubiquitin ligase Mdm2 of p53, plays a central role in the regulation of p53 level and serves as a good substrate for Akt. Mdm-2 sensitizes the p53 tumour suppressor for ubiquitin-dependent degradation by the proteasome, but, in turn p53 as a transcription factor induces Mdm-2, establishing a feedback loop (Fig. 4.19 A, B). Hypoxia or DNA damage leads to the stabilization and accumulation of transcriptionally active p53. At the molecular level, DNA damage induces the interaction of p53 with the transcriptional activator as well as with the transcriptional corepressor (Appella and Anderson, 2001; Koumenis et al., 2001; Blagosklonny, 2001; Suzuki et al., 2001; Asher et al., 2002; Ogawara et al., 2002).

Apoptosis Triggered by Chemical DNA Damage

Apoptotic DNA Fragmentation and Tissue Homeostasis

Apoptotic cell death can be triggered by different cellular stimuli, resulting in activation of apoptotic signaling pathways including the *extrinsic* Caspase Cascade pathway and the *intrinsic* Mitochondrial Apoptotic Signaling pathway. A charateristic cellular response of apoptosis is nuclear genome fragmentation creating a nucleosomal ladder. Among the multiple nucleases activated by apoptotic signaling pathways one is the DNA fragmentation factor (DFF), also known as caspase-activated DNAse (CAD). DFF/CAD is activated by the cleavage of its associated inhibitor (ICAD) through caspase proteases during apoptosis. Chromatin components such as topo II and histone H1 involved in the condensation of chromatin are supposed to interact and probably recruit CAD to chromatin. Another apoptosis activated protease is endonuclease G (EndoG). Apoptotic signaling releases EndoG from mitochondria. Mitochondria are also involved in another apoptotic signaling pathway through the release of cytochrome c induced by Bid protein to activate the caspase protease cascade. These two pathways are independent since the EndoG pathway still occurs in cells lacking DFF (Enari et al., 1998; Tang and Kidd, 1998; Liu et al., 1999; Durrieu et al., 2000; Hengartner, 2001; Li et al., 2001; Zhang and Xu, 2002).

Chemical damage to DNA triggers the cellular activation of the caspase cascade resulting in cell death. Chemical damage stimulates a sequence of events resulting in the cleavege of Bid protein similarly to the binding of the "death receptors", or initiates transition in permeability of the mitochondrial membrane. The permeability shift releases mitochondrial factors the most important of which are cytochrome c, apoptosis inducig factor (AIF) and other proteins to the cytoplasm. Cytochrome c is a key protein in electrontransport process. Cytochrome c is released in response to apoptotic signals and activates apoptotic peptidase activating factor 1 (Apaf-1). The activated Apaf-1 protease is also released from mitochondria. In the cytoplasm Apaf-1 by activating caspase-9 induces the rest of the caspase cascade. Caspases including several representatives involved in the apoptotic pathway further the apoptotic signal in the proteolytic cascade by cleaving and activating other caspases which finally degrade other cellular target proteins leading to cell death.

Checkpoint Signaling in Response to DNA Damage

Cell cycle checkpoint controls at the G1 to S transition and the G2 to M transition prevent the cell cycle from progressing when DNA is damaged. Tumour suppressor

Apoptotic Pathways

genes reduce the probability of turning a cell into a tumour cell in a multicellular organisms. An important tumour suppressor is the p53 tumour suppressor protein encoded by the *TP53* gene originally thought to be a proto-oncogene. The first discovered tumour suppressor gene was human retinoblastoma gene, which is responsible for the production of the pRb protein. Other tumour suppressor genes include the retinoblastoma susceptibility gene (RB), Wilms' tumours (WT1), neurofibromatosis type-1 (NF1), familial adenomatosis polyposis (FAP), von Hippel-Lindau syndrome (VHL), and loss of heterozygosity in colorectal carcinomas (called DCC for deleted in colon carcinoma).

The *ataxia telangiectasia* mutated (ATM) protein kinase signals the presence of double-standed DNA breaks in mammalian cells and in response phosphorylates proteins that activate DNA repair factors, initiate cell-cycle arrest and apoptosis. Two proteins are phosphorylated by ATM in response to DNA damage: the tumour suppressor p53 and the checkpoint kinase (chk1). In turn, the tumour suppressor p53 interacts with p21 which then blocks the activity of cyclin dependent kinase 2 (cdk2) and prevents the passage from G1 to S phase to avoid the harmful replication of damaged DNA. One of the targets of cdk2 is the Rb gene product, pRB protein, another tumour suppressor. When pRb is dephosphorylated it interacts with the E2F transcription factor and prevents transcription of genes required for progression through the cell cycle. The phosphorylation of pRb by cell cycle dependent kinases (cdk2, cdk4) prevents Rb from interaction with E2F, consequently the cell cycle can proceed through the G1-S checkpoint. The G2-M phase transition is also regulated by DNA damage through the cell cycle regulator cdc2.

Retinoids as Apoptotic Agents

Retinoids are derivatives of retinoic acid exerting their anti-neoplastic action through three different, though partially overlapping mechanisms: growth-inhibition, cyto-differentiation and apoptosis. Retinoid related molecules are a promising class of synthetic retinoic acid derivatives endowed with selective apoptotic activity on a large variety of leukemic and solid tumour cells (Garattini et al., 2004).

Trombospondin Induced Apoptosis in Angiogenesis

As tissues grow they require angiogenesis. Growing tissues are to be supplied with blood vessels to survive. Factors that inhibit angiogenesis are potential cancer therapeutics by blocking vessel formation in tumours and starving cancer cells. Thrombospondin-1 (TSP-1) protein inhibits angiogenesis and slows tumour growth, by inducing apoptosis of microvascular endothelial cells that line blood vessels (Jimenez et al., 2000).

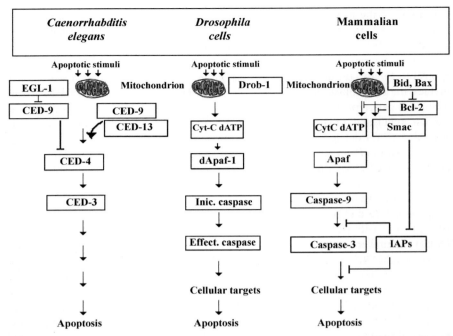

Fig. 4.23 Evolutionary conserved structures and development of regulation in apoptotic pathways. The EGL-1 gene encodes a cell death initiator protein that belongs to a group of proteins called BH3-only proteins as they contain a single BH3 domain but not the other Bcl-2 homology domains found in members of the Bcl-2 family of proteins. This BH3 domain allows EGL-1 to interact with other proteins. Abbreviations: EGL-1, egg laying abnormal-1; CED-9, CED13 prosurvival proteins; CED-4, CED-3 proapoptotic proteins; Drob-1, *Drosophila* Bax-like Bcl-2 family protein; Cyt-C, cytochrome c; dApaf-1, *Drosophila* Apaf-1/CED-4 homolog; Apaf -1, apoptotic protease activating factor 1; Inic-caspase, initiator caspase; Effect. caspase, effector caspase; Bid, Bcl-2 Interacting Domain; Bax, Bcl-2 associated X protein; Bcl-2, B-cell lymphoma 2 protein; Smac, second mitochondria-derived activator of caspases/DIABLO; IAP, inhibitor of apoptosis protein

Antiapoptotic Pathways

Based on the evolutionary conserved components of the apoptotic networks (Fig. 4.23) it is evident that in the future both apoptotic (anticancer) and antiapoptotic (e.g. cardiomyocyte cell death) therapies will be utilized. Thus, it is imperative that not only molecular pathways which initiate, but also those exerting antiapoptotic effect be deciphered. Only by doing this can the therapeutic potential be exploited.

Reverse Effects of Apoptosis-Inducing Factor (AIF) in Apoptosis and in Cell Survival

Similarly to cytochrome c, AIF (apoptosis-inducing factor) is a mitochondrial protein that is released from mitochondria upon apoptotic stimuli. While cytochrome

Antiapoptotic Pathways   261

c is involved in caspase-dependent apoptotic signaling, the release of AIF stimulates a caspase-independent apoptotic pathway by moving into the nucleus where it binds to DNA. DNA binding by AIF induces chromatin condensation and DNA fragmentation probably by recruiting nucleases. The other domain of AIF acts as an NADH oxidase, a redox enzyme. The NADH oxidase activity of AIF is independent of its DNA-binding activity and is not related to the induction of apoptosis. Actually this function of AIF protects against apoptosis rather than induces apoptosis. Lack of AIF increases sensitivity to peroxides, by a so far unknown mechanism, and suggests that AIF may act as a peroxide scavenger. Elucidating further roles of AIF such as redox enzyme will shed further light on its normal function in apoptosis and in disease (Klein et al., 2002; Lipton and Bossy-Wetzel, 2002; Ye et al., 2002).

Akt Pathway Promotes Cell Survival

The key signaling pathway responsible for maintaining the viability of erythroid cells is protein kinase B/Akt kinase (PKB/Akt) which is activated by both erythropoietin (Epo) and stem cell factor (SCF) (Miura et al., 1991; Bao et al., 1999; Uddin et al., 2000).

Akt promotes cell survival and opposes apoptosis by a variety of routes and has been shown to suppress the apoptotic death of a number of cell types induced by a variety of stimuli including cell cycle disturbance and DNA damage. Many cell-surface receptors induce production of second messengers like phosphatidylinositol 3,4,5-trisphosphate (PIP3), that convey signals to the cytoplasm from the cell surface. PIP3 signals activate the phosphoinositide 3-phosphate-dependent kinases PDK1 and PDK2 kinases, which in turn activate the kinase Akt (protein kinase B). Proteins phosphorylated by activated Akt promote cell survival.

Inactivation of p53 in Malignant Transformation of Cells

The negative control of elevated level of p53 on cell cycle progression in Fig. 4.22 indicated that reduced level or lack of p53 protein may cause anchorage independent growth as well as tumour formation. Indeed cyclin D overexpression and functional p53 inactivation induces immortalization of oral keratinocytes (Opitz et al., 2001). In normal, nontransformed cells the half-life of p53 is short (5–40 min), but significantly increased in transformed cells (Oren et al., 1981; Mora et al., 1982).

The involvement and inactivation of p53 in cellular immortalization has been well documented (Harvey and Levine, 1991; Metz et al., 1995; Wynford-Thomas, 1996; Levine, 1997). Wild type p53 by binding to DNA interacts with components of basal transcription and activates transcription of its down stream genes including, p21, mdm2 and GADD45. To the contrary mutant p53 proteins normally lack the ability to interact with TATA box-binding factors. In other cases mutant p53 proteins can still bind to DNA, but the conformational changes of p53 and its interactions with other proteins modulate various functions (Donehower and Bradley, 1993;

262   4   Apoptosis

Adler et al., 1997). p53 is known to interact with several other proteins including members of the hsp70 family (Hinds et al., 1987; Hainaut and Milner, 1992 Sugito et al., 1995).

Viral Oncoproteins Inactivate **p53**

Viral oncoproteins bind and inactivate p53. Such viral proteins are the adenovirus E1B 55-kDa and E4 34-kDa proteins which modulate the nuclear export of mRNA after virus infection (Babiss et al., 1985; Halbert et al., 1985; Pilder et al., 1986) and bind p53 antagonizing the p53-mediated transcription (Sarnow et al., 1982; Dobner et al., 1996; Querido et al., 1997). These proteins also seem to reduce the half life of p53 (Querido et al., 1997; Steegenga et al., 1998). Furthermore, the steady-state level of p53 is downregulated after transformation with the E1B 55-kDa and E4 34-kDa proteins (Moore et al., 1996; Nevels et al., 1997). Concomitantly, the E1B 55-kDa and E4 34-kDa proteins act synergistically to inactivate the transcriptional activity of p53 (Roth et al., 1998). There is no doubt that the major role of viral oncogenes is to transform cells by inactivating the p53 and pRb pathways. Other antagonists promote the intracellular degradation of p53, an activity common to oncoproteins of human papillomaviruses (HPVs) (Scheffner et al., 1990) and to the cellular murine double minute (MDM2) also known as p53-binding protein MDM2. MDM2 is a RING domain ubiquitin E3 ligase and a major regulator of the p53 tumour suppressor. MDM2 binds to p53, inactivates p53 transcription function, inhibits p53 acetylation, and promotes p53 degradation (Haupt et al., 1997; Kubbutat et al., 1997; Roth et al., 1998). MDM2 interaction with the nuclear corepressor KAP1 (KRAB domain-associated protein 1) contributes to p53 functional regulation (Wang et al., 2005).

Beside the inactivation of p53 one can also immortalize cells through the inactivation of the retinoblastoma tumour suppressor (pRb) pathway, e.g., by SV 40 large T antigen (Wright et al., 1989). Small DNA tumour viruses have developed efficient mechanisms to inactivate the function of retinoblastoma tumour suppressor (pRb). The Rb family of pocket proteins negatively regulates the progression from Go through G_1 and further to S-phase. The small pocket region in the protein is composed of two structural domains. The large pocket contains the small pocket and the C-terminal part of the Rb protein. Viral oncoproteins bind to the small pocket, while the large pocket is binding elongation factor 2 (E2F) of transcription. The reason that small DNA tumour viruses target the Rb family is that the proliferation of these viruses (adeno-, papilloma, polyoma viruses such as SV40) needs the replication machinery of the host cell. However, this machinery is available only during S-phase, while these viruses replicate in non-dividing cells. Productive infection depends on the ability of virus to initiate S-phase entry in the presence of anti-proliferative signals resulting in G_1 arrest and occasionally exit to the quiescent Go phase (Helt and Galloway, 2003).

Antiapoptotic Pathways   263

Expression of SV 40 large T antigen in primary cells leads to the inactivation of the pRb and p53 tumour suppressor pathways and allows such cells to grow beyond senescence. Although, post-senescent cells continue to loose telomeric sequences until telomeres become too short to bind the telomerase enzyme. Surviving cells maintain the telomere length through the activation of telomerase or by the involvement of a recombination based mechanism. Activation of telomerase will permit such cells with short telomeres to become immortal. Therefore, the most common method to generate immortalized cells is with oncogenic viruses such as SV 40 or human papillomavirus (HPV) (Hawley-Nelson et al., 1989; Bryan and Reddel, 1994).

p53 abnormality leads to increased genomic alterations in primary tumoir cells. These mutations are associated with the accumulation of additional genetic alterations and chromosomal abnormalities, resulting ultimately in immortalization. As already mentionaed p53 mutations occur in >50% of all human cancers. In T-cell leukemia >30% of all human T-cell leukemia virus type 1 lymphoma is caused by the functional inactivation of p53 by the HTLV-1 Tax protein, which is considered to be the transforming protein of this virus (Portis et al., 2001).

Wild-Type Level and Overexpression of p53

The nuclear p53 is expressed in normal cells and plays an essential role in the regulation of cellular proliferation. There are two mechanisms which serve the cells to give a short pulse of p53 protein in G_1 phase which is then quickly degraded. Consequently, the level of this phosphoprotein is normally low in growing cells. Its level is regulated by 1. rapid degradation by virtue of its short half-life, and 2. rapid transcription of the p53 gene which increases p53 RNA by 10- to 20-fold in late G_1 phase. Mutant forms of p53 protein in transformed cells can be distinguish from the wild-type (wt), mutant p53 proteins fail to bind to the simian virus 40 (SV40) tumour antigen, but form complexes with the heat shock protein 70 (hsp70).

Several types of malignant and premalignant tissues harbor a genetically wild-type (wt), but transcriptionally inactive form of p53, often localized in their cytoplasm (Moll et al., 1992; Tominaga et al., 1993; Isaacs et al., 1998). The effect of wild-type p53 can be counteracted by interleukin-6 and induced apoptosis in myeloid leukaemic cells (Yonish-Rouach et al., 1993). 2-methoxyestradiol, a natural byproduct of the human body can selectively posttranscriptionally induce active wild-type p53 in human cancer cell lines and induce apoptosis. This compound could be used for the induction of p53 mediated programmed cell death in cancer cells where the p53 gene is retained in native wild-type form (Mukhopadhyay and Roth, 1997). Thymocytes of p53-deficient mice are refractory to the induction of apoptosis by ionizing radiation.

Mutant p53 protein by forming oligomeric complexes with wild-type p53 may inactivate the function of the normal protein which could explain why mutant p53 is

264   4 Apoptosis

able to transform cells in the presence of wild-type protein. Another mechanism suggests that inactivation of wild-type p53 tumour suppressor can occur by electrophilic prostaglandins in a process that is unrelated to its somatic mutation or sequestration by viral oncoproteins (Moos et al., 2000).

Overexpression of wild-type p53 activity in wt p53-deficient cells induces apoptosis (Yonish-Rouach et al., 1993).

Antitapoptotic Pathways by Growth Factor Activation

Multiple antiapoptotic pathways lead from the insuline-like growth factor 1 receptor (IGF-1R) signalling to BAD phosphorylation. BAD phosphorylation seems to be good enough for survival (Downward, 1999).

The insulin-like growth factor 1 (IGF1) is produced in different tissues particularly by the liver in response to growth hormone stimulation and is an important factor in the regulation of growth and development. The chromosomal location of IGF-1R gene is 15q26.3. IGF-1R mediates cell survival and growth in response to its ligands IGF-1 and IGF-2 and leads to the phosphorylation of BAD. This tyrosine kinase receptor is widely expressed in many cell types and is a key mediator of growth. Overexpression or activation of IGF-1R may be involved in the proliferation of transformed cells, offering the inhibition of IGF-1R signaling as a strategy for the development of anticancer agents. IGF-1R activates three signaling pathways that converge to phosphorylate Bcl2-antagonist of cell death (BAD) protein and block apoptosis.

In the first pathway growth factor activation of the PI3/Akt signaling pathway culminates in the phosphorylation of the BCL-2 family member BAD, thereby suppressing apoptosis and promoting cell survival. Akt phosphorylates BAD *in vitro* and *in vivo*, and blocks the BAD-induced death. Bcl-2 family members regulate cytochrome c release.

A second pathway activated by IGF-1R involves ras oncoprotein mediated activation of the Map kinase pathway to block apoptosis.

A third pathway involves interaction of raf oncogene with mitochondria in response to IGF-1R activation. The convergence of these pathways to block apoptosis may enhance the IGF-1R response.

Stress Induction of Heat Shock Protein Regulation

Mammalian cells can respond to a variety of stresses such as heat, cold, oxidative stress, metabolic disturbance and environmental toxins through necrotic or apoptotic cell death, by increased gene expression. To the contrary, phosphorylation of heat shock proteins such as Hsp27 can protect cells against cellular stress. Heat shock proteins posses chaperons (molecular gardedames) interacting with a wide variety of proteins to exert specific effects. Hsp27 is a small heat shock protein present as monomers or forming dimers, and oligomers in the cell, with distinct activities. Oligomers exhibit molecular chaperone activity, but are disrupted upon phosphory-

Antiapoptotic Pathways

lation to form dimers and monomers. Map kinase cascades mediate Hsp27 phosphorylation. Heat stress activates the p38 kinase cascade and induces phosphorylation of Hsp27 by the downstream Map kinases. Cytokines (TNF and IL-1) can also induce Hsp27 phoshorylation through this Map kinase cascade, protecting cells in some settings against cytotoxic responses. In stressful conditions, dissociation of oligomeric Hsp27 by phosphorylation may allow lower molecular weight forms to perform other non-chaperone functions (Bitar, 2002; Bruey et al., 2000a, b; Charette and Landry, 2000; Charette et al., 2000; Garrido et al., 1999).

Hsp27 exhibits the antiapoptotic effect at different steps in the apoptotic pathway:

1. A common component of apoptotic pathways is the mitochondrial release of cytochrome c. Hsp27 induced by stress protects cells against apoptosis by preventing the release of cytochrome c and by binding to cytochrome c in the cytosol.
2. Hsp27 blocks downsteam caspase 9 activation and the subsequent activation of caspase 3, inhibiting the rest of the proteolytic caspase cascade.
3. A further role of Hsp27 in blocking apoptosis is through blocking Fas-induced apoptosis. Fas is a receptor in the TNF receptor gene family that induces apoptosis when stimulated by its cell-bound ligand, Fas-ligand (see Fas Signaling pathway). Fas also induces apoptosis through another pathway mediated by the death associated 6 protein (Daxx). Phosphorylated Hsp27 dimers block apoptosis by binding with Daxx and preventing downstream activation of the kinase Ask-1 (Apoptosis signal-regulating kinase 1).
4. The interaction of Hsp27 with actin filaments may also prevent apoptosis triggered by some agents like staurosporine that damage actin. Unphosphorylated Hsp27 monomers regulate actin filament growth by binding to the end of fibers and capping them
5. Finally, Hsp27 prevents damage to cells by reactive oxygen species (ROS), by altering the oxidative environment of the cell through the induction of glutathione expression, as well as by blocking apoptosis induced by ROS. Modulation of Hsp27 expression and phosphorylation may provide a useful means to alter cellular sensitivity to stress.

Alternative Mechanisms to Suppress Apoptosis by Cytokines

This subchapter discusses the role of cytokines which maintain cell viability by suppressing apoptosis through their interactions with genes of the apoptotic machinery and with those genes signalling for apoptosis. Cytokines are small secreted proteins produced in response to an immune stimulus mediating and regulating immune responses, inflammation, and hemopoiesis. Many cytokines induce growth and are known as growth factors. Cytokines are responsible for cell viability, and inducing cell growth. These two functions can be separately regulated. Cytokines are produced by lymphocytes (lymphokines), monocytes (monokines), leukocytes (interleukins). Cytokines may act on cells which produce them (autocrine action) or nearby cells (paracrine effect) and occasionally on distant cells (endocrine effect). Cytokines bind to specific membrane receptors and transmit signals via second

266 4 Apoptosis

messengers mainly tyrosine kinases. Response to cytokines alters gene expression (higher or lower) of membrane proteins, cell proliferation and scretion of effector molecules.

The largest group of cytokines stimulates immune cell proliferation and differentiation including:

Interleukin 1 (IL-1): activates T cells,
IL-2: stimulates proliferation of antigen-activated T and B cells,
IL-4, IL-5, and IL-6: stimulate proliferation and differentiation of B cells,
Interferon gamma (IFNγ): activates macrophages,
IL-3, IL-7 and Granulocyte Monocyte Colony-Stimulating Factor (GM-CSF): stimulate hematopoiesis.

Other groups of cytokines include interferons and chemokines. Interferons IFNα and IFNβ inhibit virus replication in infected cells, while IFNγ stimulates in antigen-presenting cells the MHC expression. Chemokines (cytokines with chemotactic activities), atract leukocytes to infection sites.

Cytokines act on their target cells by binding specific membrane receptors which have been divided into four major families based on their structures and activities:

– Hematopoetin family receptors
– Interferon family receptors
– Chemokine family receptors
– Tumour necrosis factor receptor

Here only the tumour necrosis factor (TNF) receptors are dealt with, which have four extracellular domains; they include receptors for soluble TNFα and TNFβ as well as for membrane-bound CD40 (important for B cell and macrophage activation) and Fas (which signals the cell to undergo apoptosis). T cell FasL binding to Fas leads to the activation of caspase proteases that initiate apoptosis of the cell expressing membrane Fas. Activated lymphocytes express Fas, so that FasL-positive cytotoxic T cells can regulate the immune response by eliminating activated cells. Immune deficiency linked to expression of a mutant pro-apoptotic Fas gene is characterized by over-proliferation of lymphocytes.

Cytokines contribute to the viability of normal and cancer cells by suppressing the apoptotic machinery activated by wild-type p53, or by cytotoxic agents including irradiation and compounds used in cancer chemotherapy. Cytokines can be used to decrease apoptosis in normal cells, while inhibition of cytokine activity may improve cancer therapy by enhancing apoptosis in cancer cells. The apoptosis suppressing function of cytokines is mediated by changing the balance in the activity of apoptosis inducing and suppressing genes. Apoptosis suppression is expected upstream of caspase activation in the apoptotic process. Cytokines can suppress multiple pathways leading to apoptosis, only some of which were suppressed by

Antiapoptotic Pathways 267

other agents such as some antioxidants, Ca^{2+-} mobilizing compounds and protease inhibitors (Lotem and Sachs, 1999).

Studies with p53-expressing or nonexpressing cells have shown that different cytokines, including interleukin (IL) 6, IL-3, interferon γ (IFN-γ), and granulocyte–macrophage colony stimulating factor, can suppress apoptosis induced by overexpression of wild-type p53 and by cytotoxic compounds. Some apoptotic pathways were also suppressed by certain antioxidants and protease inhibitors. The apoptosis-suppressing cytokines, antioxidants, and protease inhibitors suppressed the activation of interleukin 1β converting enzyme-like cysteine proteases (caspases), showing that these apoptosis suppressors act upstream of caspase activation. Overexpression of wild-type p53 in M1 myeloid leukemia cells induces apoptotic cell death that was suppressed by the calcium ionophore A23187 and the calcium ATPase inhibitor thapsigargin (TG). This suppression of apoptosis by A23187 or TG was associated with suppression of caspase activation but not with suppression of wild-type-p53-induced expression of *WAF-1, mdm-2,* or *FAS.* The results of Lotem and Sachs (1998) indicate that:

1. overexpression of wild-type p53 by itself or treatment of myeloid leukemic cells with cytotoxic compounds is associated with the activation of ICE-like proteases, 2. cytokines exert apoptosis by suppressing the functions upstreame of protease activation, 3. cytotoxic compounds induce additional pathways in apoptosis, 3. cytokines suppress other components of the apoptotic machinery.

One can conclude that wild-type p53-mediated apoptosis, without prior DNA damage, can be effectively suppressed by the cytokine interleukin (primarily IL-6) and to a lesser extent by interferon γ (IFN-γ) by certain antioxidants, and by some protease inhibitors.

Nicotinic Acetylcholine Receptors Against Apoptosis

Nicotinic acetylcholine receptors play an essential role in neuromuscular signaling, also expressed in non-neuronal tissues, where their function is less clear. The primary function of nicotinic acetylcholine receptors is their action as ligand-gated ion channels transducing action potentials across synapses, but they seem to have other functions. Nicotinic acetylcholine receptors in neurons alter apoptotic signaling, protecting against cell death, and through alternative signaling pathways.

The PI3 kinase/Akt pathway protects neurons against apoptotic cell death and may block apoptosis triggered by beta-amyloid fragments. Beta amyloid fragments contribute to the progression of Alzheimer's disease. Nicotinic agents may prove useful in the treatment of Alzheimer's and other neurodegenerative conditions.

Trefoil Factors for Mucosal Healing

Maintaining the integrity of the gastrointestinal tract despite the continuous presence of microbial flora and injurious agents is essential. Epithelial repair requires restitution and regeneration During restitution, epithelial cells spread and migrate

268   4 Apoptosis

across the basement membrane to re-establish surface-cell continuity, a process independent of cell proliferation. Epithelial continuity depends on a family of small abundant secreted proteins, the trefoil factors (TFFs). The trefoil factor family (TFF) comprises the gastric peptides pS2/TFF1 and spasmolytic peptide (SP)/TFF2, and the intestinal trefoil factor (ITF)/TFF3. Their fundamental action is to promote epithelial-cell restitution within the gastrointestinal tract.

Transgenic Mice for Apoptotic Research

Genetic engineering techniques also referred to as recombinant DNA technology are used to alter the genetic make-up of an organism to produce a genetically modified organism (GMO). Transgenic animals carry foreign genes inserted into their genomes. Without going into details, two basic methods of inserting genes are known: the embryonic stem cell and the pronucleus method. Transgenic animals (mice, goat, sheep, chicken, etc.) are already able to synthesize human proteins and are expected to become valuable sources of proteins for human therapy.

A further technical development was the replacement of one of the functional genes by a non-functional one ("null" allele). The mating of such heterozygous transgenic mice produces a homozygous strain known as "knockout mice" with both copies of the replaced gene being non-functional, since the gene locus has been knocked out.

Transgenic mice have provided tools for exploring important questions by inserting foreign genes into the genome of the host and expressing them correctly by the cells of the host. Transgenic mice are being used to study various aspects of apoptosis, particularly apoptosis-related gene products in primary cells. The advantage of using transgenic mice is that they may lead to different effects from similar experiments using immortalized cell lines. Moreover, trangenic experiments allow to identify the impact of these gene products on multiple aspects of physiological processes. A few transgenic experiments related to apoptosis will be mentioned.

Transgenic experiments have revealed that the SV40 large T antigen or germ line mutation can functionally inactivate p53 *in vivo*. The inactivation resulted in increased susceptibility to thymic development probably through the loss of mutation-surveillance function of p53 (Mccarthy et al., 1994). Transgenic mice have been generated expressing molecules of the Bcl-2 and Bcl-2 family members, CD95, superoxide dismutase and rhodopsin (Gil-Gómez and Brady, 1998). Transgenic mice in which Bcl-2 was overexpressed in T cells had complete protection against sepsis-induced T lymphocyte apoptosis in thymus and spleen and improved survival rate (Hotchkiss et al., 1999). It was demonstrated by transgenic experiments that survivin, an inhibitor of apoptosis protein (IAP), also crucial for mitosis and cell cycle progression, modulates caspase activation and that Fas-mediated hepatic apoptosis is regulated by survivin via mitochondrial pathways (Conway et al., 2002). In Bid-/- cells, mitochondrial dysfunction was delayed, cytochrome c was not released, effector caspase activity was reduced and the cleavage of apoptosis substrates was altered. The loss-of-function model indicates that Bid is a critical substrate *in vivo* for signalling by death-receptor agonists mediating a mitochondrial amplification loop that is essential for the apoptosis of selected cells (Yin et al., 1999).

Apoptosis Protocols

Cytotoxic assays do not distinguish between apoptosis and necrosis.
These assays were principally of two types:

- Radioactive and non-radioactive assays measure plasma membrane permeability (dying cells become leaky).
- Colorimetric assays that measure reduction in the metabolic activity of mitochondria; mitochondria in dead cells cannot metabolize dyes, while mitochondria in live cells can.

Apoptosis can be detected in populations of cells or in individual cells. Different methods have been devised to detect apoptosis:

- staining apoptotic cells
- the TUNEL (TdT-mediated dUTP Nick-End Labeling) analysis,
- ISEL (*in situ* end labeling),
- DNA laddering analysis for the detection of fragmentation of DNA in populations of cells or in individual cells,
- annexin-V analysis that measures alterations in plasma membranes,
- detection of apoptosis related proteins.

General methods for cell death staining do not distinguish between apoptosis and necrosis. These staining techniques include trypan blue staining, propidium iodide staining of living cells and fixed cells, Hoescht staining.

Staining Dead Cells

The simplest way: trypan blue stains dead cells blue.
Non-fixed cells: FDA(fluorescein diacetate)-green stains alive cells; P.I. (propidium iodide)-red stains dead cells.

TUNEL Assay

Terminal Transferase dUTP Nick End Labeling (TUNEL) is a method used to measure apoptosis-associated DNA strand breaks. At the late stage of apoptosis the fragmentation of nuclear chromatin results in a multitude of 3'-hydroxyl termini of

DNA ends. These ends can be fluorescent-tagged with deoxyuridine triphosphate nucleotides (F-dUTP). The enzyme terminal deoxynucleotidyl transferase (TdT) catalyses a template-independent addition of deoxyribonucleoside triphosphates to the 3' hydroxyl ends of double- or single-stranded DNA and generates DNA strands with exposed 3'-hydroxyl ends. Addition of the deoxythymidine analog 5-bromo-2'-deoxyuridine 5'-triphosphate (BrdUTP) to the TdT reaction serves to label the break sites. Once incorporated into the DNA, BrdU can be detected by an anti-BrdU antibody using standard immunohistochemical techniques. Non-apoptotic cells do not incorporate much of the F-dUTP because of the absence of exposed 3'-hydroxyl DNA ends. Slides can be observed under a fluorescent microscope. Images are collected and processed by using settings, to allow semiquantitative comparisons.

ISEL (In Situ End Labeling) (LEP)

The ISEL technique identifies morphologically apparent apoptotic cells and labels cells in which the DNA fragmentation has not yet progressed to clearcut nuclear fragmentation. By using biotinylated nucleotides and DNA polymerase, the DNA fragments serve as templates *in situ* to synthesize new DNA strands, which are visualized by microscopy after histochemical processing.

DNA Laddering

Apoptosis is characterized by a variety of morphological features such as loss of membrane asymmetry and attachment, condensation of the cytoplasm and nucleus, and internucleosomal cleavage of DNA. Apoptosis is associated with the fragmentation of chromatin into multiples of the 180 bp nucleosomal unit, known as DNA laddering. In DNA laddering assay, small fragments of oligonucleosomal DNA is extracted selectively from the cells whereas the higher molecular weight DNA stays associated with the nuclei. To detect fragmented DNA it has to be isolated, separated by agarose gel electrophoresis and visualized using ethidium bromide.

Immunological Detection of Low Molecular Weight DNA

The alternative method to circumvent the isolation and electrophoretic analysis of DNA is the immunological detection of low molecular weight DNA (histone-complexed DNA fragments) by the Cell Death Detection ELISA immunoassay.

Annexin V Analysis

Annexin is a common name for a group of cellular proteins also known as "lipocortin". Annexin V is a 35–36 kDa, Ca^{2+}-dependent, phospholipid binding protein. One of the earliest indications of apoptosis is the translocation of the membrane

Apoptosis Protocols

phospholipid phosphatidylserine from the inner to the outer leaflet of the plasma membrane. The translocation of phosphatidylserine precedes other apoptotic processes such as loss of plasma membrane integrity, DNA fragmentation, and chromatin condensation. Once exposed to the extracellular environment, binding sites on phosphatidylserine become available for Annexin V. Annexin V can be conjugated to biotin or to a fluorochrome e.g. FITC and used for the flow cytometric identification of cells in the early stages of apoptosis.

Fluorescence-Activated Cell Sorting (FACS)

Cells are seeded in multiple-well plates, and, after 24 h, they are treated with N-methyl-N'nitro-N-nitrosoguanidine; (MNNG) at indicated concentrations. Apoptotic cells in the medium and attached cells are collected, washed with PBS, recentrifuged, and resuspended in PBS. Cells are fixed in 70% ethanol at 4°C for 15–20 min and collected by centrifugation. The pellet is resuspended in PBS plus propidium iodide (0.02 mg/ml) and Ribonuclease A (0.25 mg/ml). Cell cycle is analysed by fluorescent activated cell sorter (FACS).

Light Scattering Flow Cytometry

The light scattering flow cytometry turned out to be a useful method at sensing the presence of apoptotic cells and apoptotic bodies. Light-scattering instruments called "scatterometer" enabled the study of various biological molecules (Liedberg et al., 1983; Shapiro, 2003). Such instruments, especially flow cytometers can be easily switched to their light scattering function and adapted for apoptotic measurements.

An important aspect of this analysis is the measurement of apoptotic DNA (ApDNA) formation, i.e. to estimate the cellular DNA degradation during the programmed cell death. This analysis which is closely related to the particle size, amplifies the apoptotic signal and shows more convincingly the change in the cell size of apoptotic cells (Bánfalvi et al., 2007).

Although, the flow cytometric DNA content in itself may indicate the presence of apoptotic cells at the G_1/G_0 border of the cell cycle, but this signal after genotoxic treatment was too low to be detected after irradiation, and in normal healthy cells practically undetectable. In damaged cells the amplification of this small signal by forward scatter flow cytometry allows a much more detailed analysis of apoptotic cells resulting in quantified data. The importance of the application of light scattering flow cytometry resides in the weakness of the comet assay which has been generally used in industrial genotoxicity *in vitro*.

Comet Assay

Due to the observed negative results, the comet assay needed to be confirmed by other methods such as the micronucleus test, sister chromatid exchange (Moller et al., 2000). Although, the comet assay is considered a suitable and fast test for DNA-damaging potential in biomonitoring studies, it is not a particularly sensitive technique to determine the effects of environmental pollution in peripheral white blood cells (Šrám et al., 1998). Other conflicting results have also been reported in the comet assay studies (Hoffmann et al., 2005) which did not indicate any difference in the radiation damage produced by X rays or γ-rays in the dose range of 0–3 Gy. In contrast, other DNA damage indicators detected significant increase in DNA damage at much lower doses (0.02 Gy) (Rössler et al., 2006). The limited use of the comet assay refers to the pressing need for the identification of apoptotic changes in cells by other methods such as light scattering measurements.

Apoptosis Proteins

These proteins can either be added to the assay to promote or to inhibit apoptosis or else they can be detected by using monoclonal antibodies raised against apoptosis proteins. These are mostly recombinant proteins, expressed in *E. Coli*, purified under native condition. Proteins can be used as input marker or positive control (Western Blotting). They are also suitable for functional study (binding assay), to identify a protein which interacts with these apoptosis proteins. Their mutants (Bcl-2, Bax) may serve as negative controls. BID and Bax samples are suitable for inducing cytochrome c release from mitochondria. Other available apoptosis proteins are: Bcl-XL, Fas cytosolic domain, Ded domain, Card domain, F-Box domain, Dap3, SKP, Bcl-G, Bak, Bad, Bik, Ced4, Daxx, Apaf-1, Grim etc.

Detection of Mono- and Oligonucleosomes

Mono- and oligonucleosomes generated during apoptosis are recognized by antibodies directed against DNA and histones in a colorimetric assay (ELISA, Cell Death Detection). Absorbance is measured with an ELISA reader. Nucleosome enrichment can be computed with respect to vehicle-treated cells.

Anti-Fas mAb

This is a simple and powerful method to induce CD95 (Fas)-mediated apoptosis of exponentially growing human cell lines by monoclonal antibodies. This method can be used as a general positive control for apoptosis detection systems.

p53 Protein Analysis

1. For Western blot analysis cells are lysed in sample buffer, boiled and loaded on 10% polyacrylamide gels containing SDS. Proteins are then transferred to

nitrocellulose membranes. The presence of p53 protein indicating apoptosis can be detected by monoclonal antibodies (Harlow et al., 1981; Offer et al., 2001).

2. For immunoprecipitation assays [^{35}S]methionine-labeled proteins are immunoprecipitated with anti-p53-specific antibodies (Harlow et al., 1981; Gannon et al., 1990; Offer et al., 2001). The complexes are precipitated with Sepharose-protein A and washed in buffer. Proteins are then separated and analyzed by SDS-PAGE.

3. Electrophoretic mobility shift assay (EMSA). In the DNA mobility shift assay radio-end-labeled DNA oligonucleotide (El-Deiry et al., 1992) is mixed with nuclear extract and anti-p53 monoclonal antibody (Harlow et al., 1981). After incubation the reaction is loaded on 4% polyacrylamide gel and electrophoresed. If the apoptotic indicator p53 is present its sequence specific binding to DNA can be traced (Offer et al., 2001).

Caspase Staining Kits

The fluorometric kits are the ideal tools for detecting activated caspases *in situ* in living cells. The kits utilize potent caspase inhibitors conjugated to FITC or rhodamine as fluorescence *in situ* markers. The caspase inhibitors are cell-permeable, nontoxic, and bind specifically and irreversibly to the activated caspases in apoptotic cells. With the FITC or rhodamine labels, activated caspases in apoptotic cells can be directly visualized by fluorescence microscopy, or analyzed by using flow cytometry or a fluorescence plate reader.

References

Adachi, S., Cross, A.R., Babior, B.M. and Gottlieb, R.A. (1997). Bcl-2 and the outer mitochondrial membrane in the inactivation of cytochrome *c* during Fas-mediated apoptosis. J Biol Chem. **272**, 21878–21882.

Adler, V., Fuchs, S.Y., Kim, J., Kraft, A., King, M.P., Pelling, J. and Ronai, Z. (1995). Jun-NH2-terminal kinase activation mediated by UV-induced DNA lesions in melanoma and fibroblast cells. Cell Growth Differ. **6**, 1437–1446.

Adler, V., Pincus, M.R., Minamoto, T., Fuchs, S.Y., Bluth, M.J., Brandt-Rauf, P.W., Friedman, F.K., Robinson, R.C., Chen, J.M., Wang, X.W., Harris, C.C. and Ronai, Z. (1997). Conformation-dependent phosphorylation of p53. Proc Natl Acad Sci USA **94**, 1686–1691.

Alderson, M.R., Tough, T.W., Davis, S.T., Braddy, S., Falk, B., Schooley, K.A., Goodwin, R.G., Smith, C.A., Ramsdell, F. and Lynch, D.H. (1995). Fas ligandmediates activationinduced cell death in human T lymphocytes J Exp Med. **181**, 71–77.

Almog, N. and Rotter, V. (1998). An insight into the life of p53: a protein coping with many function! Biochim Biophys Acta. **1378**, R43–R54.

Alnemri, E.S., Livingson, D.J., Nicholson, D.W. et al. (1996). Human ICE/CED-3 protease nomenclature. Cell. **87**, 171.

Amakawa, R., Hakem, A., Kundig, T.M., Matsuyama, T., Simard, J.J., Timms, E., Wakeham, A., Mittruecker, H.W., Griesser, H., Takimoto, H., Schmits, R., Shahinian, A., Ohashi, P., Penninger, J.M. and Mak, T.W. (1996). Impaired negative selection of T cells in Hodgkin's disease antigen CD30-deficient mice. Cell. **84**, 551–562.

274 4 Apoptosis

Ameisen, J.C. (2002). On the origin, evolution, and nature of programmed cell death: a timeline of four billion years. Cell Death Differ. **9**, 367–93.

André, N. (2003). Hippocrates of Cos and apoptosis. Lancet. **361**, 1306.

Appella, E. and Anderson, C.W. (2001). Post-translational modifications and activation of p53 by genotoxic stresses. Eur J Biochem. **268**, 2764–2772. Review.

Arends, M.J. and Wyllie, A.H., (1991). Apoptosis: mechanisms and roles in pathology. Int Rev Exp Pathol. **52**, 223–254.

Asher, G., Lotem, J., Kama, R., Sachs, L. and Shaul, Y. (2002). NQO1 stabilizes p53 through a distinct pathway. Proc Natl Acad Sci USA. **99**, 3099–3104.

Ashush, H., Rozenszajn, L.A., Blass, M., Barda-Saad, M., Azimov, D., Radnay, J., Zipori, D. and Rosenschein, U. (2000). Apoptosis induction of human myeloid leukemic cells by ultrasound exposure. Cancer Res.**60**, 1014–1020.

Babiss, L.E., Gisberg, H.S., Darnell, J.E. Jr. (1985). Adenovirus E1B proteins are required for accumulation of late viral mRNA and for effects on cellular mRNA translation and transport. Mol Cell Biol. **5**, 2552–2558.

Bacher, M., Rausch, U., Goebel, H.W., Polzar, B., Mannherz, H.G. and Aumuller, G. (1993). Stromal and epithelial cells from rat ventral prostate during androgen deprivation and estrogen treatment-regulation of transcription. Exp Clin Endocrinol. **101**, 78–86.

Baker, S.J., Fearon, E.R., Nigro, J.M., Hamilton, S.R., Preisinger, A.C., Jessup, J.M., Van Tuinen, P., Ledebetter, D.H., Barker, D.F., Nakamura, Y., White, R. and Vogelstein, B. (1989). Chromosome 17 deletions and p53 gene mutations in colorectal carcinomas. Science. **244**, 217–221.

Bálint, É. and Vousden, KH. (2001). Activation and activities of the p53 tumour suppressor protein. Br J Cancer**85**, 1813–1823.

Bánfalvi, G., Littlefield, N., Hass, B., Mikhailova, M., Csuka, I., Szepessy, E. and Chou, W.M. (2000). Effect of cadmium on the relationship between replicative and repair DNA synthesis in synchronized cho cells. Eur J Biochem. **267**, 6580–6585.

Bánfalvi, G., Klaisz, M., Ujvarosi, K., Trencsenyi, G., Rozsa, D. and Nagy, G. (2007). Gamma irradiation induced apoptotic changes in the chromatin structure of human erythroleukemia K562 cells. Apoptosis. **12**, 2271–2283.

Bánfalvi, G., Mikhailova, M., Poirier, L.A. and Chou, M.W. (1997). Multiple subphases of DNA replication in Chinese hamster ovary (CHO-K1) cells. DNA Cell Biol. **16**, 1493–1498.

Bao, H., Jacobs-Helber, S.M., Lawson, A.E., Penta, K., Wickrema, A. and Sawyer, S.T. (1999). Protein kinase B (c-Akt), phosphatidylinositol 3-kinase, and STAT5 are activated by erythropoietin (EPO) in HCD57 erythroid cells but are constitutively active in an EPO-independent, apoptosis-resistant subclone (HCD57-SREI cells). Blood. **93**, 3757–3773.

Bartek, J. and Lukas, J. (2001). Mammalian G1- and S-phase checkpoints in response to DNA damage. Curr Opin Cell Biol. **13**, 738–747.

Bartek, J., Lukas, C. and Lukas, J. (2004). Checking on DNA damage in S phase. Nat Rev Mol Cell Biol. **5**, 792–804.

Beere, H.M., Chresta, C.M., ALEJO-Herberg, A., Skladanowski, A., Dive, C., Larsen, A.K. and Hickman, J.A. (1995). Investigation of the mechanism of higher order chromatin fragmentation observed in drug-induced apoptosis. Mol Pharmacol. **47**, 986–996.

Bellairs, R. (1961). Cell death in chick embryos as studied by electromicroscopy. J Anat. **95**, 54–60.

Berman-Frank, I., Bidle, K.D. and Falkowski, P.G. (2004). The demise of the marine cyanobacteria, Trichodesmium spp., via an autocatalysed cell death pathway. Limnol Oceanogr. **49**, 997–1005.

Bernstein, E. and Hake, S.B. (2006). The nucleosome: a little variation goes a long way. Biochem Cell Biol. **84**, 505–517.

Bitar, K.N. (2002). HSP27 phosphorylation and interaction with actin-myosin in smooth muscle contraction. Am J Physiol Gastrointest Liver Physiol. **282**, G894–903.

Blagosklonny, M.V. (2001). Do VHL and HIF-1 mirror p53 and Mdm-2? Degradation-transactivation loops of oncoproteins and tumour suppressors. Oncogene. **20**, 395–398. Review.

Blagosklonny, M.V. (2002). P53: An ubiquitous target of anticancer drugs. Int J Cancer, 98, 161–166.

References

Boldin, M.P., Goncharov, T.M., Goltsev, Y.V. and Wallach, D. (1996). Involvement of MACH, a novel MORT1/FADD-interacting protease, in Fas/Apo-1 and TNF receptor induced cell death. Cell. **85**, 803–815.

Boldin, M.P., Varfolomeev, E.E., Pancer, Z., Mett, I.L., Camonis, J.H. and Wallach, D. (1995). A novel protein that interacts with the death domain of Fas/APO1 contains a sequence motif related to the death domain. J Biol Chem. **270**, 7795–7798.

Bossy-Wetzel, E. and Green, D.R. (1999). Apoptosis: checkpoint at the mitochondrial frontier. Mutation Research/DNA Repair. **434**, 243–251.

Bossy-Wetzel, E., Newmeyer, D.D. and Green, D.R. (1998). Mitochondrial cytochrome *c* release in apoptosis occurs upstream of DEVD-specific caspase activation and independently of mitochondrial transmembrane depolarization. EMBO J. **17**, 37–49.

Broker, L.E., Kruyt, F.A., Giaccone, G. (2005). Cell death independent of caspases: a review. Clin Cancer Res. **11**, 62–3155.

Brown, D.G., Sun, X.M. and Cohen, G.M. (1993). Dexamethasoneinduced apoptosis involves cleavage of DNA to large fragments prior to internucleosomal fragmentation; J Biol Chem. **268**, 3037–3039.

Bruey, J.M., Ducasse, C., Bonniaud, P., Ravagnan, L., Susin, S.A., Diaz-Latoud, C., Gurbuxani, S., Arrigo, A.P., Kroemer, G., Solary, E. and Garrido, C. (2000a). Hsp27 negatively regulates cell death by interacting with cytochrome c. Nat Cell Biol. **2**, 645–652.

Bruey, J.M., Paul, C., Fromentin, A., Hilpert, S., Arrigo, A.P., Solary, E. and Garrido, C. (2000b). Differential regulation of HSP27 oligomerization in tumour cells grown *in vitro* and *in vivo*. Oncogene. **19**, 4855–63.

Bryan, T.M. and Reddel, R.R. (1994). SV40-induced immortalization of human cells. Crit Rev Oncog. **5**, 5331–5357.

Bursch, W., Ellinger, A., Gerner, C., Fröhwein, U. and Schulte-Hermann, R. (2000). Programmed cell death (PCD). Apoptosis, autophagic PCD, or others? Ann N Y Acad Sci. **926**, 1–12.

Büttner, S., Carmona-Gutierrez, D., Vitale, I., Castedo, M., Ruli, D., Eisenberg, T., Kroemer, G. and Madeo, F. (2007). Depletion of endonuclease G selectively kills polyploid cells. Cell Cycle. **6**, 1072–1076.

Caelles, C., Helmberg, A. and Karin, M. (1994). p53-dependent apoptosis in the absence of transcriptional activation of p53-target genes. Nature. **370**, 220–223.

Cai, J. and Jones, D.P. (1998). Superoxide in apoptosis. Mitochondrial generation triggered by cytochrome *c* loss. J Biol Chem. **273**, 11401–11404.

Casiano, C.A., Ochs, R.L. and Tan, E.M. (1998). Distinct cleavage products of nuclear proteins in apoptosis and necrosis revealed by autoantibody probes. Cell Death Differ.**5**, 183–90.

Casiano, C.A. and Tan, E.M. (1996). Antinuclear autoantibodies: probes for defining proteolytic events associated with apoptosis. Mol Biol Rep. **23**, 211–216. Review.

Castedo, M., Perfettini, J.-L., Roumier, T., Andreau, K., Medema, R., Kroemer, G. (2004). Cell death by mitotic catastrophe: a molecular definition. Oncogene. **23**, 2825–2837.

Certo, M., Del Gaizo Moore, V., Nishino, M., Wei, G., Korsmeyer, S., Armstrong, S.A. and Letai, A. (2006). Mitochondria primed by death signals determine cellular addiction to antiapoptotic BCL-2 family members. Cancer Cell. **9**, 351–361.

Chakravarti, D. and Horg, R. (2003). SET-ting the stage for life and death. Cell. **112**, 589–591.

Chang, Y.C., Chang, H.W., Liao, C.B., Liu, Y.C. (2002). The role of p53, DNA repair and oxidative stress in UVC induction of PCNA expression. Ann N Y Acad Sci. **973**, 384–391.

Charette, S.J. and Landry, J. (2000). The interaction of HSP27 with Daxx identifies a potential regulatory role of HSP27 in Fas-induced apoptosis. Ann N Y Acad Sci. **926**, 126–131.

Charette, S.J., Lavoie, J.N., Lambert, H. and Landry, J. (2000). Inhibition of Daxx-mediated apoptosis by heat shock protein 27. Mol Cell Biol. **20**, 7602–7612.

Chen, Y.R., Meyer, C.F. and Tan, T.H. (1996). Persistent activation of c-Jun N-terminal kinase 1 (JNK1) in gamma radiation-induced apoptosis. J Biol Chem. **271**, 631–634.

Chinnaiyan, A.M., O'rourke, K., Lane, B.R. and Dixit, V.M. (1997). Interaction of CED-4 with CED-3 and CED-9 ± a molecular framework for cell death. Science. **275**, 1122–1126.

276 4 Apoptosis

Chinnaiyan, A.M., O'rourke, K., Tewari, M. and Dixit, V.M. (1995). FADD, a novel death domain-containing protein, interacts with the death domain of Fas and initiates apoptosis. Cell. **81**, 505–512.

Chinnaiyan, A.M., O'rourke, K., Yu, G.-L., Lyons, R.H., Garg, M., Duan, D.R., Xing, L., Gentz, R., Ni, J. and Dixit, V.M. (1996). Signal transduction by DR3, a death domain-containing receptor related to TNFR-1 and CD95. Science **274**, 990–992.

Cho, Y., Gorina, S., Jeffrey, P.D., Pavletich, N.P. (1994). Crystal structure of a p53 tumour suppressor-DNA complex: understanding tumorigenic mutations. Science. **265**, 346–355.

Cikala, M., Wilm, B., Hobmayer, E., Bottger, A. and David, C.N. (1999). Identification of caspases and apoptosis in the simple metazoan Hydra. Curr Biol. **9**, 959–962.

Clarke, P.G.H. and Clarke, S. (1996). Nineteenth century research on naturally occurring cell death and related phenomena. Anat Embryol. **193**, 81–99.

Cleary, M.L., Smith, S.D. and Sklar, J. (1986). Cloning and structural analysis of cDNAs for bcl-2 and a hybrid bcl-2/immunoglobulin transcript resulting from the t(14;18) translocation. Cell. **47**, 19–28.

Clem, R.J., Fechheimer, M. and Miller, L.K. (1991). Prevention of apoptosis by a baculovirus gene during infection of insect cells. Science. **254**, 1388–1390.

Cohen, J.J. and Duke, R.C. (1984). Glucocorticoid activation of calcium-dependent endonuclease in thymocyte nuclei leads to cell death. J Immunol. **132**, 38–42.

Cohen, G.M., Sun, X.M., Fearnhea, H., Macfarlane, M., Brown, D.G., Snowden, R.T. and Dinsdale, D. (1994). Formation of large molecular weight fragments of DNA is a key committed step of apoptosis in thymocytes. J Immunol. **153**, 507–516.

Conradt, B. and Horvitz, H.R. (1998). The *C. elegans* protein EGL-1 is required for programmed cell death and interacts with the Bcl-2-like protein CED-9. Cell. **93**, 519–529.

Conway, E.M., Pollefeyt, S., steiner-Mosonyi, M., Luo, W., Devriese, A., Lupu, F., Bono, F., Leducq, N., Dol, F., Schaeffer, P., Collen, D. and Herbert, J.M. (2002). Deficiency of survivin in transgenic mice exacerbates Fas-induced apoptosis via mitochondrial pathways. Gastroenterology. **123**, 619–631.

Cosman, D. (1994). A family of ligands for the TNF receptor superfamily. Stem Cells. **12**, 440–455.

Cossarizza, A., Kalashnikova, G., Grassilli, E., Chiappelli, F., Salvioli, S., Capri, M., Barbieri, D., Troiano, L., Monti, D. and Franceschi, C. (1994). Mitochondrial modifications during rat thymocyte apoptosis: a study at the single cell level. Exp Cell Res. **214**, 323–330.

Coux, O., Tanaka, K. and Goldberg, A. L. (1996). Structure and functions of the 20S and 26S proteasomes. Annu Rev Biochem. **65**, 801–847.

Cyrns, V. and Yuan, J. (1998). Proteases to die for. Genes Dev. **12**, 1551–1570.

Decary, S., Decesse, J.T., Ogryzko, V., Reed, J.C., Naguibneva, I., Harel-Bellan, A. and Cremisi, C.E. (2002). The retinoblastoma protein binds the promoter of the survival gene bcl-2 and regulates its transcription in epithelial cells through transcription factor AP-2. Mol Cell Biol. **22**, 7877–7888.

Deng, G. and Podack, E.R. (1995). Deoxyribonuclease induction in apoptotic cytotoxic T lymphocytes. FASEB J. **9**, 665–669.

De Ruijter, A.J., Van Gennip, A.H., Caron, H.N., Kemp, S. and Van Kuilenburg, A.B. (2003). Histone deacetylases (HDACs): characterization of the classical HDAC family. Biochem J. **370**, 737–749. Review.

Deveraux, Q.L., Roy, N., Stennicke, H.R., VAN Arsdale, T., Zhou, Q., Srinivasula, S.M., Alnemri, E.S., Salvesen, G.S. and Reed, J.C. (1998). IAPs block apoptotic events induced by caspase-8 and cytochrome *c* by direct inhibition of distinct caspases. EMBO J. **17**, 2215–2223.

Deveraux, Q., Takahashi, R., Salvesen, G.S. and Reed, J.C. (1997). X-linked IAP is a direct inhibitor of cell death proteases. Nature. **388**, 300–303.

Devireddy, L.R. and Jones, C.J. (1999). Activation of caspases and p53 by bovine herpesvirus 1 infection results in programmed cell death and efficient virus release. J Virol. **73**, 3778–3788.

Devita, V.T., Hellman, S. and Rosenberg, S.A. (2001). Cancer, Principles and Practice of Oncology (Lippincott-Raven, Philadelphia).

References

Di Gioacchino, M., Petrarca, C., Perrone, A., Martino, S., Esposito, D., Lotti, L.V., Mariani-Costantini, R. (2008). Autophagy in hematopoietic stem/progenitor cells exposed to heavy metals: biological implications and toxicological relevance. Autophagy. **4**, 537–539.

Dobner, T., Horikoshi, N., Rubenwolf, S. and Shenk, T. (1996). Blockage by adenovirus E4orf6 of transcriptional activation by the p53 tumour suppressor. Science. **272**, 1470–1473.

Donehower, L.A. and Bradley, A. (1993). The tumour suppressor p53. Biochim Biophys Acta. **23**, 181–205.

Downward, J. (1999). How BAD phosphorylation is good for survival. Nature Cell Biol. **1**, E 33–35.

Dubrez, L., Savoy, I., Hammnan, A. and Solary, E. (1996). Pivotal role of a DEVD-sensitive step in etoposide-induced and Fas-mediated apoptotic pathways. EMBO J. **15**, 5504–551.

Duckett, C.S., Nava, V.E., Gedrich, R.W., Clem, R.J., Vandongen, J.L., Gilfillan, M.C., Shiels, H., Hardwick, J.M. and Thompson, C.B. (1996). A conserved family of cellular genes related to the baculovirus IAP gene and encoding apoptosis inhibitors. EMBO J. **15** 2685–2694.

Duke, R.C. and Cohen, J.J. (1986). IL-2 addiction: withdrawal of growth factor activates a suicide program in dependent T cells. Lymphokine Res. **5**, 289–299.

Durrieu, F., Samejima, K., Fortune, J., Kandels-Lewis, S., Osheroff, N. and Earnshaw, W. (2000). DNA topoisomerase IIa interacts with CAD nuclease and is involved in chromatin condensation during apoptotic execution. Curr Biol. **10**, 923–926.

Earnshaw, W.C. (1995). Nuclear changes in apoptosis. Curr Opin Cell Biol. **7**, 337–343.

Eastman, A. (1994). Deoxyribonuclease II in apoptosis and the significance of intracellular acidification. Cell Death Differ. **1**, 7–10.

Ejercito, M. and Wolfe, J. (2004). Caspase-like activity is required for programmed nuclear elimination during conjugation in Tetrahymena. J Eukaryot Microbiol. **50**, 427–429.

El-Deiry, W., Kern, S.E., Pietenpol, J.A., Kinzler, K.W. and Vogelstein B. (1992). Definition of a consensus binding site for p53. Nat Genet. **1**, 45–49, 1992.

Ellis, H.M. and Horvitz, H.R. (1986). Genetic control of programmed cell death in the nematode *C. elegans*. Cell. **44**, 817–829.

Enari, M., Hug, H. and Nagata, S. (1995). Involvement of an ICE-like protease in Fas mediated apoptosis. Nature **375**, 78–81.

Enari, M., Sakahira, H., Yokoyama, H., Okawa, K., Iwamatsu, A. and Nagata, S. (1998). A caspase-activated DNase that degrades DNA during apoptosis, and its inhibitor ICAD. Nature. **391**, 43–50.

Enari, M., Talanian, R.V. Wong, W.W. and Nagata, S. (1996). Sequential activation of ICE-like and CPP32-like proteases during Fas-mediated apoptosis. Nature **380**, 723–726.

Evans, C.J. and Aguilera, R.J. (2003). DNase II: genes, enzymes and function. Gene. **322**, 1–15.

Fadok, V.A., Bratton, D.L., Rose, D.M., Pearson, A., Ezekewitz, R.A. and Henson, P.M. (2000). A receptor for phosphatidylserine-specific clearance of apoptotic cells. Nature. **405**, 85–90.

Fadok, V.A., Voelker, D.R., Campbell, P.A., Cohen, J.J., Bratton, D.L. and Henson, P.M. (1992). Exposure of phosphatidylserine on the surface of apoptotic lymphocytes triggers specific recognition and removal by macrophages. J Immunol. **148**, 2207–2216.

Famulski, K.S., Macdonald, D., Paterson, M.C. and Sikora, E. (1999). Activation of a low pH-dependent nuclease by apoptotic agents. Cell Death Differ. **6**, 281–289.

Fan, Z., Beresford, P.J., Oh, D.Y., Zhang, D. and Lieberman, J. (2003). Tumour suppressor NM23-H1 is a granzyme A-activated DNase during CTL-mediated apoptosis, and the nucleosome assembly protein SET is its inhibitor. Cell. **112**, 659–672.

Fesus, L., Davies, P.J.A. and Piacentini, M. (1991). Apoptosis: molecular mechanisms in programmed cell death. Eur J Cell Biol. **56**, 170–177.

Fesus, L., Madi, A., Balajthy, Z., Nemes, Z. and Szondy, Z. (1996). Transglutaminase induction by various cell death and apoptosis pathways. Experientia. **52**, 942–949. Review.

Ford, J.M., Baron, E.L. and Hanawalt, P.C. (1998). Human fibroblasts expressing the human papillomavirus E6 gene are deficient in global genomic nucleotide excision repair and sensitive to ultraviolet irradiation. Cancer Res. **58**, 599–603.

Fridman, J.S. and Lowe, S.W. (2003). Control of apoptosis by p53. Oncogene. **22**, 9030–9040.

Friedberg, E.C., Walker, G.C., Siede, W., Wood, R.D., Schultz, R.A., Ellenberger, T. (2005). DNA Repair and Mutagenesis, 2nd edition. American Society for Microbiology Press, Washington DC, p. 1118.

Fritz, G., Grösch, S., Tomicic, M. and Kaina, B. (2003). APE/Ref-1 and the mammalian response to genotoxic stress. Toxicology. **193**, 67–78.

Furukawa, Y., Nishimura, N., Furukawa, Y., Satoh, M., Endo, H., Iwase, S., Yamada, H., Matsuda, M., Kano, Y. and Nakamura, M. (2002). Apaf-1 is a mediator of E2F-1-induced apoptosis. J Biol Chem. **277**, 39760–39768.

Ganesh, L., Yoshimoto, T., Moorthy, N.C., Akahata, W., Boehm, M., Nabel, E.G., Nabel, G.J. (2006). Protein methyltransferase 2 inhibits NF-kappaB function and promotes apoptosis. Mol Cell Biol. **26**, 3864–3874.

Gannon, J.V., Greaves, R., Iggo, R. and Lane, D.P. (1990). Activating mutations in p53 produce a common conformational effect. A monoclonal antibody specific for the mutant form. EMBO J. **9**, 1595–1602.

Gannavaram, S., Vedvyas, C. and Debrabant, A. (2008). Conservation of the pro-apoptotic nuclease activity of endonuclease G in unicellular trypanosomatid parasites. J Cell Sci. **121**, 99–109.

Garattini, E., Gianni, M. and Terao, M. (2004). Retinoid related molecules an emerging class of apoptotic agents with promising therapeutic potential in oncology: pharmacological activity and mechanisms of action. Curr Pharm Des. **10**, 433–448. Review.

Garrido, C., Bruey, J.M., Fromentin, A., Hammann, A., Arrigo, A.P. and Solary, E. (1999). HSP27 inhibits cytochrome c-dependent activation of procaspase-9. FASEB J. **13**, 2061–2070.

Genini, D., Sheeter, D., Rought, S., Zaunders, J.J., Susin, S.A., Kroemer, G., Richman, D.D., Carson, D.A., Corbeil, J., Leoni, L.M. (2001). HIV induces lymphocyte apoptosis by a p53-initiated, mitochondrial-mediated mechanism. FASEB J. **15**, 5–6.

Gerace, L. and Blobel, G. (1980). The nuclear envelope lamina is reversibly depolymerized during mitosis. Cell. **19**, 277–287.

Gil-Gómez, G. and Brady, H.J.M. (1998). Transgenic mice in apoptosis research. Apoptosis. **3**, 215–228.

Ginaldi, L., De Martinis, M., Monti, D., Franceschi, C. (2005). Chronic antigenic load and apoptosis in immunosenescence. Trend Immunol. **26**, 79–84.

Glickman, M.H. and Ciechanover, A. (2002). The ubiquitin–proteasome proteolytic pathway: destruction for the sake of construction. Physiol Rev. **82**, 373–428.

Glucksmann, A. (1951). Cell deaths inn normal vertebrate ontogeny. Biol Rev. **26**, 59–86.

Green, D.R. and Reed, J.R. (1998). Mitochondria and apoptosis. Science **281**, 1309–1312.

Greenlund, L.J.S., Deckwerth, T.L. and Johnson, E.M.J. (1995). Superoxide dismutase delays neuronal apoptosis: a role for reactive oxygen species in programmed cell death. Neuron. **14**, 303–315.

Grether, M.E., Abrams, J.M., Agapite, J., White, K. and Steller, H. (1995). The head involution defective gene of *Drosophila melanogaster* functions in programmed cell death. Genes Dev. **9**, 1694–1708.

Grover, P.L. (1979). *In vitro* modification of nucleic acids by direct-acting chemical carcinogens. In: Chemical Carcinogenesis and DNA, P.L. Grover, (ed). CRC, Boca Raton. FL. Vol. I, p. 37.

Hainaut, P. and Milner, J. (1992). Interaction of heat- shock protein 70 with p53 translated *in vitro*: evidence for interaction with dimeric p53 and for a role in the regulation of p53 conformation. EMBO J. **11**, 3513–3520.

Halbert, D.N., Cutt, J.R. and Shenk, T. (1985). Adenovirus early region 4 encodes functions required for efficient DNA replication, late gene expression, and host cell shutoff. J Virol. **56**, 250–257.

Halliwell, B. and Gutteridge, J.M. (1989). Free Radicals in Biology and Medicine (Clarendon Press, Oxford).

Hampton, M.B., Zhivotovsky, B., Slater, A.F., Burgess, D.H. and Orrenius, S. (1998). Importance of the redox state of cytochrome *c* during caspase activation in cytosolic extracts. Biochem J. **329**, 95–99.

References

Hanayama, R., Tanaka, M., Miyasaka, K., Aozasa, K., Koike, M. Uchiyama, Y. and Nagata, S. (2004). Autoimmune disease and impaired uptake of apoptotic cells in MFG-E8-deficient mice. Science. **304**, 1147–1150.

Harlow, E., Crawford, L.V., Pim, D.C. and Williamson, N.M. (1981). Monoclonal antibodies specific for simian virus 40 tumour antigen. J Virol. **39**, 861–869.

Harris, C.C., (1996). P53 Tumour suppressor gene: from the basic research to laboratory to the clinican abridged historical perspective. Carcinogenesis **17**, 1187–1198.

Harris, C.C. and Hollstein, M. (1993). Clinical implications of the p53 tumour-suppressor gene. N Engl J Med. **329**, 1318–1327.

Harris, S.L. and Levine, A J. (2005). The p53 pathway: positive and negative feedback loops. Oncogene. **24**, 2899–2908.

Haupt, Y., Maya, R., Kazaz, A. and Oren, M. (1997). Mdm2 promotes the rapid degradation of p53. Nature. **387**, 296–299.

Haupt, Y., Rowan, S., Shaulian, E., Vousden, K.H. and Oren, M. (1995). Induction of apoptosis in HeLa cells by transactivation-deficient p53. Genes Dev. **9**, 2170–2133.

Harvey, D.M. and Levine, A.J. (1991). p53 alteration is a common event in the spontaneous immortalization of primary BALB/c murine embryo fibroblasts. Genes Dev. **5**, 2375–85.

Havel, A., Durzan, D.J. Stmad, M., Pec, P. and Beck, E. (eds). (1999). Programmed cell death in plant development. In Advances in Regulation of Plant Growth and Development. Peres Publisher, Prague, pp. 119–212.

Hawley-Nelson, P., Vousden, K.H., Hubbert, N.L., Lowy, D.R. and Schiller, J.T. (1989). HPV16 E6 and E7 proteins cooperate to immortalize human foreskin keratinocytes. EMBO J. **8**, 3905–3910.

Hay, B.A., Wassarman, D.A. and Rubin, G.M. (1995). *Drosophila* homologs of baculovirus inhibitor of apoptosis proteins function to block cell death. Cell. **83**, 1253–1262.

He, X., Zhang, Q., Liu, Y. and He, P. (2005). Apoptin induces chromatin condensation in normal cells. Virus Genes. **31**, 49–55.

He, J., Tohyama, Y., Yamamoto, K., Kobayashi, M., Shi, Y., Takano, T., Noda, C., Tohyama, K. and Yamamura, H. (2005). Lysosome is a primary organelle in B cell receptor-mediated apoptosis: an indispensable role of Syk in lysosomal function. Genes Cells. **10**, 23–35.

Hedgecock, E.M., Sulston, J.E. and Thomson, J.N. (1983). Mutations affecting programmed cell deaths in the nematode *Caenorhabditis elegans*. Science **220**, 1277–1279.

Helt, A.-M. and Galloway, D.A. (2003). Mechanisms by which DNA tumour virus oncoproteins target the Rb family of pocket proteins. Carcinogenesis 24, 159–169.

Hengartner, M.O. (2001). Apoptosis DNA destroyers. Nature. **412**, 27–29.

Hengartner, M.O., Ellis, R.E. and Horvitz, H.R. (1992). *Caenorhabditis elegans* gene ced-9 protects cells from programmed cell death. Nature **356**, 494–499.

Hetz, C.A., Hunn, M., Rohas, P., Torres, V., Leyton, L. and Quest, A.F.G. (2002). Caspase-dependent initiation of apoptosis and necrosis by the Fas receptor in lymphoid cells: onset of necrosis is associated with delayed ceramide increase. J Cell Sci. **115**, 4671–4683.

Hewish, D.R. and Burgoyne, L.A. (1973). Chromatin sub-structure. The digestion of chromatin DNA at regularly spaced sites by a nuclear deoxyribonuclease. Biochem Biophys Res Comm. **52**, 504–510.

Higami, Y., To, K., Ohtani, H., Masui, K., Iwasaki, K., Shiokawa, D. Tanuma, S. and Shimokawa, I. (2003). Involvement of DNase g in apoptotic DNA fragmentation in histiocytic necrotizing lymphadenitis. Virchows Arch. **443**, 170–174.

Hilt, W. and Wolf, D. H. (1996). Proteasomes: destruction as a programme. Trends Biochem Sci. **21**, 96–102

Hinds, P.W., Finlay. C.A., Frey, A.B. and Levine, A.J. (1987). Immunological evidence for the association of p53 with a heat shock protein, hsc70, in p53-plus-ras-transformed cell lines. Mol Cell Biol. **7**, 2863–2869.

Hirao, A., Kong, Y.Y., Matsuoka, S., Wakeham, A., Ruland, J., Yoshida, H., Liu, D., Elledge, S.J. and Mak, T.W. (2000). DNA damage-induced activation of p53 by the checkpoint kinase Chk2. Science. **287**, 1824–1827.

280 4 Apoptosis

Hockenbery, D.M., Oltvai, Z.N., Yin, X.-M., Milliman, C.L. and Korsmeyer, S.J. (1993). Bcl-2 functions in an anti-oxidant pathway. Cell **75**, 241–251.

Hoffmann, H., Högel, J. and Speit, G. (2005). The effect of smoking on DNA effects in the comet assay: a meta-analysis. Mutagenesis. **20**, 455–466.

Hofmann, K., Bucher, P. and Tschopp, J. (1997). The CARD domain – a new apoptotic signalling motif. Trends Biochem Sci. **22**, 155–156.

Holler, N., Zaru, R., Micheau, O., Thome, M., Attinger, A., Valitutti, S., Bodmer, J.L., Schneider, P., Seed, B. and Tschopp, J. (2000). Fas triggers an alternative, caspase-8-independent cell death pathway using the kinase RIP as effector molecule. Nat Immunol. **1**, 489–495.

Hollstein, M., Rice, K., Greenblatt, M.S., Soussi, T., Fuchs, R., Sorlie, T., Hovig, E., Smith-Sorensen, B., Montesano, R. and Harris, C.C. (1994). Database of p53 gene somatic mutations in human tumours and cell lines. Nucleic Acids Res. **22**, 3551–3555.

Hong, M.Y., Chapkin, R.S., Wild, C.P., Morris, J.S., Wang, N., Carroll, R.J., Turner, N.D. and Lupton, J.R. (1999). Relationship between DNA adduct levels, repair enzyme, and apoptosis as a function of DNA methylation by azoxymethane. Cell Growth Differ. **10**, 749–758.

Horvitz, H.R., Ellis, H.M. and Sternberg, P.W. (1982). Programmed cell death in nematode development. Neurosci Comment. **1**, 56–65.

Hotchkiss, R.S., Swanson, P.E., Knudson, C.M., Chang, K.C., Cobb, J.P., Osborne, D.F., Zollner, K.M., Buchman, T.G., Korsmeyer, S.J. and Karl, I.E. (1999). Overexpression of Bcl-2 in transgenic mice decreases apoptosis and improves survival in sepsis. J Immunol. **162**, 4148–4156.

Huang, X., Tran, T., Zhang, L., Hatcher, R. and Zhang, P. (2005). DNA damage-induced mitotic catastrophe is mediated by the Chk1-dependent mitotic exit DNA damage checkpoint. Proc Natl Acad Sci. USA **102**, 1065–1070.

Hughes, F.M. Jr., Cidlowski, J.A. (1997). Utilization of an *in vitro* apoptosis assay to evaluate chromatin degradation by candidate apoptotic nucleases. Cell Death Differ. **4**, 200–208.

Hughes, F.M.JR1., Evans-Storms, R.B. and Cidlowski, J.A. (1998). Evidence that non-caspase proteases are required for chromatin degradation during apoptosis. Cell Death Differ. **5**, 1017–1027.

Hwang, B.J., Ford, J.M., Hanawalt, P.C. and Chu, G. (1999). Expression of the p48 xeroderma pigmentosum gene is p53-dependent and is involved in global genomic repair. Proc Natl Acad Sci USA **96**, 424–428.

Irmler, M., Hofmann, K., Vaux, D.L. and Tschopp, J. (1997). Direct physical interaction between the *Caenorhabditis elegans* death proteins CED-3 and CED-4. FEBS Lett. **406**, 189–190.

Isaacs, J.S., Hardman, R., Carman, T.A., Barrett, J.C. and Weissman, B.E. (1998). Differential subcellular p53 localization and function in N- and S-type neuroblastoma cell lines. Cell Growth Differ. **9**, 545–555.

Itoh, N., Yonehara, S., Ishii, A., Yonehara, M., Mizushima, S., Sameshima, M., Hase, A., Seto, Y. and Nagata, S. (1991). The polypeptide encoded by the cDNA for human cell surface antigen Fas can mediate apoptosis. Cell. **66**, 233–243.

Jacobs-Helber, S.M., Wickrema, A., Birrer, M.J. and Sawyer, S.T. (1998). AP1 regulation of proliferation and initiation of apoptosis in erythropoietin-dependent erythroid cells. Mol Cell Biol. **18**, 3699–3707.

Jacobson, M.D., Burne, J.F., King, M.P., Miyashita, T., Reed, J.C. and Raff, M.C. (1993). Bcl-2 blocks apoptosis in cells lacking mitochondrial DNA. Nature. **361**, 365–369.

Jacobson, M.D., Weil, M. and Raff, M.C. (1996). Role of Ced-3/ICE-family of proteases in staurosporine-induced programmed cell death. J Cell Biol. **133**, 1041–1051.

Jacobson, M.D., Weil, M. and Raff, MC. (1997). Programmed cell death in animal development. Cell.**88**, 347–54. Review.

James, C., Gschmeissner, S., Fraser, A. and Evan, G.I. (1997). CED-4 induces chromatin condensation in Schizosaccharomyces pombe and is inhibited bydirect physical association with CED-9. Curr Biol. **7**, 246–252.

Jentsch, S. (1992). Ubiquitin-dependent protein degradation: a cellular perspective. Trends Cell Biol. **2**, 93–103.

References 281

Jenuwein, T. and Allis, C.D. (2001). Translating the "histone code". Science **293**, 1074–1080.

Jimenez, B., Volpert, O.V., Crawford, S.E., Febbraio, M., Silberstein, R.L. and Bouck, N. (2000). Signals leading to apoptosis-dependent inhibition of neovascularization by thrombospondin-1. Nature Med. **6**, 41–48.

Johnson, N.L., Gardner, A.M., Diener. KM., Lange-Carter, C.A., Gleavy, J., Jarpe, M.B., Minden, A., Karin, M., Zon, L.I. and Johnson, G.L. (1996). Signal transduction pathways regulated by mitogen-activated/extracellular response kinase kinase kinase induce cell death. J Biol Chem. 271, 3229–3237.

Kagi, D., Ledermann, B., Burki, K., Seiler, P., Odermatt, B., Olsen, K.J., Fodack, E.R., Zinkernagel, R.M. and Hengartner, H. (1994). Cytotoxicity mediated by T cells and natural killer cells is greatly impaired in perforin-deficient mice. Nature. **369**, 1–37.

Kanai, Y. (2003). The role of non-chromosomal histones in the host defense system. Microbiol Immunol. **47**, 553–556.

Kane, D.J., Sarafin, T.A., Anton, R., Hahn, H., Gralla, E.B., Valentine, J.S., Örd, T. and Bredesen, D.E. (1993). Bcl-2 inhibition of neuronal death: decreased generation of reactive oxygen species. Science. **262**, 1274–1277.

Karran, P. and Bignami, M. (1999). in DNA Recombination and Repair. Smith, P. J. and Jones, C. J. (eds) (Oxford Univ. Press, New York), pp. 66–159.

Kaufmann, S.H. (1989). Induction of endonucleolytic DNA cleavage in human acute myelogenous leukemia cells by etoposide, camptothecin, and other cytotoxic anticancer drugs: a cautionary note. Cancer Res. **49**, 5870–5878.

Kaufmann, S., Desnoyers, S., Ottaviano, Y., Davidson, N. and Poirier, G.G. (1993). Specific proteolytic cleavage of poly(ADP-ribose)polymerase: an early marker of chemotherapy-induced apoptosis. Cancer Res. **53**, 3976–3985.

Kelekar, A. and Thompson, C.B. (1998). Bcl-2 family proteins: The role of the BH-3 domain in apoptosis. Trends Cell Biol. **8**, 324–330.

Kerr, J. (1965). A histochemical study of hypertrophy and ischaemic injury of ratliver with special reference to changes in lysosomes. J Pathol Bacteriol. **90**, 419–435.

Kerr, J.F.R. (1971). Shrinkage necrosis: a distinct mode of cellular death. J Pathol. **105**, 13–20.

Kerr, J.F., Wyllie, A.H. and Currie, A.R. (1972). Apoptosis: a basic biological phenomenon with wide-ranging implications in tissue kinetics. Br J Cancer. **26**, 239–257.

Kharbanda, S., Pandey, F., Schofield, L., Israels, S., Roncinske, R., Yoshida, K., Bharti, A. Yuan, Z.-M., Saxena, S., Weichselbaum, R., Nalin, C. and Kufe, D. (1997). Role for Bcl-xL as an inhibitor of cytosolic cytochrome c accumulation in DNA damage-induced apoptosis. Proc Natl Acad Sci USA. **94**, 6939–6942.

Khodarev, N.N. and Ashwell, J.D. (1996). An inducible lymphocyte nuclear Ca^{2+}/Mg^{2+-} dependent endonuclease associated with apoptosis. J Immunol. **156**, 922–931.

Kihlmark, M., Imreh, G. and Hallberg, E. (2001). Sequential degradation of proteins from the nuclear envelope during apoptosis. J. Cell Sci. **114**, 3643–3653.

Kim, B.E., Roh, S.R., Kim, J.W., Jeong, S.W. and Kim, I.K. (2003). Cytochrome c-dependent Fas-independent apoptotic pathway in HeLa cells induced by delta12-prostaglandin J2. Exp Mol Med. **35**, 290–300.

Kim, C.N., Wang, X., Huang, Y., Ibrado, A.M., Liu, L., Fang, G. and Bhalla, K. (1997). Overexpression of Bcl-xL inhibits Ara-c-induced mitochondrial loss of cytochrome c and other perturbations that activate the molecular cascade of apoptosis. Cancer Res. **57**, 3115–3120.

Kitson, J., Raven, T., Jiang, Y.-P., Goeddel, D.V., Giles, K.M., Pun, K.T., Grinham, C.J., Brown, R. and Farrow, S.N. (1996). A death-domain-containing receptor that mediates apoptosis. Nature. **384**, 372–375.

Klein, J.A., Longo-Guess, C.M., Rossmann, M.P., Seburn, K.L., Hurd, R.E., Frankel, W.N., Bronson, R.T. and Ackerman, S.L. (2002). The harlequin mouse mutation downregulates apoptosis-inducing factor. Nature. **419**, 367–374.

Kluck, R.M., Bossy-Wetzel, E., Green, D.R. and Newmeyer, D.D. (1997). The release of cytochrome c from mitochondria: a primary site for Bcl-2 regulation of apoptosis. Science **275**, 1132–1136,.

Kluck, R.M., Martin, S.J., Hoffman, B.M., Zhou, J.S., Green, D.R. and Newmeyer, D.D. (1997). Cytochrome c activation of CPP32-like proteolysis plays a critical role in a Xenopus cell-free apoptosis system. EMBO J. **16**, 4639–4649.

Koken, M.H.M., Hoogerbrugge, J.M., Jaspers-Dekker, L., De Wit, J., Willemsen, R., Roest, H.P., Grootegoed, J.A. and Hoeijmakers, J.H.J. (1996). Expression of ubiquitin-conjugation DNA repair enzyme HHR6A and B suggests a role in spermatogenesis and chromatin modification. Dev Biol. **173**, 119–132.

Konat, G.W. (2003). H_2O_2-induced higher order chromatin degradation: A novel mechanism of oxidative genotoxicity. J Biosci. **28**, 57–60.

Kolter, R. (2007). Deadly priming. *Science*. 318, 578–579.

Koumenis, C., Alarcon, R., Hammond, E., Sutphin, P., Hoffman, W., Murphy, M., Derr, J., Taya, Y., Lowe, S.W., Kastan, M. and Giaccia, A. (2001). Regulation of p53 by hypoxia: dissociation of transcriptional repression and apoptosis from p53-dependent transactivation. Mol Cell Biol. **21**, 1297–1310.

Kouzarides, T. (2007). Chromatin modifications and their function. Cell **128**, 693–705.

Kreuder, V., Dieckhoff, J., Sittig, M. and Mannherz, H.G. (1984). Isolation, characterization and crystallization of deoxyribonuclease I from bovine and rat parotid gland and its interaction with rabbit skeletal muscle actin. Eur J Biochem. **139**, 389–400.

Krieser, R.J. and Eastman, A. (1998). The cloning and expression of human deoxyribonuclease II. A possible role in apoptosis. J Biol Chem. **273**, 30909–30914.

Kroemer, G. and Martin, S.J. (2005). Caspase-independent cell death. Nat Med. **11**, 725–730.

Kubbutat, M.H., Jones, S.N. and Vousden, K.H. (1997). Regulation of p53 stability by Mdm2. Nature. **387**, 299–303.

Lagarkova, M.A., Iarovaia, O.V. and Razin, S.V. (1995). Large scale fragmentation of mammalian DNA in the course of apoptosis procedes via excision of chromosomal DNA loops and their oligomers. J Biol Chem. **270**, 20239–20245.

Lane, D.P. (1984). Cell immortalization and transformation by the p53 gene. Nature. **312**, 596–597.

Lane, D.P. (1992). Cancer p53, guardian of the genome. Nature. **358**, 15–16.

Laster, S.M., Wood, J.G. and Gooding, L.R. (1988). Tumour necrosis factor can induce both apoptotic and necrotic cell lysis. J Immunol. **141**, 2629–2634.

Lazebnik, Y.A., Takahasi, A., Poirier, G.G., Kaufmann, S.H. and Earnshaw, W.C. (1995). Characterization of the execution phase of apoptosis *in vitro* using extracts from condemned-phase cells. J Cell Sci Suppl. **19**, 41M9.

Lee, J.H. and Paull, T.T. (2004). Direct activation of the ATM protein kinase by the Mre11/Rad50/Nbs1 complex. Science. **304**, 93–96.

Lee, J.H. and Paull, T.T. (2005). ATM activation by DNA double-strand breaks through the Mre11-Rad50-Nbs1 complex. Science. **308**, 551–554.

Lee, Y. S. Jang, M. S. Lee, J. S. Choi, E. J. and Kim, E. (2005). SUMO-1 represses apoptosis signal-regulating kinase 1 activation through physical interaction and not through covalent modification. EMBO Rep. **6**, 949–955.

Leist, M. and Jaattela, M. (2001). Four death and a funeral: From caspases to alternative mechanisms. Nat Rev Mol Cell Biol. **2**, 589–598.

Letai, A., Bassik, M.C., Walensky, L.D., Sorcinelli, M.D., Weiler, S. and Korsmeyer, S.J. (2002). Distinct BH3 domains either sensitize or activate mitochondrial apoptosis, serving as prototype cancer therapeutics. Cancer Cell. **2**, 183–192.

Levine, A.J. (1997). p53, the cellular gatekeeper for growth and division. Cell. **88**, 323–331.

Li, F., Srinivasan, A., Wang, Y., Armstrong, R.C., Tomaselli, K.J. and Fritz, L.C. (1997). Cell-specific induction of apoptosis by microinjection of cytochrome *c*. J Biol Chem. **272**, 30299–30305.

Li, H., Zhu, H., Xu, C.J. and Yuan, J. (1998). Cleavage of BID by Caspase 8 mediates the mitochondrial damage in the Fas pathway of apoptosis. Cell. **94**, 491–501.

Li, L.Y., Luo, X. and Wang, X. (2001). Endonuclease G is an apoptotic DNase when released from mitochondria. Nature. **412**, 95–99.

References

Liebermann, D.A., Hoffman, B. and Steinman, R.A. (1995). Molecular controls of growth arrest and apoptosis: p53-dependent and independent pathways. Oncogene. **11**, 199–210.

Liedberg, B., Nylander, C.I. and Lundstrom, L. (1983). Surface plasmon resonance for gas detection and biosensing. Sens Actuators B Chem. **4**, 299–304.

Linzer, D.I. and Levin, A.J. (1979). Characterization of a 54K dalton cellular SV40 tumour antigen present in SV40-transformed cells and uninfected embryonal carcinoma cells. Cell. **17**, 43–52.

Lyon, C.J., Evans, C.J., Bill, B.R., Otsuka, A.J. and Aguilera, R.J. (2000). The C. elegans apoptotic nuclease NUC-1 is related in sequence and activity to mammalian DNase II. Gene. **252**, 147–154.

Lipton, S.A. and Bossy-Wetzel, E. (2002). Dueling activities of AIF in cell death versus survival: DNA binding and redox activity. Cell. **111**, 147–150.

Liston, P., Roy, N., Tamai, K., Lefebvre, C., Baird, S., Chertonhorvat, G., Farahani, R., Mclean, M., Ikeda, J.E., Mackenzie, A. and Korneluk, R.G. (1996). Suppression of apoptosis inmammalian cells by NAIP and a related family of IAP genes. Nature **379**, 349–353.

Liu, X., Li, P., Widlak, P., Zou, H., Luo, X., Garrard, W.T. and Wang, X. (1998). The 40-kDa subunit of DNA fragmentation factor induces DNA fragmentation and chromatin condensation during apoptosis. Proc Natl Acad Sci USA **95**, 8461–8466.

Liu, X., Kim, C.N., Yang, J., Jemmerson, R. and Wang, X. (1996). Induction of apoptotic program in cell-free extracts: requirement for dATP and cytochrome c. Cell. **86**, 147–157.

Liu, X., Zou, H., Slaughter, C. and Wang, X. (1997). DFF, a heterodimeric protein that functions downstream of caspase-3 to trigger DNA fragmentation during apoptosis. Cell. **89**, 175–184.

Liu, X., Zou, H., Widlak, P., Garrard, W. and Wang, X. (1999). Activation of the apoptotic endonuclease DFF40 (caspase-activated DNase or nuclease). Oligomerization and direct interaction with histone H1. J Biol Chem. **274**, 13836–13840.

Lloyd, D.R. and Hanawalt, P.C. (2000). p53-dependent global genomic repair of benzo[a]pyrene-7,8-diol-9,10-epoxide adducts in human cells. Cancer Res. **60**, 517–521

Lockshin, R.A. and Williams, C.M. (1964). Programmed cell death. II. Endocrine potentiation of the breakdown of the intersegmental muscles of silkmoths. J Insect Physiol. **10**, 643–649.

Lotem, J. and Sachs, L. (1998). Different mechanisms for suppression of apoptosis by cytokines and calcium mobilizing compounds. Proc Natl Acad Sci USA. 95, 4601–4606.

Lotem, J. and Sachs, L. (1999). Cytokines as suppressors of apoptosis. Apoptosis **4**, 187–196.

Lynch, D.H., Ramsdell, F. and Alderson, M.R. (1995). Fas and FasL in thehomeostatic regulation of immune responses. Immunol Today. **16**, 569–574.

Majno, G., Joris, I. (1995). Apoptosis, oncosis and necrosis: An overview of cell death. Am J Pathol. **146**, 3–15.

Marsters, S.A., Pitti, R.M., Donahue, C.J., Ruppert, S., Bauer, K.D. and Ashkenazi, A. (1996a). Activation of apoptosis by Apo-2 ligand is independent of FADD but blocked by CrmA. Curr Biol. **6**, 750–752.

Marsters, S.A., Sheridan, J.P., Donahue, C.J., Pitti, R.M., Gray, C.L., Goddard, A.D., Bauer, K.D. and Ashkenazi, A. (1996b). Apo-3, a new member of tumour necrosis factor receptor family, contains a death domain and activates apoptosis and NF-kB. Curr Biol. **6**, 1669–1676.

Marsters, S.A., Sheridan, J.P., Pitti R.M., Brush, J., Goddard, A. and Ashkenazi, A. (1998). Identification of a ligand for the death-domain containing receptor Apo3. Curr Biol. **8**, 525–528.

Martinou, I., Desagher, S., Eskes. R., Antonsson, B., Andre, E., Fakan, S. and Martinou, J-C. (1999). The release of cytochrome c from mitochondria during apoptosis of NGF-deprived sympathetic neurons is a reversible event. J Biol Chem. **144**, 883–889.

Martins, L.M. and Earnshaw, W.C. (1997). Apoptosis: alive and kicking in 1997. Trends Cell Biol. **7**, 111–14.

Matwee, C., Betts, D.H. and King, W.A. (2000). Apoptosis in the early bovine embryo. Zygote. **8**, 57–68.

Mccarthy, S.A., Symonds, H.S. and Van Dyke, T. (1994). Regulation of apoptosis in transgenic mice by simian virus 40 T antigen-mediated inactivation of p53. Proc Natl Acad Sci USA. **91**, 3979–3983.

284   4 Apoptosis

Mccarthy, S. and Ward, W.S. (1999). Functional aspects of mammalian sperm chromatin. Hum Fertil (Camb). **2**, 56–60.

Metz, T., Harris, A.W. and Adams, J.M. (1995). Absence of p53 allows direct immortalization of hematopoietic cells by the myc and raf oncogenes. Cell. **82**, 29–36.

Mihara, M., Erster, S., Zaika, A., Petrenko, O., Chittenden, T., Pancoska, P. and Moll, U.M. (2003). p53 has a direct apoptotic role at the mitochondria. Mol Cell. **11**, 577–590.

Miura, M., Zhu, H., Rotello, R., Hartweige, A. and Yuan, J. (1993). Induction of apoptosis in fibroblasts by IL-1 beta-converting enzyme, a mammalian homolog of the *C. elegans* cell death gene ced-3. Cell. **75**, 653–660.

Miura, O., D'andrea, A., Kabat, D. and Ihle, J.N. (1991). Induction of tyrosine phosphorylation by the erythropoietin receptor correlates with mitogenesis. Mol Cell Biol. **11**, 4895–4902.

Moll, U.M., Riou, G. and Levine, A.J. (1992). Two distinct mechanisms alter p53 in breast cancer: mutation and nuclear exclusion. Proc Natl Acad Sci USA. **89**, 7262–7266.

Moller, P., Knudsen, L.E., Loft, S. and Wallin, H. (2000). The comet assay as a rapid test in biomonitoring occupational exposure to DNA-damaging agents and effect of confounding factors. Cancer Epidem Biomar Prevent **9**, 1005–1015.

Montague, J.W., Hughes, F.M. Jr. and Cidlowski, J.A. (1997). Native recombinant cyclophilins A, B, and C degrade DNA independently of peptidylprolyl cis-trans isomerase activity: Potential roles of cyclophilins in apoptosis. J Biol Chem. **272**, 6677–6684.

Moore, M., Horikoshi, N. and Shenk, T. (1996). Oncogenic potential of the adenovirus E4orf6 protein. Proc Natl Acad Sci USA. **93**, 11295–11301.

Moore, R.M., Silver, R.J. and Moore, J.J. (2003). Physiological apoptotic agents have different effects upon human amnion epithelial and mesenchymal cells. Placenta. **24**, 173–180.

Moos, P.J., Edes, K. and Fitzpatrick, F.A. (2000). Inactivation of wild-type p53 tumour suppressor by electrophilic prostaglandins. Proc Natl Acad Sci USA. **97**, 9215–9220.

Mora, P.T.K., Chandrashekaran, K., Hoffman, J.C. and Mcfarland, V.W. (1982). Quantitation of 55K cellular protein: similar amount and instability in normal and malignant mouse cells. Mol Cell Biol. **2**, 763–771.

Motyl, T., Grzelkowska, K., Zimowska, W., Skierski, J., Wareski, P., Płoszaj, T. and Trzeciak, L. (1998). Expression of bcl-2 and bax in TGF-beta 1-induced apoptosis of L1210 leukemic cells. Eur J Cell Biol. **75**, 367–74.

Muchmore, S.W., Sattler, M., Liang, H., Meadows, R.P., Harlan, J.E., Yoon, H.S., Nettesheim, D., Chang, B.S., Thompson, C.B., Wong, S.L., Ng, S.-C. and Fesik, S.W. (1996). X-ray and NMR structure of human Bcl-x(1), an inhibitor of programmed cell death. Nature **381**, 335–341.

Mukhopadhyay, T. and Roth, J.A. (1997). Induction of apoptosis in human lung cancer cells after wild-type p53 activation by methoxyestradiol. Oncogene. **14**, 379–384.

Muzio, M., Chinnaiyan, A.M., Kischkel, F.C., O'rourke, K., Shevchenko, A., Ni, J., Scaffidi, C., Bretz, J.D., Zhang, M., Ni, R.J., Gentz, R., Mann, M., Krammer, P.H., Peter, M.E. and Dixit, V.M. (1996). FLICE, a novel FADD-homologous ICE/CED-3-like protease, is recruited to the CD95 (Fas/APO-1) death–inducing signaling complex. Cell. **85**, 817–827.

Nagata, S. and Golstein, P. (1995). The Fas death factor. Science. **267**, 1449–1456.

Neame, S.J., Rubin, L.L. and Philpott, K.L. (1998). Blocking cytochrome *c* activity within intact neurons inhibits apoptosis. J Cell Biol. **142**, 1583–1593.

Nevels, M., Rubenwolf, S., Spruss, T., Wolf, H. and Dobner, T. (1997.). The adenovirus E4orf6 protein can promote E1A/E1B-induced focus formation by interfering with p53 tumour suppressor function. Proc Natl Acad Sci USA. **94**, 1206–1211.

Nezis, I.P., Stravopodis, D.J., Papassideri, I., ROBERT-Nicoud, M. and Margaritis, L.H. (2000). Stage-specific apoptotic patterns during *Drosophila* oogenesis. Eur J Cell Biol. **79**, 610–620.

Nicholson, D.W. (1996). ICE/CED3-like proteases as therapeutic targets for the control of inappropriate apoptosis. Nat Biotechnol. **14**, 297–301. Review.

References

Nicholson, D. W. (1999). Caspase structure, proteolytic substrates, and function during apoptotic cell death. Cell Death Differ. **6**, 1028–1042.

Nicholson, D.W., Ali, A., Thornberry, N.A., Vaillancourt, J.P., Ding, C.K., Gallant, M., Gareau, Y., Griffin, P.R., Labelle, M., Lazebnik, Y.A., Munday, N.A., Raju, S.M., Smulson, M.E., Yamin, T.-T., Yu, V.L. and Miller, D.K. (1995). Identification and inhibition of the ICE/CED-3 protease necessary for mammalian apoptosis. Nature. **376**, 37–43.

Nijhawan, D., Honarpour, N. and Wang, X. (2000). Apoptosis in neural development and disease. Annu Rev Neurosci. **23**, 73–87. Review.

Nishimura, K., Tsumagari, H., Morioka, A., Yamauchi, Y., Miyashita, K., Lu, S., Jisaka, M., Nagaya, T. and Yokota, K. (2002). Regulation of apoptosis through arachidonate cascade in mammalian cells. Appl Biochem Biotechnol. **102–103**, 239–250.

Norbury, C.J. and Zhivotovsky, B. (2004). DNA damage-induced apoptosis. Oncogene. **23**, 2797–2808.

Nurse, P. (1990). Universal control mechanism regulating onset of M-phase. Nature. **344**, 503–508.

Oberhammer, F., Fritsch, G., Schmied, M., Pavelka, M., Printz, D., Purchio, T., Lassmann, H. and Schulte-Hermann, R. (1993). Condensation of the chromatin at the nuclear membrane o fan apoptotic nucleus is not associated with activation o fan endonuclease. J Cell Sci. **104**, 317–326.

Oberhammer, F., Wilson, J.W., Dive, C., Morris, I.D., Hickman, J.A., Wakeling, A.E., Walker, P.R. and Sikorska, M. (1993). Apoptotic death in epithelial cells: cleavage of DNA to 300 and/or 50 kb fragments prior to or in the absence of internucleosomal fragmentation; EMBO J. **12**, 3679–3684.

Odake, S., Kam, C.M., Narasimhan. L., Poe, M., Blake, J.T., Krahenbuhl. O., Tschopp, J. and Powers, J.C. (1991). Human and murine cytotoxic T lymphocyte serine proteases: subsite mapping with peptide thioester substrates and inhibition of enzyme activity and cytolysis by isocoumarins. Biochemistry **30**, 2217–2227.

Offer, H., Zurer, I., Bánfalvi, G., Rehak, M., Falcovitz, A., Milyavsky, M., Goldfinger, N. and Rotter, V. (2001). p53 modulates base excision activity in a cell cycle-specific manner after genotoxic stress. Cancer Res. **61**, 88–96.

Ogawara, Y., Kishishita, S., Obata, T., Isazawa, Y., Suzuki, T., Tanaka, K., Masuyama, N. and Gotoh, Y. (2002). Akt enhances Mdm2-mediated ubiquitination and degradation of p53. J Biol Chem. **277**, 21843–21350.

Okuno, S., Shimizu, S., Ito, T., Nomura, M., Hamada, E., Tsujimoto, Y. and Matsuda, H. (1998). Bcl-2 prevents caspase-independent cell death. J Biol Chem. **273**, 34272–34277.

Oliverio, S., Amendola, A., DI Sano, F., Farrace, M.G., Fesus, L., Nemes, Z., Piredda, L., Spinedi, A. and Piacentini, M. (1997). Tissue transglutaminase-dependent posttranslational modification of the retinoblastoma gene product in promonocytic cells undergoing apoptosis. Mol Cell Biol. **17**, 6040–6048.

Oltvai, Z.N., Milliman, C.L. and Korsmeyer, S.J. (1993). Bcl-2 heterodimerizes *in vivo* with a conserved homolog, Bax, that accelerates programmed cell death. Cell. **74**, 609–619.

Opitz, O.G., Suliman, Y., Hahn, W.C., Harada, H., Blum, H.E., Rustgi, A.K. (2001). Cyclin D1 overexpression and p53 inactivation immortalize primary oral keratinocytes by a telomerase-independent mechanism. J Clin Invest. **108**, 725–732.

Oppermann, M., Geilen, C.C., Fecker, L.F., Gillissen, B., Daniel, P.T. and Eberle, J. (2005). Caspase-independent induction of apoptosis in human melanoma cells by the proapoptotic Bcl-2-related protein Nbk / Bik. Oncogene.**24**, 7369–7380.

Oren, M., Maltzman, W. and Levine, A.J. (1981). Post-translational regulation of the 54K cellular tumour antigen in normal and transformed cells. Mol Cell Biol. **1**, 101–110.

Oren, M. and Rotter, V. (1999). Introduction: p53 – the first twenty years. Cell Mol Life Sci. **55**, 9–11.

Ottilie, S., Wang, Y., Banks, S., Chang, J., Vigna, N.J., Weeks, S., Armstrong, R.C., Fritz, L.C. and Oltersdorf, T. (1997). Mutational analysis of the interacting cell death regulators CED-9 and CED-4. Cell Death Differ. **4**, 526–533.

Pandey, S., Smith, B., Walker, P.R. and Sikorska, M. (2000). Caspase-dependent and independent cell death in rat hepatoma 5123tc cells. Apoptosis.**5**, 265–75.

Pandey, S., Walker, P.R. and Sikorska, M. (1994). Separate pools of endonuclease activity are responsible for internucleosomal and high molecular mass DNA fragmentation during apoptosis. Biochem Cell Biol. **72**, 625–629.

Pandey, S., Walker, P.R. and Sikorska, M. (1997). Identification of a novel 97 kDa endonuclease capable of internucleosomal DNA cleavage. Biochem. **36**, 711–720.

Parant, J.M. and Lozano, G. (2003). Disrupting TP53 in mouse models of human cancers. Hum Mutat. **21**, 321–326.

Parrish, J., Li, L., Klotz, K., Ledwich, D., Wang, X. and Xue, D. (2001). Mitochondrial endonuclease G is important for apoptosis in *C. elegans*. Nature **412**, 90–94.

Peitsch, M.C., Polzar, B., Stephan, H., Crompton, T., Macdonald, H.R., Mannherz, H.G. and Tschoop, J. (1993). Characterization of the endogenous deoxyribonuclease involved in nuclear DNA degradation during apoptosis (programmed cell death). EMBO J. **12**, 371–377.

Peitsch, M.C., Polzar, B., Tschopp, J. and Mannherz, H.G. (1994). About the involvement of deoxyribonuclease I in apoptosis. Cell Death Differ. **1**, 1–6.

Peter, M., Nakagawa, J., Doree, M., Labbe, C. and Nigg, E.A. (1990). *In vitro* disassembly of nuclear lamina by cdc2 kinase. Cell. **61**, 591–602.

Petit, P.X., Lecoeur, H., Zorn, E., Dauguet, C., Mignotte, B. and Gougeon, M.-L. (1995). Alterations in mitochondrial structure and function are early events of dexamethasone-induced thymocyte apoptosis. J Cell Biol. **130**, 157–167.

Petit, P.X., Susin, S-A., Zamzami, N., Mignotte, B. and Kroemer, G. (1996). Mitochondria and programmed cell death: back to the future. FEBS Lett. **396**, 7–13.

Philpott, K.L., Mccarthy, M.J., Becker, D., Gatchalian, C. and Rubin, L.L. (1996). Morphological and biochemical changes in neurons: apoptosis versus mitosis. Eur J Neurosci. **8(9)**, 1906–15.

Pilder, S., Moore, M., Logan, J. and Shenk, T. (1986). The adenovirus E1B-55K transforming polypeptide modulates transport or cytoplasmic stabilization of viral and host cell mRNAs. Mol Cell Biol. **6**, 470–476

Pitti, R.M., Marsters, S.A., Ruppert, S., Donahue, C.J., Moore, A. and Ashkenazi, A. (1996). Induction of apoptosis by Apo-2 ligand, a new member of the tumour necrosis factor cytokine family. J Biol Chem. **271**, 12687–12690.

Plymale, D.R., Tang, D.S., Comardelle, A.M., Fermin, C.D., Lewis, D.E. and Garry, R.F. (1999). Both necrosis and apoptosis contribute to HIV-1-induced killing of CD4 cells. AIDS. **13**, 1827–1839.

Poe, M., Blake, J.T., Boulton, D.A., Gammon, M., Sigal, N.H., Wu, J.K. and Zweerink, H.J. (1991). Human cytotoxic lymphocyte granzyme B. Its purification from granules and the characterization of substrate and inhibitor specificity. J Biol Chem. **266**, 98–103.

Polyak, K., Xia, Y., Zweier, J.L., Kinzler, K.W. and Vogestein, B. (1997). A model for p53-induced apoptosis. Nature. **389**, 300–305.

Polzar, B., Zanotti, S., Stephan, H., Rauch, E., Peitsch, M.C., Irlmer, M., Tschopp, J. and Mannherz, H.G. (1994). Distribution of deoxyribonuclease I in rat tissues and its correlation to cellular turnover and apoptosis (programmed cell death). Eur J Cell Biol. **64**, 200–210.

Portis, T., Grossman, W.J., Harding, J.C., Hess, J.L. and Ratner, L. (2001). Analysis of p53 inactivation in a human T-cell leukemia virus type 1 Tax transgenic mouse model. J Virol. **75**, 2185–2193.

Punyiczki, M. and Fésüs, L. (1998). Heat shock and apoptosis. The two defense systems of the organism may have overlapping molecular elements. Ann N Y Acad Sci. **85**, 67–74. Review.

Puthalakath, H., Huang, D.C., O'reilly, L.A., King, S.M. and Strasser, A. (1999). The proapoptotic activity of the Bcl-2 family member Bim is regulated by interaction with the dynein motor complex. Mol Cell. **3**, 287–296.

Querido, E.R., Marcellus, C., Lai, A., Charbonneau, R., Teodoro, J.G., Ketner, G. and Branton, P.E. (1997). Regulation of p53 levels by the E1B 55-kilodalton protein and E4orf6 in adenovirus-infected cells. J Virol. **71**, 3788–3798.

References

287

Rabbani, A., Finn, R.M., Ausió, J. (2005). The anthracycline antibiotics: antitumor drugs that alter chromatin structure. BioEssays. **27**, 50–56.

Rathmell, J.C. and Thompson, C.B. (1999). The central effectors of cell death in the immune system. Annu Rev Immunol. **17**, 781–828.

Reed, J.C. (1997). Cytochrome *c*: can't live with it—can't live without it. Cell. **91**, 559–562.

Rehak, M., Csuka, I., Szepessy, E. and Bánfalvi, G. (2000). Subphases of DNA replication in *Drosophila* cells. DNA Cell Biol. **19**, 607–612.

Rhim, J.A., Connor, W., Dixon, G.H., Harendza, C.J., Evenson, D.P., Palmiter, R.D. and Brinster, R.L. (1995). Expression of an avian protamine in transgenic mice disrupts chromatin structure in spermatozoa. Biol Reprod. **52**, 20–32.

Roest, H.P., Van Klaveren, J., De Wit, J., Van Gurp, C.G., Koken, M.H.M., Vermey, M., Van Roijen, J.H., Hoogerbrugge, W., Vreeburg, J.T.M., Baarends, W.M., Bootsma, D., Grootegoed, J.A. and J. H. J. Hoeijmakers, J.H.J. (1996). Inactivation of the HR6B ubiquitin-conjugating DNA repair enzyme in mice causes male sterility associated with chromatin modification cell. **86**, 799–810.

Rothe, M., Pan, M.G., Henzel, W.J., Ayres, T.M. and Goeddel, D.V. (1995). The TNFR2-TRAF signaling complex contains two novel proteins related to baculoviral inhibitor of apoptosis proteins. Cell. **83**, 1243–1252.

Roth, J., Dobbelstein, M., Freedman, D.A., Shenk, T. and Levine, A.J. (1998). Nucleo-cytoplasmic shuttling of the hdm2 oncoprotein regulates the levels of the p53 protein via a pathway used by the human immunodeficiency virus rev protein. EMBO J. **17**, 554–564.

Roth, J., König, C., Wienzek, S., Weigel, S., Ristea, S. and Dobbelstein, M. (1998). Inactivation of p53 but not p73 by adenovirus type 5 E1B 55-kilodalton and E4 34-kilodalton oncoproteins. J Virol. **72**, 8510–8516.

Rössler, U., Hornhardt, S., Seidl, C., Müller-Laue, E., Walsh, L., Panzer, W., Schmid, E., Senekowitsch-Schmidtke, R. and Gomolka, M. (2006). The sensitivity of the alkaline comet assay in detecting DNA lesions induced by X rays, gamma rays and alpha particles. Radiat Prot Dosimetry. **122**, 154–159.

Rotter, V. (1983). p53, a transformation-related cellular-encoded protein, can be used as a biochemical marker for the detection of primary mouse tumour cells. Proc Natl Acad Sci USA. **80**, 2613–2617.

Rotter, V., Wolf, D. and Nicolson, G.L. (1984). The expression of transformation-related protein p53 and p53-containing mRNA in murine RAW117 large cell lymphoma cells of differing metastatic potential. Clin Exp Metastasis. **2**, 199–204.

Roy, N., Mahadevan, M.S., McLean, M., Shutler, G., Yaraghi, Z., Farahani, R., Baird, S.,Besnerjohnston, A., Lefebvre, C., Kang, X.L., Salih, M., Aubry, H., Tamai, K., Guan, X.P., Ioannou, P., Crawford, T.O., Dejong, P.J., Surh, L., Ikeda, J.E., Korneluk, R.G. and Mackenzie, A.(1995). Thegene for neuronal apoptosis inhibitory protein is partially deleted in individuals with spinal muscular atrophy. Cell. **80**, 167–178.

Ruttan, C.C. and Glickman, B.W. (2002). Coding variants in human double-strand break DNA repair genes. Mutat Res. **509**, 175–200.

Salvesen, G.S. and Dixit, V.M. (1997). Caspases: intracellular signaling by proteolysis. Cell. **91**, 443–446.

Sarafin, T.A. and Bredesen, D.E. (1994). Is apoptosis mediated by reactive oxygen species? Free Radical Res. **21**, 1–8.

Sarnow, P., Ho, Y.S., Williams, J. and Levine, A.J. (1982). Adenovirus E1b-58kd tumour antigen and SV40 large tumour antigen are physically associated with the same 54 kd cellular protein in transformed cells. Cell. **28**, 387–394.

Sastre, J., Borrás, C., GARCIA-Sala, D., Lloret, A., Pallardo, F.V. and Vina, J. (2002). Mitochondrial damage in aging and apoptosis. Ann N Y Acad Sci. **959**, 448–451.

Sattler, M., Liang, H., Nettesheim, D., Meadows, R.P., Harlan, J.E., Eberstadt, M., Yoon, H.S., Shuker, S.B., Chang, B.S., Minn, A.J., Thompson, C.B., Fesik, S.W. (1997). Structure of Bcl-x(l)-Bak peptide complex recognition between regulators of apoptosis. Science. **275**, 983–986.

Sauter, B., Albert, M.L., Francisco, L., Larsson, M., Somersan, S. and Bhardwaj, N. (2000). Consequences of cell death: exposure to necrotic tumour cells, but not primary tissue cells or apoptotic cells, induces the maturation of immunostimulatory dendritic cells. J Exp Med. **191**, 423–433.

Scaffidi, C., Fulda, S., Srinivasan, A., Friesen, C., Li, F., Tomaselli, K.J., Debatin, K.M., Krammer, P.H. and Peter, M.E. (1998). Two CD95 (APO-1/Fas) signaling pathways. EMBO J. **17**, 1675–1687.

Scheffner, M., Werness, B.A., Huibregtse, J.M., Levine, A.J. and Howley, P.M. (1990). The E6 oncoprotein encoded by human papillomavirus types 16 and 18 promotes the degradation of p53. Cell. **63**, 1129–1136.

Schou, K.B., Schneider, L., Christensen, S.T. and Hoffmann, E.K. (2007). Early-stage apoptosis is associated with DNA-damage-independent ATM phosphorylation and chromatin decondensation in NIH3T3 fibroblasts. Cell Biol Int. 2007 Sep 7; [Epub ahead of print].

Schwartz, L.M., Smith, S.W., Jones, M.E.E. and Osborne, B.A. (1993). Do All programmed cell deaths occur via apoptosis? Proc Natl Acad Sci USA. **90**, 980–984.

Schwartzman, R.A. and Cidlowski, J.A. (1993). Mechanism of tissue-specific induction of internucleosomal deoxyribonucleic acid cleavage activity and apoptosis by glucocorticoids. Endocrinology **133**, 591–599.

Shaman, J.A., Prisztoka, R., Ward, S.W. (2006). Topoisomerase IIB and an extracellular nuclease interact to digest sperm DNA in an apoptotic-like manner. Biol Reprod. 75, 741-748.

Shapiro, H.M. (2003). Practical flow cytometry 4th edn. (Hoboken, Wiley-Liss).

Shiokawa, D., Ohyama, H., Yamada, T., Takahashi, K. and Tanuma, S-I. (1994). Identification of an endonuclease responsible for apoptosis in rat thymocytes. Eur J Biochem. **226**, 23–30.

Seshagiri, S. and Miller, L.K. (1997). *Caenorhabditis elegans* CED-4 stimulates CED-3 processing and CED-3-induced apoptosis. Curr Biol. **7**, 455–460.

Sherman, M.A., Powell, D.R. and Brown, M.A. (2002). IL-4 induces the proteolytic processing of mast cell STAT6. J Immunol. **169**, 3811–3818.

Sigal, A. and Rotter, V. (2000). Oncogenic mutations of the p53 tumour suppressor: the demons of the guardian of the genome. Cancer Res. **60**, 6788–6793.

Smith, C.A., Farrah, T. and Goodwin, R.G. (1994). The TNF receptor superfamily of cellular and viral proteins: activation, costimulation, and death. Cell. **76**, 959–962.

Spector, M.S., Desnoyers, S., Hoeppner, D.J. and Hengartner, M.O. (1997). Interaction between the C. elegans cell-death regulators CED-9 and CED-4. Nature **385**, 6553–656.

Sperandio, S., D.E. Belle, I. and Bredesen, D.E. (2000). An alternative, nonapoptotic form of programmed cell death. Proc Natl Acad Sci USA. **97**, 14376–14381.

Šrám, R.J., Podrazilová, K., Dejmek, D., Mračková, G. and Pilčík, T. (1998). Single cell gel electrophoresis assay: sensitivity of peripheral white blood cells in human population studies. Mutagenesis. **13**, 99–103.

Srinivasula, S.M., Hegder, S., Aleh, A., Datta, P., Shiozaki, E., Chai, J.J., Lee, R.A., Robbins, P.D., Fernandes-Alnemri, T., Shi, Y.G. and Alnemri, E.S. (2001). A conserved XIAPinteractionmotif in caspase-9 and Smac/DIABLOregulates caspase activity and apoptosis. Nature **410**, 112–116.

Steegenga, W.T., Riteco, N., Jochemsen, A.G., Fallaux, F.J. and Bos, J.L. (1998). The large E1B protein together with the E4orf6 protein target p53 for active degradation in adenovirus infected cells. Oncogene. **16**, 349–357.

Strahl, B.D. and Allis, C.D. (2000). The language of covalent histone modifications. Nature. **403**, 41–45.

Strasser, A., Whittingham, S., Vaux, D.L., Bath, M.L., Adams, J.M., Cory, S. and Harris, A.W. (1991). Enforced BCL2 expression in B-lymphoid cells prolongs antibody responses and elicits autoimmune disease. Proc Natl Acad Sci USA. **88**, 8661–8665.

Suda, T., Takahashi, T., Golstein, P. and Nagata, S. (1993). Molecular cloning and expression of the Fas ligand, a novel member of the tumour necrosis factor family. Cell. **75**, 1169–1178.

Sugito, K., Yamane, M., Hattori, H., Hayashi, Y., Tohnai, I., Ueda, M., Tsuchida, N. and Ohtsuka, K. (1995). Interaction between hsp70 and hsp40, eukaryotic homologues of DnaK and DnaJ, in human cells expressing mutant-type p53. FEBS Lett. **358**, 161–164.

References 289

Sulston, J.E. (1976). Post-embryonic development in the ventral cord of *Caenorhabditis elegans*. Philosoph Trans Roy Soc Lond. Series B: Biol Sci. **275**, 287–297.

Sun, X.M., Snowden, R.T., Dinsdale, D., Ormerod, M.G. and Cohen, G.M. (1993). Changes in nuclear chromatin precede internucleosomal DNA cleavage in the induction of apoptosis by etoposide. Biochem Pharmacol. **2**, 187–195.

Susin, S.A., Daugas, E., Ravagnan, L., Samejima, K., Zamzami, N., Loeffler, M., Costantini, P., Ferri, K.F., Irinopoulou T., Prevost, M.C., Brothers, G., Mak, T.W., Penninger, J., Earnshaw, W.C. and Kroemer, G. (2000). Two distinct pathways leading to nuclear apoptosis. J Exp Med. **192**, 571–580.

Susin, S.A., Zamzami, N., Castedo, M., Hirsch, T., Marchetti, P., Macho, A., Daugas, E., Geuskens, M. and Kroemer, G. (1996). Bcl-2 inhibits the mitochondrial release of an apoptogenic protease. J Exp Med. **184**, 1331–1341.

Suzuki, H., Tomida, A. and Tsuruo, T. (2001). Dephosphorylated hypoxia-inducible factor 1alpha as a mediator of p53-dependent apoptosis during hypoxia. Oncogene. **20**. 5779–5788.

Szepessy, E., Nagy, G., Jenei Z., Serfozo, Z., Csuka, I., James, J. and Bánfalvi, G. (2003). Multiple subphases of DNA repair and poly(ADP-ribose) synthesis in Chinese hamster ovary (CHO-K1) cells. Eur J Cell Biol. **82**, 201–207.

Szondy, Z., Reichert, U., Bernardon, J.M., Michel, S., Tóth, R., Ancian, P., Ajzner, E. and Fesus, L. (1997). Induction of apoptosis by retinoids and retinoic acid receptor gamma-selective compounds in mouse thymocytes through a novel apoptosis pathway. Mol Pharmacol. **51**, 972–982.

Tamm, I., Wang, Y., Sausville, E., Scudiero, D.A., Vigna, N., Oltersdorf, T. and Reed, J.C. (1998). IAP-family protein survivin inhibits caspase activity and apoptosis induced by Fas (CD95), Bax, caspases, and anticancer drugs. Cancer Res. **58**, 5315–5320.

Tang, D. and Kidd, V.J. (1998). Cleavage of DFF-45/ICAD by multiple caspases is essential for its function during apoptosis. Biochem. J. **273**, 28549–52.

Tang, J.Y., Hwang, B.J., Ford, J.M., Hanawalt, P.C., Chu, G. (2000). Xeroderma pigmentosum p48 gene enhances global genomic repair and suppresses UV-induced mutagenesis. Mol Cell. **5**, 737–744.

Tarachand, U. and D'souza, S.J. (1999). Apoptosis of rat decidual cells: site specific initiation and related biochemical changes. Ind J Exp Biol **37**, 758–761.

Tata, J.R. (1966). Requirement for RNA and protein synthesis for induced regression of the tadpole tail in organ culture. Dev Biol **13**, 77–94.

Tewari, M. and Dixit, V.M. (1995). Fas- and tumour necrosis factor-induced apoptosis is inhibited by the poxvirus crmA gene product. J Biol Chem**270**, 3255–3260.

Thompson, C.B. (1995) Apoptosis in the pathogenesis and treatment of disease. Science **267**, 1456–1462.

Thornberry, N.A. and Lazebnik, Y. (1998). CASPASES: enemies within. Science. **281**, 1312–1316.

Tominaga, O., Hamelin, R., Trouvat, V., Salmon, R.J., Lesec, G., Thomas, G. and Remvikos, Y. (1993). Frequently elevated content of immunochemically defined wild-type p53 protein in colorectal adenomas Oncogene. **8**, 2653–2658.

Trauth, B.C., Klas, C., Peters, A.M., Matzku, S., Moller, P., Falk, W., Debatin, K.M. and Krammer, P.H. (1989). Monoclonal antibody-mediated tumour regression by induction of apoptosis. Science **245**, 301–305.

Tsujimoto, Y. and Croce, C.M. (1986). Analysis of the structure, transcripts, and protein products of bcl-2, the gene involved in human follicular lymphoma. Proc Natl Acad Sci USA. **83**, 5214–5218.

Uddin, S., Kottegoda, S., Stigger, D., Platanias, L.C. and Wickrema, A. (2000). Activation of the Akt/FKHRL1 pathway mediates the antiapoptotic effects of erythropoietin in primary human erythroid progenitors. Biochem Biophys Res Com. **275**, 16–19.

Ura, S., Masuyama, N., Graves, J.D. and Gotoh, Y. (2001). MST1-JNK promotes apoptosis via caspase-dependent and independent pathways. Genes Cells. **6**, 519–30.

Uren, A.G., Pakusch, M., Hawkins, C.J., Puls, K.L. and Vaux, D.L. (1996). Cloning and expression of apoptosis inhibitory protein homologs that function to inhibit apoptosis and/or bind tumour necrosis factor receptor-associated factors. Proc Natl Acad Sci USA. **93**, 4974–4978.

290 4 Apoptosis

Uren, A., O'rourke, K., Aravind, L., Pisabarro, M., Seshagiri, S., Koonin, E. and DIXIT, V. (2000). Identification of paracaspases and metacaspases: two ancient families of caspase-like proteins, one of which plays a key role in MALT lymphoma. Mol Cell **6**, 961–967.

Van Cruchten, S. and Van Den Broeck, W. (2002). Morphological and biochemical aspects of apoptosis, oncosis and necrosis. Anat Histol Embryol. **31**, 214–23. Review.

Van Loo, G., Schotte, P., Van Gurp, M., Demol, H., Hoorelbeke, B., Gevaert, K., Rodriguez, I., Ruiz-Carrillo, A., Vandekerckhove, J., Declercq, W., Beyaert, R. and Vandenabeele, P. (2001). Endonuclease G: a mitochondrial protein released in apoptosis and involved in caspase-independent DNA degradation. Cell Death Differ. **8**, 1136–1142.

Van Parijs, L. and Abbas, A.K. (1996). Role of Fas-mediated cell death in the regulation of immune responses. Curr Opin Immunol. **8**, 355–361.

Vaux, D.L., Cory, S. and Adams, J.M. (1988). Bcl-2 gene promotes haemopoietic cell survival and cooperates with c-myc to immortalize pre-B cells. Nature **335**, 440–442.

Vaux, D.L., Haecker, G. and Strasser, A. (1994). An evolutionary perspective on apoptosis. Cell. **76**, 777–779.

Vaux, D. L. and Korsmeyer, S.J. (1999). Cell death in development. Cell. **96**, 245–254.

Vaux, D.L., Weissman, I.L. and KIM, S.K. (1992). Prevention of programmed cell death in *Caenorhabditis elegans* by human bcl-2. Science. **258**, 1955–1957.

Vercammen, D., Beyaert, R., Denecker, G., Goossens, V., Van Loo, G., Declercq, W., Grooten, J., Fiers, W. and Vandenabeele, P. (1998a). Inhibition of caspases increases the sensitivity of L929 cells to necrosis mediated by tumour necrosis factor. J Exp Med. **187**, 1477–1485.

Vercammen, D., Brouckaert, G., Denecker, G., Van de Craen, M., Declercq, W., Fiers, W. and Vandenabeele, P. (1998b). Dual signaling of the Fas receptor: initiation of both apoptotic and necrotic cell death pathways. J Exp Med. **188**, 919–930.

Verheij, M., Bose, R., Lin, X.H., Yao, B., Jarvis, W.D., Grant, S., Birrer, M.J., Szabo, E., Zon, L.I. and Kyriakis, J.M. (1996). Requirement for ceramide-initiated SAPK/JNK signalling in stress-induced apoptosis. Nature. **380**, 75–79.

Vogelstein, B., Lane, D. and Levine, A.J. (2000). Surfing the p53 network. Nature. **408**, 307–310.

Voges, D., Zwickl, P. and Baumeister, W. (1999). The 26S proteasome: a molecular machine designed for controlled proteolysis. Annu Rev Biochem. **68**, 1015–1068.

Vogt, C. (1842). Untersuchungen uber die Entwicklungsgeschichte der Geburtshelerkroete (Alytes obstetricians). (Solothurn, Jent und Gassman), p. 130.

Walker, N.P.C., Talanian, R.V., Brady, K.D., Dang, L.C., Ferenz, C.R., Franklin, S., Ghayur, T., Hackett, M.C., Hamill, L.D., Herzog, L., Hugunin, M., Houy, W., Mankovich, J.A., Mcguiness, L., Orlewicz, E., Paskind, M., Pratt, C.A., Reis, P., Summani, A., Terranova, M., Welch, J.P., Xiong, L., Moèller, A., Tracey, D.E., Kamen, R. and Wong, W.W. (1994). Cystal structure of the cysteine protease interleukin-1-b-converting enzyme: A (p20/p10)2 homodimer. Cell. **78**, 343–352.

Walker, P.R., Leblanc, J. and Sikorska, M. (1997). Evidence that DNA fragmentation in apoptosis is initiated and propagated by single-strand breaks. Cell Death Differ. **4**, 506–515.

Wang, G.G., Allis, C.D. and Chi, P. (2007). Chromatin remodeling and cancer, part I: covalent histone modifications. Trends Mol Med. 13, 363–372.

Wang, C., Ivanov, A., Chen, L., Fredericks, W.J., Seto, E. and Rauscher, F.J. III, Chen J. (2005). MDM2 interaction with nuclear corepressor KAP1 contributes to p53 inactivation. EMBO J. **24**, 3279–3290.

Widlak, P., Garrard, W.T. (2006). Discovery, regulation, and action of the major apoptotic nucleases DFF40/CAD and endonuclease G. J Cell Biochem. **94**, 1078–1087.

Wilson, K.P., Black, J., Thomson, J.A., Kim, E.E., Griffith, J.P., Navia, M.A., Murcko, M.A., Chambers, S.P., Aldape, R.A., Raybuck, S.A. and Livingston, D.J. (1994). Structure and mechanism of interleukin-1-beta converting enzyme. Nature. **370**, 270–275.

White, K., Grether, M.E., Abrams, J.M., Young, L., Farrell, K. and Steller, H. (1994). Genetic control of programmed cell death in *Drosophila*. Science **264**, 677–683.

Whitten, J.M. (1969). Cell death during early morphogenesis: parallels between insect limb and vertebrate limb development. Science **163**, 1456–1457.

References   291

Wiley, S.R., Schooley, K., Smolak, P.., Din, W.S., Huang, C.-P., Nicholl, J.K., Sutherland, G.R., Davis Smith, T., Rauch, C., Smith, C. A. and Goodwin, R.G. (1995). Identification and characterization of a new member of the TNF family that induces apoptosis. Immunity. **3**, 673–682.

Williams, J.R., Little, J.B. and Shipley, W.U. (1974). Association of mammalian cell death with a specific endonucleolytic degradation of DNA. Nature. **252**, 754–755.

Woo, E.J., Kim, Y.G., Kim, M.S., Han, W.D., Shin, S., Robinson, H., Park, S.Y. and Oh, B.H. (2004). Structural mechanism for inactivation and activation of CAD/DFF40 in the apoptotic pathway. Mol Cell. **14**, 531–539.

Woods, D.B. and Vousden, K.H. (2001). Regulation of p53 function. Exp Cell Res. **264**, 56–66.

Wright, W.E., Pereira-Smith, O.M. and Shay, J.W. (1989). Reversible cellular senescence: implications for immortalization of normal human diploid fibroblasts. Mol Cell Biol. **9**, 3088–3092.

Wu, Y.-C., Stanfield, G.M., Horvitz H.R. (2000). NUC-1, a *Caenorhabditis elegans* DNase II homolog, functions in an intermediate step of DNA degradation during apoptosis. Genes Dev. **14**, 536–548.

Wu, D.Y., Wallen, H.D. and Nunez, C. (1997). Interaction and regulation of subcellular localization of CED-4 by CED-9. Science. **275**, 1126–1129.

Wyllie, A.H. (1980). Glucocorticoid-induced thymocyte apoptosis is associated with endogenous endonuclease activation. Nature. **284**, 555–556.

Wyllie, A.H. (1981). Cell death: a new classification separating apoptosis from necrosis. In: Bowen, I. and Lochshin, R.A. (eds). Cell Death in Biology and Pathology. Chapman and Hall, London. pp 9–23.

Wyllie, A.H. and Golstein, P. (2001). More than one way to go. Proc Natl Acad Sci USA. **98**, 11–13.

Wyllie, A.H., Kerr, J.F.R. and Currie, A.C. (1980). Cell death: the significance of apoptosis. Int Rev Cytol. **68**, 251–305.

Wynford-Thomas, D. (1996). Telomeres, p53 and cellular senescence. Oncology Res. **8**, 387–398.

Xue, D. and Horvitz, H.R. (1995). Inhibition of the *Caenorhabditis elegans* celldeath protease CED-3 by a CED-3 cleavage site in baculovirus p35 protein. Nature. **377**, 248–251.

Yang, J., Liu, X., Bhalla, K., Kim, C.N., Ibrado, A.M., Cai, J., Peng, T.I., Jones, D.P. and Wang, X. (1997). Prevention of apoptosis by Bcl-2: release of cytochrome c from mitochondria blocked. Science. **275**, 1129–1132.

Yee, K.S. and Vousden, K.H. (2005). Complicating the complexity of p53. Carcinogenesis **26**, 1317–1322.

Ye, H., Cande, C., Stephanou, N.C , Jiang, S., Gurbuxani, S., Larochette, N., Daugas, E., Garrido, C., Kroemer, G. and Wu, H. (2002). DNA binding is required for the apoptogenic action of apoptosis inducing factor. Nat Struct Biol. **9**, 680–684.

Yin, X.M., Wang, K., Gross, A., Zhao, Y., Zinkel, S., Klocke, B., Roth, K.A. and Korsmeyer, S.J. (1999). Bid-deficient mice are resistant to Fas-induced hepatocellular apoptosis. Nature. **400**, 886–891.

Yonehara, S., Ishii, A. and Yonehara, M. (1989). A cell-killing monoclonal antibody (anti-Fas) to a cell surface antigen co-downregulated with the receptor of tumour necrosis factor. J Exp Med. **169**, 1747–1756.

Yonish, R.E., Resnitzky, D., Lotem, J., Sachs, L., Kimchi, A. and Oren, M. (1991). Wildtype p53 induces apoptosis of myeloid leukaemic cells that is inhibited by interleukin-6. Nature. **353**, 345–347.

Yonish-Rouach, E., Resnitzky, D., Lotem, J., Sachs, L., Kimchi, A. and Oren, M. (1993). p53-mediated cell death relationship to cell cycle control. Mol Cell Biol. **13**, 1415–1423.

Yu, J., Zhang, L., Hwang, P.M., Kago, C., Kinzler, K.W. and Vogelstein, B. (1999). Identification and classification of p53 regulated genes. Proc Natl Acad Sci USA. **96**, 14517–14522.

Yuan, J. and Horvitz, H.R. (1992). The *Caenorhabditis elegans* cell death gene ced-4 encodes a novel protein and is expressed during the period of extensive programmed cell death. Development. **116**, 309–320.

Yuan, J., Shaham, S., Ledoux, S., Ellis, H.M. and Horvitz, H.R. (1993). The *C. elegans* cell death gene ced-3 encodes a protein similar to mammalian interleukin-1b-converting enzyme. Cell. **75**, 641–6452.

Yuyama, K., Yamamoto, H., Nishizaki, I., Kato, T., Sora, I. and Yamamoto, T. (2003). Caspase-independent cell death by low concentrations of nitric oxide in PC12 cells: involvement of cytochrome c oxidase inhibition and the production of reactive oxygen species in mitochondria. J Neurosci Res. **73**, 351–363.

Yung, H.W., Bal-Price, A.K., Brown, G.C. and Tolkovsky, A.M. (2004). Nitric oxide-induced cell death of cerebrocortical murine astrocytes is mediated through p53- and Bax-dependent pathways. J Neurochem. **89**, 812–821.

Zamora-Avila, D.E., Franco-Molina, M.A., Trejo-Avila, L.M., Rodríguez-Padilla, C., Reséndez-Pérez, D. and Zapata-Benavides, P. (2007). RNAi silencing of the WT1 gene inhibits cell proliferation and induces apoptosis in the B16F10 murine melanoma cell line. Melanoma Res. **17**, 341–8.

Zamzami, N., Marchetti, P., Castedo, M., Zanin, C., Vayssiere, J.-L., Petit, P.X. and Kroemer, G. (1995). Reduction in mitochondrial potential constitutes an early irreversible step of programmed lymphocyte death *in vivo*. J Exp Med. **374**, 1661–1672.

Zhang, D., Pasternack, M.S., Beresford, P.J., Wagner, L., Greenberg, A.H. and Lieberman, J. (2001). Induction of rapid histone degradation by the cytotoxic T lymphocyte protease Granzyme A. J Biol Chem. **276**, 3683–3690.

Zhang, J. and Xu, M. (2002). Apoptotic DNA fragmentation and tissue homeostasis. Trends Cell Biol. **12**, 84–89.

Zhivotovsky, B., Cederall, B., Jiang, S., Nicotera, P. and Orrenius, S. (1994). Involvement of Ca^{2+} in the formation of high molecular weight DNA fragments in thymocyte apoptosis; Biophys Biochem Res Commun. **202**, 120–127.

Zhivotovsky, B., Orrenius, S., Brustugun, O.T. and Doskeland, S.O. (1998). Injected cytochrome *c* induces apoptosis. Nature. **391**, 449–450.

Zhou, J., Ahn, J., Wilson, S.H. and Prives, C. (2001). A role for p53 in base excision repair. EMBO J. **20**, 914–923.

Zou, H., Henzel, W.J., Liu, X.S., Lutsch, G.A. and Wang, X.D. (1997). Apaf-1, a human protein homologous to *C. elegans* CED-4, participates in cytochrome c dependent activation of caspase-3. Cell. **90**, 405–413.

Chapter 5
Apoptotic Chromatin Changes

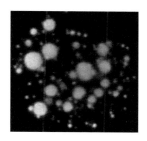

Summary

The tertiary structure of eukaryotic DNA is known as the nucleosomal organization in eukaryatic cells. Chromatin folding beyond the level of nucleosomes was not revealed. Only the end product of chromosome condensation, namely the structure of the metaphase chromosome is known. All other structures between the nucleosomal "beads on string" and the metaphase are hypothetical ones, most of them deduced from metaphase chromosomes. Consequently, early genotoxic changes at the structural level of chromatin could not be followed earlier. As far as intermediates of chromatin condensation are concerned we have characterized the major structural forms in mammalian and in *Drosophila* cells. These forms included the most decondensed veiled chromatin at the unset of S phase followed by depolarized, supercoiled, fibrous, ribboned chromatin in the first half of the S phase, chromatin bodies (earliest globular forms of chromosomes), elongated bent forms, precondensed and condensed linear forms of chromosomes in mid and lates S phase. By the time genotoxic effects led to apoptosis the whole nuclear material was degraded. Apoptotic changes have not been previously characterized at the chromatin level in interphase nuclei. This chapter analyzes the structural effects of apoptosis induced under genotoxic conditions including chemical effect, heavy metal treatment, γ and ultraviolet B (UVB) irradiation.

Genotoxic Agents

Among the DNA damaging agents eliciting stress responses are the exogenous and endogenous signals, DNA clastrogenic/mutagenic events (Nelson and Kastan, 1994) including γ-irradiation (Kastan et al., 1991; Kuerbitz et al., 1992), UV irradiation (Hall et al., 1993; Lu and Lane, 1993; Yamaizumi and Sugano, 1994), chemical exposures (Zhan et al., 1993; Tishler et al., 1993), oxidative stress (An et al., 1998; Graeber et al., 1996) and metabolite deprivation. A central role is played by p53 in the maintenance of genomic stability at the G_0–G_1 checkpoint and in the control of G_2 – M-associated DNA repair activities (Schwartz and Rotter, 1998; Gottlieb and Oren, 1998).

Changes in DNA Structure Caused by Genotoxic Agents

Common types of genotoxic agents cause changes in DNA structure, the long term effect of which can lead to malignant transformation. Structural changes involve alterations in nucleotide bases, cross-links, strand breaks, formation of bulky DNA adducts, etc. DNA damage suppresses DNA replication at checkpoints to avoid mutagenic changes being perpetuated in the genome of the next generation of cells. The primary effect of genotoxic agents is that they induce signaling cascades starting on the cell membrane, the signal is transferred to the nucleus and furthered to other cell compartments, resulting in the repair of DNA damage, alteration of cell attachment and cytoskeleton and/or resulting in cell death. Genotoxic agents comprise a first effect that inflicts damage on cellular DNA, and a second one that attracts a macromolecular cell component such as a protein.

There are several genes activated by genotoxic agents, but the responses have not been well characterized, nor have their mechanisms of induction been identified. Some genes respond to a narrow spectrum of agents corresponding to a particular DNA repair pathway, other genes are induced by a wide variety of DNA-damaging agents (Fornace et al., 1988; 1989b; Kartasova and van de Putte, 1988; Lin et al., 1990; Miskin and Reich, 1980; Devary et al., 1991), and may represent a more generalized response to stress (Applegate et al., 1991; Fornace et al., 1989a).

It is generally believed that an interaction of a reactive intermediate with cellular DNA is the initial step in the induction of mutations, neoplastic cell transformation and chemical carcinogenesis. Contrary to the considerable interest to examine these interactions, their research is hindered by the low levels of DNA adduct formation, typically ranging between $1 : 10^5$ to $1 : 10^9$ modified bases per exposure.

DNA is considered to be the primary cellular target for many cytotoxic and carcinogenic agents (Friedberg et al., 1995). DNA strand breaks constitute a major class of DNA damage, which can be induced by endogenous and exogenous agents, either directly or as intermediates in the DNA excision-repair pathways (Hoeijmakers, 2001). Frequently, strand-break termini, especially those generated by means of reactive oxygen species or as intermediates in repair of oxidative base damage, carry chemical modifications.

Chemical Mutagens Causing Air Pollution

Petrochemicals contain chemical carcinogens from industrial and vehicle emissions:

- Aromatic compounds
- Aromatic amines:
- Nitrosamines
- Alkylating agents.

Aromatic hydrocarbons can be monocyclic aromatic compounds such as the simplest benzene, polycyclic and heterocyclic compounds containing heteroatom(s). Naturally occuring polycyclic aromatic hydrocarbons come from fire (bush or forest

Genotoxic Agents   295

fires) and emitted from active volcanoes. They occur in crude oil, shale oil, and coal
tars and are formed by the incomplete combustion of coal, oil, petrol, wood, tobacco,
charbroiled meats, or other organic materials. Aromatic hydrocarbons cause primarily chromosomal aberrations and sister chromatid exchange, polycyclic compounds
are carcinogenic and teratogenic causing birth defects.

In *aromatic amines* amino ($-NH_2$) or imino ($=NH$) groups are attached to
an aromatic (benzene) ring. Aromatic hydrocarbons (e.g. aniline, o-toluidine, ani-
sidine) comprise one of the major groups of carcinogens. The detected urinary levels
of aromatic amines play a role in the bladder cancer etiology (Grimmer et al., 2000).
Occupational exposures to aromatic amines explain up to 25 percent of bladder cancers in some areas of Western countries; these estimates might be higher in limited
areas of developing countries (Vineis and Pirastu, 1997).

Nitrosamines are amines containing a nitroso ($-N=O$) group. Nitrosamines
cause cancer in different animal species. Distilling 2-naftil-amin causes bladder cancer among workers. Nitrosamines in preserved food are believed to cause stomach
cancer in humans (Jakszyn and Gonzalez, 2006).

Alkylating agents work by three different mechanisms all of which achieve the
same end result causing the disruption of DNA function and cell death. Many of
the alkylating agents are relatively inert but are made damaging by the cytochrome
P-450 oxidases (e.g. fungal toxin aflatoxin B1, mustard gas, benzopyrene in coal tar
and in tobacco smoke).

Chemical Mutagens and Carcinogens in Food

Based on this categorization chemicals can be divided into three groups. The **first**
group is naturally occurring, including plant alkaloids and plant toxins (e.g. aflatoxin
B1). Compounds in this group are usually limited to particular food materials and
are avoidable. The **second** group is composed of compounds formed by heating
food (eg. polycyclic hydrocarbons). The **third** group includes food additives and
pesticide residue contaminants.

Heavy Metals

Heavy metals are natural components of the Earth's crust. The term 'heavy metal',
in this context is inaccurate. It should be reserved for those elements with an atomic
mass of 200 or greater (mercury, thallium, lead, bismuth and the thorium series).
However, in practice the term embraces any metal exposure, which is clinically
undesirable and which constitutes a potential hazard. This raises the question: how
many elements are essential for life? The answer is known for a few laboratory
animals. Of the 90 naturally occuring elements only 26 are essential for rats and
chicken. The four most common elements: oxygen, carbon, hydrogen and nitrogen
contribute to 96% of the total body weight of mammals. The seven next most abundant elements make up almoust the remaining 4%: calcium, phosphorus, potassium,
sulfur, sodium, chlorine, and magnesium in this order. Fifteen additional trace elements are required, but their total combined amount is less than. 0.01% of the body

mass. These trace elements are: iron (transition element), cobalt, nickel, copper, zinc, vanadium, chromium, manganese, molybdenum, silicon, tin, arsenic, selenium, fluorine, iodine. Additional trace elements include: lithium, rubidium, beryllium, strontium, aluminium, boron, germanium, bromine (Schmidt-Nielsen, 1997).

Unlike complex organic pollutants most of the heavy metals cannot be degraded by microorganisms, but can be accumulated by organisms causing bioamplified toxicity through the food chain. They enter our bodies via food, drinking water and air. Some of the trace elemement heavy metals (e.g. copper, selenium, zinc) are essential in the metabolism. Heavy metals, especially those which are not included in the composition of biological systems, may produce toxic effects in organisms. Heavy metal poisoning could result from drinking-water (e.g. lead pipes), intake via the food chain, high air concentrations near emission sources. Global heavy metal discharges through the air, soil and water increased substantially over the last century. Preventive measures are important, but ultimately futile. The reality is that it is impossible in this age not to be exposed to heavy metals. It is only a matter of how much and how often. The three most important pollutant heavy metals are lead, cadmium and mercury. By the production of reactive oxygen species, heavy metals may cause alterations on proteins, DNA and cellular lipids (Leonard et al., 2004; Valko et al., 2005) leading to the generation of cell death by necrosis or apoptosis (Kang, 1997; Pulido and Parrish, 2003) and carcinogenesis (Valko et al., 2006). Due to their biological and ecological consequences heavy metals have been considered as priority pollutants (Keith and Telliard, 1979).

Heavy metals generate oxidizing radicals through the Fenton chemistry and by the Haber-Weiss-reaction leading to the hypothesis that metal induced carcinogenesis is mediated primarily by the elevated level of free radicals (see Chapter 4 and review of Kasprzak, 1995). Iron forms the most destructive free-radical of all ("hot iron"), very damaging to life. According to this view, heavy-metal-induced oxidative stress can lead to different types of DNA damage as a consequence of consumption of molecular oxygen in multiple steps of incomplete O_2 reduction, ultimately producing water. Common types of oxidative damage cause changes in DNA structure, including chromatin changes, the long-term effect of which can lead to malignant transformation. Structural changes involve alterations in nucleotide bases, cross-links, strand breaks, formation of bulky DNA adducts, etc. DNA damage suppresses DNA replication at checkpoints to avoid mutageneic changes being inherited in the genome of the next generation of cells (Hartwell and Weinert, 1989; Murray, 1992).

Radiation-Exposure

- Atomic bomb survivors, accidents of atomic power plants are exposed to fatal cancer,
- γ-irradiation and
- X-ray result in the long-term development of cancer,
- UV-induced DNA damage causing sun-induced melanoma of the non-pigmented skin.

Chromatin Changes upon Genotoxic Treatment

Since the first description of apoptosis, genetic and biochemical studies have led to a better understanding of the multiple pathways that eukaryotic cells can take to terminate their existence known as programed cell death. These studies led to the understanding of the development of various diseases such as AIDS, Alzheimer's, and Parkinson's and have provided potential targets for possible therapies. Despite of all of these developments, the mechanism of chromatin condensation, a morphological hallmark of apoptosis, remained elusive (Th'ng, 2001).

Increasing evidence suggests that the activity of genotoxic agents is the result of their manifold interactions with chromatin and its constitutive components leading either to chromatin unfolding or aggregation. The structural disruption of chromatin is thought to interfere with the metabolic processes of DNA strand separation (transcription, replication, recombination) and is likely to play an important role in the apoptosis undergone by the cells upon treatment with these drugs.

It is known that structural changes in chromatin often reflect chemical modification (Bradbury, 1992; Wolffe, 1995), phosphorylation being associated with extensive alteration. Experimental data revealed histone modification that is uniquely associated with apoptotic chromatin in species ranging from frogs to humans, namely phosphorylation of histone H2B at serine 14 (S14) correlates with cells undergoing programmed cell death in vertebrates (Cheung et al., 2003). One of the major questions to be answered in this chapter is: if chromatin changes reflect chemical modifications, then the opposite should also be valid, namely specific chemical treatment of cells should cause specific changes in chromatin structure. Based on this assumption it is of interest to investigate in more detail whether genotoxic treatment is associated with specific DNA damage manifested as changes in chromatin structure. The validity of this idea is tested in cells treated with chemicals, heavy metals and exposure to irradiation (γ and ultraviolet).

Chemically Induced Chromatin Changes

Hepatocellular tumour was induced chemically in rat by N-nitrosodimethylamine (Paragh et al., 2005). A new hepatocellular tumour cell line (HeDe, *He*patocellular *De*brecen) was established from the primary tumour induced in rats by the carcinogen N-nitrosodimethylamine. The HeDe tumour cell line caused tumour in Fisher 344 rats, in Fisher-Long-Envans hybrids, but not in Long-Evans rats. The specifity of tumour development is indicated by the fact that, although N-nitrosodimethylamine is a carcinogen in different animals, the tumour subline causes cancer only in the same strain or in the hybride strain. Since this tumour subline will cause neither tumours in other strains of rats nor in other species, it can be used safely and can serve as a model cell line for studying tumour development and treatment with antimetabolites (Trencsenyi et al., 2007).

The HeDe tumour cell line showed similar topological changes in the chromatin structure as those primary tumour cells which were isolated from the HeDe tumour (results not shown). Topological changes in the terciary structure of DNA are known to be involved in human oncology. Topoisomerases, types I and II are enzymes which exclusively serve to change the tertiary structure of DNA. Regarding these enzymes topoisomerase I is involved in embrional development and in oncology (Larsen and Gobert, 1999). Emerging evidence suggests that topoisomerase I activities are increased in human T lymphocyte proliferation. Topoisomerase I does not show significant cell cycle dependent alteration during the eukaryotic cell cycle in normal mouse and human tissues. However, this is not the case for transformed cells, where the topoisomerase I protein levels and activities may vary over more than two logs of magnitude among different tumour cells (Huang et al., 2006). Genetic studies in yeast have shown that topoisomerase I is required for chromosome condensation (Tournier et al., 1992).

Among DNA topoisomerases which are enzymes that regulate the degree of DNA supercoiling in the cell, topoisomerase I is known to play a dual role.

1. It participates in the protection of cells toward DNA damage through the formation of covalent DNA-protein complexes.
2. These DNA-protein complexes are potentially dangerous to the cell, due to their ability to mediate illegitimate recombination, leading to genomic instability and oncogenesis.

To control these two contrasting functions the cellular levels of topoisomerase I are relatively low and constant. This appears to be the case for most non-transformed cells, with a copy number of topoisomerase I molecules staying around one million per cell (Kunze et al., 1991). In transformed cells the topoisomerase I protein levels and activities are more than two orders of magnitude higher (Bronstein et al., 1996). These observations indicate that significant topological changes generated by elevated levels of topoisomerases could have structural consequences which could be manifested at the chromatin level. The validity of this idea was tested by isolating chromatin structures from nuclei of resting and hepatocellular tumour cells after reversal of permeabilization (Bánfalvi et al., 1984) and by visualizing them with fluorescent microscope. It was reasonable to assume that high degree of supercoiling will cause visible chromatin changes. Results confirm the notion that contrary to resting cells, which are in highly decondensed state, in fast growing hepatocytes transitions driven by a faster rate of supercoiling of the nuclear material (HeDe) show an inreased level of supercoiling manifested as spirated chromatin structures (Fig. 5.1).

Chromatin Structures in Nuclei of Hepatocellular Tumour Cells

Hepatocellular carcinoma cells (10^6) placed under the kidney capsule of Fisher 344 rats developed malignant tumours and resulted in death in two weeks and consisted of HeDe hepatoma cells. The most characteristic chromatin structures isolated from these tumour cells were supercoiled ribbons and fibrous structures (Fig. 5.1). The nuclei of reversibly permeabilized HeDe cells could be opened readily at the decondensed phase of chromatin condensation, making the fine chromatin veil visible

Fig. 5.1 Supercoiled chromatin structures in hepatocellular tumour cells. Hepatocellular tumours were obtained after two weeks of implantating 10^6 tumour cells under the renal capsule of rats. Tumour cells were obtained and processed, nuclei were isolated and chromatin structures were prepared. Bar, 5 μm (reproduced with permission of Trencsenyi et al., 2007)

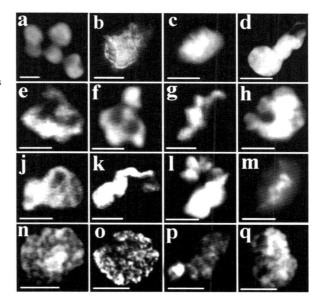

as a faint halo (Fig. 5 1 a–h). Another characteristic feature of chromatin from tumour cells is the elevated level of supercoiling observed throughout the process of chromatin condensation (Fig. 5.1 b–o). The third, and probably most important change in chromatin condensation relative to resting cells was the higher incidence of formation of apoptotic bodies visualized in Fig. 5.2.

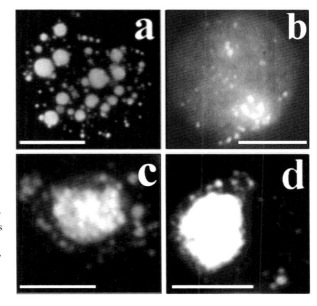

Fig. 5.2 Apoptotic changes in hepatocellular tumour cells. The isolation of tumour cells and chromatin structures corresponded to those described under Fig. 5.1. Bar, 5 μm (reproduced with permission of Trencsenyi et al., 2007)

Fluorescence Intensity Histograms of Chromatin Images

Histograms show that fluorescence intensities were higher in nuclei of tumour cells than in resting hepatocytes (Fig. 5.3). The contour length/area ratio of chromatin structures can be used as an indicator of the heterogeneity reflecting the entropy of the nuclear structures. The contour/area ratio was lower and nearly the same in resting cells reflecting a relatively uniform distribution of chromatin structures in these nuclei (Fig. 5.3 a–b). To the contrary, the contour/area ratio was nearly two times higher in tumour cells (Fig. 5.3 c–e) and almost nine times higher in apoptotic cells (Fig. 5.3 f).

Basic differences between chromatin structures of resting (Go) and hepatocellular tumour cells are of potential diagnostic importance:

1. Nuclei of resting cells contain decondensed chromatin referred to as chromatin veil. The nuclear material consisted of subdivided, six distinguishable lobes suggesting that liver chromosomes are clustered in six chromatin domains (Fig. 3.21). In decondensed continuous chromatin the chromosomes cannot be distinguished, yet.
2. Supercoiling was evident from the early stage of chromatin condensation in nuclei of hepatoma cells. The tendency of intensive supercoiling could be traced throughout the cell cycle. The rate of supercoiling is likely to be related to the length of the cell cycle. In a fast-growing spontaneous mouse mammary tumour subline the cell cycle was much shorter (11–12 h) than that of the slow-growing spontaneous mouse mammary tumour subline (15–16 h). The difference in the growth rates for fast growing tumour cells versus slow growing or resting cells was attributable to the reduction of the length of the cell cycle in tumour cells (Paveliv et al., 1978). Failure of cell growth control is thought to play a major role in the development of tumours. It is assumed that in tumour cells the increased growth rate is directly related to the high activity of topoisomerases leading to the reduced length of cell cycle. The strain caused by the shortening of the cell cycle may lead to mutations and cell death.

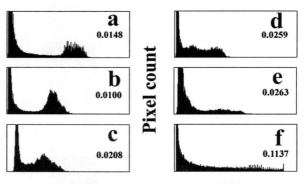

Fig. 5.3 Computer analysis of chromatin disorder. Chromatin images of nuclei from resting cells (**a, b**), from tumour cells (**c–e**) and from apoptotic cells (**f**) were subjected to computer image analysis. The number in each box expresses the contour/area ratio of the chromatin structures reflecting the disorder (entropy) inside the nucleus (reproduced with permission of Trencsenyi et al., 2007)

Heavy Metal Induced Cytotoxicity 301

3. Additionally, the cause of apoptosis may be that rapidly growing tumours need to recruit more blood vessels to meet their nutritional requirements. Indeed, data indicate that growth-inhibition of tumour cells may induce apoptosis (Huang et al., 2006). In this context it is logical to think that lagging angiogenesis, especially at the periphery of the growing tumour might be the cause of lower O_2 levels, driving anoxic tumour cells into apoptosis.

The potential diagnostic and therapeutic importance of chromatin analysis in rat hepatocytes is:

a. It was found that chromatin image analysis can be utilized to express the heterogeneity of the nuclear material as the contour/area ratio. This ratio is of diagnostic significance: it was lower in resting hepatocytes, two-times higher in tumour cells and nine-times higher in those tumour cells which underwent apoptosis.
b. The fact that the tumour induced by N-nitrosodimethylamine could be converted to HeDe cell line which maintained its agressive tumourogenic property, opens the way for individual *in vitro* antimetabolite treatment in this and potentially in any chemically induced and isolated tumours cell line. The efficiency of the individual *in vitro* anticancer treatment can be immediately tested *in vivo* by inoculating the susceptible strain of animals with the same tumour subline and treatment with the most efficient antimetabolite.

Heavy Metal Induced Cytotoxicity

The mechanisms of cell damage induced by particular heavy metals are not yet completely elucidated. Zinc, cadmium, mercury or lead have been known to impair enzyme functions by binding to sulfhydryl groups of proteins (Rothstein, 1959; Vallee and Ulmer, 1972). The ubiquitous 6 kD protein metallothionein is particularly known as a binding target for heavy metal ions due to its high cystein (about 30%) content which is found in most phyla (Hamer, 1986). High metallothionein concentrations are found in liver, kidney, intestine and pancreas. Two functions are ascribed to the binding of heavy metals to metallothioneins: maintenance of homeostasis of biochemically important metals (copper and zinc) and detoxication of cells from toxic metals, particularly cadmium and other adverse substances (Cosson, 1994; Dhawan and Goel, 1995; Liu et al., 1991; Moffatt et al., 1996; Woo and Lazo, 1997). Heavy metals are reported to cause an increase of cytoplasmic calcium (Chávez and Holguin, 1988 ; Trump et al., 1989) and to inhibit ATP synthesis (Stacey and Klaassen, 1981; Strubelt et al., 1996).

Relationship Between Replicative and Repair DNA Synthesis in Undamaged Cells During the Cell Cycle

At the molecular level we have measured replicative DNA synthesis throughout the S phase in synchronized populations of Chinese hamster ovary cells. At high resolution of elutriation multiple subphases were distinguished. These replication peaks, termed *replication checkpoints*, are distributed evenly throughout the S phase. The number of replication checkpoints seems to correspond to the chromosome number in Chinese hamster ovary and in *Drosophila* S2 cells (Bánfalvi et al., 1997a, Rehak et al., 2000). To strengthen the casual association between repair and replicative DNA synthesis the two types of DNA synthesis have been measured simultaneously in a cell-cycle-dependent manner. The comparison of cell cycle profiles of ATP-dependent replicative and ATP-independent repair synthesis in permeable CHO cells showed opposite trends. It turned out that the rates of repair and replication are inversely correlated (Bánfalvi et al., 1997b). Moreover, the cell cycle dependent fluctuation of repair activity and that of strand breaks did not coincide (Bánfalvi et al., 2000). The fluctuation of the naturally occurring strand breaks in control cells during the cell cycle was explained by replicative intermediates belonging to discontinuous DNA synthesis, such as Okazaki fragments, short replicative intermediates and the large DNA fragments accumulating in the G2 phase (Csuka and Bánfalvi, 1997).

Cadmium as a Genotoxic and Carcinogenic Agent

Cadmium (atomic mass 112.4) is one of the most common environmental metal poisons mainly due to industrial activities, particularly to zinc lead and copper smelting. These metals occur in mixed ores with cadmium, although areas of natural high cadmium content also exist. Cadmium is only weekly genotoxic, nevertheless it has been classified as carcinogen category 1 by the International Agency for Research on Cancer (1993). Cadmium is of worldwide concern because of its accumulation in the environment and long half-life estimated in humans to be between 15 and 20 years.

Cadmium is used extensively in electroplating, found in some industrial paints and may represent a hazard when sprayed. Cadmium is also present in the manufacture of some types of batteries. Cadmium emits the non-irritating CdO (brown smoke) upon heating, and thus not alarming the exposed individuals. Following long term exposure, the kidney is the critical organ (Jin et al., 1998). A substantial amount of cadmium is accumulated in the organism over a number of years. Cadmium promotes skeletal demineralization and increases bone fragility and fracture risk. Most important effects of high Cd-exposures are irreversible renal failure, immune deficiencies, bone disorders, obstructive airway disease (Bertin and Averbeck, 2006). First cases were observed in Japan in the 1940s among populations living in areas polluted with cadmium. The painful disease became known as Itai-Itai (Ouch!-Ouch!) causing proximal tubular damage, anaemia, loss of bone mineral resulting in fractures.

Heavy Metal Induced Cytotoxicity

At the cellular level, cadmium has multiple effects influencing differentiation, cell cycle progression, cell growth and it can induce apoptosis (Dong et al., 2001; Fang et al., 2002; Waisberg et al., 2003; Yang et al., 2004; Oh and Lim, 2006). Cell death resulting from cadmium intoxication has been confirmed to occur through apoptosis by morphological and biochemical changes (Hamada et al., 1997). Cd caused morphological alterations in lung epithelial cells are characteristic of apoptosis, cell nucleus shrinkage is prominent, detachment of the cell from its neighbors, cytoplasmic and chromatin condensation, and fragmentation of the nucleus into multiple chromatin bodies surrounded by remnants of the nuclear envelope (Hart et al., 1999). At molecular level cadmium affects DNA synthesis in a dose dependent manner. At concentrations higher than 1 μM cadmium inhibits DNA synthesis (Misra et al., 2003).

Biochemical and Morphological Changes Generated by Cadmium Treatment in CHO Cells

The cytotoxic effect of cadmium was tested in Chinese hamster ovary cells by mesuring the repair and replicative DNA synthesis at different stages of the cell cycle. On Cd treatment repair synthesis was elevated in certain subphases in S phase. Replicative subphases were suppressed by Cd treatment, with some of the peaks almost invisible. The number of spontaneous strand breaks measured by random oligonicleotide primed synthesis assay showed a low but cell-cycle-dependent fluctuation in control cells which was greatly increased after Cd treatment throughout the S phase. Elevated levels of oxidative DNA damage product, 8-oxodeoxyguanosine, were observed after Cd treatment, with the highest level in early S phase, which gradually declined as damaged cells progressed through the cell cycle (Bánfalvi et al., 2000).

The question emerged whether Cd as a DNA-damaging agent causing mutations leading to carcinogenesis would change the profile of multiple subphases of the two DNA-synthetic processes. The answer is that Cd changes both replicative and repair synthesis in early S phase. DNA replication was suppressed throughout the S phase as expected on DNA damage. In contrast, repair synthesis increased, and in early mid S phase was about 30-fold higher after Cd treatment than in the control population at the same subphase. Another fourfold repair activity was seen in late S phase (Bánfalvi et al., 2000).

The number of strand breaks after Cd treatment was 10–40 times higher than in control cells, showing a cell cycle dependent fluctuation. The number of strand breaks was particularly high at the beginning and at the end of S phase. In contrast, at the beginning of S phase the repair activity was low (Bánfalvi et al., 2000). This discrepancy indicated that not all of the strand breaks were repaired. To replace the catalitically active DNA polymerase at work with the one that is specific for DNA damage may postpone repair activity to the next subphase. It may well be that damaged cells undergo a selection process contributed by different types of repair activity, and those cells which miss these repair processes or cannot be repaired undergo apoptosis. In agreement with this explanation, we found that, in murine preB cells subjected to another type of genotoxic stress (γ-irradiation) augmented base excision repair in early S phase and predominantly induced apoptosis at the G2/M checkpoint (Offer et al., 2001).

The level of the carcinogenic indicator 8-oxodeoxyguanosine (8OHdG) was expected to increase upon Cd treatment. Indeed in the 30- to 60-fold overshooting, the level was highest in G1 and early S phase which gradually declined by the end of S phase (Bánfalvi et al., 2000). The decrease can be attributed to the gradual removal of modified bases by the end of S phase and/or cells are more susceptible to cadmium treatment in early S phase when the chromatin structure is more decondensed.

Inhibition of Replicative DNA Synthesis at Low Cd Concentration in CHO Cells

Cells were synchronized by centrifugal elutriation before and after cadmium (1 μM, 15 h) treatment. The characterization of synchronized populations is summarized in Tables 5.1 and 5.2.

Cells were separated into nine fractions at flow rates detailed in Tables 5.1 and 5.2 showing the relationship among cell volume, nuclear volume and nuclear diameter in elutriation fractions before and after cadmium treatment. In addition to these parameters the average C-value of each elutriation fraction was determined in untreated cells (Table 5.1) and in cells after treatment with 1 μM CdCl$_2$ (Table 5.2). Synchronization of cells by centrifugal elutriation based on the relationship between cell size and DNA content was carried out and synchrony was confirmed in each elutriated fraction by fluorescent activated cell sorter (FACS) measuring the fluorescence emitted upon propidium iodide binding. Higher C-values after cadmium treatment indicated either (a) an elevated rate of DNA synthesis and increased

Table 5.1 Characterization of synchronized populations of untreated CHO cells

Fraction number	Flow rate of elutriation (ml/min)	Elutriated cells ($\times 10^6$)	%	Average cell volume fl	nuclear volume fl	nuclear diameter μm	C-value
1	12	1.30	2.6	1070	n.d.	n.d.	2.03
2	17	4.09	8.2	1140	n.d.	n.d.	2.15
3	22	8.19	16.4	1335	220	7.1	2.41
4	27	8.37	16.7	1480	240	7.35	2.62
5	32	9.86	19.7	1705	265	7.8	2.89
6	37	8.37	16.7	1840	285	8.05	3.05
7	42	4.28	8.6	2120	335	8.6	3.34
8	47	2.05	4.1	2580	470	9.9	3.72
9	52	1.58	3.2	2945	530	10.3	3.95
Unfractionated cells	*2.0*		*4.0*	*1703*	*265*	*7.8*	

Cells subjected to elutriation: 5.0×10^7 (100%)
Number of elutriated cells: 4.8×10^7 (96%)
Loss during manipulation: 2.0×10^6 (4 %)
n. d. = not determined.
Reproduced with permission of Bánfalvi et al. (2005).

Table 5.2 Characterization of synchronized populations of Cd (1 μM) treated CHO cells

Fraction number	Flow rate of elutriation (ml/min)	Elutriated cells (×10⁶)	%	Average Cell fl	nuclear volume fl	nuclear volume μm	C-value diameter
1	12	1.11	4.44	1230	n.d.	n.d.	2.25
2	17	1.30	5.2	1320	n.d.	n.d.	2.4
3	22	1.67	6.68	1405	n.d.	n.d.	2.5
4	27	2.23	8.92	1670	278	7.9	2.85
5	32	3.35	13.4	2010	314	8.4	3.2
6	37	3.53	14.12	2230	372	9.05	3.4
7	42	7.07	28.28	2620	465	9.95	3.75
8	47	1.67	6.68	2860	515	10.3	3.9
9	52	1.21	4.84	n.d.	n.d.	n.d.	n.d.
Unfractionated cells		2.0	8.0	1948	323	8.45	

Cells subjected to elutriation: 2.5×10^7 (100%)
Number of elutriated cells: 2.3×10^7 (92%)
Loss during manipulation: 2.0×10^6 (8%)
Reproduced with permission of Bánfalvi et al. (2005).

amount of DNA, or (b) nuclear damage, leading to condensed chromatin structures. The second possibility is based on the assumption that propidium iodide uptake is directly related to chromatin compaction, similarly to the relationship between chromatin compactness and DAPI uptake observed by others (Mascetti et al., 2001). In the experiments described under Figs. 5.4, 5.5, 5.6, and 5.7 we have excluded the

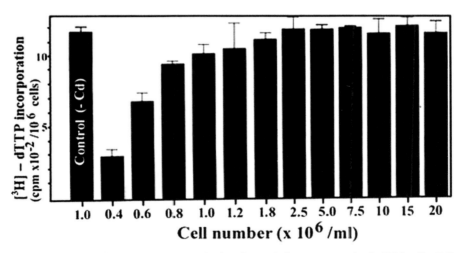

Fig. 5.4 Inhibition of replicative DNA synthesis at low cadmium concentration in CHO cells. Cell cultures representing different cell concentrations from 0.4 to 20×10^6 cells/ml were grown in the presence of 1 μM CdCl$_2$. After treatment 10^6 cells from each population were permeabilized and replicative DNA synthesis was carried out. [^{3}H]-thymidine triphosphate incorporation was measured as acid insoluble radioactivity (reproduced with permission of Bánfalvi et al., 2005)

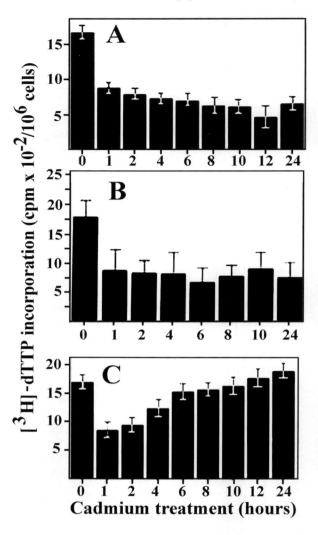

Fig. 5.5 Inhibition of replicative DNA synthesis upon cadmium treatment at different CHO cell densities. DNA synthesis was measured in permeable cells after cadmium treatment (1 μM). Cadmium treatment lasted from 0 to 24 h, at cell densities corresponding to A. 10^5, B. 2.5×10^5 and C. 5×10^5 cells/ml. [^{3}H]-thymidine triphosphate incorporation was measured as in Fig. 5.4 (reproduced with permission of Bánfalvi et al., 2005)

first possibility by measuring the rate of replicative DNA synthesis after cadmium treatment, which was either lower or equal (in dense cultures), but never higher than in untreated cells, consequently the second alternative seems to be worth of further consideration.

Cells were treated with 1 μM $CdCl_2$ at different densities ranging from 4×10^5 to 2×10^7 cell/ml. ATP-dependent replicative DNA synthesis was followed after permeabilization. Figure 5.4 shows the correlation between replicative DNA synthesis and cell concentration. The rate of DNA synthesis was measured in 10^6 cells cultured at 4×10^5, 6×10^5, 8×10^5, 10^6 and 1.2×10^6 cells/ml in the presence of Cd and 25, 57, 78, 82 and 94% of the synthetic activity of the control was detected,

Heavy Metal Induced Cytotoxicity 307

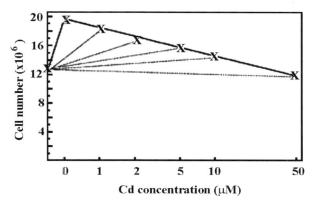

Fig. 5.6 Inhibition of cell growth at different cadmium concentration. The growth of six cell cultures was started at 1.2×10^6 cells/ml in the presence of 1, 2, 5, 10 and 50 μM $CdCl_2$, respectively. After treatment the cell number was counted in each population. In the semilogarithmic representation the abscissa is showing the cadmium concentration and the ordinate the cell number. Cell growth was registered from 0 to 10 μM $CdCl_2$ and cell death at 50 μM $CdCl_2$ concentration (reproduced with permission of Bánfalvi et al., 2005)

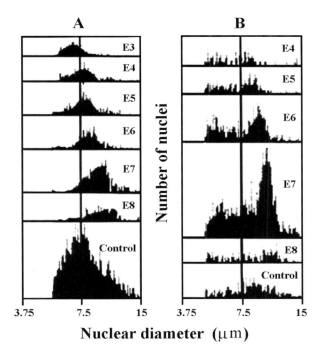

Fig. 5.7 Nuclear size analysis of CHO cells before (**A**) and after cadmium treatment (**B**). Nuclei of elutriation fractions were isolated. Nuclear size distribution of elutriation fractions was measured and nuclear diameters were expressed in micrometers Controls are nuclei isolated from cells not subjected to elutriation (reproduced with permission of Bánfalvi et al., 2005)

respectively. At higher than 2×10^6 cells/ml density there was virtually no difference between the replicative DNA synthesis of control and cadmium treated cells.

The effect of $1 \,\mu M$ $CdCl_2$ on replicative DNA synthesis was registered between 0 and 24 h treatment at three cell densities (Fig. 5.5). Maximal inhibition of DNA synthesis at cell densities was 64% (10^5 cells/ml), 56% (2.5×10^5 cells/ml) and 49% (5×10^5 cells/ml). At the lowest density (10^5 cells/ml) a gradual decrease in nucleotide incorporation was observed (Fig. 5.5A). When the cadmium treatment started in a 2.5×10^5 cells/ml culture, the initial low level of DNA synthesis was maintained between 1 and 24 h (Fig. 5.5B). In the suspension culture containing 5×10^5 cells/ml, the initial inhibition of DNA synthesis by cadmium was gradually overcome and reached the control (untreated) level in 12 h (Fig. 5.5C).

These results indicated that Cd concentration might have been lowered by the binding of metal ions to the cells. If this would be the case, higher metal ion concentrations should compensate the binding of Cd and exert an inhibitory effect on cell growth at higher cell concentrations. To prove the validity of such a stoichiometry, six identical cell cultures were grown starting at 1.2×10^6 cells/ml in the presence of 1 to $50 \,\mu M$ $CdCl_2$ (Fig. 5.6). The decreasing growth rate in the presence of 1, 2, 5 and $10 \,\mu M$ $CdCl_2$ corresponded to 80, 64, 44, 28% of the untreated cells (100%), respectively. At high Cd concentration ($50 \,\mu M$) not only complete inhibition of cell growth, but an 11% cell loss was registered, pointing to cytotoxicity, possibly to apoptosis.

Subphases of Nuclear Growth Revealed by Cadmium Treatment

The nuclear sizes of elutriation fractions have been summarized in Tables 5.1 and 5.2. These data by themselves do not reflect the nuclear size heterogeneity after cadmium treatment (Fig. 5.7B), since they are averaged values. To the contrary relative homogeneity in nuclear size was registered in nuclei isolated from elutriation fractions of untreated cells (Fig. 5.7A). Attention is called to nuclei isolated from cadmium treated cells belonging to elutriation fraction 7, since different nuclear sizes could be distinguished ranging from 6 to $10 \,\mu M$ in diameter. An even higher heterogeneity was observed in nuclei from cadmium treated, unfractionated control cells. A more detailed analysis of this unelutriated control population of nuclei from cadmium treated cells revealed at least 11 peaks (Fig. 5.8), the estimated C-values of which are: 2.04, 2.10, 2.22, 2.41, 2.53, 2.71, 2.81, 3.07, 3.24, 3.55, 3.82, respectively.

Interphase Chromatin Structures in Nuclei of Untreated Synchronized CHO Cells

After reversal of permeabilization nuclei were isolated and chromatin structures were visualized after DAPI staining by fluorescent microscopy. Decondensed veil-like structures were observed in early S phase, supercoiled chromatin later in early S, fibrous structures in early mid S phase, ribboned structures in mid S phase, continuous chromatin strings later in mid S phase, elongated prechromosomes in late S

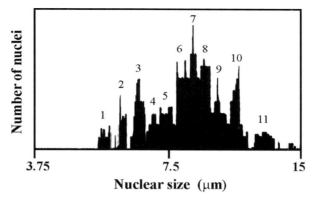

Fig. 5.8 Multiple subphases of nuclear growth upon cadmium treatment. Conditions for cadmium treatment corresponded to those under Fig. 5.7. After cadmium treatment nuclei were isolated and nuclear size analysis was carried out. The haploid genome content of eleven peaks was determined, corresponding to 2.04, 2.10, 2.22, 2.41, 2.53, 2.71, 2.81, 3.07, 3.24, 3.55, 3.82, C-values, respectively. The average haploid genom content was (C-value) was determined (Basnakian et al., 1989) (reproduced with permission of Bánfalvi et al., 2005)

phase, precondensed chromosomes at the end and after S phase (Gacsi et al., 2005). These structures have been detailed in Chapter 3. Here only a few typical chromatin structures are shown (Fig. 5.9), including the polarization of the nuclear material (Fig. 5.9A, elutriation fraction 2, 2.15 C-value), polarization causing the extrusion of spherical chromatin (Fig. 5.9B, elutriation fraction 3, 2.45 C), fibrous structures (Fig. 5.9C, elutriation fraction 4, 2.62 C), chromatin bodies (Fig. 5.9D, elutriation fraction 5, 2.89 C), ribboned structures (Fig. 5.9E, elutriation fraction 6, 3.05 C), continuous chromatin strings (Fig. 5.9F, elutriation fraction 7, 3.34 C), elongated prechromosomes (Fig. 5.9G, elutriation fraction 8, 3.72 C), precondensed chromosomes (Fig. 5.9H, elutriation fraction 9, 3.95 C).

Changes in Chromatin Structure upon Cd Treatment

Apoptotic Bodies in Early S Phase (2.2–2.5 C)

Chromatin structure analysis revealed that Cd treatment disturbed the normal process of chromatin condensation during the S phase. Apoptotic bodies were observed in early S phase (Fig. 5.10, elutriation fraction 2–3, 2.4–2.5 C-value). The formation of apoptotic bodies seems to be related with premature chromosome condensation represented by elutriation fraction 2 and 3 (upper part Fig. 5.10 C, E, F). Attention is called to the lack of polarization and extrusion of the nuclear material. Polarization of nuclear material and extrusion of looped chromatin veils (interphase chromosomes) are typical structures seen in elutriation fractions 2–3 of untreated cells.

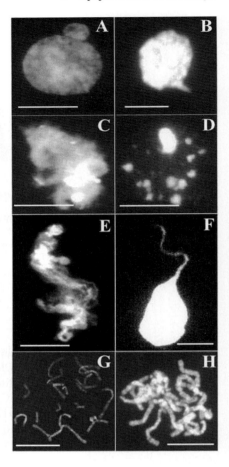

Fig. 5.9 Intermediates of chromatin condensation in untreated CHO cells. Permeabilization of cells from elutriation fractions was followed by the restoration of membrane structures, colcemid treatment and isolation of chromatin structures. Regular structures seen under fluorescent microscope: fraction 2, polarization of the nuclear material (**A**), fraction 3, extrusion of spherical chromatin (**B**), fraction 4, fibrous chromatin (**C**), fraction 5, chromatin bodies (**D**), fraction 6, ribboned, supercoiled chromatin (**E**), fraction 7, chromatin strings (**F**), fraction 8, elongated prechromosomes (**G**) approaching metaphase (**H**). Bars 5 μM each (reproduced with permission of Bánfalvi et al., 2005)

Ribboned Chromatin Margination in Early Mid S Phase (2.5–3.0 C)

After Cd treatment the absence of veiled chromatin, the lack of extrusion of polarized chromatin and the absence of supercoiled loops seem to prevent the formation of chromatin bodies typical to this stage of the cell cycle. Rather the early condensation of chromatin resulted in ribboned chromatin arranged in a semicircular manner (Fig. 5.10).

Perichromatin Fibers and Chromatin Bodies in Mid S Phase (3.0–3.4 C)

The condensation from the thick fibers (300 nm) to chromatidal ropes (700 nm) seen in elutriation fraction 5, Fig. 5.11 A–C (upper panels) is another indication of the premature condensation of the interphase chromatin. The presence of chromatin structures, such as chromatin bodies belong to the normal intermediates of this stage of the cell cycle (Fig. 5.11 B, D, E, upper panels) while the fibers are early

Fig. 5.10 Cadmium induced chromatin changes in early S phase. CHO cells were treated with cadmium and elutriated as described in Table 5.2. Reversible permeabilization, colcemid treatment and isolation of chromatin structures were the same as in Fig. 5.9. Fluorescent microscopy after DAPI staining was carried out. Typical chromatin structures in elutriation fractions 2–3 were apoptotic bodies (**A–F**, *upper panels*) Elutriation fraction 4, ribboned chromatin arranged in a semicircular manner (**A–F**, *lower panels*). Bars 5 μM each (reproduced with permission of Bánfalvi et al., 2005)

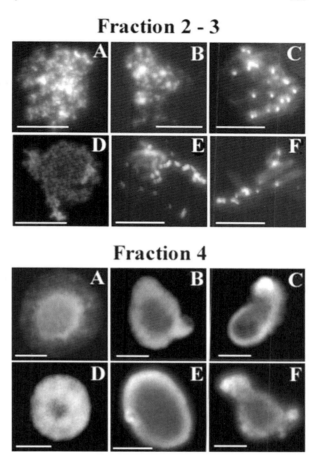

intermediates (Fig. 5.11 F, upper panels). To summarize the changes of chromatin condensation in elutriation fraction 5, major changes occured in this fraction containing delayed, normal and premature forms among the intermediates.

Intra-Nuclear Inclusions, Perichromatin Granules and Elongated Forms of Chromosomes in Late S Phase (3.4–3.9 C)

Reversible permeabilization allowed to open cellular and nuclear membranes any time during the cell cycle in untreated cells and upon gamma irradiation (Nagy et al., 2004). After cadmium treatment an increasing stickyness of the nuclear material was observed in elutriation fractions 6, 7 and 8, obscuring the events of chromatin condensation. In elutriation fraction 6 nuclear disruptions, small holes in the nuclear membrane, the extrusion of nuclear material, and elongated chromosomes were found (Fig. 5.11, lower panels).

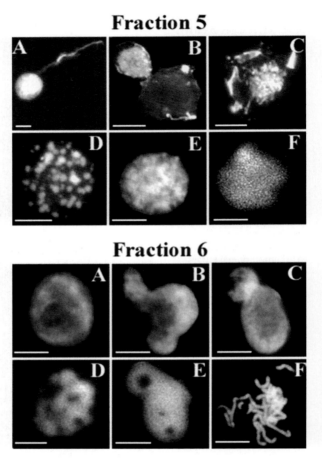

Fig. 5.11 Cadmium induced chromatin changes in mid S phase. Cells were treated and chromatin structures were isolated as described in Table 5.2. Fluorescent microscopy after DAPI staining was carried out. Typical chromatin structures in elutriation fraction 5: thin and thick chromatin fibres (**A–C**), chromatin bodies (**B, D, E**), fibrils (**F**). Fraction 6 (*lower panels*) cadmium induced chromatin changes in late mid S phase, disruptions (**B, C**), holes in the nuclear membrane (**A, D, E**), the extrusion of the nuclear material (**B, C**), and elongated chromosomes (**F**). Bars 5 μM each (reproduced with permission of Bánfalvi et al., 2005)

The ejection of some of the chromatin bodies, intra-nuclear inclusions and the formation of clusters of large-sized perichromatin granules were seen in elutriation fraction 7 (Fig. 5.12, upper panels). The most severe losses and disruptions of the nuclear membrane were observed in elutriation fraction 8 (Fig. 5.12, lower panels). The disruption of the nuclear membrane did not lead to distinguishable intermediates, rather it obscured the picture (Fig. 5.12 A–E, lower panels) resulting in large nuclei with typical large holes and sticky elongated and condensed chromosomal forms (Fig. 5.12 F–J, lower panels).

To summarize the effects of cadmium on DNA synthesis and chromatin condensation during the cell cycle, an inverse correlation was observed between the increasing cell density and diminishing rate of replicative DNA synthesis. Cell growth and DNA synthesis at higher cell concentrations could be suppressed by increasing Cd concentration.

Heavy Metal Induced Cytotoxicity 313

Fig. 5.12 Cadmium induced chromatin changes in late S, G and M phases. Cells were treated and chromatin structures were isolated. Chromatin structures in elutriation fraction 7 with expelled chromatin bodies (**A, B**), elongated nuclei with disrupted membranes (**B, C**), intra-nuclear inclusions (**A, D, E, F**), large-sized perichromatin granules (**E, F**). Elutriation fraction 8: disrupted nuclear membrane with large holes (**A–E**), elongated chromosomal forms (**F, G**), sticky elongated chromosomes (**H, J**). Bars 5 μM each (reproduced with permission of Bánfalvi et al., 2005)

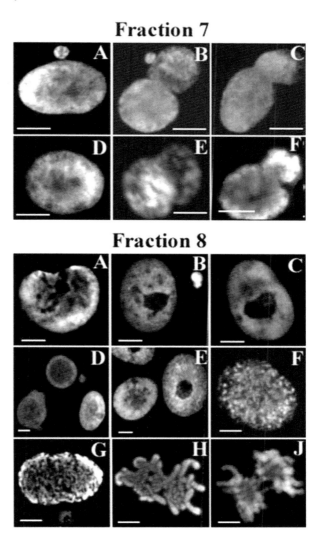

The notion that small apoptotic bodies interfere with chromatin condensation seems to be supported by their appearance and by the lack of supercoiled structures in early S phase. Decreased rate of supercoiling leads to increased nuclear size and to the disruption of the nuclear membrane, often seen as large holes inside the nucleus, especially when nuclear envelop is damaged.

Reversible permeabilization was used to open the nucleus any time during the cell cycle (Nagy et al., 2004; Gacsi et al., 2005). The relationship between cell growth and cadmium concentration is pointing to a stoichiometric binding of cadmium to cells. Cadmium binding seems to interfere with permeabilization as indicated by the stickiness of the nuclear membrane upon Cd treatment. It is assumed that Cd may compete with the binding of bivalent cations to transmembrane proteins. The stoichiometric Cd binding to the nuclear membrane i. contributes to a loose, disruptive nuclear envelop, ii. prevents its continuous growth and iii. results in discontinuous nuclear growth revealed as stop sites. The peaks of nuclear growth expressed in C-values coincide with the replication checkpoints (Bánfalvi et al., 1997a) which are likely to correspond to the subphases of chromosome replication (Rehak et al., 2000).

As cells progress through the S phase the visibility of intermediates of chromatin condensation decreases relative to untreated cells and the condensing chromatin remains locked inside the nucleus. The stickyness of chromatin structures, the formation of gradually increasing holes in the nuclear membrane indicate that cadmium binding to the components of the membrane may be the primary cause of its toxicity. This idea seems to be supported by ultrastructural studies of others related to the localization of cadmium in nuclear inclusions (Bell et al., 1979), leading to the transformation of cytoplasmic organells upon cadmium treatment (Ord et al., 1988; Hirano et al., 1991), cytoplasmic vacuolation, deterioration of cellular membrane, extrusion of nuclei (Gill and Pant, 1985), membrane leakage (Morselt et al., 1983), cytoplasmic invaginations, extensive disruption of the nuclear membrane (Ree et al., 1982). The consequence of cadmium binding is the disruption of nuclear membrane followed by the activation of cytoplasmic DNases, caspase-activated deoxyribonuclease (CAD), which are known to promote apoptosis *in vitro* and *in vivo* (Kimura et al., 2004; Nagata et al., 2003). DNA strand breaks were measured in CHO cells after $0.5\,\mu M$ CdCl$_2$ treatment and the random oligonucleotide primed synthesis (ROPS) assay showed 10–40-times increase during the cell cycle (Bánfalvi et al., 2000). Consequently, the idea that the disruptive nuclear membrane activates death endonucleases leading to apoptosis is consistent with the degradation of chromosomal DNA and subsequent chromatin condensation during apoptosis (Woo et al., 2004).

In addition to the binding of cadmium to the nuclear membrane, the inreasing stickiness may prevent the continuous growth of the nucleus and the discontinuous nuclear growth may be the cause of the nuclear disruptions. The distribution of peaks of the nuclear growth after cadmium treatment corresponded to C-values of 2.04, 2.10, 2.22, 2.41, 2.53, 2.71, 2.81, 3.07, 3.24, 3.55, 3.82, showing similarities to the number and position of the eleven replicative peaks in CHO cells at 2.05, 2.12, 2.2, 2.45, 2.6, 2.8, 2.95, 3.15, 3.3, 3.45 and 3.85 C-values published earlier (Bánfalvi et al., 1997a). These C-values indicate that nuclear growth takes place in several subphases following the path of chromosome replication similarly to the subphases of DNA replication (Bánfalvi et al., 1997a; Rehak et al., 2000).

We confirm the observations of others regarding the toxic effects of cadmium and followed the structural changes of chromatin in a cell cycle dependent manner. These changes include:

(a) the formation of apoptotic bodies in early S phase, noted by others as cell death being preceeded by chromatin condensation and DNA fragmentation (el Azzouzi et al., 1994),

Heavy Metal Induced Cytotoxicity

(b) the lack of fibrous and supercoiled chromatin and the appearance of ribboned chromatin in the form of perichromatin semicircles in early mid S phase, described by others as chromatin margination (Ree et al., 1982),
(c) the presence of perichromatin fibrils and chromatin bodies in mid S phase, observed as augmentation of perichromatin fibrils (Cervera et al., 1983),
(d) intra-nuclear inclusions, elongated forms of premature chromosomes, the extrusion and disrupture of nuclear membrane later in mid S phase, referred to as intra-nuclear inclusions (Hirano et al., 1991), deterioration of cellular membrane, extrusion of nuclei (Gill and Pant, 1985), disruption of the nuclear membrane (Ree et al., 1982),
(e) excluded chromatin bodies and formation of clusters of large-sized perichromatin granules in late S phase, mentioned as large-sized perichromatin granules, increase in chromatin margination (Ree et al., 1982), the appearance of clusters of 29 to 35-nm granules and the accumulation of perichromatin granules (Cervera et al., 1983),
(f) large extensive disruptions and holes in the nuclear membrane and the clumping of incompletely folded chromosomes at the end of S phase and in M phase, are likely to correspond to nuclear chromatin clumping (Di Sant'Agnese et al., 1983), extensive disruption of this membrane (Ree et al., 1982), membrane leakage leading to cell death (Morselt et al., 1983).

After cadmium treatment most severe changes occurred in late mid S phase (fraction 5 and 6) between 3.2 and 3.4 C-values, followed by the accumulation of cells in fraction 7 (28%), and by a sudden increase in nuclear size (9.95 µm). This is an indication that the most severe effect of cadmium on chromatin condensation takes place in late mid S phase at 3.2–3.4 C-values.

The final conclusion regarding Cd binding is that it

(i) contributes to a loose disruptive nucleus,
(ii) preventing its gradual increase during S phase and
(iii) resulting in discontinuous nuclear growth revealed as stop sites.

Contradictory Results of Cadmium Treatment in Different Mammalian Cells

The induction of apoptotic cell death by cadmium was investigated in different mammalian cell lines and great differences in the cytotoxicity of cadmium were found, which could not be explained so far. Rat glioma C6 cells turned out to be sensitive with an IC50-value of 0.7 µM, while human adenocarcinoma A549 cells were relatively resistant with an IC50-value of 164 µM CdCl$_2$ (Watjen et al., 2002). The idea that glioma cell proliferation is controlled predominantly by cell size-dependent mechanisms (Rouzare-Dubois et al., 2004) and the observation that adenocarcinoma is the most common form of non-small cell lung cancer (Kreuzer et al., 1999) may be an indication that the occurence of apoptosis after cadmium treatment may depend on cell size.

316   5 Apoptotic Chromatin Changes

Contradictory results show that at concentrations higher than 1 μM cadmium inhibits DNA synthesis (Misra et al., 2003), but very low cadmium concentrations stimulate DNA synthesis and cell proliferation (Von Zglinicki et al., 1992). Cadmium induces several genes involved in the stress response to pollutants or toxic agents. Several cell cycle regulating genes are over-expressed after exposure to cadmium, and many proteins are up-regulated such as proteins of the RAS signaling pathway and kinases of the RAS pathway. In conformity with the effetcs of cadmium at different concentrations the previous subchapter described that by increasing the density of the CHO cell culture, a diminishing inhibitory effect on replicative DNA synthesis was measured. Flow cytometry revealed increasing cellular and nuclear sizes after 1 μM $CdCl_2$ treatment in CHO cells (Bánfalvi et al., 2005) contradicting one of the most characteristic morphological changes of apoptosis known as cell shrinkage (Watjen et al., 2002).

Resolving Contradictory Results Caused by the Cadmium Treatment

The conflicting observations regarding the cytotoxicity of Cd raised several questions:

1. Is cadmium toxicity dependent on cell culture density, cell size or cell cycle stage?
2. What is the reason that low cadmium concentrations stimulate, higher cadmium concentrations inhibit DNA synthesis?
3. Why does Cd cause inrease in cellular and nuclear size or result in morphological changes leading to cellular and nuclear shrinkage known as apoptosis?

Recent studies served as a basis to tackle these questions and led to the idea that Cd concentration might have been lowered by its binding to the cells (Bánfalvi et al., 2005). The practical relevance of the confirmation of a stoichiometric relationship between toxicity and cell culture density could be the elimination or reduction of cadmium toxicity, by increasing the number of protective cells and/or metal binding proteins. This subchapter confirms the validity of the hypothesis regarding the stoichiometric relationship between cadmium toxicity and cell culture density. Differences in Cd toxicity were found during the cell cycle with higher cytotoxicity in early and late S, G2 and M phase than in mid S-phase.

The shift from replicative to repair DNA synthesis in murine preB cells was confirmed and apoptotic chromatin changes were visualized which took place after cadmium treatment during the cell cycle. Murine pre-B-cells grown in the presence of lower (1 μM) or higher (5 μM) concentration of cadmium chloride were synchronized by separating them into 13 fractions by centrifugal elutriation. The rate of DNA synthesis after cadmium treatment determined in permeable cells was dependent on cell culture density during cadmium treatment. Cell cycle analysis revealed a shift in the profile of DNA synthesis from replicative to repair DNA synthesis upon cadmium treatment. The study of the relationship between cell culture density and cell diameter at lower and higher cell densities in the presence of 1 μM cadmium chloride concentration showed that:

Heavy Metal Induced Cytotoxicity 317

(a) at 5×10^5 cell/ml or lower densities cells were shrinking indicating apoptotic changes,
(b) at higher cell culture densities the average cell size increased,
(c) the treatment of cells with low $CdCl_2$ concentration (1 μM) at higher cell culture density ($>5 \times 10^5$ cell/ml) did not change significantly the average cell diameter.

At 5 μM cadmium concentration and higher cell culture densities ($>5 \times 10^5$ cell/ml) the average cell size decreased in each elutriated fraction. Most significant inhibition of cell growth took place in early S phase (2.0–2.5 C value). Apoptotic chromatin changes in the chromatin structure of murine preB cells after cadmium treatment were seen as large extensive disruptions, holes in the nuclear membrane and stickiness of incompletely folded chromosomes, similarly to observations in CHO cells.

Effect of Cadmium on Replicative DNA Synthesis in Permeable Murine preB Cells

When murine pre-B-cells were treated with 1 μM $CdCl_2$ at three cell culture densities, more than 60% inhibition of replicative DNA synthesis was obtained at 10^5 cells/ml and a gradual decrease in nucleotide incorporation was observed between 1 and 24 h. When the same Cd treatment was applied at culture density of 2.5×10^5 cells/ml, the initial lower level of DNA synthesis ($>50\%$) was maintained throughout the 24 h treatment. In cell cultures containing 5×10^5 cell/ml, the initial inhibition ($\approx 40\%$) of DNA synthesis gradually decreased and reached the control level of DNA synthesis in 12 h. These results are similar to those observed in CHO cells (Bánfalvi et al., 2005) and indicate that intracellular Cd concentration is likely to be lowered by its binding to metallothionein and other proteins (Nordberg et al., 1994). If this would be the case one would expect a significant difference between the toxic effects of lower and higher metal ion concentrations and apoptotic effects could be eliminated or at least reduced by increasing cell culture density. To test this possibility we have determined the cell culture density (10^5 cells/ml) and the corresponding cadmium concentration causing 50% inhibition in DNA replication. Gradual decrease in replicative DNA synthesis upon 0.2, 0.5, 1, 2 and 5 μM $CdCl_2$ treatment was observed, causing 8, 21, 27, 39 and 51% inhibition, respectively (Fig. 5.13).

Effect of Cadmium Concentration on Cell Growth During the Cell Cycle

Two concentrations of cadmium (1 and 5 μM) have been chosen to determine which stage of the cell cycle is most vulnerable to heavy metal treatment. Murine preB cell cultures were grown at an initial density of 1.2×10^5 cells/ml in the absence and in the presence of 1 and 5 μM $CdCl_2$. Cell growth measured in a cell cycle dependent manner showed a significantly higher inhibitory effect of cadmium in mid-S phase (Fig. 5.14). Detailed analysis of cadmium inhibition on cell

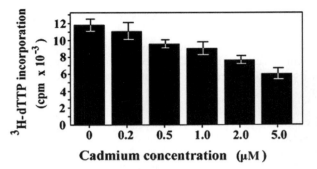

Fig. 5.13 Inhibition of replicative DNA synthesis at different cadmium concentration. Cell growth of six cell cultures each started at 10^5 cell/ml each in the presence of 0.2, 0.5, 1, 2 and 5 μM $CdCl_2$, respectively. Cadmium treatment lasted for 15 h. After treatment 10^6 cells from each population were permeabilized and replicative DNA synthesis was measured. [^{3}H]-thymidine triphosphate incorporation was measured as acid insoluble radioactivity (reproduced with permission of Bánfalvi et al., 2007)

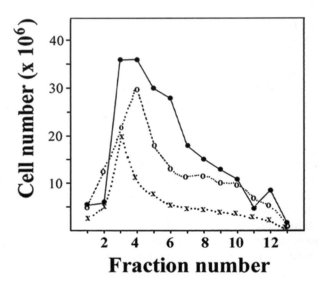

Fig. 5.14 Elutriation profiles of exponentially growing cells before and after cadmium treatment. Three suspension cultures (1 liter each) of murine pre-B-cells (at 1.2×10^5 cells/ml) were grown in spinner flasks for 15 h in RPMI medium supplemented with 10% fetal bovine serum. Control (untreated) cells were harvested at 2.2×10^5 cells/ml by centrifugation at 600 g for 5 min at 5°C and resuspended in 20 ml elutriation medium containing RPMI medium and 1% fetal bovine serum. Control cells (2.2×10^8) were loaded into a Beckman JE6B elutriator rotor and elutriation was carried out (•–•). The same experiment was carried out with the exception that cells were grown to 1.7×10^5 cell/ml in the presence of 1 μM $CdCl_2$ and 1.7×10^8 cells were loaded into the elutriation chamber (o...o). In the third experiment cells were grown in the presence of 5 μM $CdCl_2$, harvested at 8×10^4 cells/ml and 8×10^7 cells were subjected to elutriation (x...x). (Reproduced with permission of Bánfalvi et al., 2007)

Heavy Metal Induced Cytotoxicity

Table 5.3 Cadmium induced inhibition of cell growth during the cell cycle of murine preB cells

| Fraction number | Cell cycle phase | Elutriated | | | | | |
| | | Control | | 1 μM CdCl$_2$ | | 5 μMCdCl$_2$ | |
		$\times 10^6$ cells	%	$\times 10^6$ cells	%	$\times 10^6$ cells	%
1	discarded	5.5	2.54	5.1	2.36	2.2	1.01
2	early S	6	2.78	12.5	5.78	5.2	2.41
3	early S	36	16.67	22	10.19	20.3	9.39
4	early mid S	36.5	16.90	30.3	14.02	11.8	5.46
5	early mid S	30	13.68	17.5	8.10	7.5	3.47
6	mid S	28	12.96	13.5	6.25	5.2	2.41
7	mid S	18	8.33	12	5.56	4.8	2.22
8	late mid S	16	7.41	12.5	5.78	4.8	2.22
9	late mid S	14	6.48	11	5.09	4.0	1.85
10	late S	12	5.56	10.5	4.86	3.8	1.76
11	late S, G2, M	5	2.31	7.5	3.47	2.4	1.11
12	late S, G2, M	8	3.70	6	2.78	2.0	0.93
13	late S, G2, M	1	0.46	0.5	0.23	0.2	0.09
Elutriated total:		216×10^6 100%		160.9×10^6 61.9%		74.2×10^6 34.4%	

Loaded control population (2.16×10^8) was taken as 100%.
Percentage of cell number in each fraction is expressed as % of the total population.
Control cells subjected to elutriation: 2.2×10^8.
Loss during manipulation 4×10^6 (1.8%).
Cadmium treated cells (1 μM) subjected to elutriation: 1.7×10^8.
Loss of Cd treated (1 μM) cells during manipulation: 9.1×10^6 (5.4%).
Cadmium treated cells (5 μM) used for elutriation: 8×10^7.
Loss of Cd treated (5 μM) cells during manipulation: 5.8×10^6 (7.3%).
Reproduced with permission of Bánfalvi et al. (2007).

growth is summarized in Table 5.3. Treatment with 1 and 5 μM CdCl$_2$ caused an overall 38 and 66% cell growth inhibition, respectively. Most significant reduction in cell number occured in elutriation fraction 7, corresponding to mid S phase.

Elutriation fractions of control (untreated) and cadmium treated (1 μM) cells (Fig. 5.14) were subjected to flow cytometric analysis (Fig. 5.15). These measurements revealed that C-values of untreated cells increased from 2C to 4C (Fig. 5.15A), while the C-values of cadmium treated cells were higher in each elutriated fraction (Fig. 5.15B). One would expect that the higher C-values reflect increased DNA content, coming from the elevated rate of DNA synthesis. Consequently we have measured replicative and repair DNA synthesis in elutriated fractions after Cd treatment to find an explanation to the increased fluorescence.

The fact that the smaller glioma cells were more sensitive to cadmium treatment than the larger adenocarcinoma cells (Watjen et al., 2002) raised the question whether or not the cytotoxic effect of cadmium toxicity depends on cell size within the same species. The cell size was measured in elutriated fractions following toxic treatment (5 μM CdCl$_2$). The relationship between cell cycle distribution versus cell diameter in control untreated and treated cells is shown in

Fig. 5.15 Flow cytometry of murine pre B cells before (**A**) and after (**B**) 1 μM CdCl$_2$ treatment. The DNA content of each elutriation fraction is indicated as haploid genome content (C-value) on the abscissa and cell number is given on the ordinate. Controls are representing unelutriated cells before and after cadmium treatment (reproduced with permission of Bánfalvi et al., 2007)

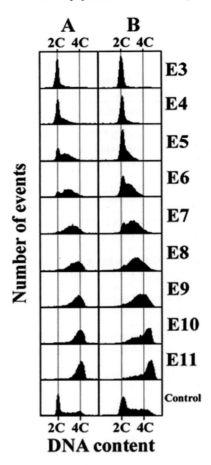

Fig. 5.16. The size of untreated cells is significantly higher in each elutriated fraction (Fig. 5.16A, C) than the size of Cd treated and thus apoptotic cells (Fig. 5.16B, C). Smaller cells seem to be more vulnerable in early S phase (Fig. 5.16C, fractions 2, 3). However, the curve showing the limited increase in cell size during the cell cycle also indicates that cadmium has a more toxic effect in late S, G2 and M phase (Fig. 5.16C, fractions 8–12) than in mid S phase (Fig. 5.16C, fractions 4–6). Consequently a direct relationship between cell size and cadmium toxicity can be ruled out.

Replicative and Repair DNA Synthesis upon Cd Treatment During the Cell Cycle

DNA synthesis in permeable preB cells was measured in elutriated fractions under replicative and repair synthesis conditions after cadmium (5 μM) treatment (Fig. 5.17). The simultaneous measurement of replicative and repair DNA synthesis

Heavy Metal Induced Cytotoxicity

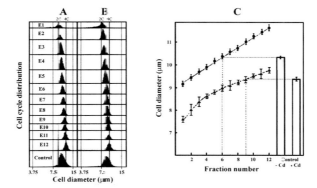

Fig. 5.16 Cell cycle analysis of murine pre-B-cells before (**A**) and after 5 μM CdCl$_2$ treatment (**B**). Cell cycle distribution is given on the ordinate and cell diameter on the abscissa. The DNA content of each elutriation fraction (E1–E12) is indicated by the haploid genome content (average C-value from 2C to 4C). The continuity of the normal cell growth is disrupted by cadmium treatment in elutriation fractions (C). Cell growth in elutriated fractions in the absence (●–●) and in the presence (▲–▲) of 5 μM CdCl$_2$. Controls are representing unelutriated cells before and after cadmium treatment (reproduced with permission of Bánfalvi et al., 2007)

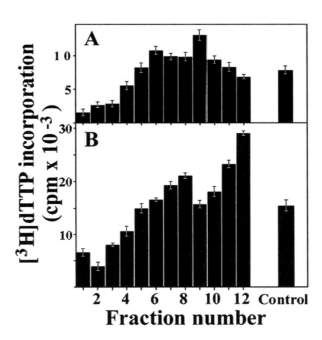

Fig. 5.17 Replicative and repair DNA synthesis in permeable cells of elutriated fractions after cadmium treatment. [^{3}H]-thymidine triphosphate incorporation was measured under replicative (**A**) and repair conditions (**B**) after cadmium chloride (5 μM) treatment. Control, unelutriated cells (reproduced with permission of Bánfalvi et al., 2007)

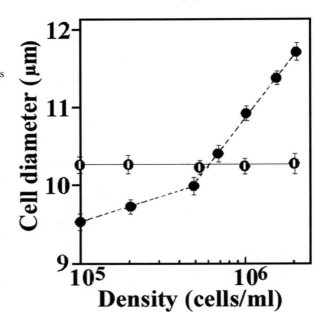

Fig. 5.18 Cadmium induced changes in cell size at different cell densities. Logarithmic growth of murine preB cell cultures was allowed to continue at different cell densities (10^5, 2×10^5, 5×10^5, 10^6, 1.5×10^6 and 2×10^6 cells/ml) in the presence and absence of 5 μM CdCl$_2$ (reproduced with permission of Bánfalvi et al., 2007)

showed a typical biphasic, but reduced replicative DNA profile (Fig. 5.17 A), and an increased repair DNA synthesis throughout the cell cycle (Fig. 5.17 B). The shift in the two peaks indicate that replicative and repair DNA synthesis are temporally distinct processes. In CHO cells cadmium induced oxidative damage reduced ATP-dependent nucleotide incorporation reflecting DNA replication but ATP-independent repair synthesis was elevated in certain subphases of the cell cycle (Bánfalvi et al., 2000).

Relationship Between Cell Density and Cadmium Toxicity

To test the validity of a stoichiometric relationship between cell density and apoptosis, six murine cell cultures with different cell densities were grown in the presence of 5 μM CdCl$_2$ and cell diameters were measured after 15 h of treatment. In Fig. 5.18 the shrinking of cells can be observed at cell densities $\leq 5 \times 10^5$ cells/ml, while at higher than 5×10^5 cells/ml the cell diameter was increasing.

Interphase Chromatin Structures in Untreated preB Cells

Interphase chromatin structures in healthy murine pre-B-cells have been described earlier at lower resolution of centrifugal elutriation by collecting eight fractions (Nagy et al., 2004; Bánfalvi et al., 2006). When higher resolution was used (13 fractions) the same chromatin structures were visualized after DAPI staining (Fig. 5.19). These structures include veil-like nuclear material with some polarization of the chromatin in early S phase (Fig. 5.19 A–D), fibrous structures in early mid S phase

Fig. 5.19 Intermediates of chromatin condensation in untreated pre-B-cells. Regular structures after DAPI staining seen under fluorescence microscope: fractions 2–3, polarized nuclear material (**a–d**); fractions 4–6, extrusion of fibrillar chromatin (**e–h**); fractions 7–8, formation of chromatin bodies (**j–m**); fractions 9–10, linearization of chromatin bodies (**n–q**); fractions 11–12, precondensed chromosomes (**r–u**). Bars 5 μm each (reproduced with permission of Bánfalvi et al., 2007)

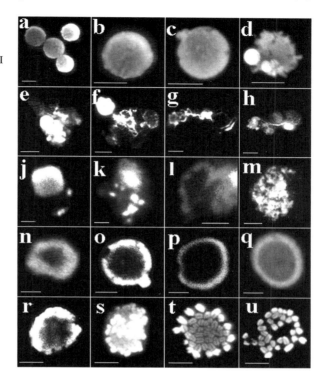

(Fig. 5.19 E–H), chromatin bodies in mid S phase (Fig. 5.19 J–M), linearization of chromatin later in mid S phase (Fig. 5.19 N–Q), precondensed chromosomes in late S, G2 and M phases (Fig 5.19 R–U). As exponentially growing cell cultures in early logarithmic phase were used, the contribution of G2 and M phase cells (fractions 11–13) to the total cell population was only 6.5% (Table 5.3).

Apoptotic Changes in Chromatin Structure After Cadmium Treatment of Murine preB Cells

The cell size distribution after cadmium treatment already indicated the high vulnerability of cells in early S-phase. Fluorescent images of chromatin changes confirmed this notion by the appearance of apoptotic bodies in early S-phase (Fig. 5.20, Fraction 2) which is related to premature chromatin condensation. The polarization of the nuclear material and extrusion of the nuclear material of looped chromatin veil are typical structures later in early S phase (Fig. 5.20, fraction 3). Chromatin margination after cadmium treatment in early mid S phase and the absence of supercoiled loops were found preventing the formation of chromatin bodies (Fig. 5.20, fraction 4).

Reversible permeabilization allowed to open the nucleus any time during the cell cycle in untreated cells. After cadmium treatment the increasing stickiness

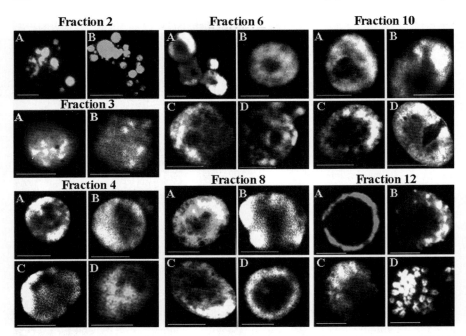

Fig. 5.20 Cadmium induced chromatin changes in S phase. Murine preB cells (2 × 10^5/ml) were subjected to 1 μM CdCl$_2$ treatment followed by reversible permeabilization and isolation of chromatin structures. Fluorescent microscopy after DAPI staining was carried out. Early S phase: Elutriation Fraction 2–4. Mid S phase: Elutriation Fraction 6–8. Typical changes in elutriation fraction 6: disruptions (**A, D**) and holes in the nuclear membrane (**B–D**) and in elutriation fraction 8: large-sized perinuclear granules and holes in the middle of the nucleus (**A–D**). End of S, G2 and in M phase: Elutriation fraction 10: disrupted nuclear membrane with large holes (**A–D**). Elutriation fraction 12: linear arrangement of condensing chromosomes (**A–C**), sticky precondensed chromosomes (**D**). Bars 5 μm each (reproduced with permission of Bánfalvi et al., 2007)

of the nuclear material prevented the opening of the nucleus and obscured the events of chromatin condensation. The formation of perinuclear granules (Fig. 5.20, fraction 6) and ribboned chromatin arranged in a semicircular manner is evident (Fig. 5.20, fraction 8).

Most typical changes of cadmium treatment are the severe disruptions of the nuclear membrane observed in fraction 10 (Fig. 5.20, fraction 10). These disruptions did not allow to distinguish among the final stages of chromatin condensation, resulting in nuclei with big holes inside them and sticky imperfectly condensed chromosomes (Fig. 5.20, fraction 12).

Conclusions: Biochemical and Morphological Effects of Cd

Earlier observation that cadmium has an effect on the relationship between cell culture density and the rate of DNA replication suggested that cell growth and

DNA synthesis at higher cell concentration could be suppressed by increasing Cd concentration. The connection between cell growth and cadmium concentration pointed to the stoichiometric binding of cadmium to CHO cells (Bánfalvi et al., 2005). An extention of this idea was that the cytotoxicity of cadmium could be eliminated or at least minimized by increasing cell culture density. This observation raised the question whether only specific cadmium-binding (metallothionein) proteins are involved in the protective mechanism against the toxic cadmium ion, or non-specific binding may also be involved. Nephrotoxic studies after cadmium exposure indicate that pretreatment with cadmium effectively induced the metallothionein synthesis. Cadmium was mainly bound to metallothionein and other low molecular weight proteins in pretreated mammals, while in non-pretreated animals the major part of Cd was bound to high molecular weight proteins (Nordberg et al., 1994). These experiments are in conformity with experiments regarding the possibility of using inert proteins to lower cadmium toxicity.

The distribution of peaks of nuclear growth after cadmium treatment showed similarities to the number and position of the replicative DNA synthesis peaks in CHO cells. Nuclear growth takes place in several subphases (Bánfalvi et al., 2005) which are likely to correspond to the number of subphases of chromosome replication and to the number of chromosomes (Bánfalvi et al., 1997a). Further coincidence, namely the binding of cadmium to membrane proteins and the fact that nuclear membranes could not be permeabilized after cadmium treatment also favour the idea that cadmium concentration could be lowered by its adsorption to membrane structures including the nuclear envelop. An explanation why cadmium treatment prevented the opening of the nuclear membrane could be that by damaging the cell membrane, cadmium increases the permeability of the cells interfering with reversible permeabilization and causing stickiness of the nuclear material obscuring chromatin structures.

It has been shown by others that lower and higher cadmium concentrations have adverse cellular effects (Von Zglinicki et al., 1992; Hart et al., 1999; Misra et al., 2003). These contradictory results can be explained by cadmium binding, lowering the inhibitory metal ion concentration to the subtoxic level which may then have an opposite stimulatory effect.

Genotoxicity on metaphase chromosomes in mammalian cells *in vitro* were found only at highly cytotoxic concentrations of cadmium. Reduced metaphase number and shortening of metaphase chromosomes were detected in McCoy fibroblast cells exposed to 100 µM $CdSO_4$. The cell nucleus was labeled within one hour exposure to [109]Cd, which suggested an early nuclear involvement in Cd-induced cell damage (Fighetti et al., 1988). Two modes of action are likely to be predominant: the induction of oxidative DNA damage, best established and the interaction with DNA repair processes, leading to enhanced genotoxicity (Hartwig, 1995; Bánfalvi et al., 2000; Hartwig et al., 2002).

However, these changes do not exculde interphase chromatin damages taking place before chromosome condensation. In fact characteristic nuclear changes induced at 1 µM cadmium chloride concentration including apoptotic bodies in early S phase, smaller holes in mid S phase and larger holes and disruption in the nuclear membrane in late S phase. Such toxic chromatin changes were visible throughout the interphase of the cell cycle in CHO cells (Bánfalvi et al., 2005). Such holes and disruptions in nuclei are regarded as diagnostic symptoms of cadmium toxicity. These chromatin changes seem to be adequate endpoints for genotoxic effect-biomonitoring in cadmium exposed cells and organisms.

Regarding the questions addressed at the beginning of this subchapter it is logical to assume that: 1. There is no direct relationship between cell size and cadmium toxicity since the shrinkage of small cells in early S and large cells in late S, G2 and M phases is distinct from those in mid S phase. The density of the cell culture influences cadmium toxicity, high cell culture density reduces cytotoxicity probably by binding and lowering the metal ion concentration. 2. The contradiction between the lower stimulatory Cd concentrations on DNA synthesis versus inhibitory effect of higher Cd concentrations could be explained by the shift from replicative to repair DNA synthesis. At low Cd concentration (0.2–5 μM) the replicative DNA synthesis is gradually inhibited, but this inhibition is less than the increase in repair DNA synthesis and the overall rate of DNA synthesis will be higher than in normal untreated cells. Several studies have shown that Cd interferes with DNA repair. Exceeding the tolerable level of Cd concentration repair synthesis is unable to cope with DNA damages. Consequently high Cd concentrations will inhibit both replicative and repair synthesis and cells undergo apoptosis. 3. The increase in cellular and nuclear size at lower concentrations of Cd is likely to be the reflection of an increased metabolic load manifested in the induction of both cytoplasmic and nuclear processes to overcome the toxic effect. Apoptotic levels of Cd toxicity result in the complete breakdown of metabolic activity seen as cellular and nuclear shrinkage.

Effect of Ionizing Radiation on Chromatin Structure

Gamma Irradiation-Induced Apoptosis in Murine preB Cells

This subchapter compares chromatin intermediates of normal untreated cells with chromatin changes after γ irradiation (4 Gy) in a cell cycle dependent manner. Especially those changes are of importance which are related to apoptosis. The results to be presented show that after γ-irradiation the cell number is reduced mainly in G_0–G_1 and early S phases as compared to non-irradiated cells. Cellular and nuclear volumes are higher than average throughout the cell cycle after irradiation. Apoptotic changes also manifested as apoptotic bodies, the number and size of which were related to the progression of the cell cycle. Significant changes in chromatin structure and nuclear volume were found in early S phase. In damaged cells the condensation of nuclear material was prevented at the fibrous stage of chromatin condensation.

Regeneration of Cellular Membrane After γ-Irradiation

The kinetics of cell regeneration was followed after permeabilization in non-irradiated and in irradiated murine pre-B cells and found that this process slowed down after 4 h. Cells have lost their ability to incorporate [³H]-thymidine during permeabilization, but after 8 h of regeneration 72% of the non-irradiated cells and 65% of irradiated cells regained the capacity to incorporate [³H]-thymidine. After 4 h of incubation 62% of the non-irradiated and 56% of the irradiated cells were able to exclude the dye molecules, corresponding to data obtained by [³H]-thymidine incorporation. In subsequent experiments a 4 h regeneration period was used after permeabilization (Nagy et al., 2004).

Early S Phase Block After γ-Irradiation

To obtain cells at different stages of the cell cycle, exponentially growing murine preB cells in suspension culture were subjected to centrifugal elutriation. Twelve synchronized fractions were collected. Conditions for centrifugal elutriation were the same as described in Table 5.3 except that genotoxicity included gamma irradiation rather than Cd treatment. Cellular fractions obtained by centrifugal elutriation were subjected to FACS analysis. The growth profile and the C values of subpopulations confirmed that the cells were in early exponential phase. Two third of the cells were elutriated in the first 6 fractions belonging to early S phase and only one third in the rest of the fractions in mid S, late S and G2 – M phases.

The profile of DNA content in non-irradiated cells showed a gradual increase in DNA content from 2.03 to 3.86 C-value (Fig. 5.21 A). To the contrary, in cells exposed to 4 Gy γ-irradiation the DNA content showed a biphasic distribution in each elutriated fraction (Fig. 5.21 B) pointing to some nuclear heterogeneity in the C-value of each subpopulation of cells. Overlapping regions of the elutriation profiles of non-irradiated cells (dark gray) and irradiated cells (light gray) are shown in black (Fig. 5.21 C). It was observed that after γ-irradiation the increase in DNA content was smaller in each irradiated subpopulation than in the non-treated control cells. The C-value in elutriation fraction 8 (2.41 C) was not higher than in fraction 7 (2.49 C) suggesting that the growth in DNA content is completely blocked in fraction 8. This observation raised the question whether irradiated cells were properly synchronized. Forward and side scatter analysis was carried out which confirmed the synchrony of irradiated cells.

Change in Cell and Nuclear Sizes After γ-Irradiation

Elutriation is based on the relationship between cell size and DNA content. An even closer relationship is expected between the nuclear size and the DNA content. To determine these relationships, cell size, cell volume and nuclear volume were registered in each elutriated fraction before and after irradiation (Fig. 5.22). The cell number was lower in irradiated subpopulations belonging primarily to Go–G1 (elutriation fraction 1–2; 2.0 C-value) and early S-phase (fractions 3–6; 2.03–2.15 C),

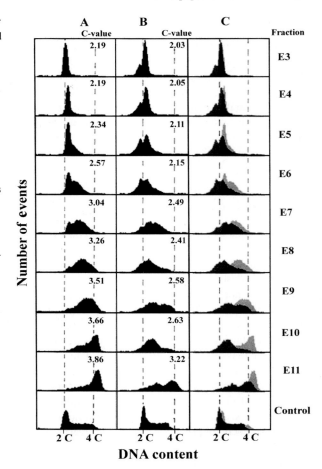

Fig. 5.21 Characterization of cell populations synchronized by centrifugal elutriation. Murine preB cells were grown, irradiated and synchronized by elutriation. Cell cycle patterns of various elutriated fractions (E3–E11 and non-elutriated control) of non-irradiated (**A**) and irradiated pre-B (**B**) cells were analyzed by FACS using propidium iodide staining. Overlapping regions of elutriation profiles were derived from panel **A** (shown in *black*) and from panel **B** (in *dark gray*) and are shown in combined form in panel **C**. Cell-cycle analysis is given on the abscissa and cell number on the ordinate. The DNA content of each elutriated fraction is expressed in C value (reproduced with permission of Nagy et al., 2004)

than in the non-irradiated control population. Irradiation resulted in increased cell number being in early mid S phase (fractions 7–8; 2.4–2.5 C) and in mid S phase (Fig. 5.22 A, fraction 11–12; 3.2–3.3 C). The cell volume was higher in each fraction containing irradiated cells, than in the non-irradiated populations of synchronized cells (Fig. 5.22 B). The last fraction (fraction 13) of the non-irradiated population was not regarded as synchronous, since the cell volume decreased. To the contrary, the largest irradiated cells which belonged to fraction 13 seemed to be synchronized with respect to cell size. The nuclear size of each irradiated elutriation fraction was larger than that of the non-irradiated populations of cells (Fig. 5.22C). After irradiation the higher than expected increase in nuclear size was 3–6% at the onset of S phase (in fractions 3, 4) and in mid S phase (fractions 10, 11). Major changes with respect to nuclear size occurred in early S phase (fractions 6–9), particularly in fraction 8 where the nuclear volume was the highest among all fractions. The nuclear volume in fraction 8 (2.4 C) was three times higher than in non-irradiated

Effect of Ionizing Radiation on Chromatin Structure 329

Fig. 5.22 The effect of γ-irradiation on cell number, cell volume, and nuclear volume in elutriated fractions. Elutriation fractions (described in Fig. 5.21) were analyzed with respect to their cell number (**A**), cell volume (**B**) and nuclear volume (**C**). *Black* columns indicate non-irradiated cells, striped columns correspond to values obtained from irradiated cells (reproduced with permission of Nagy et al., 2004)

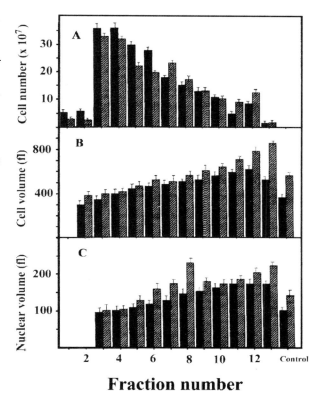

elutriation fraction 1 and 60% more than in elutriation fraction 8 (3.86 C) of non-irradiated cells.

Fibrous, Supercoiled Chromatin and Precondensed Chromosomal Structures in Non-irradiated Murine preB Cells

Reversible permeabilization allowed to restore cellular membranes and to open the nucleus any time during the cell cycle. The regeneration process was also followed by fluorescent microscopy. In one of the control experiments the isolation of nuclei from permeable cells was carried out without reversal of permeabilization resulting in a sticky network of the nuclear material, pointing to the importance of regeneration of nuclear and cellular membranes to visualize chromatin structures.

The change in shape from fibrillar and supercoiled chromatin to elongated and then to condensed forms of chromosomes in non-irradiated preB cells is illustrated in Fig. 5.23. In its decondensed form (fraction 4) at 2.2 C, chromatin appears as a long veil-like structure turned around itself with some polarized regions in the nucleus (Fig. 5.23 A). In elutriation fraction 6 (2.57 C) the fibrous structure of chromatin veil is forming round chromatin bodies and supercoils which are left handed

330   5 Apoptotic Chromatin Changes

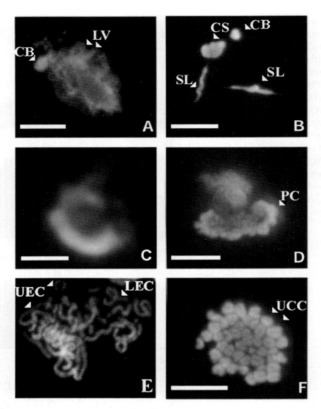

Fig. 5.23 Intermediates of chromatin condensation in non-irradiated murine preB cells. Permeabilization of cells from elutriated fractions was followed by restoration of membrane structures, colcemid treatment and isolation of chromatin structures. The following regular structures were seen: fraction 4 (2.19 C-value), looped veil (**A**); fraction 6 (2.57 C), chromatin bodies and super-coiled loops (**B**); fraction 7 (3.04 C), linear connectivity between supercoiled loops (**C**); fraction 8 (3.26 C), precondensed early chromosomes (**D**); fraction 9 (3.51 C), elongated, U and L-shaped prechromosomes (**E**); and fraction 10 (3.66 C), *semi circled* (U shaped) condensed prechromosomes (**F**). Abbreviations: CB, chromatin body; LP, looped veil; SL, supercoiled loop; PC, precondensed chromosome; UEC, U-shaped elongated chromosome; LEC, looped elongated chromosome; UCC, U-shaped condensed chromosome. Bars 5 μm each (reproduced with permission, Nagy et al., 2004)

turns (Fig. 5.23 B). The fluorescent image of a stained nucleus from elutriation fraction 7 (3.04 C) shows the linear connection of supercoiled loops (Fig. 5.23 C). Precondensed chromosomes are linked to each other and lined up in a semicircle with the two ends open. The string of chromosomes is arranged in a semicircular, helical fashion inside the nucleus (Fig. 5.23 D). In late S phase (3.3–4.0 C) distinct forms of chromosomes became visible. Supercoiled loops form chromatin bodies turn into semicircles during their condensation resembling horseshoe-like arrays. In Fig. 5.23 E the chromatin is seen as a string of clearly distinguishable chromatin bodies and precondensed chromosomes. Some continuity seems to be maintained even during the formation of these precondensed chromosomal forms (elutriation fraction 10, 3.66 C). In elutriation fraction 11 (3.86 C) chromosomes are bent, compact structures, but did not reach the metaphase stage (Fig. 5.23 F). Linear metaphase chromosomes were seen at the end of the condensation process (not shown). A similar condensation pattern was observed in CHO cells and in other mammalian cells including murine preB cells (Figs. 3.19 and 3.20). A more detailed description of the intermediates of chromatin condensation in CHO and other mammalian cells was communicated (Gacsi et al., 2005; Bánfalvi et al., 2006). In the following experiments emphasis is placed primarily on changes in chromatin structure caused by γ-irradiation.

Apoptotic Changes in Chromatin Structure After γ-Irradiation

Fluorescent images of chromatin structures isolated from synchronized murine preB cells showed that γ-irradiation disturbed the condensation process throughout the S phase. Changes are manifested mainly as condensed chromatin particles (apoptotic bodies), the number of which depends on the stage of the cell cycle (Fig. 5.24).

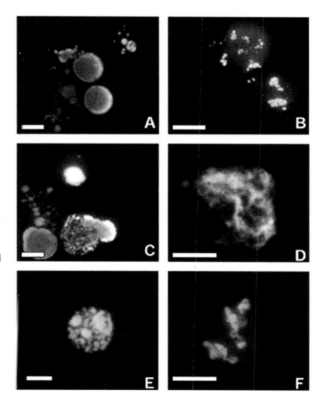

Fig. 5.24 Intermediates of chromatin condensation in γ-irradiated murine preB cells. Cells were irradiated with 4Gy, incubated for 2 hours, elutriated and chromatin structures were isolated. Apoptotic bodies in nuclei from fractions 4, 5 and 6 (A, B and C, respectively). Fibrous chromatin network (fraction 7, D). Incomplete transition from fibrous chromatin to precondensed chromosomes (fractions 8 and 9; E and F, respectively). Bars 5 μm each (reproduced with permission, Nagy et al., 2004)

In the early S phase premature condensation of chromatin manifested as multiple distinct loci inside the nucleus. In elutriation fractions 4, 6 and 7 many small condensed apoptotic particles were observed representing early S phase with an average C-value between 2.0 and 2.4 C. (Fig. 5.24 A, B, C). The delayed progression of chromatin condensation through the S phase is characterized by the lower than expected DNA content in elutriation fractions 4–7. Less damage was observed with respect to the number of apoptotic bodies in elutriation fractions 8–11. Most dramatic changes occurred in elutriation fraction 8, the C-value of which corresponded to fraction 5 (2.34 C) of non-irradiated cells. This stage of the cell cycle (around 2.4 C) seems to be the point when thin condensing chromatin fibers turn to thick supercoiled chromatin structures (Fig. 5.24 D). Due to the delay in cell cycle, most of the γ-irradiated cells belonging to elutriation fractions 10 and 11 did not reach the stage of precondensed forms of chromosomes (Fig. 5.24 E, F).

A more detailed microscopic analysis of elutriation fraction 8 indicated that the fibrous chromatin structure of irradiated cells (Fig. 5.25 A–D) corresponded to the stage of chromatin condensation seen in fractions 4 and 5 (2.19 and 2.34 C) of non-irradiated cells (Fig. 5.23 A). Figure 5.25 shows that in elutriation fraction 8 of irradiated cells the nucleus consists mainly of thin fibrous chromatin unable to turn

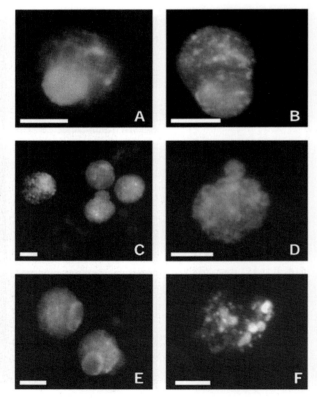

Fig. 5.25 Accumulation of fibrous chromatin structures in mid S phase in γ-irradiated preB cells. Typical structures of chromatin condensation in fraction 8 of irradiated cells are visualized. Fibrous structures with polarized condensed regions in nuclei (**A–C**). Fibrous structures containing recognizable U-shaped precondensed chromosomes (**D, E**). Fibrous structure with a few distorted precondensed chromosomes (**F**). Bars 5 μm each (reproduced with permission of Nagy et al., 2004)

into supercoiled precondensed chromosomes. After irradiation chromatin condensation seems to be blocked in early S phase, causing the accumulation of irradiated subpopulations of cells with enormous nuclear size primarily in fractions 7 and 8.

Another analysis of chromatin structures isolated from nuclei of irradiated cells is seen in Fig. 5.26. These experiments were carried out in the absence of colcemid treatment to assure that even if the rate of chromatin condensation is lower, colcemid would not interfere with the condensation process. In early S phase (elutriation fractions 3 and 4), many small apoptotic bodies were observed (Fig. 5.26 A–D). In elutriation fractions 5 and 6, the size of apoptotic bodies was increased and their number reduced (Fig. 5.26 E–H). Apoptotic changes manifested as larger apoptotic bodies in elutriation fraction 7 with some development of fibrous chromatin obscuring the background (Fig. 5.26 J, K). Elutriation fraction 8 was associated with the prevalence of fibrous chromatin structures (Fig. 5.26 L, M). In elutriation fraction 9, fibrous chromatin was turning into supercoiled loops (Fig. 5.26 N, O). Elutriation fraction 10 showed an incomplete conversion of fibrous chromatin into supercoiled loops (Fig. 5.26 P) or folded back chromosomal forms (chromatin bodies) (Fig. 5.26 Q). Elutriation fraction 11 contained incompletely folded precondensed chromosomes (Fig. 5.26 R, S), and in elutriation fraction 12 U-shaped semicircles (Fig. 5.26 T) or incompletely folded metaphase chromosomes (Fig. 5.26 U).

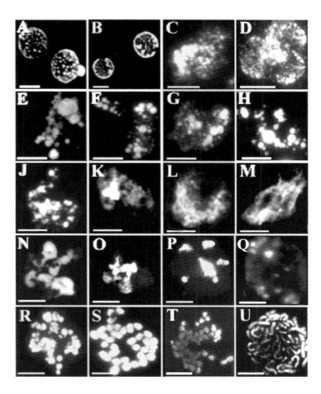

Fig. 5.26 Chromatin structures isolated from irradiated cells in the absence of colcemid treatment. Permeabilization of cells from elutriated fractions was followed by restoration of membrane structures. Chromatin structures from nuclei of elutriation fraction. A–D: Elutriation fractions 3 and 4. E–H: Elutriation fractions 5 and 6. J, K: Elutriation fraction 7. L, M: Elutriation fraction 8. N–O: Elutriation fraction 9. P, Q: Elutriation fraction 10. R, S: Elutriation fraction 11: incompletely folded chromosomes (R, S). T, U: Elutriation fraction 12. Bars 5 μm each (reproduced with permission of Nagy et al., 2004)

334 5 Apoptotic Chromatin Changes

Conclusions. To summarize the results, murine preB cells have been used for γ-irradiation and studies focused on structural changes related to chromatin condensation taking place during S phase. It was shown previously that after the same 4 Gy dose of γ-irradiation in preB cells up to a 2–3-fold increase in apoptosis was evident. The percentage of apoptosis at the onset of S phase was low (5–6%), increased in mid S phase to 9–10% and reached its highest level in late S and G2–M phases (12–13%) (Offer et al., 2001).

After gamma irradiation apoptotic changes in S phase were manifested by the "stalling" of DNA replication in early S phase. Simultaneously, there was a swelling of the nucleus, reaching an anormous size at ∼2.4 C. Morphologically, the condensation of irradiated chromosomes appeared to be blocked in a spread-out membrane or veil-like form, apparently unable to resolve from fibrillar to precondensed chromosomal form. To the question why the large swelling of the nucleus coincides with the apparent stalling of DNA replication a reasonable answer could be that DNA is more vulnerable in its decondensed state. Indeed it was described by others that in the early stages of chromosome condensation cells were especially vulnerable (Mikhailov and Rieder, 2002). It is assumed that double stranded DNA breaks release their torsional energy accumulated during the supercoiling of the chromatin. As a result apoptotic bodies are formed, the size of which is dependent on the compactness of the chromatin structure. According to this view in early S phase when chromatin is in a relatively decondensed state, smaller loops and fibrils form smaller apoptotic bodies, while at later stages of S phase larger apoptotic bodies are generated. To explain the increase of nuclear volume after gamma irradiation an analogy is used which is based on a physical law. Although, no analogy is perfect, it may help to understand why gamma irradiation causes the swelling of the nucleus and the cell especially in those fractions where DNA is decondensed by gamma irradiation. Boyle's law states that in a closed system when the amount and temperature of a gas are not changing PV = constant (where P is pressure and V stands for volume). In the nucleus at a certain stage of the cell cycle P is taken as the torsional pressure (energy) and V as the nuclear volume. To keep at this stage PV constant in the closed nucleus and to avoid its disruption the decrease of the supercoiled energy caused by strand breaks is counteracted by the increase of the nuclear volume. The increase of the nuclear size after γ irradiation suggests that most of the torsional energy is released when chromatin returns to its decondensed fibrillar form. Upon gamma irradiation the condensation process is blocked between the transitions from fibrillar to supercoiled ribbon structures at 2.4 C-value. It is assumed that the release of the supercoiled energy at strand breaks is manifested as the swelling of the nucleus coincident with the apparent stalling of DNA replication.

Gamma Irradiation-Induced Apoptosis in Radiation Resistant Human Erythroleukemia K562 Cells

Double-stranded DNA breaks are generally accepted to be the most biologically significant lesions caused by ionizing radiation. As γ-irradiation-induced cytotoxicity

Effect of Ionizing Radiation on Chromatin Structure 335

is not necessarily correlated with DNA strand break formation and repair capacity, the differences of γ-irradiation tolerance in cells such as chronic lymphocytic leukemia cells are explicable in terms other than DNA strand break formation or repair (Mylliperkiö et al., 1999). Consequently, alternative solutions have to be found to explain the radiation resitance of certain leukemia cells.

The mechanisms involved in radiation-induced cellular injury and death remain incompletely understood. The tumour suppressor p53 has been implicated in gamma irradiation-induced apoptosis. There are two distinct cell cycle responses to γ-radiation: the p53 independent, and the p53 dependent G1 arrest (Miyashita et al., 1994). The mechanism of the Bcl-2 anti-apoptotic activity regulated by p53 is still not fully understood and it is disputed whether or not the radiation resistance is related to the bypass of the p53-dependent mitochondrial apoptotic pathway. Formerly, the development of radiation resistance has been attributed to the abnormal or absent p53 function (Chen et al., 1995; Haas-Kogan et al., 1996). However, a strong correlation between functional p53 and radiation sensitivity has not been satisfactorily established (Haas-Kogan et al., 1996; Yount et al., 1996). The picture regarding the radiation resistance of tumour cells was further blurred by the finding that drug resistance (CAM-DR) contributes to the protection of K562 cells against γ-irradiation (Damiano et al., 2001). It is thus likely that besides the p53 apoptotic pathway there might be other molecular players which contribute to the radiation resistance in experimental systems and may influence the response of tumour cells to radiation therapy.

It was reported earlier that in murine preB 70Z/3 cells after γ-irradiation p53 functioned as a modulator that determined apoptosis in a cell cycle-specific manner. Augmentation of the G_1/G_0 and reduced G_2/M associated base excision repair after γ-irradiation confirmed our notion that these patterns of regulation were regulated by p53 (Offer et al., 2001). DNA damage and increased repair activity suppress DNA replication at checkpoints to avoid mutagenic changes being perpetuated in the next generation of cells (Hartwell and Weinert, 1989; Murray, 1992; Kastan et al., 1992). Apoptotic chromatin changes followed in a cell cycle dependent manner after γ-irradiation proved that the progression of the cell cycle was arrested in the fibrillary chromatin stage corresponding to early S phase at 2.4 C value (Nagy et al., 2004), close to the G_1-G_0 checkpoint as described in the previous subchapter. Ultraviolet B irradiation arrests the cell cycle somewhat earlier and closer to the G_1/G_0 checkpoint between 2.2 and 2.4 C-values (Ujvarosi et al., 2007).

To get an answer to the question whether or not in radiation resistant cells the p53 pathway is blocked in the early stage of the cell cycle the radiation resistant K562 erythroleukemia cell line was used to iniciate apotosis in S-phase cells. As far as the radiation resistance of K562 cells is concerned, it was explained by the inhibition of the apoptotic pathway induced by diverse apoptotic stimuli through the blockage of mitochondrial release of cytochrome c and blockage of caspase-3 activation (Amarante-Mendes et al., 1998).

The next subchapters describe that in the radiation resistant K562 cell line γ-irradiation arrests chromatin condensation in the early fibrillary stage corresponding to 2.4 C similarly to murine preB cells (Nagy et al., 2004) and results in the formation of many small chromatin corpuscules resembling or being apoptotic

336   5 Apoptotic Chromatin Changes

bodies. Forward light scatter flow cytometry has been used to estimate the proportion of apoptotic cells present in various elutriation fractions after irradiation. Results indicate that the resistance of K562 cells to γ-irradiation is either independent of the apoptotic chromatin function of p53, or is due to the bypass of the early events of the p53-dependent apoptotic pathway.

Characterization of K562 Cell Populations Obtained by Centrifugal Elutriation

In an attempt to induce apoptosis in human erythroleukemia K562 cells the same dose of γ-irradiation (4 Gy) was applied as described earlier for murine preB cells (Nagy et al., 2004). Cell cycle enriched populations of K562 cells were obtained by elutriation (Table 5.4). The autenticity of the cell cycle position of each fraction was confirmed by FACS analysis before γ-irradiation and after exposing cells to 4 Gy (Fig. 5.27 A, B). Flow cytometric profiles were used to calculate the C-value of each elutriated fraction. C-values of control unirradiated cells inreased gradually from 2C to 4C (Fig. 5.27 A). In cells exposed to γ-irradiation the DNA content increased as cells progressed through the S phase, but showed the presence of small apoptotic cells (Ap) and a lower DNA content in each fraction from 2.0 to 3.7 C-values (Fig. 5.27 B). The same C-values of fractions 3 and 4 (2.38 and 2.37) indicate that after irradiation the cell cycle was arrested at around 2.4 C-value. After γ-irradiation the cell number did not change significantly (Fig. 5.27 C).

Cells were grown from an initial cell culture density of 1.5×10^5/ml in 1500 ml RPMI medium for 15 h (37°C, 5% CO_2) and harvested at 2.4×10^5/ml by centrifugation (600 g, 5 min, 4°C). Cells (1.5×10^8) were resuspended in 20 ml RPMI medium containing 1% fetal bovine serum and introduced in the elutriation chamber at 19 ml/ml flow rate. Another portion of cells (1.5×10^8) was irradiated with a dose

Table 5.4 Fractions of control and gamma-irradiated K562 cells obtained by centrifugal elutriation

Fraction-number	Flow rate ml/min	Non-irradiated cells		Gamma-irradiated cells	
		Cell number ($\times 10^6$)	C-value	Cell number ($\times 10^6$)	C-value
1	21.5	19.2	2.07	17.6	2.06
2	23.5	15.8	2.18	13.8	2.11
3	25.5	16.7	2.46	19.9	2.38
4	27.6	18.1	2.73	15.8	2.37
5	29.7	17.9	2.95	18.2	2.52
6	31.9	18.4	3.21	19.8	2.85
7	34.0	17.9	3.47	19.1	3.13
8	36.1	13.2	3.69	12.3	3.42
9	38.3	8.9	3.95	9.0	3.70

Recovered by elutriation: 1.48×10^8 cells (98.6%); 1.45×10^8 (96.6.0%)
Loaded to elutriation chamber: 1.5×10^8 cells (100%); 1.5×10^8 (100%)
Loss of cells during elutriation : 2×10^6 cells (1.3%)
Further loss of cells due to radiation: 3×10^6 cells (2.0%)

Effect of Ionizing Radiation on Chromatin Structure 337

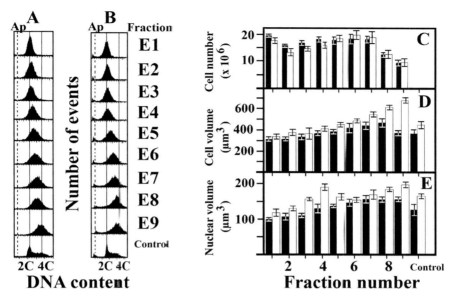

Fig. 5.27 Characterization of human erytroleukemia K562 cell populations synchronized by centrifugal elutriation. K562 cells were exposed to 4 Gy γ-irradiation, incubated for 2 h. Cell cycle patterns of various elutriated fractions (1–9) in unirradiated (**A**) and irradiated K562 cells (**B**) were analyzed by FACS using PI staining. The control population was not elutriated. Elutriated fractions were analyzed with respect to their cell number (**C**), cell volume (**D**) and nuclear volume (**E**). *Black columns* indicate unirradiated cells, *white columns* correspond to values obtained from irradiated cells. DNA content is expressed in C-values. 2C, diploid genome content; 4C, tetraploid DNA content; Ap, small apoptotic cells (reproduced with permission of Bánfalvi et al., 2007)

of 4 Gy γ-irradiation. After further growth for 2 h elutriation was performed in a Beckman J21 centrifuge at 20°C and 1850 rpm. C-values and cell numbers are averages of two measurements (reproduced with permission of Bánfalvi et al., 2007).

The flow cytometric profiles of cell fractions from unirradiated cells being in early logarithmic growth exhibited a normal pattern of cell cycle progression without any sign of apoptosis (Fig. 5.27 A). In cell fractions obtained 2 h after irradiation the appearance of an apoptotic subpopulation containing smaller cells was observed in the sub-G_1 marker window. The dotted line in Fig. 5.27 B helps to estimate the shift of apoptotic DNA from left to the right as irradiated cells progress through the S phase. To magnify the differences among subpopulation of apoptotic cells (Ap) seen at the sub-G_1 marker position, the flow cytometric profiles of irradiated cells were subjected to forward light scatter analysis (Fig. 5.28).

Changes in Cell Size and Nuclear Size After Irradiation of K562 Cells

Cells seeded at an intial density of 1.5×10^5 cells/ml, were grown for 15 h then divided into two portions. Control cells were not irradiated, the other part was subjected to 4 Gy γ-irradiation followed by 2 h incubation. Cell number of nonapoptotic

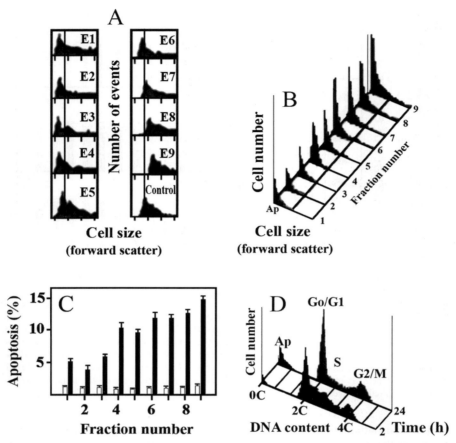

Fig. 5.28 Apoptotic changes detected by the forward scattering of the light in gamma-irradiated K562 cells during the cell cycle. After propidium iodide staining different fractions of irradiated K562 cells shown in Fig. 5.27 B were subjected to forward light scatter analysis in the sub-G$_1$ marker window (**A**). To follow the increase in apoptotic cell size a vertical line was placed on the main peak of the irradiated, non-elutriated control (**B**). After irradiation of cells elutriated fractions were incubated for 24 h and then tested for the increase in apoptotic cells by forward scatter analysis. Cells were stained with PI and analyzed for cell cycle patterns and forward scatter analysis of apoptotic cells was carried out. Panel C. Percentage of apoptotic cells of non-treated (□) and of γ-irradiated (■) cell fractions. Panel D. Equal number of γ-irradiated cells from non-elutriated populations were cultured at 37°C. One aliquot of cells was incubated for 2 h, the other for 24 h, then stained with PI and was subjected to forward light scatter analysis. The C-values and major phases of the cell cycle are indicated. Ap, apoptotic cells (reproduced with permission of Bánfalvi et al., 2007)

Effect of Ionizing Radiation on Chromatin Structure 339

cells in elutriated fractions was determined. After irradiation a small reduction in cell number occured in early and mid S phase, while the cell number did not change significantly in late S phase relative to unirradiated cells (Fig. 5.27 C). The total loss off cell number due to irradiation (2%) is negligable. Based on the relationship between cell volume and DNA content the cell cycle distribution versus cell volume in control untreated and irradiated cells is shown in Fig. 5.27 D. The volume of irradiated cells is higher in each elutriated fraction than the average size of the control fractions. The most significant increase in cell volume was observed in late S phase (Fig. 5.27 D). An even closer relationship between the nuclear volume and the DNA content was used to visualize the gradual increase of nuclear size of elutriated control fractions which was smaller in each fraction than those of irradiated populations (Fig. 5.27 E). After irradiation a higher than expected increase in nuclear size was measured at the beginning of S phase in fractions 1–5, corresponding to a nuclear expansion of 18, 25, 42, 52 and 20%, respectively. Major increase in nuclear volume took place in fraction 4 (52%). Moderate increase of nuclear size relative to control cells was measured in fractions 6–9 (5–15%). The average nuclear volume of non-elutriated irradiated cells was 31% higher than that of the unirradiated control population. The average expansion of nuclear volume was somewhat less ($\sim$10%) than in murine preB cells caused by the same dose of γ-irradiation (Nagy et al., 2004).

Flow Cytometric Forward Scatter Analysis

An important aspect of this analysis was the measurement of apoptotic DNA (Ap-DNA) formation, i.e. the estimate of the cellular DNA degradation during the programmed cell death. This analysis which is closely related to the particle size, amplifies the apoptotic signal and shows more convincingly the increase in the cell size of apoptotic cells during the cell cycle (Fig. 5.28 A). The forward light scatter size analysis was repeated by incubating cells in culture medium for 24 h after irradiation. The quasi 3D representation of forward light scatter analysis in Fig. 5.28 B shows the progression of apoptotosis during 24 h from fraction 1 to 9. The flow cytometric data served to estimate the DNA content of distinct apoptotic cell fractions, which increased from $\sim$ 0.05 to $\sim$1 C-value.

The flow cytometric forward scatter analysis was used to determine the proportion of apoptotic cells relative to unirradiated ones after 24 h incubation. Apoptosis was low in unirradiated cells ($\sim$1%) and increased significantly in irradiated cells from 5 to 13–14%. (Fig. 5.28 C). To estimate the average value of apoptosis, non-elutriated control populations were also subjected to γ-irradiation and grown in culture medium for 2 and 24 h, respectively. The number of apoptotoc cells seen in the sub-G_1 marker window increased from 4.8 to 15.6% (Fig. 5.28 D). Besides the accumulation of apoptotic cells, the increase in G_1/G_0 and G_2/M phase and the decrease of S phase cells were observed (Fig. 5.28 D). These changes confirm the radiation-induced cell cycle arrest.

Intermediates of Chromatin Condensation in Unirradiated K562 Cells

Since chromatin condensed either into a unique chromatin mass in one cell line reflecting apparently an early apoptosis and in another cell line the late apoptosis with apoptotic bodies already visible (Buendia et al., 1999), we have used the technique that is based on the reversible permeabilization of cells to visualize large scale chromatin changes. Using this approach, our studies in a variety of cell types with different chromosome numbers (Chinese hamster ovary, Indian muntjac, murine pre-B, and human erythroleukemia K562 cells) showed that chromosome condensation follows in mammalian cells during the normal cell cycle a common pathway that includes several intermediate chromosomal structures (Bánfalvi et al., 2006). The shape and density change from the veil-like chromatin in early S to fibrillar and supercoiled chromatin in mid S, is followed by the formation of elongated and condensed forms of chromatin in late S phase are illustrated in Fig. 5.29. Since these geometric forms have been described earlier (Fig. 3.19 and Fig. 3.20) (Bánfalvi et al., 2006), here the similar structures are presented only to demonstrate how much these physiological chromatin forms differ from those isolated from cells irradiated with 4 Gy in early S (Fig. 5.30) and late S phase (Fig. 5.31).

Apoptotic Chromatin Changes After γ-Irradiation in K562 Cells

The aim was to test whether γ-irradiation causes characteristic chromatin changes in K562 cells which are known to be resistant to apoptosis-inducing agents including

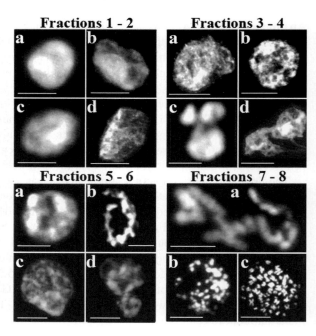

Fig. 5.29 Intermediates of chromatin condensation in unirradiated K562 cells. Permeabilization of cells from elutriated fractions was followed by the restoration of membrane structures, and chromatin structures were isolated. Regular structures included: decondensed, veil-like chromatin (fractions 1–2), fibrous, looped chromatin (fractions 3–4), linear connectivity between supercoiled loops (fractions 5–6), precondensed, early chromosomes (fractions 7–8). Bars, 5 μm each (reproduced with permission of Bánfalvi et al., 2007)

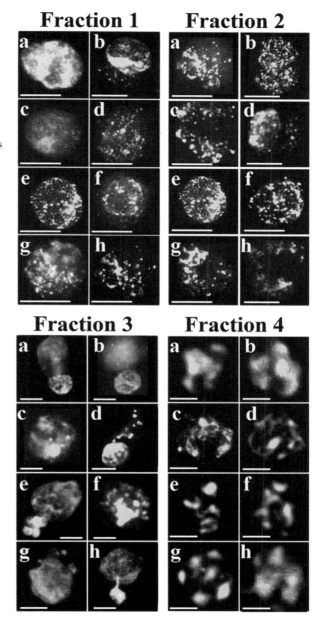

Fig. 5.30 Intermediates of chromatin condensation in γ-irradiated K562 cells in early S phase. Cells were irradiated with 4 Gy, incubated for 2 h and elutriated, and chromatin structures were isolated. Apoptotic chromatin structures from nuclei of elutriation fractions 1–4. Bars 5 μm each (reproduced with permission of Bánfalvi et al., 2007)

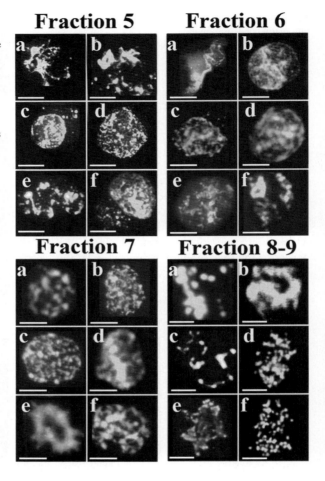

Fig. 5.31 Intermediates of chromatin condensation in γ-irradiated K562 cells in late S phase. Cells irradiated with 4 Gy, were incubated for 2 h and elutriated, and chromatin structures were isolated. Apoptotic chromatin structures from nuclei of elutriation fractions 5–9. Bars 5 μm each (reproduced with permission of Bánfalvi et al., 2007)

ionizing radiation. Fluorescent images of chromatin structures showed that γ-irradiation blocked the condensation process throughout the S phase in murine preB cells (Nagy et al., 2004). The question was whether similar apoptotic changes are manifested in K562 cells after γ-irradiation. The forward scatter flow cytometry already indicated that γ-irradiation has a significant effect on the cellular and nuclear size distribution of K562 cells and causes the accumulation of apoptotic cells. Fluorescent microscopic imaging of chromatin structures which has been proved to be a sensitive method (Nagy et al., 2004) provided further evidence for the genotoxic effect of irradiation (Figs. 5.30 and 5.31). The generally observed effect of γ-irradiation can be summarized by the disruption of the nuclear material into smaller (early S phase) and larger apoptotic bodies (late S phase).

Effect of Ionizing Radiation on Chromatin Structure 343

Early S Phase

The disruption of the nuclear material into many small apoptotic bodies 2 h after γ-irradiation was observed at the unset of the S phase (Fig. 5.30, Fraction 1–2, 2.0–2.1 C-value). The polarization of the nuclear material and the extrusion of the highly decondensed chromation veil are typical structures seen in early S phase (Fig. 5.30, Fraction 3, 2.38 C-value).

Early Mid S Phase

In nuclei of irradiated cells the disruption of decondensed chromatin was seen (Fig. 5.30, Fraction 4) The observation that the cell cycle was arrested in Fraction 4 is confirmed by the unaltered C-values at 2.37–2.38 (Table 5.4). It did not escape our attention that the disrupted chromatin separated into six major fluorescent intensities within individual nuclei, an indication that human chromosomes might be clustered in these chromatin domains (Fig. 5.30, Fraction 4 a, b, e, f, g, h). This idea is supported by the finding that the nuclear material of hepatocytes isolated from resting (G_0) hepatocytes of rats also consisted of six, subdivided, well distinguished chromatin lobes (Trencsenyi et al., 2007) suggesting that rat and human chromosomes are clustered in six major chromatin domains.

Mid S Phase

After the early S phase block, the cell cycle continued with some delay in γ-irradiated cells (Fig. 5.31, Fraction 5) and was characterized by its lower (2.52 C) than the normal C-value (2.95 C) of unirradiated cells (Table 5.4). Supercoiling gradually changes the shape of fibrillary chromatin to fibrous structures, which are disrupted by γ-irradiation.

Later in Mid S Phase

In unirradiated mammalian cells the formation of round globular bodies referred to as chromatin bodies was regarded as the appearance of the earliest visible forms of interphase chromosomes (Bánfalvi et al., 2006). In γ-irradiated K562 cells the apoptotic bodies were not very well distinguishable form the chromatin bodies but the heterogeneous distribution of nuclear material which dominated the picture indicated apoptosis (Fig. 5.31, Fraction 6).

Late S Phase

The formation of a thick continuous chromatin fiber seen in unirradiated cells was prevented by γ-irradiation, the nuclear material was disintegrated (Fig. 5.31, Fraction 7). Elutriation Fractions 8 and 9 represent the final stage of chromatin condensation in normal cells between 3.70 and 4.0 C value. Due to the delayed cell cycle most of the γ-irradiated cells belonging to the last elutriated fractions did not

344 5 Apoptotic Chromatin Changes

reach the precondensed stage of chromosome condensation. Disrupted nuclear parts and stuck-together chromosomes are regarded as large apoptotic bodies (Fig. 5.31, Fraction 8 a-c). Condensed chromosomes close to metaphase isolated from irradiated cells tended to stick together (Fig. 5.31. Fraction 9 e, f).

Conclusions: effects of γ *irradiation.* One can conclude from the results to the possible relationship between the checkpoint surveillance mechanisms of the cell cycle and DNA damage in radiation resistant K562 cells in S-phase:

(a) The cell cycle distribution as an indicator of radiation resistance of K562 cells revealed negligible loss of cells, a relatively low number of early apoptotic cells (4–5%) and an increased number of late apoptotic cells (14–15%) in synchronized fractions of irradiated cells.

(b) The C-values of elutriation fractions showed that after irradiation the cell cycle was blocked in early S phase corresponding to 2.4 C-value. The cell cycle was arrested at the same C-value in murine preB cells (Nagy et al., 2004). Based on this congruency one can predict that γ-irradiation will have a similar effect in other mammalian cells.

(c) Two hours after γ-irradiation the same cellular changes took place in K562 as in murine preB cells. The most significant increase in the nuclear size of K562 cells took place in fraction 4 (2.37 C-value) identical with the early cell cycle checkpoint and similar to the cell cycle checkpoint after UV-irradiation between 2.2 and 2.4 C-value (Ujvarosi et al., 2007). The effect of UV irradiation on chromatin structure will be reviewed in the next subchapter.

(d) Twenty-four h after irradiation the K562 cell population showed an enrichment in G_1/G_0 and G_2/M phase cells, an increased apoptotic cell number and reduced cell number in S phase. These experiments strengthen the view that γ-irradiation inhibits the S phase and arrests cell cycle near the G_1/G_0 and at the G_2/M checkpoints. The cell cycle arrest at the G_1/G_0 checkpoint was explained by the induction of the p53 mitochondrial pathway directed apoptosis in murine preB cells (Offer et al., 2001; Nagy et al., 2004). The same apoptotic changes taking place in K562 cells suggest that radiation resistance is not due to the inhibition of the p53 pathway.

(e) γ-irradiation causes the same apoptotic chromatin changes in radiation resistant K562 cells as described earlier in radiation sensitive murine preB cells (Nagy et al., 2004). Chromatin changes induced by cadmium treatment were markedly different from those seen after γ-irradiation, but cadmium-induced chromatin changes were the same in different cells. Apoptotic chromatin changes in chromatin structure after cadmium treatment were seen as large extensive disruptions, holes in the nuclear membrane and stickiness of incompletely folded chromosomes (Bánfalvi et al., 2005; Bánfalvi et al., 2007). After γ-irradiation the cellular and nuclear sizes increased, the DNA content was lower in each elutriated subpopulation of cells, the progression of cell cycle was arrested in early S phase, the chromatin condensation was blocked between the fibrillary chromatin and precondensed elongated chromosome forms and the number and size of apoptotic bodies were inversely correlated with the progression of the

Effect of Ionizing Radiation on Chromatin Structure345

cell cycle (Nagy et al., 2004). These differences suggest that typical chromatin injuries can be categorized based on the assessment of injury-specific changes. This categorization could be done preferably during S-phase when chromatin is in its decondensed state and DNA might be very vulnerable especially if the repair is not proficient.

Regular and temporally distinct chromatin structures follow a general pathway of chromatin condensation in unirradiated cells (Bánfalvi et al., 2005; 2006; Gacsi et al., 2005). These chromatin structures differ significantly from those large-scale chromatin disruptions which are characteristic to γ-irradiated murine preB (Nagy et al., 2004) and K562 cells, and are unlike to those chromatin changes which were caused by ultraviolet irradiation (Ujvarosi et al., 2007). The difference is easily explained by the fact that γ-irradiation induces many DNA double-standed breaks, UV-B induces only few. Typical genotoxic chromatin changes observed after various treatments (γ-, UV-irradiation, Cd treatment) were absent in non-treated cells excluding the possibility of artifact formation. Further experiments have to be directed to compare normal and tumour cell types with respect to chromatin condensation and possible correlations with radioresistance. It may well be that some radioresistant cells have more densely packed or epigenetically modified chromatin related to radioresistance.

One cannot disregard the explanation related to the formation of apoptotic bodies after γ-irradiation. The condensation process of irradiated K562 cells seems to be most affected in its veil-like form. Indeed it was found by others that chromatin is particularly vulnerable in the early steps of chromatin condensation (Mikhailov and Rieder, 2002). It is logical to assume that double stranded breaks caused by γ-irradiation release the torsional energy and result in the formation of apoptotic bodies. The size of the apoptotic bodies seems to depend on the compactness of the chromatin, with smaller loops and fibrils forming smaller apoptotic bodies in a relatively decondensed state and larger, but less numerous apoptotic bodies later in S phase. The increase of nuclear size seems to counteract the release of the increased torsional energy caused by the strand breaks. Upon γ-irradiation the condensation is blocked at the fibrillary stage, between the transition from fibrillar to supercoiled fibrous ribbon structure at 2.37 C value. This probably means that the chromatin is in its most decondensed state at this C-value and somewhat more condensed in G_1/G_0 (2.0 C). This explanation also implies that the physical force of γ-irradiation by releasing the torsional energy stored in the chromatin structure would be the immediate cause of the formation of apoptotic bodies. Although, we still name these structures apoptotic, contrary to our belief that the apoptotic pathway joins in after these DNA damages probably via the p53 pathway, and apoptosis is completed by caspases only at the end of the pathway. This would also mean that after γ-irradiation the formation of apoptotic bodies could not be prevented in radiation resistant cells since γ-irradiation acts directly on DNA, generating apoptotic bodies or simply fragmented DNA is accumulating after irradiation during S-phase.

The idea that radiation resistance can be among the downstream factors of the apoptotic pathway is supported by experiments carried out in radiation sensitive

346   5 Apoptotic Chromatin Changes

Jurkat and radiation resistant K562 cells (Rexford et al., 1999). Jurkat cells displayed increased levels of long term C16 ceramide accumulation, which mediates the release of cytochrome c form mitochondria, whereas K562 cells did not. Further supporting evidence came from the increase of caspase-3 activity in Jurkat cells, whereas caspase-3 activity in K562 cells remained unchanged. As the connection between ceramide and Fas/Fas ligand and other DISC family members in apoptotic signaling is unclear, ceramide *per se* may not participate in the early phases of the apoptotic response (Rexford et al., 1999).

Due to the uncertainties in the p53 dependent pathway, the p53 independent pathway also deserves consideration in the development of radiation resistance of tumour cells. That this might be the case is supported by the evidence provided by the exposure of p53-deficient lymphoblastic leukemia cells to 3–96 Gy causing p53-independent cell death in a dose and time-dependent fashion. By electron microscopic and other criteria, this cell death was classified as apoptosis. At lower levels of irradiation, apoptosis was preceded by the accumulation of cells in the G2/M phase of the cell cycle. Expression of Bcl-2 and Bax were not detectably altered after irradiation (Strasser-Wozak et al., 1998).

The protein composition of different cell types is a further possibility to be mentioned. Experimental data indicate that distinct classes of chromosomal proteins afford the DNA with different levels of protection against ray-induced DNA double strand breaks. Thus, chromatin domains that differ in tertiary structure and protein composition may also differ in their susceptibility to DNA double strand breaks induced by ionizing radiation (Elia and Bradley, 1992).

Finally, the light scattering flow cytometry for the detection of the apoptotic cells deserves to be mentioned, as it turned out to be a useful method at sensing the presence of apoptotic cells and apoptotic bodies. Light-scattering instruments called "scatterometers" enabled the study of various biological molecules (Liedberg et al., 1983; Shapiro, 2003). Such instruments, especially flow cytometers can be easily switched to their light scattering function and adapted for apoptotic measurements.

Preapoptotic Chromatin Changes Induced by Ultraviolet B Irradiation in Human Erythroleukemia K562 Cells

Besides revealing intermediates of the normal supranucleosomal organization of DNA it was shown that structural aberrations take place in chromatin organization upon genotoxic treatments. Abnormal chromosomal forms were found after cadmium treatment and gamma irradiation leading to apoptosis (Nagy et al., 2004; Bánfalvi et al., 2005). Cadmium treatment is known to cause disruptions and large holes on the nuclear membrane (Bánfalvi et al., 2005). Experiments with gamma irradiated cells revealed that ionizing radiation generates apoptotic bodies the size of which depends on the cell cycle, with many small apoptotic bodies in early S phase. Larger but less numerous chromatin bodies were seen at the end of S phase,

Effect of Ionizing Radiation on Chromatin Structure   347

correlating with the observation that in the most dense chromatin structure, the metaphase chromosomes, ionization leading to DNA breaks did not cause significant visible changes (Nagy et al., 2004).

Ultraviolet B radiation is known to be a potent agent for the induction of programmed cell death. There is a general consensus that UV-induced apoptosis is the consequence of nuclear DNA damage (primarily pyrimidine dimers) sensed by DNA associated protein kinases which activate the tumour suppressor protein p53. UV irradiation activates both the external death receptor and the intrinsic mitochondrial apoptotic pathway. Yet, the mechanistic aspects of UV-induced apoptosis remain ill-defined. Human leukemia HL-60, myelomonocytic U937, T-lymphoblastoid Molt-4 and Molt-3 cells were found to rapidly undergo apoptosis after short periods of UV-irradiation, whereas prolonged exposure to UV radiation induced a more rapid form of cell death which was suggestive of necrosis. Pre-erythroid K562 and B-lymphoblastoid Daudi cell lines proved to be more resistant to the death-inducing properties of UV-irradiation by comparison (Martin and Cotter, 1991). DNA fragmentation pattern indicative of endogenous endonuclease activation was not detected immediately after UV-irradiation. Lymphoid cells showed moderate damage after UV-irradiation, as assessed by DNA double-stranded breaks, while normal granulocytes and myeloid leukemia blasts expressed a sharp increase in double-stranded breaks (Bogdanov et al., 1997). Ultraviolet rays induced chromosomal giant DNA fragmentation, which was followed by internucleosomal DNA fragmentation associated with apoptosis in rat glioma cells (Higuchi et al., 2003). Degradation into an oligonucleotide ladder appeared as early as 2 h after UV irradiation in hybridoma cells (Winter et al., 1998). These results, especially the observation that lethal doses of UV irradiation generate high molecular weight (100–800 kbp) DNA fragments prior to ladder-formation of internucleosomal DNA fragmentation associated with apoptosis through caspase activation indicate the involvement of large scale chromatin changes in DNA structure induced by UV irradiation (Higuchi et al., 2003).

Next subchapters deal with biochemical and morphological changes taking place in regenerating non-irradiated K562 cells and in cells exposed to low (2 and 5 J/m^2), moderate (15 J/m^2) and high (25 J/m^2) doses of ultraviolet B (UVB) irradiation after reversal of permeabilization. Biochemical measurements showed partial recovery of replicative DNA synthesis at lower and moderate UVB doses and stagnant suppression of DNA replication at high UVB dose. Large-scale chromatin changes generated by ultraviolet (UV) B irradiation were visualized throughout the cell cycle. These preapoptotic chromatin changes preceded the apoptotic disintegration of the nuclear material.

Biochemical and Morphological Changes Generated by UV-Irradiation in Human Erythroleukemia K562 Cells

Experiments on cell permeability were discussed in Chapter 3 and served as a basis to follow the two types of DNA synthesis and chromatin condensation in reversibly

348   5   Apoptotic Chromatin Changes

permeabilized K562 cells. Before going into experimental details a scheme is given for experiments on UV-irradiation.

Non-irradiated and UV-irradiated cells were used to test:

A. Biochemical changes including:

-   viability of intact, non-irradiated cells (Fig. 5.32 A),
-   regeneration of non-irradiated cells after permeabilization (Fig. 5.32 A),
-   recovery of [^{3}H]thymidine incorporation after reversal of permeabilization,
-   dose dependent recovery of [^{3}H]thymidine incorporation,
-   decreasing incorporation of [^{3}H]TTP after reversal of permeabilization,
-   the absence of apoptotic morphology at cellular level in irradiated cells after reversal of permeabilization (Fig. 5.32 B, flow cytometry),
-   regeneration of UV irradiated cells after reversal of permeabilization (Fig. 5.32 C),
-   DNA content during the cell cycle of non-irradiated and irradiated populations of synchronized cells (Table 5.6, Fig. 5.33),
-   ATP-dependent replicative and ATP-independent repair DNA synthesis (Table 5.5)
-   replicative synthesis in non-irradiated cells,
-   repair synthesis in non-irradiated cells,
-   replicative DNA synthesis in irradiated cells,
-   repair synthesis in irradiated cells.

B. Morphological changes:

-   interphase chromatin structures in non-irradiated cells (Fig. 5.34),
-   interphase chromatin structures in irradiated cells (Fig. 5.35),
-   final stage of chromosome condensation in non-irradiated and irradiated cells (Fig. 5.36).

Cell Viability and DNA Synthesis in Permeable Cells

The viability of intact cells was measured by the trypan blue exlusion test and expressed as percentages of the total cells. At the beginning of the experiment less than 2% of intact K562 cells were able to take up trypan blue dye molecules. After 10 h of growth we counted approximately 5% less viable cells in the same control population (Fig. 5.32 A). Human erythroleukemia K562 cells were made permeable with hypotonic buffer in the presence of 4.5% dextran T-150. K562 cells became permeable rapidly, more than 95% of the cells took up trypan blue within 5 min. Simultaneously, [^{3}H]thymidine incorporation was measured in permeable cells (Fig. 5.32 A). The kinetics of cell regeneration was followed and showed that reversal of permeabilization was fast at the beginning and slowed down after 4 h. Permeable cells which were originally unable to incorporate [^{3}H]thymidine, gradually regained their capacity to incorporate the radioactive nucleoside. After 10 h approximately 80% of

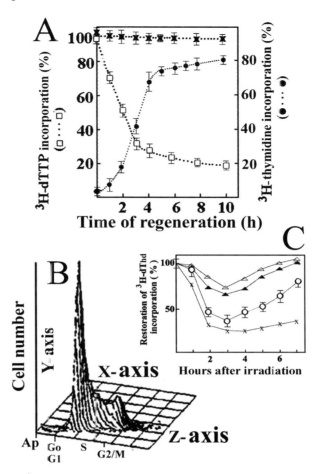

Fig. 5.32 Reversible permeabilization and inhibition of DNA replication after UV irradiation of K562 cells. A. The viability of intact K562 cells was counted during the incubation by the trypan blue exclusion test and expressed as percentages of the total cells (**x–x**). K562 cells were permeabilized and then allowed to regenerate. [^{3}H]-thymidine (●–●) and [^{3}H]dTTP incorporation (□–□) were measured from 0 to 10 h. [^{3}H]-thymidine incorporation was expressed as percentages of the radioactive nucleoside incorporation of intact cells. [^{3}H]dTTP incorporation was expressed as percentages of [^{3}H]dTTP uptake in freshly permeabilized cells. B. Permeable cells were subjected to flow cytometry 3 h after regeneration and tested for apoptotic cells. C. Rate of DNA synthesis relative to control as a function of time after UV irradiation. [^{3}H]dThymidine ([^{3}H]dThd) incorporation was followed after permeabilization, applying the following UV doses: 2 J/m^2 (Δ–Δ), 5 J/m^2 (▲–▲), 15 J/m^2 (o–o) and 25 J/m^2 (x–x) (reproduced wit permission of Ujvarosi et al., 2007)

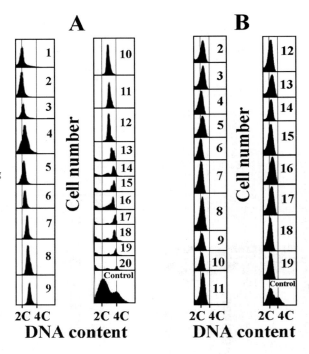

Fig. 5.33 Flow cytometric characterization of K562 cell populations synchronized by centrifugal elutriation before and after UV irradiation. K562 cells were grown, UV irradiated with 24 J/m^2 and elutriated. Cell cycle patterns of various elutriated fractions (1–20 and non-elutriated control) of non irradiated (**A**) and irradiated K562 cells (**B**) were analyzed by FACS using propidium iodide staining (reproduced with permission of Ujvarosi et al., 2007)

the cells regained their membrane integrity. The opposite tendency was observed when [^{3}H]dTTP incorporation was measured. Permeable cells gradually lost their ability to incorporate the four deoxyribonucleotides (dNTPs) including [^{3}H]dTTP (Fig. 5.32 A).

Effect of Ultraviolet Irradiation on Replicative and Repair DNA Synthesis in Permeable K562 Cells

The rate of DNA synthesis of fibroblasts was inhibited by approximately 60% after irradiation with 5.2 J/m^2 of 254 nm UV light (Cleaver, 1982). This observation served as an orientation point to determine the radiosensitivity of K562 cells. Since K562 cells have lower radiation sensitivity (Klein et al., 1976; Martin and Cotter, 1991), higher UV doses have been applied. Exponentially growing human K562 cells treated with 15 J/ m^2 280 nm ultraviolet light at 10^5 cells/ml cell culture density were grown further for 3 h, then permeabilized. After these treatments 65% inhibition of ATP-dependent replicative DNA synthesis was measured. Repair synthesis was measured in the absence of ATP. Table 5.5 demonstrates the ATP requirement for replicative DNA synthesis in non-irradiated cells. The ATP independent repair synthesis after UV irradiation was almost as high as the ATP-dependent replicative synthesis in non-irradiated cells indicating that UV treatment induced DNA repair. The difference between the ATP-dependent replicative synthesis and

Effect of Ionizing Radiation on Chromatin Structure

Table 5.5 ATP-dependent DNA synthesis in K562 cells UV irradiated prior to permeabilization

Incubation mixture	DNA synthetic activity in cells	
	Non-irradiated	Ultraviolet irradiated
	(%)	(%)
Complete	100	35.2
– dNTPs (dATP, dGTP, dCTP)	8.6	9.2
– ATP	19.7	31.6

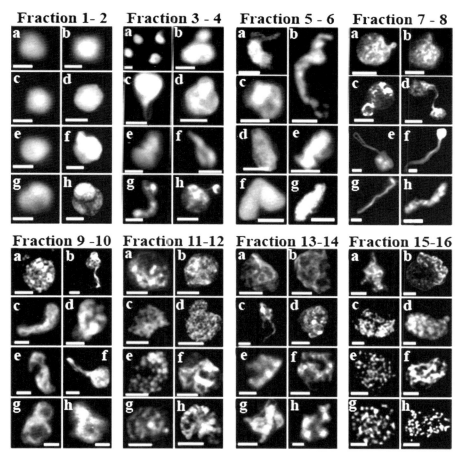

Fig. 5.34 Intermediates of chromatin condensation in non-irradiated K562 cells. Permeabilization from subpopulations of cells after elutriation was followed by reversal of permeabilization to restore membrane structures, colcemid treatment and isolation of chromatin structures. Regular structures included: fractions 1–2, chromatin veil; fractions 3–4, polarized chromatin veil; fractions 5–6, looped veil; fractions 7–8, chromatin ribbon; fractions 9–10, supercoiled ribbon; fractions 11–12, chromatin bodies; fractions 13–14, early elongated chromosomal forms; 15–16, precondensed early chromosomes. Odd fractions: a-d, even fractions e–f, each. Bars 5 μm each (reproduced with permission of Ujvarosi et al., 2007)

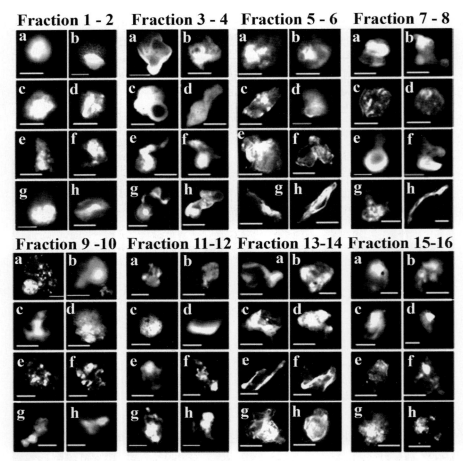

Fig. 5.35 Changes in chromatin structure after UV-irradiation of K562 cells. Cells were irradiated with 24 J/m^2, incubated for 3 h, then elutriated, resealed after permeabilization, treated with colcemid and chromatin structures were isolated. Fibrillary chromatin network covering and blurring chromatin structures from fraction 1 to 16. Incomplete transition from fibrillary to fibrous chromatin without chromosome formation. Bars 5 μm each (reproduced with permission of Ujvarosi et al., 2007)

ATP independent repair synthesis after UV irradiation clearly indicates, that replication is almost completely blocked after irradiation and the measured rate of residual DNA synthesis can be attributed to DNA repair.

[^{3}H]dTTP incorporation was measured in permeable cells. DNA synthesis was carried out in control, non-irradiated and ultraviolet-irradiated cells. Results are expressed as percentages of the activity of the complete incubation mixture containing unirradiated cells.

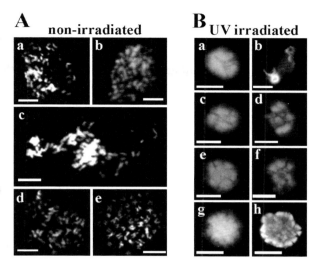

Fig. 5.36 Comparison of chromatin condensation in non-irradiated K562 cells and irradiated (24 J/m^2) cells in late S. The last fractions of non-irradiated and irradiated cells were permeabilized, regenerated, subjected to colcemid treatment and chromatin structures were isolated. **A**. Complete chromosome formation in non-irradiated cells. **B**. Segregation of nuclear material without chromosome formation in irradiated cells. Bars 5 μm each (reproduced with permission of Ujvarosi et al., 2007)

Flow Cytometric Analysis After UV Irradiation

Exponentially growing K562 cells were subjected to increasing fluences of UV irradiation (6, 12 and 24 J/m^2). Figure 5.32 B shows the flow cytometric profile of the cell population which was subjected to 24 J/m^2 UV dose. Irradiation of cells was followed by incubation in growth medium for further 3 h. The lack of small apoptotic cells in Fig. 5.32 B shows that apoptosis did not develop in such a short period of time after irradiation.

Restoration of Nucleoside Incorporation After Irradiation in Intact K562 Cells

K562 cells (10^6) were irradiated with increasing doses of UV light from 2 to 25 J/m^2. [^{3}H]Thymidine ([^{3}H]dThd) incorporation was measured in intact non-irradiated cells and after irradiation. The UV dose dependent decrease of DNA synthesis was measured (Fig. 5.32 C). The rate of DNA synthesis of cells irradiated with lower (2 and 5 J/m^2) doses of UV light showed a graual recovery after 3 h. Higher dose of UV irradiation (15 J/m^2) could be partially overcome (80%) by regenerating cells, while cells exposed to high dose of UV light (25 J/m^2) were unable to regenerate their DNA synthesizing capacity.

Cell Cycle Analysis in Intact and UV Irradiated Cells

The flow cytometric profiles of fractions of non-irradiated cells were averaged to assay the DNA content (Fig. 5.33 A). The DNA content of exponentially growing non-irradiated cells increased from 2.07 to 3.88 C-values (Table 5.6). Typical chromatin structures belonging to fractions isolated from non-irradiated cells will be described in the next paragraphs. The separation of irradiated cells (24 J/ m^2) was

354   5   Apoptotic Chromatin Changes

Table 5.6 Elutriation of synchronized K562 cells

		Non-irradiated cells		Ultraviolet-irradiated cells	
Fraction #	Flow rate ml/min	C-value	Name of chromatin structure	C-value	Name of chromatin structure
2	21.5	2.07.	n.s.	n.d.	decondensed fibrillary
4	23.5	2.20	decondensed chromatin	n.d.	polarized fibrillary
6	25.5	2.49	supercoiled chromatin	n.d.	fibrous
8	27.6	2.74	fibrous chromatin	n.d.	coiled fibrous
10	29.7	2.97	chromatin fiber	n.d.	coiled fibrous
12	31.9	3.24	chromatin bodies	n.d.	coiled fibrous
14	34.0	3.50	elongated prechromosomes	n.d.	coiled fibrous
16	36.1	3.72	precondensed chromosomes	n.d.	coiled fibrous
18	38.3	3.88	condensed chromosomes	n.d.	coiled fibrous

Cells were grown from an initial cell culture density of 1.5×10^5/ml in 1200 ml RPMI medium for 15 h (37°C, 5% CO_2) and harvested at 2.4×10^5/ml by centrifugation (600 g, 5 min, 4°C). Fifty percent of the cells were resuspended in 20 ml RPMI medium containing 1% fetal bovine serum and introduced in the elutriation chamber at 19 ml/ml flow rate. The other half of the cell population was irradiated with 24 J/m^2 ultraviolet light and then subjected to elutriation. Elutriation was performed in a Beckman J21 centrifuge at 20°C and 1850 rpm. Cell number loaded into the elutriation chamber: 1.8×10^8 sejt.
n.d. = not determined
n.s. = not specified.
Reproduced with permission of Ujvarosi et al. (2007).

carried out in the same way, but their C-values could not be determined due to their variation. The estimated C-values of fractions of irradiated cells ranged between 2.2 and 2.4 C-values (Fig. 5.33 B), corresponding to fibrillary chromatin structures seen in non-irradiated cells.

Interphase Chromatin Structures in Non-irradiated Cells

Reversible permeabilization was used to restore cellular membranes which allowed to open the nucleus throughout the cell cycle. Control experiments included cells: a. without permeabilization, b. after permeabilization without reversal of permeabilization as described earlier (Nagy et al., 2004).

The gradual change in shape of chromatin structures from 2.0 to 3.7 C-values is summarized in Fig. 5.34 (Elutriation fractions 1–16). Decondensed chromatin appears as a veil-like fibrillary structure (Fraction 1–2), which becomes polarized with oval or elongated shapes (Fraction 3–4). Elongated chromatin gradually turns to supercoils (Fractions 5–6). The fibrillary structure changes to fibrous forms (Fractions 7–8) showing the linear connection between early decondensed chromosomes (Fractions 9–10). Supercoiled loops form chromatin bodies, which are regarded as the earliest visibile forms of interphase chromosomes (Fraction 11–12). The continuity is maintained in precondensed bent chromosomal forms (Fraction 13–14). Bent forms open up to linear structures, which did not reach the compactness of metaphase chromosomes (Fraction 15–16).

Changes in Chromatin Structure After UV Irradiation

Chromatin changes occured only occasionally at lower doses of UV light (6, $12 \, J/m^2$), and due to their infrequency we did not visualize them. High UV light dose ($24 \, J/m^2$) manifested mainly as an increased fibrillary cloud covering the condensing chromatin structure. As a result chromatin structures became blurred (Fig. 5.35).

At the unset of S-phase fibrillary structures are more decondensed than in the same fraction of non-irradiated cells and polarization of chromatin starts earlier (Fig. 4, Fraction 1–2). Elongated chromatin forms appear earlier (Fig. 5.35, Fractions 3–4), but the elongation process seems to be arrested (Fig. 5.35, Fraction 5–6). Fibrous structures covered with a fine fibrillary network dominate the pictures from fraction 5 without significant progression in supercoiling which would lead to more compact structures. Although, small apoptotic bodies were observed occasionally (Fig. 5.35, Fraction 9 and 10), but contrary to gamma irradiation (Nagy et al., 2004) they are not typical structures of UV irradiation. Neither chromatin bodies, nor other typical chromosomal forms could be observed after UV irradiation. Primitive, early forms of chromatin condensation were covered with faint fibrillary chromatin. Chromatin condensation did not reach the stage of visible chromosomes (Fractions 11–16).

Final Stage of Chromatin Condensation After UV Irradiation

Fluorescent images of chromatin structures of the last elutriation fractions were compared in non-irradiated and irradiated cells (Fig. 5.36). In the last fraction of non-irradiated (normal) cells chromatin condensation reached its final stage, occasionally metaphase (Fig. 5.36A). After UV irradiation ($24 \, J/m^2$) only partial segregation of chromosome domains could be seen, fully condensed chromosomes were not visible. The stickiness of the fine fibrillary network covering the incompleteley folded primitive chromosomes is assumed to prevent the opening of the nucleus (Fig. 5.36B).

Concluding remarks. To summarize the results of biochemical changes upon Ultraviolet B irradiation in K562 cells these studies proved that replicative DNA synthesis in permeable K562 cells is:

a. an ATP-dependent process,
b. reversible permeabilization maintains cell viability,
c. UV irradiation inhibits replicative DNA synthesis and favours repair DNA synthesis,
d. UV irradiation and reversal of permeabilization do not immediately cause apoptotic shrinkage of cells.

Earlier experiments on DNA replication in ultraviolet-irradiated HeLa cells have shown that the extent of inhibition varied with time. At doses less than $10 \, J/m^2$ the rate was initially depressed but later showed some recovery. At higher doses,

356 5 Apoptotic Chromatin Changes

a constant, low rate of synthesis was seen for at least 6 h (Edenberg, 1976). Our experiments confirm these observations, and the higher UV resistance of K562 cells. The lower UV doses which allowed recovery turned out to be 2, 5 and 15 J/m² for K562 cells. After these treatments the repair mechanisms of K562 cells were able to remove at least partially DNA damages, while at high dose (25 J/m²) K562 cells were unable to recover and deemed to undergo apoptosis. This probably means that DNA damages of short exposures or low concentrations of genotoxic agents can either be repaired or cause delayed apoptotic death.

Summarizing the Apoptotic Changes of Genotoxic Treatments

Morphological studies after genotoxic treatments suggest that the consequences of various chromatin injuries can be categorized based on the assessment of injury-specific chromatin changes. In a broader sense one can characterize external apoptotic changes caused by genotoxic agents and classify them according to their structural chromatin changes. Since the potential to distinguish among different chromatotoxic effects is of diagnostic significance, we have started to determine and systematize the effects of cadmium treatment (Bánfalvi et al., 2005), gamma irradiation (Nagy et al., 2004) and UV irradiation (Ujvarosi et al., 2007). Genotoxic treatments may have multiple effects on different cell lines, therefore two different cell lines (Chinese hamster ovary cells and murine preB cells) were used for cadmium treatment and have seen the same large extensive disruptions and holes in the nuclear membrane and sticky incompletely folded chromosomes typical for cadmium treatment (Nagy et al., 2004; Bánfalvi et al., 2007).

Preapoptotic changes upon γ-irradiation in CHO cells manifested as: (a) The cellular and nuclear sizes increased. (b) The DNA content was lower in each elutriated subpopulation of cells. (c) The progression of the cell cycle was arrested in the early S phase at 2.4 C value. (d) The chromatin condensation was blocked between the fibrillar chromatin and precondensed elongated chromosomal forms. (e) The number and size of apoptotic bodies were inversely correlated with the progression of the cell cycle, with many small apoptotic bodies in early S phase and less but larger apoptotic bodies in late S phase (Nagy et al., 2004). Similar observations were made in K562 cells after gamma irradiation (Bánfalvi et al., 2007).

To correlate the presented data with those of others it was reported that UV can induce G_1 or G_2 cell cycle arrest in human keratinocytes and neuroblastoma cells (Gujuluva et al., 1994; Ceruti et al., 2005), while we have observed that chromatin condensation after UV irradiation was arrested in early S phase between 2.2 and 2.4 C-values. This discrepancy can be resolved by the fact that the C-value of G1 (2.0 C) is close to our observation (2.2–2.4 C). However, the small initial rate of DNA replication indicates that replication has started, thus the C-value is higher than 2.0 C. The G2 arrest means that chromosomes could not enter metaphase and condensed chromosomes were not formed after UV irradiation. Indeed, metaphase chromosomes were not visible after irradiation, but this can be explained by the S phase block which is probably maintained throughout the S phase and in G2 phase. Results are in conformity with the view that UV irradiation generates first giant DNA fragments (Higuchi et al., 2003) which could be seen as a fibrillary cromatin cloud. The view of Higuchi et al. (2003) is shared that the absence of apoptotic cells is an indication of the formation of high molecular weight DNA which preceeds ladder-formation of internucleosomal DNA fragmentation associated with apoptosis through caspase activation.

Morphological Similarities and Differences Between UV and Gamma Irradiation

The effects of gamma and UV irradiation on mammalian cells are in compiled Table 5.7.

By summarizing the morphological effects the similarities and differences between ultraviolet light and gamma irradiation are accentuated:

1. UV irradiation did not cause significant changes in cellular or nuclear size,
2. the DNA content expressed in C-value was lower in each synchronized cell population after UV irradiation,
3. the progression of cell cycle was arrested somewhat earlier in S phase (between 2.2 and 2.4 C), than after gamma irradiation (2.4 C),
4. UV irradiation blocked chromatin condensation at its fibrillary stage, nuclear structures were blurred and covered with fibrillary chromatin,
5. although, some apoptotic bodies were seen in mid S phase, they are not typical to UV rather to gamma irradiation,
6. the lack of metaphase chromosomes indicates that UV damage may release fibrillary chromatin at any stage during chromatin condensation and prevent the folding process.

Finally, data on the genotoxic effects of chemical (Trencsenyi et al., 2007), cadmium (Bánfalvi et al., 2005), gamma (Nagy et al., 2004) and ultraviolet light irradiation support the notion that preapoptotic events can be categorized by fingerprinting the injury-specific chromatin changes.

Table 5.7 The effect of gamma and UV irradiation on cells

Biochemical and morphological changes	Gamma irradiation	UVB irradiation
Replicative DNA synthesis	inhibited	inhibited
Repair DNA synthesis	elevated	elevated
Cellular size	increased	increased
Nuclear size	increased	increased
Apoptotic cells (shrinkage)	many	none
C-values during S phase	lowered	uniformly low
Arrest in S phase	2.4 C	2.2–2.4 C
Chromatin stage	fibrous	fibrillary
Apoptotic bodies	many	few
Metaphase chromosomes	visible	invisible

References

Amarante-Mendes, G.P., Naekyung, K.C., Liu, L., Huang, Y., Perkins, C.L., Green, D.R. and Bhalla, K. (1998). Bcr-Abl exerts its antiapoptotic effect against diverse apoptotic stimuli through blockage of mitochondrial release of cytochrome c and activation of caspase-3. Blood. **91**, 1700–1705.

An, W.G., Kanekal, M., Simon, M.C., Maltepe, E., Blagosklony, M.V. and Neckers, L.M. (1998). Stabilization of wild-type p53 by hypoxia-inducible factor 1 alpha. Nature. **392**, 405–408.

Applegate, L.A., Lauscher, P. and Tyrrell, R.M. (1991). Induction of hemeoxygenase: a general response to oxidant stress in cultured mammalian cells. Cancer Res. **51**, 974–978.

Bánfalvi, G., Gacsi, M., Nagy, G., Kiss, Z.B. and Basnakian, A.G. (2005). Cadmium induced apoptotic changes in chromatin structure and subphases of nuclear growth during the cell cycle in CHO cells. Apoptosis. **10**, 631–642.

Bánfalvi, G., Littlefield, N., Hass, B., Mikhailova, M., Csuka, I., Szepessy, E. and Chou, M.W. (2000). Effect of cadmium ont he relationship between replicative and repair DNA synthesis in synchronized CHO cells. Eur J Biochem. **267**, 6580–6585.

Bánfalvi, G., Mikhailova, M. Poirier, L.A. and Chou, M.W. (1997a). Multiple subphases of DNA replication in Chinese hamster ovary (CHO-K1) cells. DNA Cell Biol. **16**, 1493–1498.

Bánfalvi, G., Nagy, G., Gacsi, M., Roszer, T. and Basnakian A.G. (2006). Common pathway of chromosome condensation in mammalian cells. DNA Cell Biol. **25**, 295–301.

Bánfalvi, G., Poirier, L.A., Mikhailova, M. and Chou, M.W. (1997b). Relationship of repair and replicative DNA synthesis to cell cycle in Chinese hamster ovary (CHO-K1) cells. DNA Cell Biol. **16**, 1155–1160.

Bánfalvi, G., Sooki-Toth, A., Sarkar, N., Csuzi, S. and Antoni, F. (1984). Nascent DNA synthesized reversibly permeable cells of mouse thymocytes. Eur J Biochem. **139**, 553–559.

Bánfalvi, G., Ujvarosi, K., Trencsenyi, G., Somogyi, C., Nagy, G. and Basnakian, A.G. (2007). Cell culture density dependent toxicity and chromatin changes upon cadmium treatment in murine pre-B-cells. Apoptosis. **12**, 1219–1228.

Basnakian, A., Bánfalvi, G. and Sarkar, N. (1989). Contribution of DNA polymerase delta to DNA replication in permeable cells synchronized in S phase. Nucleic Acids Res. **17**, 4757–4767.

Bell, S.W., Masters, S.K., Ingram, P., Waters, M. and Shelburne, J.D. (1979). Ultrastructure and x-ray microanalysis of macrophages exposed to cadmium chloride. Scan Electron Microsc. **3**, 111–121.

Bertin, G. and Averbeck, D. (2006). Cadmium: cellular effects, modifications of biomolecules, modulation of DNA repair and genotoxic consequences (a review). Biochimie. **88**, 1549–1559.

Bogdanov, K.V., Chukhlovin, A.B., Zaritskey, A.I., Frolova, O.I. and Afanasiev, B.V. (1997). Ultraviolet irradiation induces multiple DNA double-strand breaks and apoptosis in normal granulocytes and chronic myeloid leukaemia blasts. Brit J Haemat. **98**, 869–872.

Bradbury, E.M. (1992). Reversible histone modifications and the chromosome cell cycle. Bioessays. **14**, 9–16.

Bronstein, I.B., Vorobyev, S., Timofeev, A., Jolles, C.J., Alder, S.L. and Holden, J.A. (1996). Elevations of DNA topoisomerase I catalytic activity and immunoprotein in human malignancies. Oncol Res. **8**, 17–25.

Buendia, B., Santa-Maria, A. and Courvalin, J.C. (1999), Caspase-dependent proteolysis of integral and peripheral proteins of nuclear membranes and nuclear pore complex proteins during apoptosis. J Cell Sci. **112**, 1743–1753.

Ceruti, J.M., Scassa, M.E., Flo, J.M., Varone, C.L. and Canepa, E.T. (2005). Induction of p19INK4d in response to ultraviolet light improves DNA repair and confers resistance to apoptosis in neuroblastoma cells. Oncogene. **24**, 4065–4080.

Cervera, J., Alamar, M., Matinez, A. and Renau-Piqueras, J. (1983). Nuclear alterations induced by cadmium chloride and L-canavanine in HeLa S3 cells. Accumulation of perichromatin granules. J Ultrastruct Res. **82**, 241–263.

Chávez, E. and Holguin, J.A. (1988). Mitochondrial calcium release as induced by Hg^{2+}. J Biol Chem. **263**, 3582–3587.

References 359

Chen, P., Lavarone, A., Fick, J., Edwards, M., Prados, M. and Israel, M.A. (1995). Constitutional p53 mutations associated with brain tumours in young adults. Cancer Genet Cytogenet. **82**, 106–115.

Cheung, W.L., Ajiro, K., Samejima, K., Kloc, M., Cheung, P., Mizzen, C.A., Beeser, A., Etkin, L.D., Chernoff, J., Earnshaw, W.C. and Allis, C.D. (2003). Apoptotic phosphorylation of histone H2B is mediated by mammalian sterile twenty kinase. Cell. **113**, 507–517.

Cleaver, J.E. (1982). Normal reconstruction of DNA supercoiling and chromatin structure in Cockayne syndrome cells during repair of damage from ultraviolet light. Am J Hum Gent. **34**, 566–575.

Cosson, R.P. (1994). Heavy metal intracellular balance and relationship with metallothionein induction in the liver of carp after contamination by silver, cadmium and mercury following or not pretreatment by zinc. BioMetals. **7**, 9–19.

Csuka, I. and Bánfalvi, G. (1997). Analysis of 5'-termini of early intermediates of Okazaki fragments accumulated after emetine treatment in mice. DNA Cell Biol. **16**, 979–984.

Damiano, J.S., Hazlehurst, L.A. and Dalton, W.S. (2001). Cell adhesion-mediated drug resistance (CAM-DR) protects the K562 chronic myelogenous leukemia cell line from apoptosis induced by BCR/ABL inhibition, cytotoxic drugs, and γ-irradiation. Leukemia. **15**, 1232–1239.

Devary, Y., Gottlieb, R. A., Lau, L F. and Karin, M. (1991). Rapid and preferential activation of the c-jun gene during the mammalian UV response. Mol Cell Biol. **11**, 2804–2811.

Dhawan, D. and Goel, A. (1995). Further evidence for zinc as a hepatoprotective agent in rat liver toxicity. Experimental Molecular Pathology **63**, 110–117.

Di Sant'Agnese, P.A., Jensen, K.D., Levin, A. and Miller, R.K. (1983). Placental toxicity of cadmium in the rat: an ultrastructural study. Placenta. **4**, 149–163.

Dong, S., Shen, H.M. and Ong, C.N. (2001). Cadmium-induced apoptosis and phenotypic changes in mouse thymocytes. Mol Cel Biochem. **222**, 11–20.

Edenberg, H.J. (1976). Inhibition of DNA replication by ultraviolet light. Biophys J. **16**, 849–860.

el Azzouzi, B., Tsangaris, G.T., Pellegrini, O., Mauel, Y., Benveniste, J. and Thomas, Y. (1994). Cadmium induces apoptosis in a human T cell line. Toxicology. **88**, 127–139.

Elia, M.C. and Bradley, M.O. (1992). Influence of chromatin structure on the induction of double strand breaks by ionizing radiation. Cancer Res. **52**, 1580–1586.

Fang, M.Z., Mar, W. and Cho, M.H. (2002). Cadmium affects genes involved in growth regulation during two-stage transformation of Balb/3T3 cells. Toxicology. **177**, 253–265.

Fighetti, M.A., Miele, M., Montella, A., Desole, M.S., Congiu, A.M. and ANANIA, V. (1988). Possible involvement of nuclei in cadmium-induced modifications of cultured cells. Arch Toxicol. **62**, 476–478.

Fornace, A.J.Jr., Alamo, I. Jr. and Hollander, M.C. (1988) DNA damage inducible transcripts in mammalian cells. Proc Natl Acad Sci USA. **85**, 8800–8804.

Fornace, A.J.Jr., Nebert, D.W., Hollander, M.C., Luethy, J.D., Papathanasiou, M., Fargnoli, J. and Holbrook, N.J. (1989a) Mammalian genes coordinately resulted by growth arrest signals and DNA-damaging agents. Mol Cell Biol. **9**, 4196–4201.

Fornace, A.J.Jr., Zmudka, B.Z., Hollander, M.C. and Wilson, S.H. (1989b). Induction of fi-polymerase mRNA by DNA damaging agents in Chinese hamster ovary cells. Mol Cell Biol. **9**, 851–853.

Friedberg, E.C., Walker, G.C. and Siede, W. (1995). DNA Repair and Mutagenesis, ASM Press Washington DC. pp. 1–58.

Gacsi, M., Nagy, G., Finter, G., Basnakian, A.G. and Bánfalvi, G. (2005). Condensation of interphase chromatin in nuclei of synchronized chinese hamster ovary (CHO-K1) cells. DNA Cell Biol. **24**, 43–53.

Gill, T.S. and Pant, J.C. (1985). Erythrocytic and leukocytic responses to cadmium poisoning in a freshwater fish, *Puntius concaonius Ham.* Environ Res. **36**, 327–337.

Gottlieb, T.M. and Oren, M. (1998). p53 and apoptosis. Seminars in Cancer Biology. **8**, 359–368.

Graeber, T., Osmanian, C., Jacks T., Housman, D., Koch, C., Lowe, S.W. and Giaccia, A.J. (1996). Hypoxia-mediated selection of cells with diminished apoptotic potential in solid tumours. Nature. **379**, 88–91.

360   5 Apoptotic Chromatin Changes

Grimmer, G., Dettbarn, G., Seidel, A. and Jacob, J. (2000). Detection of carcinogenic aromatic amines in the urine of non-smokers. Sci Total Environ. **247**, 81–90.

Gujuluva, C.N., Baek, J.H., Shin, K.H., Cherrick, H.M. and Park, N.H.(1994). Effect of UV-irradiation on cell cycle, viability and the expression of p53, gadd153 and gadd45 genes in normal and HPV-immortalized human oral keratinocytes. Oncogene. **9**, 1819–1827.

Haas-Kogan, D.A., Yount, G., Haas, M., Levi, D., Kogan, S.S., Hu, L., Vidair, C., Deen, D.F., Dewey, W.C. and Israel, M.A. (1996). p53-dependent G1 arrest and p53-independent apoptosis influence the radiobiologic response of glioblastoma. Int J Radiat Oncol Biol Phys. **36**, 95–103.

Hall, P.A., Mckee, P.H., Menage, H.D-P., Dover, R. and Lane, D.P. (1993). High levels of p53 protein in UV-irradiated normal human skin. Oncogene. **8**, 203–207.

Hamada, T., Tanimoto, A. and Sasaguri, Y. (1997). Apoptosis induced by cadmium. Apoptosis. **2**, 359–367.

Hamer, D.H. (1986). Metallothionein. Annu Rev Biochem. **55**, 913–951.

Hart, B.A., Lee, C.H., Shukla, G.S., Shukla, A., Osier, M., Eneman, J.D. and CHIU, J.F. (1999). Characterization of cadmium-induced apoptosis in rat lung epithelial cells: evidence for the participation of oxidant stress. Toxicology. **133**, 43–58.

Hartwell, L.H. and Weinert, T.A. (1989). Checkpoint: controls that ensure the order of cell cycle events. Science. **246**, 629–634.

Hartwig, A. (1995). Current aspects in metal genotoxicity. Biometals. **8**, 3–11.

Hartwig, A., Asmuss, M., Ehleben, I., Herzer, U., Kostelac, D., Pelzer, A., Schwerdtle, T., Bürkle, A. (2002). Interference by Toxic Metal Ions with DNA Repair Processes and Cell Cycle Control: Molecular Mechanisms. Environ Health Perspect. **110** Suppl. 5, 797–799.

Higuchi, Y., Mizukami, Y. and Yoshimoto, T. (2003). Ultraviolet ray induces chromosomal giant DNA fragmentation followed by internucleosomal DNA fragmentation associated with apoptosis in rat glioma cells. Ann N Y A C. **1010**, 326–330.

Hirano, T., Ueda, H., Kawahara, A. and Fujimoto, S. (1991). Cadmium toxicity on cultured neonatal rat hepatocytes: biochemical and ultrastructural analyses. Histol Histopathol. **6**, 127–133.

Hoeijmakers, J.H. (2001). Genome maintenance mechanisms for preventing cancer. Nature. **411**, 366–374.

Huang, L., Dong, L., Chen, Y., Qi, H. and Xiao, D. (2006). Effects of sinusoidal magnetic field observed on cell proliferation, ion concentration, and osmolarity in two human cancer cell lines. Electromagn Biol Med. **25**, 113–126.

International Agency for Research on Cancer (1993). Berryllium, cadmium, mercury and exposures in the glass manufacturing industry. International Agency for Research on Cancer Monographs on the Evaluation of Carcinogenic Risks to Humans. Lion: IARC. Scientific Publications. **5**, 119–237.

Jin, T., Lu, J.and Nordberg, M. (1998). Toxicokinetics and biochemistry of cadmium with special emphasis on the role of metallothionein. Neurotoxicology. **19**, 529–535.

Jakszyn, P. and Gonzalez, C.A. (2006). Nitrosamine and related food intake and gastric and oesophageal cancer risk: a systematic review of the epidemiological evidence. World J Gastroenterol. **12**, 4296–4303.

Kang, Y.Y. (1997). Cellular and molecular mechanisms of metal toxicities. In: E.J. Massaro (ed), Handbook of Human Toxicology, CRC Press, Boca Raton, FL, pp. 256–284.

Kartasova, T. and van de Putte, P. (1988). Isolation, characterization, and UV stimulated expression of two families of genes encoding polypeptides of related structure in human epidermal keratinocytes. Mol Cell Biol. **8**, 2195–2203.

Kasprzak, K.S. (1995). Possible role of oxidative damage in metal induced carcinogenesis. Cancer Invest. **13**, 411–430.

Kastan, M.B., Onyekwere, O., Sidransky, D., Vogelstein, B. and Craig, R.W. (1991). Participation of p53 protein in the cellular response to DNA damage. Cancer Res. **51**, 6304–6311.

Kastan, M.B., Zhan, Q., El-Deiry, W.S., Carrier, F., Jacks, T., Walsh, W.V., Plunkett, B.S., Vogelstein, B. and Fornace, A.J. (1992). A mammalian cell cycle checkpoint pathway utilizing p53 and GADD45 is defective in ataxia-telangiectasia. Cell. **71**, 587–597.

References

Keith, L.H. and Telliard, W.A. (1979). Priority pollutants: a perspective view, Environmental Science Technology. **13**, 416–423.

Kimura, Y., Sugimoto, C., Matsukawa, S., Sunaga, H., Igawa, H., Yamamoto, H., ITO, T., Saito, H. and Fujieda, S. (2004). Combined treatment of cisplatin and overexpression of caspase-activated deoxyribonuclease (CAD) promotes apoptosis *in vitro* and *in vivo*. Oral Oncol. **40**, 390–399.

Klein, E., Ben-Bassat, H., Neumann, H., Ralph, P., Zeuthen, I., Polliack, A. and Vanky, F. (1976). Properties of the K562 cell line, derived from a patient with chronic myeloid leukemia. Int J Cancer. **18**, 421–431.

Kreuzer, M., Kreienbrock, L., Muller, K.M., Gerken, M. and Wichmann, E. (1999). Histologic types of lung carcinoma and age at onset. Cancer Phila. **85**, 1958–1965.

Kuerbitz, S.J., Plunkett, B.S., Walsh, W.V. and Kastan, M.B. (1992). Wild-type p53 is a cell cycle checkpoint determinant following irradiation. Proc Natl Acad Sci USA. **89**, 7491–7495.

Kunze, N., Yang, G.C., Dolberg, M., Sundarp, R., Knippers, R. and Richter, A. (1991). Structure of the human type I DNA topoisomerase gene. J Biol Chem. **266**, 9610–9616.

Larsen, A.K. and Gobert, C. (1999). DNA topoisomerase I in oncology: Dr Jekyll or Mr Hyde. Pathol Oncol Res. **5**, 171–178.

Leonard, S.S., Harris, G.K. and Shi, X. (2004). Metal-induced oxidative stress and signal transduction, Free Radical Biology and Medicine. **12**, 1921–1942.

Liedberg, B., Nylander, C.I. and Lundstrom, L. (1983). Surface plasmon resonance for gas detection and biosensing. Sens Actuators B Chem. **4**, 299–304.

Lin, C.S., Goldthwait, D.A. and Samols, D. (1990). Induction of transcription from the long terminal repeat of Moloneymurine sarcoma provirusby UV irradiation, X-irradiation, and phorbol ester. Proc Natl Acad Sci USA. **87**, 36–40.

Liu, J., Kershaw, W.C. and Klaassen, C.D. (1991). The protective effect of metallothionein on the toxicity of various metals in rat primary hepatocytes. Toxicol Appl Pharmacol. **107**, 27–34.

Lu, X. and Lane, D.P. (1993). Differential induction of transcriptionally active p53 following UV or ionizing irradiation: Defects in chromosome instability syndromes? Cell. **75**, 765–778.

Martin, S.J. and Cotter. T.G. (1991). Ultraviolet B irradiation of human leukemia HL-60 cells *in vitro* induces apoptosis. Int J Radiat Biol. **59**, 1001–1016.

Mascetti, G., Carrara, S. and Vergani, L. (2001). Relationship between chromatin compactness and dye uptake for *in situ* chromatin stained with DAPI. Cytometry. **44**, 113–119.

Mikhailov, A. and Rieder, C.L. (2002). Cell cycle stressed out of mitosis. Curr Biol. **12**, 331–333.

Miskin, R. and Reich, E. (1980). Plasminogen activator: induction of synthesis by DNA damage. Cell. **19**, 217–224.

Misra, U.K., Gawdi, G. and Pizzo, S.V. (2003). Induction of mitogenic signalling in the 1LN prostate cell line on exposure to submicromolar concentrations of cadmium. Cell Signal. **15**, 1059–1070.

Miyashita, T., Krajewski, S., Krajewski, M., Wang, H.G., LIN, H.K., Liebermann, D.A., Hoffman, B. and Reed, J.C. (1994). Tumour suppressor p53 is a regulator of bcl-2 and bax gene expression *in vitro* and *in vivo*. Oncogene. **9**, 1799–1805.

Moffatt, P., Plaa, G.L. and Denizeau, F. (1996). Rat hepatocytes with elevated metallotionein expression are resistant to N-methyl-N'-nitro-N-nitrosoguanidine cytotoxicity. Toxicol Appl Pharmacol **136**, 200–207.

Morselt, A.F., Peereboom-Stegeman, J.H., Jongstra-Spaapen, E.J. and James, J. (1983). Investigation of the mechanism of cadmium toxicity at cellular level. I. A light microscopical study. Arch Toxicol. **52**, 91–97.

Murray, A.W. (1992). Creative blocks: cell-cycle checkpoints and feedback control. Nature. **359**, 599–604.

Mylliperkiö, M.H., Koski, T.R., Vilpo, L.M. and Vilpo, J.A. (1999). Gamma-irradiation-induced DNA single- and double-strand breaks and their repair in chronic lymphoid leukemia cells of variable radiosensitivity. Hematol Cell Ther. **41**, 95–103.

Nagata, S., Nagase, H., Kawane, K., Mukae, N. and Fukuyama, H. (2003). Degradation of chromosomal DNA during apoptosis. Cell Death Differ. **10**, 108–116.

Nagy, G., Gacsi, M., Rehak, M., Basnakian, A.G., Klaisz, M. and Bánfalvi, G. (2004). Gamma irradiation-induced apoptosis in murine pre-B cells prevents the condensation of fibrillar chromatin in early S phase. Apoptosis. **9**, 765–776.

Nelson, W.G. and Kastan, M. (1994). DNA strand breaks-the DNA template alteration that triggers p53 dependent DNA damage response pathways. Mol Cell Biol. **14**, 1815–1823.

Nordberg, G.F., Jin, T., Nordberg, M. (1994). Subcellular targets of cadmium nephrotoxicity: cadmium binding to renal membrane proteins in animals with or without protective metallothionein synthesis. Environ Health Perspect. **102** Suppl. 3, 191–194.

Offer, H., Zurer, I., Bánfalvi, G., Rehak, M., Falcovitz, A., Milyavsky, M., Goldfinger, N. and Rotter, V. (2001). p53 modulates base excision activity in a cell cycle-specific manner after genotoxic stress. Cancer Res. **61**, 88–96.

Oh, S.H. and Lim, S.C. (2006). A rapid and transient ROS generation by cadmium triggers apoptosis via caspase-dependent pathway in HepG2 cells and this is inhibited through N-acetylcysteine-mediated catalase upregulation, Toxicol Appl Pharmacol. **212**, 212–223.

Ord, M.J., Bouffler, S.D. and Chibber, R. (1988). Cadmium induced changes in cell organelles: an ultrastructural study using cadmium sensitive and resistant muntjac fibroblast cell lines. Arch Toxicol. **62**, 133–145.

Paragh, G., Foris, G., Paragh, G. Jr., Seres, I., Karanyi, Z., Fulop, P., Balogh, Z., Kosztaczky, B., Teichmann, F. and Kertai, P. (2005). Different anticancer effects of fluvastatin on primary hepatocellular tumours and metastases in rats. Cancer Lett. **222**, 17–22.

Pavelic, Z.P., Porter, C.W., Allen, L.M., Mihich, E. (1978). Cell population kinetics of fast- and slow-growing transplantable tumours derived from spontaneous mammary tumours of the DBA/2 Ha-DD mouse. Cancer Res. **38**, 1533–1538.

Pulido, M.D. and A.R. Parrish, A.R. (2003). Metal-induced apoptosis: mechanisms. Mutation Research. **533**, 227–241.

Ree, K., Rugstad, H.E. and Bakka, A. (1982). Ultrastructural changes in the nucleus of a human epithelial cell line exposed to cytotoxic agents. The effect of PUVA and cadmium. Acta Pathol Microbiol Immunol Scand [A]. **90**, 427–435.

Rehak, M., Csuka, I., Szepessy, E. and Bánfalvi, G. (2000). Subphases of DNA Replication in *Drosophila* Cells. DNA and Cell Biol. **19**, 607–612.

Rouzaire-Dubois, B., Malo, M., Milandri, J.B. and Dubois, J.M. (2004). Cell size-proliferation relationship in rat glioma cells. Glia. **45**, 249–257.

Rexford, L., Christopher, T.Jr., Matsko, M., Lotze, M.T. and Amoscato, A.A. (1999). Mass Spectrometric Identification of Increased C16 ceramide levels during apoptosis. J Biol Chem. **274**, 30580–30588.

Rothstein, A. (1959). Cell membranes as site of action of heavy metals. Fed Proc. **18**, 1026–1035.

Schmidt-Nielsen, K.S. (1997). Food and fuel. In Animal Physiology, Cambridge University Press. pp. 154–158.

Schwartz, D. and Rotter, V. (1998). p53-dependent cell cycle control: response to genotoxic stress. Semin Cancer Biol. **8**, 325–36.

Shapiro, H.M. (2003). Practical Flow Cytometry (4th ed.) (Wiley-Liss, Hoboken, NJ).

Stacey, N.H. and Klaassen, C.D. (1981). Comparison of the effects of metals on cellular injury and lipid peroxidation in isolated rat hepatocytes. J Toxicol Environ Health. **7**, 139–147.

Strubelt, O., Kremer, J., Tilse, A., Keogh, J., Pentz, R. and Younes, M. (1996). Comparative studies on the toxicity of mercury, cadmium and copper toward the isolated perfused liver. Journal of Toxicology and Environmental Health **47**, 267–283.

Strasser-Wozak, E.M.C., Hartmann, B.L., Geley, S., Sgonc, R., Böck, G., Oliveira Dos Santos, A., Hattmannstorfer, R., Wolf, H., Pavelka, M. and Kofler, R. (1998). Irradiation induced G2/M cell cycle arrest and apoptosis in p53-deficient lymphoblastic leukemia cells without affecting Bcl-2 and Bax expression. Cell Death Differ. **5**, 687–693.

Th'ng, J.P.H. (2001). Histone modifications and apoptosis: Cause or consequence? Biochem Cell Biol. **79**, 305–311.

References

Tishler, R.B., Calderwood, S.K., Coleman, C.N. and Price, B.D. (1993). Increases in sequence specific DNA binding by p53 Following treatment with chemotherapeutic and DNA damaging agents. Cancer Res. 53, 2212–2216.

Tournier, M.F., Sobczak, J., DE Nechaud, B. and Duguet, M. (1992). Comparison of biochemical properties of DNA-topoisomerase I from normal and regenerating liver. Eur J Biochem. 210, 359–364.

Trencsenyi, G., Kertai, P., Somogyi C., Nagy, G., Dombradi, Z., Gacsi, M. and Bánfalvi, G. (2007). Chemically induced carcinogenesis affecting chromatin structure in rat hepatocarcinoma cells. DNA Cell Biol. 26, 649–655.

Trump, B.F., Berezesky, I.K., Smith, M.W., Phelps, P.C. and Elliget, K.A. (1989). The relationship between cellular ion deregulation and acute and chronic toxicity. Toxicol Appl Pharmacol. 97, 6–22.

Ujvarosi, K., Hunyadi, J., Nagy, G., Pocsi, I. and Bánfalvi, G. (2007). Preapoptotic chromatin changes induced by ultraviolet B irradiation in human erythroleukemia K562 cells. Apoptosis. 12, 2089–2099.

Valko, M., H. Morris, H. and M.T. Cronin, M.T. (2005). Metals, toxicity and oxidative stress. Current Medicinal Chemistry 12, 1161–1208.

Valko, M., Rhodes, C.J. Moncol, J., Izakovic, M. and Mazur, M. (2006). Free radicals, metals and antioxidants in oxidative stress-induced cancer, *Chemico-Biological Interactions* 160, 1–40.

Vallee, B.L. and Ulmer, D.D. (1972). Biochemical effects of mercury, cadmium and lead. Annu Rev Biochem. 41, 91–128.

Vineis, P. and Pirastu, R. (1997). Aromatic amines and cancer. Cancer Causes Control. 8, 346–355.

Von Zglinicki, T., Edwall, C., Ostlund, E., Lind, B., Nordberg, M., Ringertz, N.R. and Wroblewski, J. (1992). Very low cadmium concentrations stimulate DNA synthesis and cell growth. J Cell Sci. 103, 1073–1081.

Waisberg, M., Joseph, P., Hale, B. and Beyersmann, D. (2003). Molecular and cellular mechanisms of cadmium carcinogenesis. Toxicology. 192, 95–117.

Watjen, W., Cox, M., Biagioli, M. and Beyersmann, D. (2002). Cadmium-induced apoptosis in C6 glioma cells: mediation by caspase 9-activation. Biometals. 15, 15–25.

Winter, D.B., Gearhart, P.J. and Bohr, V.A. (1998). Homogeneous rate of degradation of nuclear DNA during apoptosis. Nucleic Acids Res. 26, 4422–4425.

Wolffe, A. (1995). Chromatin, Structure and Function. Academic Press Inc New York.

Woo, E.J., Kim, Y.G., Kim, M.S., Han, W.D., Shin, S., Robinson, H., Park, S.Y. and Oh, B.H. (2004). Structural mechanism for inactivation and activation of CAD/DFF40 in the apoptotic pathway. Mol Cell. 14, 531–539.

Woo, E.S. and Lazo, J.S. (1997). Nucleocytoplasmic functionality of metallothionein. Cancer Res. 57, 4236–4241.

Yamaizumi, M. and Sugano, T (1994). UV-induced nuclear accumulation of p53 is evoked through DNA damage of actively transcribed genes independent of the cell cycle. Oncogene. 9, 2775–2784.

Yang, P.M., Chiu, S.J., Lin, K.A. and Lin, L.Y. (2004). Effect of cadmium on cell cycle progression in Chinese hamster ovary cells. Chem Biol Interact. 149, 125–136.

Yount, G.L., Haas-Kogan, D.A., Vidair, C.A., Haas, M., Dewey, W.C. and ISRAEL, M.A. (1996). Cell cycle synchrony unmasks the influence of p53 function on radiosensitivity of human glioblastoma cells. Cancer Res. 56, 500–506.

Zhan, Q., Carrier, F. and Fornace, A.J. (1993). Induction of cellular p53 activity by DNA damaging agents and growth arrest. Mol Cell Biol. 13, 4242–4250.

Abbreviations

A23187 mobile divalent cation ionophore (Mn^{2+}, Ca^{2+}, Mg^{2+})

Ac-DEVD Caspase-3 inhibitor

ACTH Adrenocorticotrope hormon

Ac-YVAD Caspase-1 Inhibitor

ADC AIDS-Dementia-Complex

AIDS Acquired Immune Deficiency Syndrome

AIF Apoptosis-inducing factor

Akt 1 (PKB, B/Akt) Protein kinase B involved in cellular survival pathways, inhibiting *apoptotic* processes

AKT family AKT1, AKT2, AKT3

Allopurinol 4-hidroxipirazolo(3,4-d) pyrimidine

AMP Adenosine-5'-monophosphate

ANA Arabinonucleic acid

ANT Adenine nucleotide translocator

AP-1 Activator protein 1, transcription factor

Ap-DNA apoptotic DNA

Apaf-1 Apoptosis peptide activating factor 1

APL Acute promyelocytic leukemia

Apo2L or TRAIL Tumour necrosis factor-related apoptosis-inducing ligand

Apo3 see DR3

Apo3 ligand TWEAK

ApoE Apolipoprotein E protein

APE1/Ref-1 human AP-endonuclease

APOE Apolipoprotein E gene

AraC Cytosine arabinoside (compound)

ASK1 Apoptosis signal-regulating kinase 1

ATL Adult T-cell leukemia

ATM *Ataxia Telangiectasia* Mutated

ATP Adenosine-5'-triphosphate

ATR A-T and Rad3-related

AZT Azidothymidine

Bad Bcl2-antagonist of cell death

Bax Bcl2-associated X protein

B/Akt see Akt 1, PKB

BBC3 Bcl2 binding component 3

Bcl B cell leukaemia

Bcl-2 B-cell leukemia 2 protein

Bcl-X(L) Bcl X-long, an antiapoptotic protein from Bcl2 family

Bcl-X(S) Bcl leukemia X-short

Bcl-W a Bcl-2 family member that is anti-apoptotic

BDNF Brain-derived neurotrophic factor

BER Base excision repair

bFGF Basic Fibroblast Growth Factor

BH1,2,3,4 Bcl-2 homology 1,2,3,4

Bid Bcl2 Interacting Domain

Bik BCL2-interacting killer (apoptosis-inducing)

Bim BH3-only polypeptide BIM (also known as Bcl2-like 11)

Bir Baculovirus inhibitor of apoptosis

BirC 2 Baculoviral IAP repeat-containing 2

BMK1/ERK5 Big MAP kinase-1/extracellular signal regulated kinase 5

Bok Bcl-2-related ovarian killer

bp Base pair

BrdUTP 5-bromo-2'-deoxyuridine 5'-triphosphate

Abbreviations

BRUCE BIR repeat containing ubiquitin-conjugating enzyme

CAD/DFF40 Caspase Activated DNase

cAMP Cyclic adenosine monophosphate

CaP Prostate cancer

CARD Caspase Recruitment Domains

CARDIAK Card containing interleukin-1 beta converting enzyme (ICE) associated kinase (RIP$_2$, RICK, CCK)

CB Cajal body, or small nuclear ribonucleoprotein (snRNP) foci

CBP CREB binding protein, transcriptional coactivator

CD4 Cluster of differentiation 4, glycoprotein expressed on the surface of T helper cells, regulatory T cells, monocytes, macrophages, and dendritic cells

CD8 Cluster of differentiation 8 transmembrane glycoprotein serving as a co-receptor for the T cell receptor

CD95 Fas receptor mediates apoptotic signaling by Fas-ligand expressed on the surface of other cells

CDC2 Cell division cycle 2 regulator

CDK2 Cyclin-dependent kinase 2

CDK4 Cyclin-dependent kinase 4

CDK6 Cyclin-dependent kinase 6, cell division protein kinase 6, MGC59692, PLSTIRE, Serine/threonine-protein kinase

CDKN1A (Cip1) see p21

cDNA complementary DNA

CDNB 1-Chloro 2,4-dinitrobenzene

C/EBP CAAT/enhancer binding proteins

Ced1 transmembrane protein inducing cell death in *C. elegans,* similar to human SREC

Ced2 gene encodes an Src homology (SH) 2 and 3 containing adaptor in *C. elegans*

Ced3 and **Ced4** proteins induce the death of *C. elegans* cells

Ced9 protein prevents cell death of *C. elegans* cells

C. elegans *Caenorhabditis elegans*

CEM Human acute lymphoblastic T cell leukemia

cIAP1 see p21and CDKN1A**and cIAP2** two closely related members of inhibitor of apoptosis (IAP)

Cip1(CDKN1A) see p21

Cip/Kip family (CKIs) of cyclin-dependent kinases inhibitors

Cisplatin Cis-Diamino dichloroplatin

CHK1 Checkpoint kinase 1

CHK2 kinase Checkpoint kinase 2, a tumour suppressor playing role in DNA damage signaling

CK Creatine Kinase

CK2 Casein kinase II

CML Chronic myeloid leukemia

CMT Cell-mediated cytotoxicity

c-myc Transcription factor

COX Citochrome oxidase

CPP32 Caspase 3

CrmA cytokine response modifier gene encoding the cowpox viral caspase inhibitor protein, a specific inhibitor of the interleukin-1β converting enzyme

CTL Cytotoxic T lymphocytes (CD8$^+$)

CVA Cerebrovascular accident, stroke

Cyt-C Cytochrome c

DAB Diaminobenzidine

Dad-1 an endogenous programmed cell death suppressor in *Caenorhabditis elegans*

dApaf-1 *Drosophila* Apaf-1/CED-4 homolog

Daxx Death-associated protein 6

DCC Deleted for colon carcinoma

DD Death domain

DED Death effector domain

DFF40/CAD DNA fragmentation factor also known as Caspase Activated DNase

DFF 45/ICAD DNA Fragmentation Factor (45 kDa), human homolog to ICAD (from mouse)

DIABLO protein binds to IAP and antagonizes antiapoptotic effect, analogous to proapoptotic *Drosophila* molecules, Grim, Reaper, and HID

Dicer a member of the RNase III family of dsRNA specific endonucleases

Abbreviations

DISC Death Inducing Signal Complex

DMSO Dimethyl sulfoxide

DNA Deoxyribonucleic acid

DNA-PK DNA dependent Proteine Kinase

DNase DNA degrading enzyme

DOC Deoxicholate

DR Death receptor

DR3 (Ws1, Apo3, LARD, TRAMP, TNFSFR12), a member of the death domain-containing tumour necrosis factor receptor (TNFR) superfamily

DR3L Death receptor 3 ligand or TWEAK

Drob-1 *Drosophila* Bax-like Bcl-2 family protein

DTT Dithiothreitol

E2F1 E2F transcription factor 1. Synonymes: PBR3, PRB-binding protein E2F-1, RBAP-1, RBBP3, RBBP-3, RBP3, retinoblastoma-associated protein 1, retinoblastoma-binding protein 3

EDTA Ethylene diamine tetracetic acid

EGTA Ethylene glycol tetraacetic acid

EGF Epidermal growth factor

EGL-1 Egg laying abnormal-1, one of the more than 40 genes affecting egg laying in *Caenorhabditis elegans*

eIF$_2$ eukaryotic initiation factor 2

ELISA Enzyme-linked Immuno Sorbent Assay

EMSA Electrophoretic mobility shift assay

EndoG Endonuclease G

Epo Erythropoetin

ER Endoplasmic reticulum

ERK Extracellular signal-Regulated Kinase

ES Embryonic stem cell

FACS Fluorescence-Activated Cell Sorter

FAD Flavine adenine dinucleotide

FADD/MORT1 Fas Associated Death Domain protein

Fak Focal adhesion kinase

FAK Focal Adhesion Kinase

Fas Cell-surface receptor

FasL Fas death ligand

FAP familial adenomatosis polyposis

FAP Fas associated phosphatase

F-dUTP Fluorescent-tagged with deoxyuridine triphosphate nucleotides

FI TC Fluorescein isothiocyanate

FLICE FADD homologous ICE/CED3-like protease

FLIP FLICE-inhibitory protein

FRET Fluorescense resonance energy transfer

GADD45 Growth Arrest and DNA Damage

GEF Guanine Exachange Factor

GEM Gemini of coiled body

Gln Gluthamine

GMO Genetically modified organism

GNEF – Guanine nucleotide exchange factor

GPCR G Protein Coupled Receptor

Grim a cell death gene in *Drosophila*

GSH Glutathione reduced form

GSSG Glutathione oxidized form

GST Glutation-S-transferase

GTP Guanosine triphosphate

Gy Gray is the SI unit of absorbed radiation dose

H_2O_2 Hydrogen peroxide

hAPE/Ref-1 human apurinic/apyrimidinic endonuclease/redox factor-1

HAT Histone acetyl transferase

HB-EGF Heparin Binding Epidermal Growth Factor

HDAC Histone deacetlylase

Hid – Head involution defective *Drosophila* gene, a direct molecular target of Ras-dependent survival signaling

Abbreviations

HIF-1 Hypoxia-inducible factor-1

HIV-1 Human Immunodeficiency Virus-1, that causes AIDS

HMG2 one of two isozymes of hydroxymethyl glutaryl-CoA (HMG-CoA) reductase that convert HMG-CoA to mevalonate

HPLC High Performance Liquid Chromatography

HPV Human papillomavirus

HR6B Ubiquitin-conjugating DNA repair enzyme

Hrk Harakiri, BCL2 interacting protein (contains only BH3 domain)

HSP family heat shock proteins (from Hsp10 to Hsp110) with chaperon function

HTLV-1 Human T-cell leukemia virus type 1 lymphoma

HX Hypoxanthine

IAP-1 Inhibitor of apoptosis proteins 1

IAP-2 Inhibitor of apoptosis proteins 2

IC50 Quantitative measure indicating how much substance is needed to inhibit growth (or DNA replication in this book)

ICAD Inhibitor of Caspase-Activated DNase

ICE Interleukin-1 β-Converting Enzyme (caspase 1)

I-FLICE Inhibitor of FLICE

IFNs interferons (types I-III)

IFN-γ Interferon-γ

IGF-1 Insuline-like Growth Factor 1

IGFBP Insulin-like Growth Factor Binding Proteins

IGF-1R Insuline-like Growth Factor 1 Receptor

I kappa Bα NF-kappa B inhibitor alfa

IKK IκB kinase

IL-1 Interleukine-1

IL-2 Interleukine 2

IL-6 Interleukine-6

ILMCE Inducible lymphocyte Ca^{2+}/Mg^{2+}-dependent endonuclease

ISEL (LEP) *In Situ* End Labeling

Jak2 Tyrosine kinase

JNK c-Jun N-terminal kinase

JURKAT human, peripheral blood, leukemia, T cell

KAP1 KRAB domain-associated protein 1

kb Kilobase

kDa Kilodalton

kDNA Kinetoplast DNA

LMNA Lamin A

LTα Lymphotoxin alpha (TNF superfamily, member 1)

LTβ Lymphotoxin beta (TNF superfamily, member3). Synonyms: LT-beta, Lymphotoxin-beta, p33, TNFC, TNF-C, TNFSF3, Tumour necrosis factor C, Tumour necrosis factor ligand superfamily member 3

LTRs Long terminal repeats

Lupus erythematosus a chronic autoimmune disease

MAPK Mitogen-activated protein kinase

MAPKK MAPK Kinase

MAPKKK MAPKK Kinase

MCL1 Myeloid cell leukemia sequence 1 (human)

MDM Monocyte-derived macrophages

MDM2 Murine Double Minute2, p53-binding protein

MEKK MAPK/ERK kinase

MFG-E8 Milk fat globule epidermal growth factor

MFGM Milk fat globule membrane

MHC Major histocompatibility complex

MKK Mitogen-activated protein kinase kinase

MLK Mixed lineage kinases

MMP Metaloproteinases from cellular matrix

Mn-SOD Mn-dependent superoxide dismutase

Mort1/FADD – a cytosolic adaptor protein which is critical for signalling from 'death receptors'

MST1 Macrophage stimulating 1 (hepatocyte growth factor-like) also known as HGFL; NF15S2; D3F15S2; DNF15S2

Abbreviations

mRNA Messenger RNA

Mtd/Bok pro-apoptotic regulator abundantly present in bursa of Fabricius

Myc marvelously complex (oncogene)

NAD Nicotinamide adenine dinucleotide

NAIP Neuronal apoptosis inhibitor protein

Nbk/Bik inducer of apoptosis in the presence of mutant p53 and low levels of Bax

NEM N-ethylmaleimide

NEMO NF-kB Essential Modifier

NER Nucleotide excision repair

NF1 Neurofibromatosis type-1

NF-1 Nuclear Factor 1

NF-κB Nuclear Factor kappa B

NGF Nerve growth factor

NIK NF-κB induced kinase

NK cells Natural killer cells

NLS Nuclear location sequence

NM23-H1 a DNA nicking DNAse

NO• Nitric oxide (radical)

NO$_2$• Nitrogendioxide radical

NOS Nitric oxide synthase (e-NOS, NOS endotelial; iNOS, NOS inducible; nNOS, NOS neuronal)

Noxa a pro-apoptotic Bcl2 family protein

NUF7 Nuclear filament-related

Nur 77 Nuclear receptor 77

O$_2$•$^-$ – Superoxide anion (radical)

OD Optical density

OH• Hydroxyl radical

Omi/HtrA2 a mitochondrial serine protease released to antagonize IAPs

ORF Open reading frame

p21 (Cip1, CDKN1A, WAF1) Cyclin-dependent kinase inhibitor 1A

p27 (KIP1) belongs to the family of cell cycle regulators cyclin-dependent kinase inhibitors (CDKI), which bind to "cyclin-CDK" complexes and cause cell cycle arrest in the G1 phase

p21/WAF1/Cip1 Inhibidor of CDK

p38 MAPK p38 mitogen-activated protein kinase

p53 protein 53 (*TP53*), a *transcription factor* regulating the *cell cycle* and functioning as a *tumour suppressor*

PALA N-phosphoacetyl-L-aspartate

PARP Poly(ADP-ribose) polymerase

PBS Phosphate buffered saline

PCA Perchloric acid

PCAF Histone acetylatransferase (p300/CBP-associated factor)

PDK1 3'-phosphoinositide-dependent kinase

PG Prostaglandin

PGF2alfa Prostaglandin F2alpha

PI Propidium iodide

PI3-K Phosphatidylinositol 3-kinase-type protein

PIGs p53-inducible genes

PIP2 Phosphatidylinositol 3,4,5-trisphosphate, the substrate for cleavage with phospholipase C.

PKA Protein kinase A

PKB (Akt1, B/Akt) Protein kinase B or Akt is a serine/threonine kinase

PKC Phospholipid-dependent protein kinase

PKCδ Protein kinase C delta

PLA(2) Phospholipase A (2)

PML Promyelocytic leukemia (protein)

PNA Peptide nucleic acid

PP2A Phosphate protein 2A

pp32 tumour suppressor protein

pRb human retinoblastoma protein

PS Phosphatidylserine

pS2/TFF1 gastric peptide

Abbreviations

pS2/TFF3 intestinal trefoil factor

PT Permeability Transition

PUMA p53 upregulated modulator of apoptosis

Raf – MAP kinase kinase kinase activated by Ras

RAIDD RIP-associated ICH-1/CED-3 homologous protein with DD

Ras – an oncoprotein implicated in development of cancer, small GTPase coupling the activation of growth factor tyrosine kinases to Raf at the beginning of the MAP kinase pathway

Ras-GNEF (Sos, Son-of-sevenless) – nucleotide exchange factor that activates Ras

RCC1 (regulator of chromosome condensation) – Ran guanine nucleotide exchange factor (GNEF) associated with chromatin maintains high nuclear concentration of RanGTP throughout the cell cycle

RB Retinoblastoma susceptibility gene

RES Rough endoplasmic reticulum

RHD Rel homology domain

Rho Homologoue to Ras proto-oncoprotein

RING Really interesting new gene

RING finger domains are defined by the consensus sequence with the Cys and His zinc binding residues

RIP Receptor Interacting Protein

RNA Ribonucleic acid

RNAi molecules capable of mediating RNA interference

RNS Radical Nitrogen Species

ROO• Peroxil radical

ROS Radical or Reactive Oxigen Species

RP-2 STK19P Serine/threonine kinase 19

RP-8 PDCD2 Programmed cell death protein 2

Rpr Reaper, cell death gene in *Drosophila*

rRNA Ribosomal RNA

RSK Ribosomal S6 kinase

SAM S-adenosylmethyonine

SARs Scaffold attached regions

SCF Stem cell factor

SDS Sodium dodecil sulfate

SDS-PAGE Sodium dodecil sulfate polyacrylamide gel electrophoresis

SER Smooth endoplasmic reticulum

SERCA, Sarco(Endo)plasmic **R**eticulum **Ca^{++}-ATP**ase – endoplamic reticulum Ca^{++} pump

sGp Small G-proteins

SGP-2 Sulfated glycoprotein-2

SH2 domain – adapter domain that binds to peptides (containing phosphotyrosine) and is part of many signaling proteins

siNA Short interfering nucleic acid

SINEs Short interspersed nuclear elements

SINV Sindbis virus

siRNA Short interfering RNA

Smac Second mitochondria-derived activator of caspases)/DIABLO

S/MARs Scaffold/matrix attachment regions

SMC Structural maintainance of chromosomes

SMN Survival motor neuron

SNAP-25 soluble N-ethyl maleimide senstive attachment proteins that interact with SNARE

SNARE SNAP receptor (tSNARE, target SNARE; vSNARE, vesicular SNARE)

snRNPs Small nuclear ribonucleoproteins

SOD Superoxide dismutase

SODD Silencer of death domains

Sos (Son-of-Sevenless) proteins that are missing the seven transmembrane α-helices

SP/TFF2 Spasmolytic peptide (trefoil factor)

SRC a Protooncogene tyrosine kinase

SREC Scavenger Receptor from Endothelial Cells, similar to Ced1 in *C. elegans*

STAT Signal transducers and activators of transcription

Abbreviations

STAT3 Signal transducer and activator of transcription 3, mediates the expression of genes in response to cell stimuli, plays role in many cellular processes including cell growth

STAT6 a survival factor in prostate cancer progression regulating the transcriptional program

STH Somatotropin (growth hormone)

SUMO-1 *Small ubiqutin-like modifier-1*

SV40-LT SV40 large T antigen

TAF Transcription initiation factor TFIID

TBS Tris buffered saline

tBid truncated Bid

Tc cells Cytotoxic T cells

TCA Trichloro acetic acid

TCR T cell receptor

TdT Deoxinucleotidyl terminal transferase

TFIID General transcription factor

TFFS Trefoil factor family

TG Thapsigargin, a calcium ATPase inhibitor

TGFα Transforming Growth Factor alfa protein

TGFβ Transforming growth factor beta proteins (*TGF-β1*, *TGF-β2* and *TGF-β3*)

TIMP Metalloproteinase inhibitor 1 precursor

Tm Melting temperature

*TNF Tumour necrosis factor

TNFL TNF ligand

TNF-R1 Tumour Necrosis Factor Receptor 1, also known as p55R, CD120a, FPF; p55; p60; TBP1; TNF-R; TNFAR

TNF-α Tumour necrosis factor α

TPA Tetradecanoil phorbolacetate, tumour promoter

TRADD TNF-Receptor Associated Death Domain

TRAF-1 TNF-receptor associated factor 1

TRAF-2 TNF-receptor associated factor 2

TRAIL (Apo2L) Tumour necrosis factor-related, apoptosis-inducing ligand

TRIS base 2-amino-2-(hidroximethyl)-1,3-propanediol

tRNA Transfer RNA

TRPM/SGP-2 Testosterone repressed message/sulfated

TSH Tyroid stimulating hormone

tTG Tissue Transglutaminase

TUNEL TdT-mediated dUTP Nick-End Labeling

TWEAK TNF-related weak inducer of apoptosis, a potent skeletal muscle-wasting cytokine

UBC Ubiquitin-conjugating enzyme

UVB Ultraviolet B irradiation

uPA Urokinase-type plasminogen activator

VDAC Voltage-Dependent Anion Channel

vFLIP Viral FADD like ICE inhibitory protein

WAF-1 Mediator of p53 tumour suppression

WAP Whey acidic protein

WT Wilms tumour

XDH Xantine dehidrogenase

XIAP X human chromosome-linked inhibitor of apoptosis protein

XO Xantine oxidase

XOR Xantine oxidoreductase

Z-VAD-FMK Caspase inhibitor

* For details regarding abbreviations of the TNF ligand and receptor superfamily members see: Meager, A. (2004). Assays for cytotoxicity. Methods Mol Biol. **249**, 135–152.

Glossary

30 nm fiber – compacted filament of fibrillary chromatin consisting of nucleosomes.

Acetylcholine – neurotransmitter in neuromuscular junctions and other synapses.

Acrocentric chromosomes – those chromosomes, specifically human 13, 14, 15, 21 and 22, that are able to take part in Robertsonian translocations, i.e. when two chromosomes fuse, usually at the centromere.

Acute myocardial infaction (MI, Heart attack) – complete blockage of blood flow in a coronary artery.

Acute promyelocytic leukemia (APL) – characterized by a block in the promyelocyte stage of myeloid development, comprising approximately 10% of myelogenous leukemias.

Adenocarcinoma – malignant epithelial tumour.

Adenoma – benign epithelial tumour (of glandular organization).

Agonist – ligand that activates its receptor.

Alternative splicing – production of more than one gene product (mRNA, protein) from a gene by excluding certain exons or including introns in mRNA.

Amino terminal – free amino end of an oligo- or polypeptide.

Amyloid beta (Aβ or Abeta) peptide of 39–43 amino acids. main constituent of amyloid plaques in the brains of Alzheimer's disease patients.

Allostery – modulation of enzyme activity involving conformational change of protein upon binding of a modulating agent.

Alzheimer's disease – a progressive and fatal brain disease, destroying brain cells, causing problems with memory and thinking.

Aneuploidy – missing or excess chromosomes caused during mitosis or meiosis.

Antiapoptotic – signal or protein protecting against apoptosis.

Antimetabolites – similar to body molecules but slightly different (analogous) in structure.

380   Glossary

Antagonist – ligand exerting inhibition on its receptor.

Apoptotic protease activating factor (Apaf-1) – this protein is activated by cytochrome c released by mitochondria in the intrinsic pathway of apoptosis. Cytochrome in turn binds to procaspase 9 to form the apoptosome.

Apoptin viral protein induces apoptosis in the absence of p53, is stimulated by Bcl-2.

Apoptosis – type I programmed cell death triggered by external stimuli or internal signals manifested as biochemical and morphological changes.

Apoptotic bodies – membrane enclosed remnants of apoptotic cells.

Apoptosomes – caspases involved in the initiation of apoptosis are activated by interaction with large complexes of scaffolding and activating proteins called apoptosomes. Apoptosomes are ring-like structures with seven spokes containing Apaf-1 and procaspase 9 which start the intrinsic (mitochondrial) pathway of apoptosis.

Arabinosylcytosine – nucleotide analogue antibiotic, acts by blocking DNA synthesis.

Archeae (archaebacteria) – the third major domain of organisms, as distinct from true bacteria (eubacteria or prokaryotes) and eukaryotes.

Arachidonate cascade a series of complex biosynthetic pathways producing lipid mediators derived from the oxygenation and/or hydroxylation of polyunsaturated arachidonic acid.

Aromatic compounds – consist of conjugated rings (such as benzene) with delocalized pi electron clouds.

Asp – aspartic acid, amino acid.

Assembly proteins (AP1, AP2) – clathrin constituents that regulate the assembly of clathrin coat.

Ataxia telangiectasia – primary immune deficiency.

Ataxia telangiectasia **mutated (ATM)** – in response to DNA damage this protein kinase activates downstream checkpoint kinases (Chk1 and Chk1).

ATR (ATM and RAD3 related) kinase – responds to the accumulation of single stranded DNA by activating downstream checkpoint kinases (Chk1 and Chk1).

Attenuation – trasncriptional regulation to abort RNA synthesis.

Autoimmunity – the failure of an organism to recognize its own constituent parts ("self") from foreign causing an immune response against its own cells and tissues.

Autonomously replicating sequences (ARS) – short DNA sequences (100–150 bp) acting as replication origins in yeast cells.

Glossary

Autophagy – intracellular degradation of substrates in membrane-bounded compartments.

AZT – azidothymidine, a drug used against human immunodeficiency virus (HIV), the virus that causes AIDS.

Bacteroides – genus of rod-shaped Gram negative bacteria.

Baculoviruses a family of large rod-shaped pathogens that attack insects and other arthropods.

Baculovirus IAP repeat – See BIR.

Bak – Bcl-2 family protein activating the intrinsic pathway of apoptosis by inserting into the inner membrane of mitochondria and releasing pro-apoptotic factors.

Barr body – inactivated, heterochromatized X chromosome.

Base excision repair – the most commonly used pathway to remove incorrect bases (such as uracil formed by the spontaneus oxidative deamination of cytosine) or damaged bases (e.g. methyladenine). Replaces reduced, alkylated or deaminated bases in DNA.

Bax – Bcl-2 family protein activating the intrinsic pathway of apoptosis by inserting into the inner membrane of mitochondria and releasing pro-apoptotic factors.

B-box type family protein – similar to zinc finger protein.

Bcl-2 – apoptosis is controlled through proteins, known as the Bcl-2 family of proteins.

Benign tumour – cells grow and remain together, complete cure can be reached by removing the tumour mass surgically.

BH domain (Bcl-2 homology domain) – conserved sequences defining Bcl-2 family members, present in Bcl-2 family members (BH3), promoting interactions among family members.

BIR domain – is present in all IAP (Inhibitor of Apoptosis) family proteins. BIR (baculovirus IAP repeat) is a baculovirus inhibitor of apoptosis, a caspase inhibitor.

Bloom syndrome – rare autosomal recessive disorder characterized by telangiectases and photosensitivity, growth deficiency of prenatal onset, variable degrees of immunodeficiency, increased susceptibility to neoplasms.

Box(es) – group of nucleotides forming usually consensus sequences with recognizable functions.

Bursa of Fabricius (*Bursa cloacalis* or *Bursa fabricii*) – site of hematopoiesis, a specialized organ that is necessary for B cell development in birds.

Caspase activated DNase (CAD) – nuclease in apoptotic cells degrading chromatin into nucleosomal fragments of $\approx$200 base pairs.

CAD domain – extracellular part of cadherin adhesion proteins.

Cadherin – adhesion proteins binding to similar cadherins on other cells.

Caenorrhabditis elegans – small nematode worm, genetic model organism in developmental biology and apoptosis.

Cajal body (CB) or sn ribonucleoprotein (RNP) foci – suggested to be temporarily associated with chromatin containing actively transcribing genes.

Calpains – Ca^{++}-activated cysteine proteases that cleave intracellular proteins involved in cell motility and adhesion.

Cancer – cells that do not follow the rules of cell proliferation and occupy the place of normal tissues.

Carbohydrates (saccharides) – organic polyhydroxy aldehyde and polyhydroxy ketone compounds.

Carcinogenesis (creation of cancer) – is the process by which normal cells are transformed into cancer cells.

Carcinoma – cancer of epithelial origin.

Cartilage – a type of dense consecutive tissue, composed of collagen and/or elastin fibers, supplying smooth surfaces for the movement of bones.

Caspase recruiting domain (CARD) – protein interaction domain of adapter proteins and caspases involved in cell death.

Caspases (*Cysteine Aspatate Specific ProteASEs*) – ICE-like cysteine proteases involved in the activation and implementation of apoptosis.

Caryolysis – lysis and disappearance of nucleus.

Caryorexis – convolution and breakdown of nucleus.

CD40 – a cell surface transmembrane 45 kDa glycoprotein receptor expressed on B lymphocytes and certain epithelial cells.

CD95 (APO-1/Fas) – is a prototype death receptor characterized by the presence of an 80 amino acid death domain in its cytoplasmic tail.

Cdc25 – protein phosphatases that remove inhibitory phosphatases from Cdk-cyclin complexes to trigger kinase activation.

Cdk (cyclin-dependent kinase) – in association with a cyclin subunit regulates cell cycle progression.

cDNA – DNA copy of mRNA.

Cell death abnormal (Ced) – mutation in *C. elegans* genes affecting programmed cell death.

Glossary

CED-3 protein – shows homology to the family of interleukin-1β-converting enzyme (ICE) which are cysteine proteases involved in the apoptotic signaling through the type 1 TNF receptor.

Cell cycle – growth and division of cells consisting of phases G1, S. G2 and M.

Cellulose – polysaccharide macromolecule consisting of β-D-glucose units.

Centrifuge – instrument generating force to sediment particles in a liquid sample.

Centromere – chromosome locus containing specific DNA sequences that regulates chromosome movement during mitosis and meiosis.

Ceramide – sphingolipide backbone converted to glucosylceramide and sphingomyelin in the Golgi apparatus.

Chaperones – molecular "gardedames", that assist the unfolding of other proteins in the cytosol and help their translocation across the membrane in the lysosomal lumen.

Checkpoint – when activated, blocks progression through the cell cycle temporally or occasionally permanently.

Chemical mutagens – aromatic hydrocarbons, aromatic amines, nitrosamines, alkylating agents (mustard gas), many of them are relatively inert but made damaging by cytochrome P-450 oxidases (fungal toxin aflatoxin B1, benzopyrene in coal tar and in tobacco smoke)

Chk (Chk 1 and 2, checkpoint kinases) – protein kinases activated by ATM and RAD3 related (ATR) kinases to phosphorylate Cdc25A protein phosphatase and other proteins and to block the progression of the cell cycle.

Chondroma, chondrosarcoma – benign and malignant tumours of cartilage.

Chromatin – complex of nucleic acids (DNA or RNA) and proteins (mainly histones) constituting eukaryotic chromosomes; located in the cell nucleus.

Chromatin remodeling – dynamic modulation of chromatin structure occurring throughout the cell cycle, a key component in the regulation of gene expression, apoptosis, DNA replication and repair and chromosome condensation and segregation.

Chromatid – single chromosomal molecule containing DNA and its proteins.

Chromodomain (chromatin modification organizer) – domain of 50 amino acids binding to histone 3 trimethylated on lysine.

Chromonema fiber – theoretical structure 100–300 nm in diameter supposed to be part of the higher-order chromatin structure.

Chromosome – independent inheritable unit of mitosis and meiosis consisting of DNA and attendant proteins.

384   Glossary

Chromosome cycle – replication and division of chromosomes into two daughter cells.

Chronic inflammation – an inflammatory immune response of prolonged duration that eventually leads to tissue damage.

Cis-acting – elements or modulators exerting their effects within the same DNA molecule (e.g. promoters and upstream elements affecting transcription rates).

Comet-assay – single cell based technique allowing to detect and quantitate DNA damages.

Cytokines – signalling compounds produced by animal and plant cells to communicate with one another.

Cytotoxicity – the cell-killing property of a chemical (food, cosmetic, pharmaceutical) or mediator cell (cytotoxic T cell).

CKI (cyclin-dependent kinase inhibitor) – proteins regulating negatively cell cycle kinases (Cdks) to block the progression of the cell cycle.

Codon – three successive nucleic acid bases in mRNA specifying a particular amino acid in the polypeptide chain during protein synthesis on a ribosome.

Coiled coil – structural motif in proteins, alpha-helices coiled together like the strands of a rope.

Colchicin – alkaloide of autumn crocus that inhibits microtubule assembly and blocks chromosome condensation at metaphase.

Common ancestor – true evolutionary relationship, the origin of life some 3.3–4 billion years ago.

Conformational change – change in the shape of a macromolecule owing to the free rotation around single bonds.

Consensus sequence – sequence of nucleotides compiled by the comparison of homologous regions of many genes and giving the most frequently occuring sequence.

Constitutive heterochromatin – inactive chromatin remaining condensed throughout the cell cycle due to the permanent binding to special proteins and histone modifications.

Core histone – histones H2A, H2B, H3 and H4 forming the disk shaped octameric core of nucleosomes.

Coupled reactions – in two consecutive chemical reactions which share a common intermediate, the product of the first reaction is the reactant in the second reaction and energy is transfered from one side to the other. The energy carrier is regularly ATP.

Covalent modification – reversible regulation of enzyme activity by phosphorylation and dephosphorylation, acetylation and deacetylation, methylation

Glossary

and demethylation as the most common but not the only means of covalent modification.

Crossover – physical breakage and reunion of DNA strands on different chromosomes serving the exchange of DNA sequences during recombination.

C-value – haploid genome content per cell.

Cyanobacteria – photosynthesizing blue-green algae with both types of photosystems and manganese enzyme, catalysing the splitting of water and producing carbohydrates and molecular oxygen.

Cyclins – subunits of cyclin-dependent kinases that undergo cycling patterns of accumulation and degradation during the cell cycle.

Cyclin-dependent kinase – See Cdk.

Cytochrome c – heme protein that is part of the electron transport chain and when released from mitochondria triggers intrinsic (mitochondrial) apoptosis.

Cytochrome p450 – consisting of enzymes involved in the biotransformation of xenogenes (foreign substances) in the smooth endoplasmic reticulum.

Cytokines – diverse family of protein hormones and growth factors produced by phagocytes in response to immune stimuli, inflammation, hemopoesis.

Cytokine receptor – cell surface membrane recptors for cytokines linked to JAK kinases inside the cell.

Cytokinesis – cell division at the end of mitosis.

Dark energy – hypothetical form permeating through the space and increasing the rate of expansion of the universe.

Dark matter (astrophysics) – hypothetical form of matter of unknown composition emitting or reflecting electromagnetic radiation.

Daxx – a nuclear protein involved in apoptosis and transcriptional repression, interacting with the death receptor Fas, with promyelocytic leukemia protein (PML), and several transcriptional repressors.

Death domain (DD) – domain of protein interactions in cell death signaling pathways, including Fas receptors and FADD adapters.

Death effector domain (DED) – interaction of adapter proteins, e.g. FADD, caspase 8 and 10 prodomains or certain inhibitors of apoptosis.

Death inducing signaling complex (DISC) – complex of proteins that forms upon triggering of CD95. The DISC consists of an adaptor protein and initiator caspases and is essential for induction of apoptosis.

de novo **chromosome abnormality** – occurring in an individual and not inherited from the parents.

Deoxiribonucleotide – building block of DNA consisting of sugar (deoxyribose), phosphate and base (A, G, C or T).

Dephosphorylation – enzymatic removal of phosphate moiety from protein side chains of amino acids.

Diabetes – a disease in which the blood glucose, or sugar levels are too high.

Dictyostelium discoideum – slime mold, model organism to study chemotaxis, cell motility and differentiation.

Diploid chromosome number (2n) – total chromosome number in a diploid organism, including pairs of chromosomes of maternal and paternal origin.

DISC – see death-inducing signaling complex.

Disjunction – separation of chromosomes or chromatids in meiosis or mitosis.

DNA – deoxyribonucleic acid, informational macromolecule consisting of deoxynucleotide (dNMP) units constituting the genetic information of organisms.

DNA damage checkpoints – pathways that detect DNA damage by either blocking the progression of the cell cycle or triggering apoptosis.

DNA laddering – endonuclease activation during apoptosis degrades genomic DNA at internucleosomal linker regions and produces ≈ 180 base-pair long DNA fragments.

DNA repair mechanisms – the three major DNA repairing mechanisms: base excision, nucleotide excision and mismatch repair.

DNA replication – complete synthesis of complementary strands of DNA double helix to duplicate the genome.

DNA replication checkpoints – mechanisms to detect unreplicated DNA and stabilize stalled replication forks for DNA repair.

DNA topoisomerase IIα – mitotic scaffold enzyme catalyzing the alteration of tertiary structure of DNA (topology) by passing double stranded DNA through another duplex.

Dolichol phosphate – pyrophosphorylated unsaturated isoprenoid alkohol serving as substrate in the core glycosylation and subsequent transfer of oligosaccharides to the asparagine residues of proteins in the lumen of the endoplasmic reticulum.

Down syndrome – human aneuploidy with three copies of chromosome 21.

Drosophila melanogaster – fruit fly, genetic model organism for studying development.

EF2 – family of mammalian transcription factors that regulate genes involved in cell cycle progression and can trigger apoptotic cell death.

Glossary

Effector caspases – belong to "downstream" caspases activated by the cleavage of initiator caspases during the intracellular proteolysis leading to apoptosis.

Eicosanoids – lipid seconder messengers originating from arachidonic acid.

Electric potential – voltage difference across a membrane, calculated by the Nernst equation and expressed in mV.

Electron transport pathway – biochemical pathway in the inner membrane of mitochondria and bacterial plasma membrane by passage of electrons and creating a proton gradient across the membrane generating chemiosmotic energy for ATP synthesis.

Electrostatic interaction – attraction between oppositely charged atoms.

Embrionic stem cells – cells of the first embrionic cell divisions, precursors of the entire embrio.

Endergonic reaction – chemical reaction that consumes energy (nonspontaneous).

Endocytosis – the capture of extracellular materials enclosed in membrane-bound vesicles that invaginate and are released inside the cytoplasm.

Endoplasmic reticulum (ER) – intracellular compartment collecting proteins synthesized on ribosomes on its surface, modifying and delivering proteins into the secretary pathway. The lipid bilayer of ER is continuous with the outer nuclear membrane sharing its functions.

Enhancers – regulator DNA sequences found mainly in eukaryotes helping to upregulate transcription of neighbouring genes, they can also exert their effect over long distances of intervening DNA located upstream or downstream of the gene in either direction.

Enthalpy (H, heat content) – internal energy (U) of a system plus the product of its pressure (p) times its volume (V) $(H = U + pV)$.

Entropy – the measure of molecular disorder in a thermodynamic system.

Eosinophyl – type of white blood cells the large granules of which stain with eosin. Eosinophyl granulocytes are active against parasites.

Epigenetic change – change of gene expression without the change of DNA, e.g. inactivation of X chromosome as Barr body.

Epigenetic trait – modification of DNA or proteins associated with DNA by enzymes which provide an inheritable property.

Erythropoietin (EPO) – cytokine produced by the kindney, initiates red blood cell production. EPO, and transfusion blood was used for doping. There is now an accurate urine test that can detect the differences between normal and synthetic EPO.

Escherichia coli – Gram-negative colon bacterium, genetic model organism.

Euchromatin – transcriptionally active chromatin, contains most of the structural genes to be transcribed.

Eukaryote – organism with the genome packaged within the nucleus. Most of the organsims apart from bacteria and certain algae (cyanobacteria) are eukaryotes.

Evolutionary convergence – organism evolved similar structures in response to similar ways of life.

Exergonic reaction – free energy is released during the chemical reaction which can be used to do biological work, the free energy of the final state is lower than the initial state (spontaneous reaction). Enzymes increase the low rate of exergonic reactions.

Exons – parts of eukaryotic split genes coding for RNA which will be spliced into mature RNA molecules.

Extrinsic pathway of cell death – initiated when cell surface ligands bind to and activate their receptors triggering apoptosis in target cells.

Fas-associated protein with a death domain (FADD) – also known as MORT1, is a common adapter protein in both CD95-mediated and TNF-R-mediated apoptosis that promotes association of activated Fas receptor with procaspase 8, triggering extrinsic cell death.

Fanconi anaemia – inherited bone marrow failure.

Fas receptor (Apo1 or CD95) – a death receptor of the tumour necrosis factor (TNF) (TNF) family of receptors carrying a death domain at the cytoplasmic end. Mediates apoptotic signaling by Fas-ligand expressed on the surface of other cells.

Fas ligand – trimeric cell surface protein binding to Fas on a target cell, initiating the extrinsic pathway of apoptosis.

Flow cytometry – light scattering, light excitation, and emission of fluorochrome molecules to generate specific multi-parameter data from particles and cells in the size range of 0.5–40μm in diameter.

Fluorescence – emission of light from a molecule after excitation by shorter-wavelength, high energy photons.

Fluorochrome – fluorescent dyes used to stain biological specimens.

Free energy – thermodinamic energy available for work.

Fusion protein (in molecular biology) – fusion of two DNA coding sequences of two proteins to produce a hybride protein.

Gel electrophoresis – using an electric field for the separation of molecules based on their size and charge in a gel matrix.

Gemini of coiled body (GEM) – immunolocalization using survival of motor neurons protein (SMN) monoclonal antibodies show several intense dots in HeLa cell, similar in size (0.1–1.0 μm) to coiled bodies.

Glossary

Gene – Definitions: a. section of DNA coding for a functional RNA or protein product (containing introns), b. the hereditary unit consisting of a sequence of DNA that occupies a specific location on a chromosome, c. recent official definition (Guidelines for Human Gene Nomenclature): a DNA segment that contributes to phenotype/function. In the absence of demonstrated function a gene may be characterized by sequence, transcription or homology.

Genetic code – the correspondence between nucleotide triplets in mRNA and aminoacids in a polypeptide.

Genetic diversity – refers to any variation in the nucleotides, genes, chromosomes, or whole genomes of organisms.

Genetic likage map – linear arrangement and relative positions of genes along a chromosome.

Genetic marker – specific DNA sequence for identification and monitoring by examining the phenotype of cells carrying this sequence.

Genome – the entire complement of DNA in an organism.

Genomics – the study of an organism's entire genome.

Genotype – the genetic constitution of an organism that results from the arrangement of the DNA within the cell or organelles.

Genotypic variation – the variation that exists between the genetic constitution of different individuals.

Germ cell – the egg and sperm cells also kown as the reproductive cells. Each mature germ cell is haploid, a human germ cell has a single set of 23 chromosomes (22 autosomes (non-sex chromosomes) and one sex chromosome).

Germline – the line (sequence) of germ cells that have genetic material that may be passed to a child.

Granulocyte Monocyte Colony-Stimulating Factor (GM-CSF), stimulates hematopoiesis.

Guanine nucleotide exchange factor (GNEF) – also known as Son-of-Sevenless (Sos), which binds to the SH3 domains of the adapter protein Grb2. through its SH2 domain, associates with receptor tyrosine kinases (e.g. Src) to transmit the signal from the autophosphorylated receptor to Ras.

G_0 phase – cell cycle stage of nondividing (resting, quiescent, often terminally differentiated) cells which will not grow and replicate and will not re-enter the cell cycle. G_0 cells can occasionally enter G_1 and go through the cell cycle under certain circumstances (e.g. stimulation of T lymphocytes). The G_0 terminology in such cases is confusing.

G_1 phase (first gap phase) – cell cycle stage between mitosis and DNA replication.

G₂ checkpoint – cell cycle checkpoint controlling the G_2 phase to block mitotic entry if DNA is damaged or replication of DNA is incomplete.

G₂ phase (second gap phase) – cell cycle stage between the completion of DNA replication and mitosis.

Golgi apparatus – secretory membrane compartment in the cell for the processing of glycoproteins and sorting molecules.

G-protein – GTPase subunits (G_α, G_β and G_γ) of trimeric GTPases.

Granzymes – exogenous serine proteases released by cytotoxic T cells and natural killer cells, by inducing apoptosis in virus-infected cells destroy infected cells.

Grb2 adapter protein consisting of two SH3 and one SH2 domains binds to the autophosphorylated protein tyrosine kinase (PTK) leading to the activation of Sos, GDP-GTP exchanger of Ras and consequently to a cascade of phosphorylation resulting in the activation of mitogen-activated protein kinase MAPK).

Growth cycle – increase of cellular mass during cell growth.

Gynecomastia – increased breast tissue.

Haploid chromosome number (n) – numbers of chromosomes donated by the mother or by the father in a diploid cell or organism. Haploid cells in animals are gametes (sperm and egg cells).

Heat-shock genes – become active after sudden increase of temperature, and dominate the pattern of gene expression after heat shock, but are inactive at normal temperatures.

Helicases – DNA unwinding enzymes using the energy of ATP to dissociate the complementary strands of DNA, remove secondary structures or bound proteins from nucleic acids.

Hereditary factors – genetic links determined in twins suggest that up to 80 percent of body shape is hereditary. Other risk factors for high cholesterol include obesity, high blood pressure and smoking.

Herpes virus (*Herpes simplex virus* 1 and 2, *HHV-1 and HHV-2*) – two strains of the herpes virus family, causing painful and lifelong infections in humans either as cold sores or as genital herpes.

Heterochromatin – transcriptionally inert, highly condensed regions in chromatin rich in histon H3 trimethylated on lysine 9.

Heterogenous nuclear RNA (hnRNA) – incompletely processed precursors of large RNAs containing transcribed introns present in nuclei of eukaryotic cells and include precursors of functional species such as mRNA.

Heterozigous – the condition in a diploid organism in which a genetic marker has two different forms on two homologous chromosomes.

Glossary

Hexose – a six carbon sugar.

Histones – chief protein components of chromatin.

Histone code hypothesis – a proposal that postranslational modification of histones determine the functional activity of a particular region of chromatin. Together with other modifications such as DNA methylation they are part of the epigenetic code.

Hodgkin's lymphoma – spread of disease from one lymph node group to another, one of the most curable forms of cancer (by radiation and combined chemotherapy).

Hogness box – See TATA box.

Homeobox (HOX) gene – class of genes for transcription factors specifying the development of embrionic segments.

Homology – similarity of form seen in the biological world, explained by common descent; features are homologous if they are inherited from a common ancestor.

Homologous chromosomes – pairs of chromosomes in diploid organisms one of which is donated by the mother and the other by the father.

Homozygous – the condition of a diploid organism in which a particular genetic marker has the same sequence on both maternal and paternal chromosomes.

Housekeeping genes – universally required by most tissues of a multicellular organism (e.g. genes coding for enzymes of ubiquitous metabolic pathways, such as glycolysis).

Hsp70 – protein chaperones (molecular gardedames) using the energy of ATP hydrolysis to drive the cyclic binding and release of short hydrophobic segments to promote protein folding.

Hsp90 – chaperones that maintain steroid receptors in "open" position to bind their receptors.

Hydrogen bonds – weak polar interaction between a H atom (with a partial positive charge) and an oxygen or nitrogen (with a partial negative charge) contributing in the stabilization of secondary structures of informational macromolecules.

Hydrophilic – molecules or groups of atoms in molecules favouring interactions with water.

Hydrophobic – molecules or groups of atoms in molecules repelling interactions with water.

Inhibitor of apoptosis protein (IAP) – family of proteins inhibiting caspases through a motif of ≈ 80 amino acids. This motif is also konwn as baculovirus IAP repeat (BIR).

Inhibitor of CAD (ICAD) – chaperone that folds caspase-activated DNase (CAD) in its inactive form until ICAD is cleaved by caspase and releases the active DNase (CAD).

Interleukin-1β-converting enzyme (ICE) – a mammalian cysteine protease, homologoues to the CED3 cysteine aspartase in the nematode *Caenorhabditis elegans.*

ICE-like cysteine proteases – collectively called caspases.

Image processing – optical and computation methods to reduce noise, average or make 3D reconstructions of micrographs.

Informational macromolecules – DNA, RNA, protein.

Inhibitor of CAD – See ICAD.

Initiator (DNA sequences) – in promoters near the transcription site of genes.

Initiator (tumour) – makes tumour, alters DNA (carcinogen).

Initiator caspases – autoactivated by association with scaffolding cofactors and propagating apoptosis by zymogen activation (cleavage) of effector caspases.

Inositol-1,4,5-trisphosphate (IP$_3$) – cyclohexanol containing phosphate esters at 1, 4 and 5 positions. IP$_3$ is a signal transducing molecule in cells.

Integral membrane protein – protein embedded in the membrane bilayer.

Integrins – heterodimeric cell surface receptors, mediating adhesion of cells to the extracellular matrix as well as to other cells.

Interferons (IFNs) – natural proteins produced by the cells of vertebrates in response viruses, parasites and tumour cells. Signaling: through the JAK-STAT and IFN signaling pathway.

Interchromatin granules – intranucleolar accumulation of factors for RNA processing.

Interchromosomal domain – region of nucleoplasma between adjacent chromosome territories.

Interleukins (IL1 – ≈IL35) cytokines sectered signaling molecules found first in leukocytes (white blood cells).

Interleukin converting enzyme (ICE) – enzyme participating in protease cascade. Activation of ICE occurs *in vivo* in CD4$^+$ lymphocytes from HIV-1-infected individuals, accounting for the increased susceptibility of CD4$^+$ cells in Fas mediated apoptosis.

Interphase – proceeds in three stages, G1/S/G2, preceded by the previous cell cycle of mitosis and cytokenesis.

Intrinsic pathway of cell death (mitochondrial pathway) – apoptosis triggered by the release of pro-apoptotic factors from mitochondria and regulated by Bcl-2 family members.

Glossary

Introns – regions of eukaryotic split genes that are removed from immature hnRNA molecules by splicing and thus not contributing to the final functional RNA product.

Inversions – a portion of the chromosome has broken off, turned upside down and reattached, therefore this portion of genetic material is inverted.

IP$_3$ receptor – IP$_3$ activates channels to release calcium from the endoplasmic reticulum. PIP$_2$ is the substrate for cleavage with phospholipase C, a membrane-bound enzyme activated through α1 adrenergic receptors. Products of this reaction are inositol triphosphate (IP$_3$) and diacylglycerol (DAG).

Ischemia – lack of blood supply caused by thrombosis, embolism, reduced blood circulation, delayed angiogenesis.

H1 histone – See linker histone.

Hepatitis – injury to liver characterised by the presence of inflammatory cells in the liver tissue.

JAK (just another kinase) – family of tyrosine kinases associated with cytokine receptors.

"Junk" DNA (pseudogenes) – is widely used as descriptor of noncoding DNA, originally used to refer to pseudogenes, i.e. defunct copies of protein-coding genes.

Kaposi's sarcoma – a cancer causing patching abnormal tissue to grow under the skin, in the mouth, nose, throat or in other organs.

Karyotype – an organism's profile of chromosomes, a complete set of chromosomes of a cell or an organism.

Kinetochore – surface protein structure of the centromeric chromatin that binds the chromosome to microtubule polymers from the mitotic spindle and directs chromosomal movement in mitosis and meiosis.

Kinetoplast – independently replicating structure near the base of the flagellum in certain protozoans.

"Knockout" animal – transgenic animal, genetically engineered to exhibit mutations in specific genes.

Lagging strand – replication in direction opposite to the replication fork movement generating a series of short discontinuous DNA segments called Okazaki fragments.

Lamin A (LMNA) – gene on chromosome 1 encoding a key protein of the membrane surrounding the cell nucleus.

Lampbrush chromosomes – actively transcribed animal chromosomes showing visible loops during the diplotene stage of meiosis.

Lariat – lasso shaped RNA parts spliced out from hnRNA to form mRNA.

Leading strand – DNA strand synthesized continuously in the direction of replication fork movement.

Ligand – molecule that binds to the receptor.

LINES – see long interspersed nuclear elements.

Linkage analysis – distance of genes which determines the frequency at which two gene loci become separated during chromosomal recombination.

Linker DNA – between adjacent nuclesomes.

Linker histone (H1 histone) – binds to linker DNA and participates in the formation of the 30 nm thin fibril in chromatin condensation.

Lipocortin – an enzyme activated by glucocorticoids, inhibits the activity of phospholipase A2.

Long interspersed nuclear elements (LINEs) – class of autonomous retrotransposons with a consensus sequence of 6–8 kb making up about 20% of the human genome. LINEs have two open reading frames (ORFs). ORF2 encodes reverse transcriptase.

Long terminal repeats (LRTs) – transposable genetic elements characterized by flanking long terminal repeat, similar to retroviruses and contain a group specific antigen (*gag*).

Lymphocyte – white blood cells (T and B) of the adaptive immune system.

Lysosome – membrane-bound organelle containing acid hydrolases for the digestion of degradable contents.

Macroautophagy – engulfment of large cytoplasmic parts including glycogene granules, ribosomes and other organelles into an autophagic vacuole.

Macrophage stimulating 1 (MST1 also known as HGFL; NF15S2; D3F15S2; DNF15S2) an upstream kinase of the JNK and p38 MAPK pathways whose expression induces apoptotic morphological changes such as nuclear condensation.

Major histocompatibility complex (MHC) – the most gene-dense region of the mammalian genome playing an important role in the immune system, autoimmunity, and reproductive success.

Malignant transformation – cells acquired the ability to invade other tissues, leading to secondary tumour (metastasis) formation.

MAP-2 – high molecular weight protein belonging to the tau family and associated to microtubules.

MAP kinase – cytoplasmic serine/threonine kinases activated by phosphorylation then moving to the nucleus to activate transcription factors required for cell growth and division.

Glossary 395

MAP kinase kinase – MAP kinases are activated by kinases phosphorylating serine and threonine residues.

MAP kinases kinase kinase – MAP kinase kinase is activated by phosphorylation through MAP kinase kinase kinase.

Matrix attachment regions (MARs) – See scaffold attachment regions.

Meiosis – two coupled eukaryotic cell divisions to maintain proper chromosome number during sexual reproduction. These cell divisions result in gametes (eggs and sperm cells), each containing half the number of chromosomes as normal (diploid) cells.

Melanoma – derived from epithelial pigment cells frequently continue to produce pigment and once it has metastasized widely, is often impossible to eliminate.

Membrane potential – See electric potential.

Mesosome – infolding of bacterial cell membrane.

Messenger RNA (mRNA) – RNA molecules which code for proteins and are translated on ribosomes.

Metabolism – all the chemical reactions that occur in living organisms to maintain life.

Metamorphosis – change in taxon during development. Physical development of animals after birth or hatching.

Metaphase – third phase of mitosis when all chromosomes are attached to both spindle poles and alingned near the equator of the mitotic spindle.

Metaphase checkpoint (spindle checkpoint) – delays the unset of anaphase until each chromosome is properly attached to the mitotic spindle.

Micronucleus assay – genotoxicity causes the formation of micronuclei during cell division. Cytochalasin-B (an inhibitor of actins) addition allows to distinguish between mononucleated (not dividing) and binucleated (dividing) cells.

Microtubule – cylindrical polymer of α- and β- tubulin serving as tracks for movements by positive (kinesin) and negative (dynein) motor proteins.

Microtubule-associated proteins (MAPs) – proteins regulating microtubule properties.

Milk fat globule epidermal growth factor factor 8 (MFG-E8) – involved in the engulfment of apoptotic cells by macrophages.

Mismatch repair – process that removes errors occuring during DNA replication.

Mitochondria – symbiotic proteobacterium organelle of eukaryotic cells specialized for terminal oxidation and oxydative phosphorylation from ADP to ATP. Also involved in citric acid cycle, fatty acid oxidation and by releasing cytochrome c in apoptosis.

Mitochondrial pathway of cell death – See intrinsic pathway of cell death.

Mitosis (M phase) – cell division phase consisting of prophase, metaphase, anaphase, and telophase stages resulting in the partitioning of chromosomes and other cellular components to two daughter cells that have the normal amount of chromosomes.

Mitotic spindle – framework of microtubules that segregates chromosomes during mitosis.

Model organisms – research on a small number of organisms played a pivotal role in advancing understanding of numerous biological processes. Mammalian models: rat, mouse, chick. Non-mammalian models: archeae "extremophyl" bacteria), microbes (*E. coli*), fruit fly (*Drosophila melanogaster*), for malaria research (*Plasmodium falciparum* and *Plasmodium vivax)*, round worm (*Caenorhabditis elegans*), retroviruses (pneumonia, leukemia, and AIDS), yeast (*Saccharomyces cerevisiae, Schisosaccharomyces pombe*), zebrafish (*Daniao rerio*), frog (*Xenopus laevis*) and amoeba (*Dictyostelium discoideum*). Plant model: *Arabidopsis thaliana*, a small flowering plant that belongs to the Brassica (mustard) family. Note: viruses are not organisms.

Monocyte – white blood cell, precursor of osteoclasts (cells specialized for bone resorption) and tissue macrophages.

Monokines – soluble mediators of immune responses produced mainly by monocytes and macrophages.

Monte Carlo method – a computational algorithm relying on repeated random sampling used to simulate physical, mathematical and biological systems.

Morphogenesis – see ontogeny, ontogenesis.

Most Recent Common Ancestor – of all living humans, a human within historical times. "Mitochondrial Eve" is only defined relative to AD 2000.

Motor end plate – See neuromuscular junction.

Maturation promoting factor, M phase-promoting factor (MFP) – causing interphase cells to enter mitosis, shown to be an active Cdk with a cyclin partner.

Messenger RNA see mRNA.

Mus musculus **(the mouse)** – mammalian model organism with highly developed genetics.

Mutation – any permanent alteration in the genomic DNA sequence.

Myc oncogene – activated product of myc oncogene is a transcriptional factor, promoting the expression of genes involved in cell cycle progression; represses the expression of Cdk inhibitors, deregulates both cell growth and death check points, and in a permissive environment, rapidly accelerates the affected clone through the carcinogenic process.

Glossary

Myeloid leukemia – in almost every case of this cancer reciprocal translocation takes place between the arm of chromosome 22 and the end of chromosome 9, also known as Philadelphia chromosome.

Necrosis – irreversible cell injury (leakage of cell membranes, destruction of cellular contents and lysosomal enzymes, local inflammation) causing cell death.

Negative control – regulation of transcription by factors which are normally present and prevent RNA synthesis. Transcriptional activation takes place by the removal of these repressors.

Negative selection – potentially harmful lymphocytes with T cell receptors recognize self-antigens leading to apoptosis.

Nematode – roundworm.

Neoplasm – cell proliferation without control.

Neuromuscular junction – synapse between skeletal muscle cell and motor nerve cell (neuron).

Neurotransmitter – small organic molecules used for chemical communication between neuron and between nerve cells and other cells.

Neutrophyl – white blood cell phagocyting and destructing bacteria.

Nondisjunction – erroneous separation of chromosomes resulting in aneuploidy.

Nonhistone proteins – proteins in chromatin and chromosomes that are not histones.

Notochord – rod-shaped body found in embryos of all chordates, composed of cells defining the primitive axis of an embryo.

N-terminal tail – the N-terminal end of core histones (H2A, H2B, H3, H4) containing about 30 amino acids which regulate chromatin compaction.

Nuclear pore complexes – channels bridging the nuclear membrane providing communication between the nucleus and cytoplasm during interphase.

Nuclear receptor – lipid soluble ligands (steroid hormones) diffusing in cells activate transcription factors which enter then into the nucleus and regulate gene expression.

Nucleic acids – informational macromolecules (DNA, RNA) consisting of nucleotides linked by phosphodiester bridges.

Nucleolus – nucleolar subdomain containing ribosomal genes and specialized for the assembly of ribosomal subunits.

Nucleolar-organizing regions (NORs) – remnants of nucleolar fibrillar centers associated with rRNA genes seen in condensed mitotic chromosomes.

Nucleoplasm – region enclosed within the nucleolar envelope.

Nucleoside – pentose (ribose or deoxyribose) attached to the nucleic acid base through an N-glycosidic bond.

Nucleosome – DNA-protein complex of 167 base pairs of DNA wrapped around the histone core particle consisting of two copies each of histones H2A, H2B, H3 and H4.

Nucleosome remodeling complex – enzyme complex using ATP energy to reassemble nucleosomes at another location on DNA.

Nucleotide – building block of nucleic acids consiting of three subunits: sugar, base and 1–3 phosphates.

Nucleotide excision repair – involved in the removal of a variety of bulky DNA lesions such as UV induced cyclobutane pyrimidine dimers and photoproducts.

Neurodegenerative disease – a condition in which cells of the brain and spinal cord are lost.

Mosaicism – abnormal chromosome division resulting in two or more kinds of cells, each containing different numbers of chromosomes (chromosome mosaicism).

Nephroblastoma – see Wilms tumour.

Nucleus – largest membrane-bounded subcellular compartment in eukaryotic cells containing genomic (cellular) DNA and the machinary for DNA and RNA synthesis and RNA processing.

Null allele – mutant copy of a gene that lacks the wild type gene's normal function.

Okazaki fragments – short DNA segments of discontinuous DNA replication on the lagging strand.

Oncogenes – especially those of viral genes causing the transformation of normal cells to cancerous tumour cells when their protein products are overexpressed and activated. These products belong to the components of the pathway of signal transduction and regulate cell growth and proliferation.

Oncosis – cell death marked by cellular swelling.

Ontogeny (ontogenesis, morphogenesis) – development of an organism from fertilized egg to mature form.

Operator – regulator region of DNA between the TATA box (Pribnow box in bacteria, Goldberg-Hogness box in eukaryotes) and protein-coding region subject to negative control, to which the repressor binds.

Operon – section of DNA containing regions of regulation and structural gene(s). Common in bacteria, rare in eukaryotes.

Organelles – specialized subunits within a cell with specific functions, separated within their own lipid membrane. See organelles of eukaryotic cells in Table I/1, page 8.

Osteoporosis – weak bones that are likely to break, common in older women.

Glossary

Outer nuclear membrane – lipid bilayer continuous with endoplasmic reticulum and sharing its function.

Oxidative phosphorylation – biochemical reactions in mitochondria and some bacteria, utilizing the energy of nutrient breakdown to produce ATP from ADP.

p21 protein – blocks cell cycle progression by binding to Cdk1-cyclin A complex. The expression of p21 is stimulated by p53 in response to DNA damage.

p53 transcription factor – activated in response to DNA damage, induces the expression of proteins hampering cell cycle progression or initiates apoptosis.

PAK - p21 activated kinase – both the apoptosis and DNA synthesis are mediated by the p21 activated kinase PAK 3, a serine-threonine kinase.

Palindrome – DNA sequence which reads the same in both directions, taking account of the two antiparalel strands.

Paradigm (scientific) – key model, pattern, fundamental property of a theory, from Greek word: either παράδειγμα (*paradeigma*) meaning "pattern" or "example", or παραδεικνύναι (*paradeiknunai*) meaning "to demonstrate".

Paralogue (in genetics) – one of the two genes of a pair of genes that derive from the same ancestral gene.

Paraptosis – cell death that involves cytoplasmic vacuolation, mitochondrial swelling and absence of caspase activation without condensation or fragmentation of the nuclei.

Partial pressure – defined as the pressure of a single gas in the mixture as if that gas alone occupied the container (e.g. the partial pressure of oxygen in air is 0.21, corresponding to 21%) (Dalton's law).

Pathogenicity – the ability to produce disease in a host organism.

Pentoses – five carbon sugars (ribose, arabinose, xylose, lyxose).

Peptide bond – amine bond between amino acids in peptides and proteins.

Perforin – cytolytic protein found in the granules of CD8 T-cells and NK cells. Upon degranulation, perforin inserts in the target cell's plasma membrane, forming a pore.

Perichromatin fibrils – nucleoplasmic structures on the surface of condensed chromatin.

Phagocytosis – the ingestion of large, solid particles (bacteria, foreign bodies, remnants of dead cells) by cells.

Phenotype – manifestation (appearance, macromolecular composition) of the genotype of an organism.

Philadelphia chromosoma – see myeloid leukemia.

Phosphatidylcholine – glycerophospholipid (phosphatidic acid) with a choline head group.

Phosphatidylinositol – glycerophospholipid with an inositol (cyclohexanol) head group which can be phosphorylated on C3, C4 and C5.

Phosphatidylinositol 4,5-bisphosphate (PIP_2) – phosphatidylinositol esterified with phosphate at C4 and C5.

Phosphodiester bond – phosphate esterified to two hydroxyl groups linking different molecules (in DNA and RNA) or forming cyclic phosphodiesters (cAMP, cGMP).

Phospholipase C – activated by a signal cascade, catalyzing the cleavage of PIP_2 to yield the signal molecules diacylglycerol and inositol trisphosphate (IP_3).

Photolysis of water – during photosynthesis water is split by means of light energy (sunlight), hydrogen binds to an acceptor (CO_2) and oxygen is released into the atmosphere.

Photosynthesis – a process in plants, some bacteria and some protists to reduce carbon dioxide to carbohydrates, $NADP^+$ to NADPH by photolysis and to produce energy (ATP) and oxygen using sunlight.

Phosphorylation – formation of ester bond between a phosphate and the hydroxyl group of an amino acid, sugar, lipid or other molecule.

Plasma state – in the fourth state of matter (besides the three states being solid, liquid and gas) the negatively charged electrons are freely streaming through the positively charged ions. The plasma state is not related to the plasma of blood, the most common usage of the word in biology.

Plasmids – usually small, circular double-stranded DNA molecules found in bacteria in addition to the cellular (genomic) DNA, used as vectors in DNA cloning.

Polymorphism – sequence variations in the genom and multiple states for a single property varying from person to person.

Polypeptide (protein) – polymer of amino acids linked by peptide bonds.

Polyploidy – chromosome abnormality with one or more entire extra sets of chromosomes.

Positive control – transcription increased by the addition of an activating, mostly protein factor.

Positive selection – apoptotic pathway of lymphocytes ineffective in immune responses since they eypress T-cell receptors that fail to interact with an individual's major histocompatibility complex glycoproteins.

Posttranslational targeting (protein targeting) – directing completed polypeptides inside the cell or outward to the particular location for use.

Glossary

Retinoblastoma protein (pRb or Rb) – a tumour suppressor protein disfunctional in several cancer cells.

Preapoptotic – chromatin condensation occurs before mitochondrial membrane permeabilization and caspase activation.

Prebiotics – 1st meaning: before the existence of biotics (living microorganisms), 2nd meaning: prebiotics are indigestible carbohydrates. In this book the first meaning is used.

Prebiotic evolution – abiogenesis, the most important events leading to the origin of life on Earth.

Precursor – inactive compound that can be turned into an active form.

Primordial soup – theory suggesting that that life began in sea from chemicals forming nucleotides and amino acids which then evolved into macromolecules and species.

Proapoptotic – signal or protein triggering apoptosis.

Procaspase – caspases synthesized as inactive zymogens.

Product oncogene domains (PODs) – one class of nuclear bodies containing the promyelocytic leukemia (PML) and several other transcription factors including CREB binding protein, transcriptional coactivator (CBP), involved in protein metabolism as well as transcriptional regulation.

Prokaryote – simple cellular life-form without a distinct nucleus, including bacteria and certain blue-green algae (cyanobacteria).

Programmed cell death – active cellular process leading to cell death in response to developmental signals, environmental effects, physiological damage.

Proliferation – expansion of cell number by cell growth and divison.

Promoter – regulator DNA sequence required to form preiniciation complex for transcription.

Promyelocytic leukemia gene (PML) (human) – the protein encoded by this gene is present in a set of nuclear organelles (PML nuclear bodies) as a member of the tripartite motif (TRIM) family.

Pronucleus – haploid nucleus of a sperm or egg before fusion of the nuclei in fertilization.

Prophase – first subphase of mitosis defined by chromosome condensation inside the nucleus accompanied by cytoplasmic changes of formation of microtubules.

Proteasome – a large barrel-shaped enzyme complex tagged with chains of ubiquitin for the selective, energy requiring degradation of proteins to short peptides.

Proteasome core complex – with a sedimentation coefficient of 20S, contains 2 copies each of 14 different polypeptides.

Protein – one or more polypeptides folded into functional three-dimensional structure.

Protein domain – independently folded functional part of a protein.

Protein folding – linear polypeptide turning into a specific three-dimensional structure.

Protein kinase – catalyses the formation of phosphate esters on hydroxyl groups of amino acid residues (tyrosine, threonine, serine) in proteins.

Protein kinase A (PKA) – the formation of phosphate esters is regulated by cAMP binding to the regulatory subunit of the enzyme.

Protein kinase C (PKC) – the formation of phosphate esters is regulated by the binding of diacyglycerol (DAG) or other lipids and calcium to the regulatory subunit of the enzyme.

Protein kinase C delta (PKCdelta) – delta isoform of protein kinase C, implicated in many important cellular processes, including regulation of apoptotic cell death.

Protein phophatase – catalyses the removal (hydrolysis) of phosphate from the ester bond of proteins.

Protein synthesis – See translation.

Protein targeting – See posttranslational targeting.

Proteomics – large-scale study of proteins, their structures and functions.

Protooncogenes – normal, nonpathogenic genes, when mutated or otherwise altered, become oncogenes; many of the protooncogenes code for growth factors, growth factor receptors, genetic regulatory proteins, or other signal transduction molecules.

Pseudogenes ("Junk" DNA) – show strong base sequence homology to functional genes but are non-functional (eg. ribosomal genes, Ig genes). DNA sequences likely to be derived from gene transcripts that have been reverse transcribed and inserted back into the chromosome.

PTB domain (phosphotyrosine binding domain) – adapter domain found in multiple proteins that binds particular peptides containing phosphotyrosine.

Guanine nucleotide exchange factor (GNEF) – protein that stimulates the exchange (replacement) of nucleoside diphosphates for nucleoside triphosphates and associates with chromatin throughout the cell cycle.

Reaper (rpr) – cell death gene in *Drosophila*.

Receptor – macromolecule that binds selectively to its partner molecule (ligand) initiating cellular response.

Receptor interacting protein (RIP) – binds to the cytoplasmic segment of Fas (along with FADD) and activates the death pathway.

Glossary

Receptor mediated endocytosis – ligand taken up by the cell due to its binding to to the receptor that undergoes endocytosis.

Reciprocal translocation – when segments from two different chromosomes have been exchanged.

Recombination – the crossing over and physical exchange of DNA strands between homologous chromosomes during meiosis. As a consequence previously unlinked genes may become linked or *vice versa*.

Redox reaction – oxidizing one substrate while reducing another.

Reductive power – in catabolism a set of chemical reactions where nutrients are converted to smaller molecules while energy and reductive power is produced in the form of ATP and NADH or NADPH.

Repetitive DNA – sequences present in thousands of copies in eukaryotic genomes.

Replication foci – thousands of replication sites in eukaryotic nuclei during S-phase.

Replication fork – replication bouble consisting unwound DNA strands and two replication forks for bidirectional replication with the replication machinaries on both forks.

Replicon – region of chromosomal DNA replicated from a single origin of replication.

Repressor – protein binding to the operator region of the promoter DNA to prevent transcription. See also negative control.

Retinoblastoma – a childhood cancer arising from immature retinal cells in one or or both eyes.

Retinoids – chemical compounds related to vitamin A.

Restriction point – checkpoint in late G1 phase that blocks cells from commiting to proliferation. Proliferation takes place when nutrients and mitogens are present in sufficient concentration.

Retrotransposons – long terminal repeats (LTRs), long interspersed nuclear elements (LINEs), short interspersed nuclear elements (SINEs).

Reverse genetics – gene function studied by engineering mutations into cloned regions of genes. The genetic screen for new genes is often referred to as forward genetics as opposed to reverse genetics, the term for identifying mutant alleles in genes that are already known.

Reverse transcriptase – an enzyme that functions as an RNA-dependent DNA polymerase encoded by retroviruses, copying the viral RNA into DNA.

Rhamnose – methylpentose, 6-deoxy-hexose, a naturally occurring deoxy sugar.

Ribonucleotide – building block of RNA consisting of sugar (ribose), phosphate and base (A, G, C or U).

Ribonucleotide reductase – catalyses the reduction (deoxygenation) of the ribose ring to yield 2'deoxyribose from ribonucleoside diphosphate substrates (NDP $\rightarrow$ dNDP conversion).

Ribose – five-carbons sugar, component of RNA.

Ribosome – complex of ribosomal RNA and multiple proteins catalysing the synthesis of polypeptides.

Ribosomal RNAs (rRNAs) – types of RNA found in ribosomes in prokaryotes (5S, 16S and 23S), and eukaryotes (5S, 5.8S, 18S and 28S rRNAs).

Ribozymes – RNA enzymes, a special class of biological catalysts independent of proteins.

Ring chromosome – a portion of a chromosome broken off and forming a circle or ring. This can happen with or without loss of genetic material.

Ring finger domain (Zn- finger Ring type) – a cysteine-rich sequences structurally diverse, all contain two interleaved Zn^{2+} binding sites.

Ribonucleic acid (RNA) – informational macromolecule consisting of ribonucleotide (NMP) building blocks.

RNA interference (RNAi) – an mRNA cleaved by RNAi is no longer protected by a 5'cap and a 3'poly(A) tail, the free ends of mRNA are rapidly degraded by cellular RNases.

RNA polymerase – enzyme that transcribes DNA template to complementary RNA.

RNA silencing – double-stranded (ds) RNA is processed into small interfering (si) RNAs of 21 to 25 nucleotides by Dicer, a member of the RNase III family of dsRNA specific endonucleases.

rNDPs – ribonucleoside diphosphates (ADP, GDP, CDP, UDP).

Robertsonian translocation – when two chromosomes fuse, usually at the centromere, creating a translocation. Only certain chromosomes, called acrocentric chromosomes, are capable of participating in this kind of translocation.

Rough endoplasmic reticulum (RES) – subdomain of endoplasmic reticulum (ER) associated with ribosomes which synthesize secretion proteins and insert them into the ER.

Saccharomyces cerevisiae – budding (baker) yeast, genetic model organism to study basic cell biology.

Sarcoma – cancer from connective tissue or muscle cell.

Glossary

Satellite DNA – clusters of DNA repeats in discrete areas of chromosomes (e.g. at centromeres).

Scaffold attached regions (SARs) – regions of DNA associated with nuclear matrix and chromosome scaffold.

SC35 Speckles – identified immunocytologically in the nuclei with an antibody against spliceosomal protein SC35.

Schizosaccharomyces pombe – fission yeast, genetic model organism to study cell cycle.

Scleroderma (pigmentosum) – a chronic disease with excessive deposits of collagen in the skin or other organs.

Second messengers – small molecules, calcium ions and lipids carrying biochemical signals inside cells.

Secondary structure – formed primarily by hydrogen bonds in polypeptides (α-helix, β-pleated sheet) and nucleic acids (e.g. A-, B-, Z-DNA).

Sepsis – a life-threatening illness in response to the body's response to a bacterial infection. The immune system overreacts the normal processes in the blood resulting in small blood clots form, blocking blood flow to vital organs leading to organ failure.

Seven-helix receptor – a class of receptor proteins composed of seven transmembrane α-helices coupled to cytoplasmic trimeric protein, as opposed to Sos (Son-of-Sevenless) proteins that are missing the seven transmembrane α-helices.

Sex chromosomes – carrying genes defining the sex of an organism.

SH2 domain – adapter domain that binds to peptides (containing phosphotyrosine) and is part of many signaling proteins.

Short interspersed terminal repeats (SINEs) – are non-autonomous retrotransposons of about 300 base pairs making up about 13% of the human genome, that exploit the enzymatic retrotransposition machinery of LINEs.

Side chain – of amino acids is a group on the α-carbon.

Signal peptidases – enzymes cleaving signal peptides from the N-terminal of proteins that are translocated accross the membrane of the endoplasmic reticulum.

Signal transduction – reactions converting an external stimulus to internal one(s).

Sindbis virus (SINV) – member of the *Togaviridae* family, isolated from mosquitos, causes epidemic polyarthritis and rash.

Singlet oxygen – metastable (the lowest excited) state of molecular oxygen.

Sister chromatids – two identical condensed DNA molecules packaged by chromatin proteins forming the metaphase chromosome.

Small nuclear RNAs (snRNAs) – a family of small RNAs (11S) found in eukrayotic nuclei, complexed with small nuclear ribonucleoproteins (snRNPs) forming splicesosomes that recognize intrones and exons and are involved in the processing of heterogeneous nuclear RNAs to functional mRNAs.

Small *u*biqutin-like *mo*difier-*1* (SUMO-1) – small protein covalently linked to other proteins, and changes the properties of the linked proteins.

Structural maintainance of chromosomes (SMC) proteins – ATPases as part of the condensin and cohesin complexes are essential for mitotic chromosome structure, are involved in sister chromatid pairing, DNA replication and regulation of gene expression.

Smooth endoplasmic reticulum (SER) – subdomain of endoplasmic reticulum lacking ribosomes, involved in detoxication, drug metabolism, steroid biosynthesis and calcium homeostasis.

SNAP receptor (SNARE) – proteins that participate in the fusion of carriers with the acceptor compartment.

Somatic cells – of the body compose the tissues, organs, and parts of an individual other than the germ (sex) cells.

Son-of-Sevenless (Sos) – GDP-GTP exchanger of Ras.

S phase (synthetic phase) – cell cycle stage when DNA is replicated.

Sphingomyelin – sphingolipid with choline or ethanolamine head group.

Splicing – the processing of heterogeneous nuclear RNA (hnRNA) leading to the maturation by removing specific regions (introns) and rejoining exon sequences.

Signal transducers and activators of transcription (STAT) – a family of transcription factors that regulate cell growth and differentiation, phosphorylated by JAK kinases and enter the nucleus.

Stem cell – by asymmetrical cell division produces itself and a second cell that is able to differentiate into specialized cells.

Stroke (cerebrovascular accident, CVA) – ischemic heart disease, clinical symptome of a rapid loss of brain function due to a disturbance in the blood vessels supplying blood to the brain.

Substrate – a molecule upon which an enzyme acts.

Subunit – macromolecular building block of a larger structure.

Sugar puckering – five-membered furanose rings in ribo- and deoxyribonucleotides are never planar, they can be puckered in either envelope, but primarily to twist (half-chair) forms. In nucleotides the five-membered sugar ring is a symmetrically substituted, and the oxygen as heteroatom creates energy tresholds which limit the pseudorotation leading to preferred puckering modes, the two principal ones are $C_{3'}$-endo and $C_{2'}$-endo conformations.

Glossary 407

Supercoiling – extra tension in DNA generated by revolving around the axis.

Superoxide dismutase – dismutes the superoxide anion radical $(O_2\bullet^-)$ to O_2 and hydrogen peroxide (H_2O_2), outcompetes damaging reactions of superoxide, protects cell from superoxide toxicity.

Survivin – an inhibitor of apoptosis protein (IAP), localizes to the inter-mitochondrial membrane space in tumour cells and accelerates tumourigenesis.

SV40 large T antigen (SV40-LT) – a viral oncoprotein transactivating viral and cellular promoters and induces DNA synthesis in quiescent cells.

S-values – annotations given to macromolecules (e.g. 5S, 16S, 23S rRNA) which reflect their rate of sedimentation through a sucrose gradient and is given in S (Svedberg) units.

Symbiosis – close relationship and interaction of different biological species.

Synapse (chemical) – specialized space (junction) through which nerve cells (neurons), nerve and muscle, nerve and glandular cells communicate. Electrical and immunological synapses exist as well.

TATA box – promoter element for transcription by RNA polymerase II, consisting of the TATAAAA consesus sequence recognized by the TATA box binding protein (TBP). Known as Goldberg-Hogness box in eukaryotes and as Pribnow box in prokaryotes.

Tau family – belongs to the microtubule-associated proteins (MAP2, MAP4, tau) abundant in nerve cells of the central nervous system and less common elsewhere, containing microtubule-binding motifs that stabilize microtubules. Tau proteins found in Alzheimer's disease can facilitate each other to form amyloid lesions at least under laboratory conditions.

Tax protein – viral protein is considered to play a central role in the process leading to adult T-cell leukemia (ATL).

Taxon – designating an organism or group of organisms.

Telomere – specialized DNA sequences located at the end of the chromosome.

Telomerase – specialized reverse transcriptase containing RNA and protein subunits responsible for maintaining DNA sequences at telomeres.

Terminator – DNA sequence indicating the termination of transcription at the 3' end of a gene.

Tetrahymena – a common freshwater ciliate.

TNF-receptor associated death domain (TRADD) – associated with the cytoplasmic domain of TNFR1 induces two major responses to TNFR1 activation: apoptosis and NF-κB induction of transcriptional activation in the nucleus.

Thrombosis – the formation of a blood clot or thrombus inside a blood vessel, obstructing the flow of blood through the circulatory system.

Thrombospondin 1 – inhibits angiogenesis and modulates endothelial cell adhesion, motility, and growth, induces apoptosis of endothelial cells.

TNF-related weak inducer of apoptosis (TWEAK, Apo3 ligand), a potent skeletal muscle-wasting cytokine.

Toad – tailless amphibians (mainly *Bufonidae*), related to and resembling the frogs but more terrestrial with a broader body and rougher, drier skin.

Tobacco smoke – far the most important enviromental cause of cancer, revealed by the decreased total cancer death rate by 30% in North-America and Europa.

TP53 – a tumour suppressor gene that is named after a protein called tumour protein 53 (TP53 or p53). See also p53.

TRADD – See TNF-receptor associated death domain.

Trans-acting – transcription factors which are not part of the gene in question and are up- or down regulating transcription.

Transcription – synthesis of RNA from the template DNA strand.

Transcription factors – specific proteins associating with promoter sequences necessary for the accurate initiation of transcription of specific DNA sequences but are not integral parts of RNA polymerases.

Transfer RNA (tRNA) – a family of adapter molecules carrying amino acids to the site of protein synthesis on the ribosome.

Transformed cells – lacking cell growth control, growing caotically regardless whether they are touching neighbouring cells or not.

Transforming growth factor-β (TGF β including TGF-β1, TGF-β2 and TGF-β3) – activate serine/threonine kinase receptor pathways promoting differentiation of mesenchymal cells.

Transforming growth factor beta superfamily – includes the TGF families, inhibins, activin, anti-müllerian hormone, bone morphogenetic protein, decapentaplegic and Vg-1.

Transgenic – organism which carries foreign genes, inserted deliberately in its genome using recombinant DNA technology.

Translation – protein synthesis taking place on ribosomes, the sequence of nucleotides in mRNA specifies the sequence of amino acid in a polypeptide.

Translocation – Breakage and removal of a large segment of DNA from one chromosome, followed by the segment's attachment to a different chromosome.

Transposable elements (transposons) – discrete DNA sequences dispersed throughout the genome that move from one location to another within the genome. Transposable elements were discovered by Barbara McClintock in the 1940s and 1950s. She received the Nobel Prize in 1983.

Glossary

Transposons – See transposable elements.

Trefoil factors family – peptides are the secretory products of mucous epithelia playing role in cytoprotection, anti-apoptosis.

Tripartite motif (TRIM) family – includes three zinc-binding domains, a RING, a B-box type 1 and a B-box type 2, and a coiled-coil region.

Tumour necrosis factor (TNF) – inflammatory protein activating trimeric receptors.

Tumour promoter – non mutagenic, but promotes cancer, enriches the tissue in irreversibly damaged cells (e.g. tetradecanoil phorbolacetate – TPA).

Tumour suppressor – gene, the products of which are negative regulators of cell proliferation, its inactivation predisposes to cancer.

Ubiquitin – protein usually tagged for selective destruction in proteolytic complexes called proteasomes by covalent attachment of ubiquitin, a small, compact protein that is highly conserved.

Unique genes – are present only once per haploid genome. Most structural (protein-encoding) genes of eukaryotes and all those of prokaryotes fall into this category.

Upstream elements – DNA sequence upstream (in the 5' end direction) of the core promoter affecting the rate of transcription.

van der Waals interaction – distance dependent attraction or repulsion of closely spaced atoms.

Werner syndrome – premature aging disease.

Western blotting – transfer of proteins from a gel after electrophoresis to a membrane (nitrocellulose) for subsequent analysis with an antibody of the protein to be found.

Wild type – the typical form of an organism, strain or gene.

Wilms tumour – or nephroblastoma is a pediatric kidney cancer arising from pluripotent embryonic renal precursors.

Viral infection – caused by the presence of a virus in the body.

Xanthine – chemical found in caffeine, theobromine, and theophylline encountered in tea, coffee, and the colas.

Xeroderma pigmentosum – human genetic defect of DNA repair mechanism, inability to correct errors caused by UV-irradiation in the epidermal cells, increased incidence of skin cancer.

Zymogen (proenzyme) – an inactive enzyme precursor.

Index

A

Angiogenesis, 259, 301
Antiapoptotic pathways, 203
Apoptosis, 15, 17, 21, 24, 37, 87, 106, 112, 127, 135, 142, 148, 164, 169, 174, 203–273, 293, 296, 297, 301, 303, 304, 308, 314–316, 322, 326, 334–337, 339, 340, 343–347, 353, 356, 357
Apoptosis inducing signals, 229–237
Apoptosis protocols, 203, 269–272
Apoptotic chromatin changes, 218, 231, 293–352
Apoptotic pathways, 203, 206, 215, 222, 223, 226, 237–268, 335, 336, 345, 347
Arguments supporting RNA World, 9

B

Biosynthesis, 18, 23, 37–39, 45, 50–54, 57, 59, 60, 110, 111, 228
Biotinylated DNA, 140, 143, 149
Bond types, 31, 49, 50
Building blocks, 3, 9, 31–61, 64, 184, 191, 242
Building units of chromosomes, 191

C

Cadmium, 158, 174, 296, 301–309, 311–327, 344–346, 356, 357
Cancer cells, 38, 100, 112, 224, 226, 228, 230, 231, 251, 252, 256, 259, 263, 266
Caspases, 204–206, 209, 213, 214, 216–219, 221–229, 234–236, 238, 239, 241–251, 256, 258, 260, 261, 265–269, 273, 314, 335, 345–347, 357
Cell cycle analysis, 182, 188, 316, 321, 328, 353
Cell types, 6, 22, 91, 129, 141, 158, 161, 215, 216, 219, 225, 227, 228, 231, 240, 257, 261, 264, 340, 345, 346
Cellular responses, 24–26, 246, 254, 258

Cell viability, 265, 348, 355
Chemical carcinogenesis, 294
Chemical inducers of apoptosis, 203, 229
Chemical mutagens, 294, 295
Chromatin models, 87, 92, 136, 193
Chromosome arrangement, 129, 166
Chromosome models, 135
Chromosomes, 8, 25–27, 31, 62, 63, 69, 80, 81, 87, 91–96, 99–110, 112–114, 125–127, 129–139, 141–147, 150–158, 160–163, 165–195, 215, 217, 233, 237, 251, 252, 293, 298, 300, 302, 309, 311–315, 317, 323–325, 329–334, 340, 343, 344, 347, 348, 351–357
Common pathway of condensation, 158–163
Computer analysis, 25, 26, 193, 300
Core information processes, 16–24

D

Death receptors, 223–225, 238, 243, 244, 249, 250, 258, 269, 347
Degradation, 37, 57–60, 87, 205, 209, 213–216, 219, 222, 231, 234–237, 240, 242, 252, 253, 258, 262, 263, 271, 314, 339, 347
DNA content, 62, 97–99, 106, 167–169, 182, 188, 271, 304, 319–321, 327, 328, 332, 336, 337, 339, 344, 348, 353, 356, 357
DNA structure, 4, 27, 31, 35, 62–114, 217, 243, 294, 296, 347

E

Early chromatin models, 87
Effect of ionizing radiation, 326–357
Epigenetic processes, 16
Euchromatin, 90, 93, 101, 126, 127, 139, 150, 161, 183, 233

Index

F

Fluorescence microscopy, 137, 140, 170, 273, 323
Free radical formation, 10, 22
Function, 7, 11, 23, 36, 59, 86, 87, 96–98, 103, 107, 110, 111, 125, 144, 168, 184, 191, 206, 218, 223, 224, 228, 233, 235, 252, 255, 257, 261–263, 266–269, 271, 295, 335, 336, 346, 349

G

Gamma irradiation, 174, 236, 254, 293, 296, 297, 304, 311, 326, 327, 329, 331, 334–337, 339, 340, 342–346, 356, 357
Genotoxic agents, 100, 218, 222, 293, 295–297
Genotoxic changes, 25, 293

H

Heavy metal induced cytotoxicity, 301–326
Hepatocellular tumour cells, 164, 297–300
Heterochromatin, 90, 102, 126–128, 139, 174
Histone code hypothesis, 127–129

I

Injury-specific changes, 345, 356, 357
Intermediates of chromatin condensation, 26, 135, 137, 141–148, 155, 158–160, 166, 170, 180, 181, 183, 187, 190, 293, 310, 314, 323, 330, 331, 340–342, 351
Interphase chromatin structures, 135, 141, 147, 163, 165, 169, 180, 187, 189, 308, 322, 348, 354
Irradiation, 25, 104, 174, 221, 236, 238, 239, 254, 266, 271, 293, 296, 297, 304, 311, 326–329, 331, 333–340, 342–350, 352, 353, 355–357

L

Linear connection of chromosomes, 127, 151, 153, 155–157, 175, 184, 190, 354

M

Mammalian and *Drosophila* cells, 125, 187, 293
Metabolism, 7, 11, 17, 21, 25, 31, 50, 97, 99, 107, 109, 111, 127, 219, 228, 230, 232, 245, 296

N

Necrosis, 203, 205, 212–228, 230, 235, 243, 245–248, 266, 269, 296, 347
Nomenclature, 45, 50, 51, 100

O

Oxygen toxicity, 21, 22

P

p53, 17, 25, 112, 204–206, 209, 214, 223, 224, 226, 227, 229, 232, 233, 238, 239, 251–259, 261–264, 267, 268, 272, 273, 293, 335, 336, 344–347
Permeable cells, 25, 140–144, 169, 174, 306, 316, 321, 348–350, 352
Plectonemic chromatin model, 91, 92, 136, 193, 195
Pre-RNA World, 9, 18, 41
Primary information, 15, 24
Principles of genetic information, 12–15

R

Regulation, 16, 57, 60, 70, 75, 86, 93, 107, 111, 126–128, 135, 158, 206, 213, 216, 223, 225, 232, 233, 235, 242, 245, 252, 253, 255–257, 260, 262–264, 335
Replicative and repair synthesis, 26, 233, 303, 320, 326
Retinoids, 205, 259
Reversal of permeabilization, 25, 92, 125, 140–149, 155, 156, 159, 166, 170, 174, 183, 190, 194, 298, 308, 311, 313, 323–325, 329, 340, 347–349, 351, 354
RNA World, 8–12, 16–20, 23, 26, 27, 41

S

Stress, 21, 72, 73, 97, 113, 208, 210, 213, 224, 226, 233, 235, 238–240, 244, 252–254, 256, 264, 265, 293, 294, 296, 304, 316
Structural levels, 25, 63, 87, 133, 164, 293
Structure of nucleotides, 31
Subphases of nuclear growth, 308, 309
Synchronized cells, 139, 140, 142, 146, 148–153, 162, 166, 169, 176, 178, 180, 188, 328, 348, 357

T

Transition to DNA Empire, 1–26

U

UV irradiation, 293, 344, 345, 347–353, 355–357